Democracy in America

Democracy in America

ALEXIS DE TOCQUEVILLE

Introduction by Bruce Frohnen

GATEWAY
EDITIONS

Originally published as two volumes in 1835 and 1840 and translated from French by Henry Reeve. This 2024 edition reproduces Reeve's revised 1862 text with an introduction from the 1889 reissue published by Longmans, Green, and Co.

Gateway Editions™ and Regnery® are imprints of Skyhorse Publishing, Inc.®, a Delaware corporation.

Visit our website at www.regnery.com.
Please follow our publisher Tony Lyons on Instagram @tonylyonsisuncertain.

10 9 8 7 6 5 4 3 2 1

Library of Congress Cataloging-in-Publication Data is available on file.

Cover design by John Caruso

Print ISBN: 978-1-68451-538-7
eBook ISBN: 978-1-68451-572-1

Printed in the United States of America

Contents

First Part

Second Part

Second Part, First Book: Influence of Democracy on the Progress of Opinion in the United States

Second Part, Second Book: Influence of Democracy on the Feelings of the Americans

Second Part, Third Book: Influence of Democracy on Manners, Properly So Called

Second Part, Fourth Book: Influence of Democratic Opinions and Sentiments on Political Society

Note on This Reeve Edition

By Bruce Frohnen

This volume reproduces the revised 1862 edition (published in 1889) of Tocqueville's *Democracy in America,* translated by Henry Reeve. In addition to Tocqueville's masterful analysis of America and democratic culture, it includes Reeve's introduction and biographical sketch of the author. It also includes Tocqueville's report to the Academy of Moral and Political Sciences on Cherbuliez's book, *On Democracy in Switzerland.* In this essay, Tocqueville examines the ways in which the old aristocracies of Switzerland, through their violent opposition to the inevitable rise of democracy, ended by encouraging democratic revolution, to the detriment of democratic freedom. Finally, this volume contains Tocqueville's eloquent speech before the French Chamber of Deputies on January 27, 1848, urging his fellow legislators to prevent the oncoming socialist revolution by reforming the spirit of French laws and institutions.

Republication of Tocqueville's masterpiece of political philosophy requires no justification. However, some might question the need to reproduce Reeve's nineteenth-century translation, particularly when several twentieth-century translations are or soon will be available.[1] The Reeve translation itself has undergone significant revisions by Francis Bowen and again by Phillips Bradley.[2] Hasn't it been rendered obsolete by subsequent scholarship?

No. The Reeve present edition is a unique, contemporaneous translation of Tocqueville's work, providing an early- to mid-nineteenth-century English version of an early- to mid-nineteenth-century French work. The result is not merely historically interesting, but also highly useful to anyone who views great books not as mere collections of facts but as great literature, to be looked to as a source of nuanced arguments concerning human nature and the social order. In addition, Reeve brought to his translation a wealth of knowledge and experience within the circle of Tocqueville's English friends and a facility with language that well qualified him to become, as he did, translator of the bulk of Tocqueville's works into English.

A number of scholars have faulted the Reeve translation because it does not live up to standards of scholarship emphasizing literal, word-for-word translation. But the value of such standards is itself open to question. Indeed, as recently as 1946, Henry Steele Commager presented an abridged edition of *Democracy in America,* stating, "I have used here the translation by Henry Reeve, which, while perhaps not as accurate as that by the American Francis Bowen, seems to me to catch more the spirit and style of the original."[3]

Some may denigrate notions of "spirit and style" that subtract from scientistic accuracy. In justifying his correction of the Reeve translation, Bowen went so far as to call it "utterly inadequate and untrustworthy," deeming it "generally feeble, inelegant, and verbose, and too often obscure and incorrect." According to Bowen, such extensive revisions were required to render the Reeve translation acceptable that, particularly in the first volume, his [Bowen's] might almost be deemed a new translation.[4]

Unfortunately for those who placed their faith in Bowen's translation, however, that retranslation itself garnered severe criticism. Phillips Bradley, in justifying his own corrections to the Bowen edition, noted that "in many passages Bowen adopted an almost slavishly literal translation without much consideration of the context." And "Bowen took other and more substantial liberties with Tocqueville's style—and meaning. He altogether suppressed portions or the whole of sentences and paragraphs, especially in the footnotes and appendices." Bradley also found "a great many archaisms . . . reflecting the change in usage over the past eighty years."[5]

Thus Bradley revealed that Bowen's claims to accuracy were themselves inaccurate. Moreover, Bradley put forth a persuasive argument that the search for absolute accuracy is bound to fail and to cause problems of its own. Those who aim at absolute accuracy will find in language a moving target. Thus Bradley, in correcting Bowen's corrected translation, wrote "the present text reflects . . . no more than one individual's effort to make *Democracy in America* more comprehensible to the contemporary reader."[6]

Faced with the reality of shifting usage and meaning, many academics have given up the search for accuracy altogether. But Reeve's example shows that good, faithful translation is possible—though based on criteria and knowledge less narrow and instrumentalist than those prized by academics seeking reimport the supposed exactitude of physics and other "hard" sciences into humanistic realms. Translation requires not pseudoscientific tools of measurement and a narrow search for exact corollaries but wisdom and a prudent understanding of the relative usage of the two languages under study.

Reeve was part of a learned tradition, very much alive throughout the nineteenth century, in which the educated classes generally, and scholars in particular, were expected to know several languages, ancient and modern. The average university-educated person had read important works, from Thucydides to Montesquieu, in the original. Translation remained important, however, as laymen rarely retained the mastery of foreign languages necessary to make reading books in the original pleasant or worth their time and effort once they had left the university. But what widespread knowledge of languages provided was a comfort with translation that has become all too rare. Because the original was accessible to many people, translators were not given the responsibility or the power to decree the final word on the text. Rather, the translator's job was to produce a book that rendered the spirit of the original readable and persuasive in translation. Thus, in praising his friend's translation of his work, Tocqueville wrote to Reeve, "I should have written sooner, but I wanted to be able to speak about your translation. I have not yet seen the whole of it, but from what I have seen, I can honestly assure you that I am very much pleased with it. You have rendered my thoughts, in their most delicate shades, with a fidelity and clearness that seem to me perfect. As for the style, of which I am less able to judge, Mme. De Tocqueville assures me that it is excellent—clear, simple, and, in short, exactly what it ought to be for a book on political philosophy."[7] This almost casual attitude on the part of the author toward his translator might scandalize some academics today. But it is indicative of the trust that quickly developed between Tocqueville and Reeve, and that was fitting, given the level of learning among the educated of the time.

This trust between men of learning justifies the republication of this important translation of one of the world's greatest achievements in political philosophy. The eminent American man of letters Russell Kirk referred to Tocqueville as "perhaps the only social thinker of the first rank since the end of the eighteenth century."[8] Kirk also wrote that Tocqueville, after Burke, had "no peer as a critic of society."[9] Having read and reread, and having both learned and taught from *Democracy in America,* I am inclined to agree. But, thanks to Regnery's republication of this valuable translation the reader can, as he ought to, judge for himself.

One final note to the reader: the endnotes in brackets are Reeves's. The others are Tocqueville's.

Preface to This Edition

BY THE TRANSLATOR

At an advanced time of my life another edition is required of the translation of this work, which I published fifty-four years ago. It is gratifying to find that this long period of time has rather increased than diminished the interest which is felt in the writings of M. de Tocqueville and the respect entertained for his opinions. Not many writers on speculative politics have stood the test of time or acquired a permanent authority in political science; but in the opinion of the present generation, the works of M. de Tocqueville stand not far below those of Montesquieu and Burke, and they may be ranked with those of the late Sir Henry Maine, to which they bear a striking resemblance.

If the principal object of M. de Tocqueville had been to describe the political and social institutions of the United States, as he saw them nearly sixty years ago, his work would now be obsolete, for that country has undergone changes in the last half-century surpassing the furthest extent of human foresight. The population of the North American States has quadrupled. Their territory has extended from the Atlantic to the Pacific. The number of the sovereign States of the Union has been raised to forty-two, each of them possessing an independent legislature and executive, subject only to the terms of the Federal compact. The rural townships of New England, in which M. de Tocqueville studied the discreet and orderly action of democratic institutions, have given place to huge and turbulent cities. The equality of conditions has been modified by the acquisition of enormous wealth, and the increase of a proletariate, swelled by a vast immigration from Europe. The negroes are emancipated. The Indians have dwindled away. The fertile valley of the Mississippi, which was a desert in 1832, is already peopled by millions of citizens. The power of the Eastern and maritime States is passing to the West: and the immense development of the material and industrial resources of the country has changed the aspect of the Union, and, in some respects, the very life of the nation.

But the true object of M. de Tocqueville's researches was not so much to describe the condition of the American people and their institutions, as to forecast the effects upon mankind of the progress of democracy, which he regarded as an irresistible revolution, transforming, destroying, or perhaps regenerating the condition of the world. The theme of his book is not America, but Democracy. In America, as he says, in the introductory chapter which supplies a key to the whole work, he saw more than America: he sought the image of 'Democracy itself, with its inclinations, its characters, its prejudices, and its passions, in order to learn what we have to fear or to hope from its progress.' His observations on the peculiar condition of American society are no more than illustrations of the great problem of the age. The main drift and purpose of all he wrote was not so much to criticise America as to study the future of Europe, and especially of France. In America the establishment and progress of Democracy was easy, natural, and unopposed, for it had free scope in a new country and an uninhabited territory. In France it has been accomplished by a series of revolutions and by the ruin of the institutions of a thousand years: and what has been substituted for them?

These are the considerations which press even more keenly on the minds of men at the present time than they did when this book was written. They apply not only to France, but to ourselves, and to every State in Europe. The onward movement is everywhere felt. The ultimate result is everywhere obscure. But nowhere is the subject discussed with more wisdom, sublety, and impartiality than in these pages. Hence the application of the work (as the elder writers would have termed it) is direct and permanent, for it concerns not the past but the present and the future, not of one country, but of the civilized world.

By a singular coincidence this present year is the centenary of two of the principal events of modern history—the final adoption of the Federal Act of Union by the United States, and the commencement of the French Revolution. In both countries a democratic system of government was established, in America by an admirably regulated constitution, in France by the destructive overthrow of the monarchy. We have before our eyes the instructive lesson of a hundred years. In America democratic institutions have raised a people to great prosperity and power; in France they have plunged a great nation into a state bordering on anarchy, for although the strong administrative system of the Empire still holds society together, the traditions of government have, at present, ceased to exist, and the existence of the State is insecure.

The contrast is striking, but not more striking than the conflict of opinions which everywhere exists as to the ultimate consequences of the democratic revolution. There are those who believe, with the first members of the National Assembly of 1789, that a period of reconstruction and regeneration has arrived: that we are witnessing a new birth of society, or as some of them enthusiastically expressed it, 'a new heaven, and a new earth!' That the conditions of human life will be changed by the transfer of power to popular election: and that the voice of the people will solve the most intricate questions of government. But even the successful experience of America has not altogether realized the hopes of the more ardent apostles of the democratic party, since it has shown that equality of rights does not mean equality of conditions, and that popular election, the grand instrument of democracy, is liable to be so manipulated and controlled by artifice or by accident, that the freedom of individual opinion may be crushed by organization as effectively as by tyranny or corruption. There are those, on the other hand, who are more impressed by the destructive powers of uninstructed numbers and impatient poverty; who do not believe that the government of a great nation can be carried on with success upon an unstable basis, shaken by the passions or delusions of men without experience of affairs of State, guided more by party or by personal interests than by an enlightened devotion to the welfare of all classes; and who fear that if supreme power is wrested from the hands of the wisest and ablest members of the community law will be less respected, order less maintained, and the fundamental principles of social life attacked.

These adverse tendencies of a democratic age are both of them impartially indicated and discussed in these volumes. They might be further illustrated by numerous events which have occurred since these volumes were written. But the principal object of the writer was to point out by what means, in his opinion, the advantages of democracy may be secured and its dangers averted. Whatever be the form of a government its success depends on the wisdom and rectitude of those who are placed at the head of affairs, and the more a nation advances in freedom the more essential it becomes that its statesmen should not be swayed by popular agitation but governed by sound principles of action. The conclusion to be drawn from this work is that the fate of democratic society will be determined, for better or worse, by the moral qualities of those who are called upon to control its excesses and to direct its course. They will hold in their hands the future destinies of the world.

Henry Reeve
Foxholes, Christchurch,
April 1889

Introductory Notice

BY THE TRANSLATOR

FIFTY-EIGHT YEARS have now elapsed since the author of this book visited the United States of America; fifty-four years since the publication of the first portion of his commentary on the American Constitution; forty-eight years since the publication of the concluding portion of his task, which forms the second volume of the present edition. Revolutions, both political and social, of extraordinary magnitude and frequent recurrence, have, in this interval, changed the condition of many of the greatest States in either hemisphere, and modified the aspect of society itself. Nor can any man, to whom it is given to look back through this long vista of life, recall without emotion the scenes he has witnessed in the moving drama of the world. Yet the demand for this work is still considerable, and I am induced to publish a cheap and popular edition of it, for time would already have consigned it to oblivion if it had not a lasting claim upon the interest of mankind. Experience has demonstrated the profound sagacity with which the youthful author analysed the great political and social problems of the age; and at every page the reader meets with some searching intuition, which, seen by the light of subsequent events, seems to bear the mark of prophetic power. The tendency of democracy in France to the re-establishment of absolute government—the tendency of democracy in America to a disruption of the Federal Union, ending in Civil War and the subjugation of one portion of the country by the other, in spite of the wisdom which had framed the American Constitution and the circumstances which favoured its duration—the tendency of democracy to change, and in some cases to lower, the condition of society in other countries—are here not only indicated, but described with a precision which could hardly be surpassed after the occurrence of these events. Many of the facts referred to in these volumes have been modified by subsequent events; but such is the soundness of M. de Tocqueville's political principles, and the accuracy of the judgments based upon them, that every opinion is as worthy of consideration as at the time when it was first formed, and may serve as a guide to futurity itself. The value of this book consists, not in the careful and exact description of the institutions and resources of the United States at the time the author visited America, for these have undergone changes and a prodigious increase; but rather in the study of that great social and political revolution, which is gradually moulding and transforming in our age the institutions of the civilised world. It is this which has given to this work an importance and a significance scarcely understood at the time when it was first published, and has made it a textbook of every political student; yet not to the political student only, for in the Second Part the author combats, in the spirit of a Christian and a philosopher, the sceptical and materialist tendency of democratic ages.

I avail myself of the present occasion to prefix to these pages some faint memorials of their illustrious author, which have already appeared, for the most part, in another place. The most conspicuous monument of his wisdom and his virtues is to be found in the work which laid the basis of his fame; but even his writings derive additional lustre from the purity and elevation of his character. To some extent the 'Memoir and Correspondence of M. de Tocqueville,' published by M. Gustave de Beaumont, and ably rendered into English by another hand, have contributed to make known to the world qualities which, while he lived, lay folded and hallowed in the recesses of friendship and of domestic life; and it may not be out of place to connect a brief notice of his career with this edition of his first literary production.

M. de Tocqueville exclaimed in early youth to his intimate friend, who has since become his biographer:—'Il n'y a pas à dire, c'est l'homme politique qu'il faut faire en nous.' His studies, his journeys, his pursuits, were already directed to a life of political action. He engaged in politics with matchless ardour, and with an ambition the more intense that it was absolutely free from the slightest taint of personal interest. He pursued this noble enterprise for fifteen years, in the contests of parliamentary debate, in the paroxysms of revolution, in the ranks of a Constituent Assembly, in the service of the President of the Republic, and in the direction of the Department of Foreign Affairs. He witnessed the catastrophe which extinguished the liberties of his country, and realised the darkest of his own marvellous predictions; but subjection to despotic power wasted him like an incurable disease, and amongst the causes which doubtless contributed to exhaust his delicate and sensitive frame was the ever-recurring thought that he who survives the freedom and the dignity of his country has already lived too long. After the Revolution of February 1848 a thick darkness settled over the history of the French nation. Men learned to whisper their opinions. The former divisions of party appear ludicrous and mischievous, when they are measured by that great chasm which yawns between Imperial despotism and constitutional freedom. Those who, like M. de Tocqueville himself, have actually written a record of the political events in which they took part, bury their manuscripts or deposit them in foreign countries, till better times shall vindicate the rights of history.

Yet a knowledge of M. de Tocqueville's birth, parentage, and connexions, is required to explain the true bearing of his political opinions; and this is the chief result which can be drawn from so uneventful a biography. It is not, however, an unimportant result, if it removes a misconception which has very generally prevailed as to the spirit and design of his principal writings. Because M. de Tocqueville based his literary and political reputation on the study of democracy and democratic institutions, it was hastily inferred that these institutions were the object of his own predilections. Because he described with perfect impartiality the means by which the American people appeared to have succeeded in combining a highly democratic state of society with a free and regular government, it was supposed than M. de Tocqueville carried a love of democracy to the length of republicanism. Even among some of his intimate friends an opinion existed that his political principles had in them something extreme and revolutionary, and his own family, ardently attached to the royalist party in France, were half alarmed at the audacity and the fame of the most illustrious member of their house. The truth is, that this celebrated book had the singular good fortune to find equal favour in the eyes of opposite parties. It was hailed with equal satisfaction by the ardent friends of democracy and by those who dread the exclusive predominance of democratic power. The former were gratified by M. de Tocqueville's admission of the preponderance of this great element in modern societies, and by his prediction of its future dominion over the world; the latter were no less struck by the acuteness with which he pointed out its tendency to favour absolute government, and to degrade the noblest faculties of man. His doctrine of the universal extension of social equality was applauded by Mr. Mill and Mr. Grote; his doctrine of the tyranny of democratic majorities was quoted with extraordinary effect by Sir Robert Peel, when he was laying the foundations of the great party of conservative resistance, after the popular movement of 1832. But no party objects whatever entered into the mind of M. de Tocqueville himself. Even in this controversy, which may be said to have formed the business of his life, because he saw more clearly than any other man that the fate and freedom of the world depend on it, he maintained an inviolable impartiality, the more difficult and meritorious that his personal sympathies inclined to the cause of aristocracy, although the result of his profound political observations led him to believe that the cause of aristocratic government was irreparably lost, and that democracy must hereafter be mistress of the world. This apparent contradiction was perfectly well explained by himself in a letter to his friend Stoffels, which deserves to be cited. Stoffels had imagined that the tendency of his theories was radical and almost revolutionary. M. de Tocqueville replied, that his love of liberty was tempered by so great a respect for justice, and so genuine a love of law and order, that he might fairly pass for a Liberal of a new sort, not to be confounded with most of the democrats of the time. The following sentences contain his own view of the book he had just published:—

> 'The political object of the work is this: I have sought to show what a democratic people is in our days, and by this delineation, executed with rigorous accuracy, my design has been to produce a twofold effect on my contemporaries. To those who make to themselves an ideal democracy, a brilliant vision which they think it easy to realise, I undertake to show that they have arrayed their picture in false colours; that the democratic government they advocate, if it be of real advantage to those who can support it, has not the lofty features they ascribe to it; and, moreover, that this government can only be maintained on certain conditions of intelligence, private morality, and religious faith, which we do not possess; and that its political results are not to be obtained without labour. To those for whom the word "democracy" is synonymous with disturbance, anarchy, spoliation, and murder, I have attempted to show that the government of democracy may be reconciled with respect for property, with deference for rights, with safety to freedom, with reverence to religion; that if democratic government is less favourable than another to some of the finer parts of human nature, it has also great and noble elements; and that perhaps, after all, it is the will of God to shed a lesser grade of happiness on the totality of mankind, not to combine a greater share of it on a smaller number, or to raise the few to the verge of perfection. I have undertaken to demonstrate to them that whatever their opinion on this point may be, it is too late to deliberate; that society is advancing and dragging them along with itself towards equality of conditions; that the sole remaining alternative lies between evils henceforth inevitable; that the question is not whether aristocracy or democracy can be perpetuated, but whether we are to live under a

> democratic society devoid indeed of poetry and greatness, but at least orderly and moral, or, under a democratic society, lawless and depraved, abandoned to the frenzy of revolution, or subjected to a yoke heavier than any of those which have crushed mankind since the fall of the Roman Empire. I have sought to calm the ardour of the former class of persons, and, without discouragement, to point out the only path before them. I have sought to allay the terrors of the latter, and to bend their minds to the idea of an inevitable future, so that with less impetuosity on the one hand, and less resistance on the other, the world may advance more peaceably to the necessary fulfilment of its destiny. This is the fundamental idea of the book; an idea which connects all its other ideas in a single web, and which you ought to have discerned more clearly than you have done. There are, however, as yet very few persons who understand it. Many people of opposite opinions are pleased with it, not because they understand me, but because they find in my book, considered on one side only, certain arguments favourable to their own passion of the moment. But I have confidence in the future, and hope the day will come when everybody will see clearly what a few only perceive at present.'
>
> —Tocqueville Correspondence, vol. i. p. 427.

In a letter to one of his English friends, he expresses with greater precision his own personal connexion with the subject:—

> 'People want to make me a party man, which I am not. They ascribe to me passions when I have only opinions—or rather but one passion, the love of freedom and human dignity. All forms of government are in my eyes but means to satisfy this sacred and lawful passion of man. Democratic and aristocratic prejudices are alternately ascribed to me. I should perhaps have had these or those had I been born in another century or in another country; but the accident of my birth has easily enabled me to defend myself against either tendency. I came into the world at the end of a long revolution, which, after having destroyed the former state of things, had created nothing lasting in its place. Aristocracy was already dead when I began to live, and democracy was not yet in existence. No instinct, therefore, impelled me blindly towards one or the other. I was an inhabitant of a country which had been for forty years

trying everything and stopping definitively at nothing. I was not easily addicted to political illusions. Belonging myself to the old aristocracy of my country, I had no natural hatred or jealousy of aristocracy; nor had I any natural love of it, for people only attach themselves to what is in existence. I was near enough to judge it with knowledge, far enough to judge it without passion. The same may be said of the democratic element. No interest gave me a natural or necessary propensity to democracy; nor had democracy inflicted on me any personal injury. I had no particular motive to love it or to hate it, independently of my own reason. In a word, I was so well balanced between the past and the future, that I did not feel myself naturally and instinctively drawn towards one or the other, and it was no great effort to me to take a tranquil survey of both sides.'

—Tocqueville Correspondence, vol. ii. p. 70.

The maintenance of this state of philosophical impartiality, widely remote from indifference, was one of the great objects of M. de Tocqueville through life, and it is one of the finest qualities of his writings. He was, as an ingenious writer expresses it, essentially 'binocular;' he saw correctly, because he saw the object in two positions at once, the angle of one point of vision correcting the obliquity of the other. But this singular rectitude of judgment must be attributed to the skill with which he preserved the balance between his sympathies and his understanding, rather than to the absence of those passions to which other men are more apt to yield.

The family of Clerel, or, as it was anciently spelt, Clarel, has been established for many centuries in the peninsula of the Cotentin, on the Norman coast, and the village and lands of Tocqueville give them their territorial designation. The Clerels figure in the roll of Battle Abbey, among the companions of the Conqueror; for an extraordinary number of the gallant Norman adventurers who overran Britain, and filled the world with their exploits, drew their first breath in some manor-house of this district. Tradition indeed relates that the village of Tocqueville owed its name to a Norman chief, or sea-rover, called Toki, whose tumulus may still be seen on the high ground above the château: and certainly this point commands a vast range of sea and land of no common historic interest;—hard by, Barfleur, now a neglected port, but once famous in the annals of English royalty and English wars; to the east, the Hogue; to the west, Cherbourg. On this spot the seigneurs of Tocqueville have dwelt for many generations, leading the life of the country gentlemen of France before the Revolution, always ready to pay their debt to their country with their blood, for their descendant relates in one of these letters that his grandfather and his great-uncle perished on the field of battle or died of their wounds; seeking their amusements in field sports or in the neighbouring county town of Valognes; proud of their gentle descent, though not entitled to be ranked among the highest order of the French nobility. Their actual residence at Tocqueville dates from about 250 years ago. Before that time the Clerels lived on an estate at Rampan, near St. Lô, and the family was known as Clerel de Rampan. Several of the Seigneurs de Rampan figure in the annals of the Parliament of Rouen in the seventeenth century; and as the spirit and learning of the French provincial magistracy—the old Parliamentary spirit—was the very salt of the nation before the Revolution of 1789, it may be said that Alexis de Tocqueville inherited the qualities for which this order of men was justly conspicuous. But when he himself went to the bar, an old country neighbour, well versed in Norman pedigrees, the Countess de Blangy, who had inherited the domain of the Abbé St. Pierre in the same district, said to the young stagiaire, 'Souvenez-vous, Monsieur, que votre famille a toujours été de la noblesse d'épée.' She was right in point of fact. The Clerels had always been soldiers, and long before 1789 the family bore the title of Count. That title, subsequently conferred by Louis XVIII. on the father of Alexis, was no more than the recognition of an ancient distinction. It is still borne by the elder brother and representative of the House, but Alexis himself always refused to adopt it, and he mentions in one of his letters to Madame Swéchine, that titles had long ago lost in his estimation and in France all meaning and all value.

The Château de Tocqueville consisted originally of what would be termed, north of the Tweed, a 'peel' flanked by a huge tower of enormous solidity, and this part of the edifice is probably as old as the battle of Agincourt. Such was the type of the Norman manor-house of the fifteenth century. But when the gentry of the Cotentin had ceased to dread the incursions of English marauders, their houses expanded, and in the reign of Louis XIII. the château was considerably enlarged. A quadrangle was built, which served partly for the residence of the family and partly for farm buildings, the windows looking out on the farm-yard in the middle. A large dovecote, though now guiltless of pigeons, still marks the ancient seignorial right of the lord to keep his pigeons at the expense of his peasantry; and a stain over the door indicates the spot from which the Revolution of '93 tore the escutcheon of the family. The quadrangle has made way for the convenience of a modern approach, and the old château has assumed the elegance of a mansion of the nineteenth century; but every stone of it tells of the past. Alexis de Tocqueville came into possession of this residence by a family arrangement in 1837. He speaks of it in one of his letters at that time as 'mon pauvre vieux Tocqueville,' a sort of big farmhouse, which had not been inhabited for half a century. Indeed at that time the floors were gone, and the roof was in danger, though happily the old 'girouette féodale' still turned on the big tower. But its aspect was speedily changed; it became for the next twenty years the scene of uninterrupted domestic happiness, and of never-failing rural interests, a repose after the contests of political life, a retreat in the dark hour of national adversity, and the scene of literary labour, of liberal hospitality, of counsel and consolation to all who needed or asked for them.

At an early age the father of Alexis entered into possession of this inheritance, then surrounded with all its seignorial rights, and contracted a marriage with Mdlle. Lepeletier de Rosambo, a granddaughter of M. de Malesherbes. This connexion with a house so distinguished as that of the Lamoignons proves the consideration at that time enjoyed by the Clerels of Tocqueville. M. de Tocqueville's connexion with the old Marquise d'Aguesseau was also by his mother's side, Madame d'Aguesseau being one of the three daughters in whom the Lamoignon family expired. One of her sisters married Count Molé's father, and the other M. Feydeau de Brou. The paternal grandfather of Alexis de Tocqueville married Mdlle. de Damas Crux, whence the Duke de Damas was his great-uncle.

After the execution of the King M. de Malesherbes returned to his countryseat. And it was at this very time and under these distressing and alarming circumstances at the Château de Malesherbes in 1793 that the Count de Tocqueville married his granddaughter. Barely six months had passed after the marriage, Malesherbes still living on his estate with the several branches of his descendants, when his eldest daughter and her husband, M. de Rosambo, were torn from him by the revolutionary emissaries. A few days later Malesherbes himself and all the other members of his family were also seized; and on the 22nd April, 1794, he was sent to the scaffold with his daughter, his granddaughter, recently married to M. de Chateaubriand, and her husband, the elder brother of the well-known statesman and writer. They were executed before his eyes, and his own death instantly followed that of those he loved. M. and Madame de Tocqueville, she being a sister of Madame de Chateaubriand, were arrested at the same time, and remained for several months in the Conciergerie, until they were liberated by the fall of Robespierre. I remember to have heard that the first thing they did after their liberation was to drive about Paris for a whole day in a hackney coach, partly for the enjoyment of the sense of freedom, and partly from the confusion of mind produced by the scenes they had witnessed and the perils they had escaped. They returned, however, to their family mansion: the plate had been buried, and was saved; a service of Dresden china had also been buried in another part of the grounds, but the clue to the hiding-place was lost, and it has never been rediscovered. The Tocquevilles never emigrated; they therefore retained their landed property, and continued to live peaceably upon it. In 1805 Alexis, their third son, was born in Paris, but soon afterwards, being still an infant, he was brought to Tocqueville in a panier slung across a horse, with his nurse on a pillion. In those primitive times, scarcely fifty years ago, there was no such thing as a road for wheeled carriages from the mansion of a country gentleman to the village, or even from the village to the chief town of the department.

I relate these details (which I heard from M. de Tocqueville himself) because, independently of the interest they may possess, they serve to show the influence of the Revolution on the last and present generations of the French. In the higher ranks of society, more especially, there is hardly a family in which events of the deepest tragic interest have not occurred within living memory; and if the actual witnesses of those dreadful scenes have now almost disappeared, their children received from them in early life impressions which no time can efface. When Alexis de Tocqueville was born, less than eleven years had elapsed since the most illustrious members of his mother's family had perished on the scaffold. The age of martyrs was still near. Is it yet over? Tocqueville himself was wont to say that he lived in a country where no man could foretell with certainty whether he should die in his bed or on the block. These traditions doubtless contributed to produce on a mind, naturally so sensitive and so reflective, impressions of which he was himself scarcely conscious. His family was ardently royalist, and might be compared to a high Tory family on this side the water; with some change of conditions, their prejudices and disposition of the mind were the same. His education was scanty, having been conducted by an Abbe Lesueur, whose death, during his absence in America, he affectionately deplored. But that which was not scanty and not deficient was the high principle, the lofty conception of truth and duty, the unselfish dignity, with which his father, like himself, was completely imbued. On the Count's death, in 1856, Alexis wrote to M. de Corcelles, one of his most intimate and highly-valued friends:—'You are right. If I am worth anything, I owe it above all to my education, to those examples of uprightness, simplicity, and honour which I found about me in coming into the world and as I advanced in life. I owe my parents much more than existence.'

The following anecdote, related by himself, recalls these impressions of his early life:—

> 'That sort of idolatry of royalty which ennobled obedience, and made men capable of acts of self-sacrifice, not only to the principle of government, but to the person of the sovereign, may be said to be gradually disappearing entirely from the world. In some countries, as in France,

> not a trace of it remains. I met with it again in your narrative, and the more kindly as the scenes to which it belongs carry me back to the earliest days of my childhood. I remember even now, as if it were still before me, one evening, in a château where my father was then living, and where some family rejoicings had brought together a large number of our near relations. The servants had retired. We were all sitting round the hearth. My mother, who had a sweet and touching voice, began to sing an air well known in our civil disturbances, to words relating to Louis XVI, and his death. When she ceased every one was in tears, not for the personal sufferings they had undergone, not even for the loss of so many of our own blood on the field of civil war and on the scaffold, but for the fate of a man who had died fifteen years before, and whom most of those present had never seen. But that man had been the King.'

Alexis de Tocqueville was ten years old at the Restoration in 1815, and his father became successively prefect at Metz, at Amiens, and at Versailles. He was also raised, very deservedly, to the rank of a peer of France. These mutations had some effect on the earlier career of his son. In 1822 he gained the prize of rhetoric at the academy of Metz; and in 1827 he entered the profession of the magistracy, as Juge Auditeur at Versailles. In the interval he had made a tour in Italy, of which some record has been preserved. Probably he had then never heard of the celebrated passage in Gibbon's Memoirs, where that great historian relates that the idea of his 'Decline and Fall' came into his mind as he sate amidst the ruins of the Capitol and heard the voices of the barefooted friars singing vespers in the Temple of Jupiter. But a similar vision seems to have passed over the mind of another youthful traveller on the same spot; as Tocqueville describes in his journal a procession of barefooted friars mounting the steps of the Ara Cœli, whilst a shepherd calls his goats browsing in the Forum, the past history of Rome rises before him, and he traces the extinction of her greatness to the day when her liberties fell beneath the sceptre of imperial power.

The following years were eagerly devoted to extend the range of his education, as well as to qualify himself for his legal functions; but it is easy to perceive that his ambition would never have contented itself with the honours of the bench, and, in those days more especially, the whole youth of France were launched with inconceivable energy in historical researches, in literary controversies, in philosophical theories, which called forth the full powers of a mind earnest in the pursuit of all knowledge. In political affairs he took as yet no part, but his sympathies were entirely on the side of the Liberal party, whilst his remarkable foresight enabled him to discern the perils of the monarchy. In August, 1829, on the formation of the Polignac Ministry, a year before the celebrated Ordinances, he wrote:—

> 'These ministers can neither summon a new chamber with the present law of election, nor pass a new law of election in the existing chambers. They are launched then on the plan of coups d'état, of laws by ordinance; that is, the question lies between the royal power and the popular power, a conflict in closed lists, a conflict in which, in my opinion, the popular power only stakes its present, but the royal authority will stake both present and future. If this ministry falls, the crown will suffer much from its fall; for it is the creation of the crown, and it will cause securities to be taken hereafter, which will still further restrict a power already too limited. God grant that the House of Bourbon may not one day repent what has just been done!'
>
> —Tocqueville Correspondence, vol. ii. p. 6.

The Revolution, which in 1830 realised these sinister predictions, was a severe, if not a fatal blow to the hopes of a man of five-and-twenty entering on public life with M. de Tocqueville's prospects and opinions. It was not only that his personal chances of advancement in the world were at an end, and that his family, deeply imbued with the passions of the Royalist party, viewed with horror a new form of popular government. These considerations had small weight with a mind alike disinterested and independent. But it became manifest in 1830 that the passions of the French Revolution had slumbered, but were not extinct. Another experiment had failed—another form of government had been overthrown. To use an expression of his own, 'The Revolution has not stopped. It no longer, indeed, brings to light any great novelties, but it still keeps everything afloat. The mighty wheel turns and brings nothing up, but it seems that it will turn for ever.' What then was this blind but irresistible force which swept before it in ever-recurring paroxysms the institutions, the orders, the government of the country? Not merely the love of freedom, for freedom has existed in England for nearly two hundred years, without any grave perturbation of social order, and it has existed for seventy years in the United States, combined with a purely democratic state of society. Nor indeed had the love of freedom acquired any permanent hold over the French people. They adored it in 1789, they were indifferent to it in 1800; and the same phenomenon has since been repeated. One of the last passages which has been preserved from M. de Tocqueville's pen describes his countrymen in the following words:—

> 'Accustomed though we be to the fleeting inconsistency of men, there is something astonishing in so vast a change in the moral inclinations of a people; so much selfishness succeeding to so much patriotism, so much indifference to so much passion, so much fear to so much heroism, so great a scorn for that which had been so vehemently desired and so dearly purchased. A change so complete and so abrupt cannot be explained by the customary laws of the moral world. The temperament of our nation is so peculiar that the general study of mankind fails to embrace it. France is for ever taking by surprise even those who have made her the special object of their researches; a nation more apt than any other to comprehend a great design and to embrace it, capable of all that can be achieved by a single effort of whatever magnitude, but unable to abide long at this high level, because she is ever swayed by sensations and not by principles, and that her instincts are better than her morality; a people civilised among all civilised nations of the earth, yet, in some respects, still more akin to the savage state than any of them, for the characteristic of savages is to decide on the sudden impulse of the moment, unconscious of the past and careless of the future.'

This inconstancy in the pursuit of political objects, this inability to estimate the true value of such objects or to retain them, and lastly the malignant passions which the Revolution had arrayed against all social, intellectual, and moral superiority, were the evil powers which Alexis de Tocqueville resolved to combat and to resist. The shock of the Revolution of 1830 was scarcely needed to teach him that a deep gulf lay fixed between the principles to which he was immutably attached, and the dreams which his countrymen were determined madly and vainly to pursue. He was led, or rather compelled, to the study of democratic institutions not by any natural sympathy with popular agitation or any illusion as to the results of it, but by consternation at the ravages it had already made, and by a deep-seated dread of its furthest consequences. Throughout his writings, throughout his parliamentary career, throughout his correspondence, the conviction may be traced that modern democracy tends to the establishment of absolute power, unless it be counteracted by a genuine love and practice of freedom. The modern theory of democracy is not so much a love of freedom as the love of a particular kind of power. Democratic power differs in its origin, but not at all in its nature, from other forms of absolutism. It is as impatient of control, as liable to overleap the restraint of law, as much addicted to flatterers and abuses, as the most arbitrary monarchy or the corruptest oligarchy.

He perceived that freedom itself could with difficulty be practised or maintained in countries where high principles were giving way to low interests; where the spirit of personal dignity and independence was crushed by the government and hated by the masses; where, to use his own illustration, the impulses of savage life prevailed over the laws of civilisation, and revolution triumphed over tradition. He perceived, too, that as the ruling principle of democracies is the principle of interest, so the principle of aristocracies, if they are to last, must be that of duty. It is apparent from what we have already said of his descent and education, that he belonged by nature to a chosen order of men. Indeed, the extreme delicacy of his physical organisation, the fastidious refinement of his tastes, the exquisite charm of his manners, made him the very type of a high-bred gentleman; and if these were in him the outward signs of distinction, not less was he ennobled by the very soul of chivalry, by that purity and simplicity of character which are the truest nobility, and by a combination of manly virtues with an almost feminine grace—qualities which Englishmen are wont to trace to an ideal perfection in the person of Sir Philip Sidney.

Conceive such a man placed by fate on the brink of the French Revolution, stripped of the traditions of the past by one blast of that great convulsion, robbed by another blast of the hopes of the future, hating with an equal hatred the abominations of the Ancien Régime, the crimes of the Revolution, and the iron yoke of the French Empire, whether imposed by the military genius of one Napoleon or by the civil craft of another; and all this time, viewing with almost superhuman penetration and with patriotic despondency the gradual decline of the French people from that standard of moral dignity and public spirit which could alone enable them to fulfil the generous aspirations of their forefathers! Well aware of the difficulty, perhaps the impracticability, of so great an enterprise, he never ceased to contend for those genuine principles of liberty which could alone, as he thought, preserve society and civilisation from the greatest calamities.

Such were the views, still probably indistinct, which led the young 'Juge Auditeur' to throw up his office at Versailles, and in the company of M. Gustave de Beaumont to proceed in 1831 to the United States. A mission was given them by Count Montalivet to examine the Penitentiary System, then recently introduced in America: they performed this part of their duty conscientiously; but the real motive of their journey was to examine the political institutions of the American people, and the result of it is the book entitled 'Democracy in America.'

M. de Tocqueville was not thirty years old when his great work appeared. He woke one morning, like Byron, and found himself famous. 'I feel,' said he in a letter to his friend Stoffels, written in February 1835, 'like a lady of the Court of Napoleon, whom the Emperor took it into his head to make a Duchess. That evening, as she heard herself announced by her new title when she came to Court, she forgot to whom it belonged, and ranged herself on one side to let the lady pass whose name had just been called. I assure you this is just my case. I ask myself if it be I that they are talking about? and when the fact is established, I infer that the world must consist of a poor set of people, since a book of my making, the range of which I know so well, has had the effect this appears to produce.' His first interview with Gosselin, the publisher, was by no means flattering. That great man consented with some hesitation to strike off an edition of five hundred copies, and Tocqueville remarked that it was rather a humiliating condition of the profession of authors to have to treat one's bookseller as if he were a superior being. Nine months afterwards the tables were turned. 'I went yesterday to see Gosselin, who received me with the most expansive countenance in the world, exclaiming, "Ah, çal mais il paraît que vous avez fait un chef-d'œuvre!" Is not this the tradesman all over?' The success of the book was indeed prodigious. It was instantly translated into all languages. It has become a text-book of constitutional law in the United States, where the English translation has run through numberless editions. It shortly afterwards opened to Tocqueville the doors of the French Institute, and eventually of the Academy. M. Royer-Collard affirmed that since Montesquieu, nothing like it had appeared. Even the compositors and readers in the printing-office testified their interest in the production of it.

Soon after the publication of his first two volumes in 1835, M. de Tocqueville paid a visit (though not his first visit) to England. He was received by many Englishmen with attention and hospitality, which soon ripened into cordial friendship and the deepest mutual regard. Indeed, no inconsiderable portion of the collection of letters now given to the public mark the strong attachment and the sedulous interest with which he kept up his connexions in English society. Perhaps, indeed, there was no society now in existence to which he may be said so naturally to have belonged, as that which he met with in this country. In the polished circles of Lansdowne House and Holland House, his manners and his powers of conversation ensured him a cordial reception; he found there not only the easy citizenship of good breeding, but the same deep interest in the progress of mankind, and the same ardent attachment to every great and free object, which had become the ruling passion of his life. His own ideal of social excellence and political greatness lay precisely in the combination of aristocratic tastes with popular interests, and in that independence of position and character which is never more complete than when it is united to a high sense of the duties and obligations of property and station. Twenty years elapsed before he revisited England, and was again received with all the honours that could be paid by society to one of the most eminent and interesting men of the time. But during the whole of that interval his intimacy with his English friends had been strengthened and increased, partly by correspondence, and partly by their visits to his own country house in Normandy. His confidence and his affection were not easily given; they were given to few; but when given, his friends became a portion of himself; none of them was ever in the faintest degree slighted, or neglected, or forgotten; between them and him, each in his respective manner, there was entire communion; not one of them ever broke from that charmed circle, nor did the vicissitudes of life at all affect the unalterable tenderness of his regard. It is not less interesting to us to know that the first and only object of his affections, who became his wife, and who in that name comprised the strongest and purest ties of human existence—his constant companion, counsellor, and friend—with whom no place was solitary to him, and without whom no society was attractive—was an Englishwoman, who brought him for her portion that best of gifts, the comfort and the trust of English domestic life.

In 1837, when Alexis de Tocqueville had not been long settled in the old family château of his house, he came forward as a candidate for the representation of the arrondissement of Valognes in his own department. His reception was not very flattering. A trace of the old revolutionary prejudices lingered in the neighbourhood; a cry of pas de nobles was got up: his opponent, a retired cotton-spinner who had built a big house, said: 'Prenez garde! il va vous ramener les pigeons,' pointing to the mighty dovecote of Tocqueville Manor; and, in short, the aristocratic though liberal candidate was defeated. He was himself surprised at the intensity of the democratic passions which sent up the large Norman farmers to vote against him. 'My opponents admit,' said he, 'that I have none of the prejudices they ascribe to the nobility; but there is something in the head of these fellows against us which resembles the instinctive aversion of the Americans to men of colour.' So that by a curious contradiction, at the very moment when the 'Democracy in America' was in everybody's hands, and generally regarded as a vindication of democratic institutions, the democracy of his own country rejected the author for his aristocratic descent.

It is true that his opponent also had the support of the Government, and that by M. de Tocqueville's own act and choice. When Tocqueville's name was first announced as a candidate, Count Molé, then Prime Minister of France, gave orders that he should have all the support the Government could afford him, and this without the slightest pre-engagement or even inquiry as to the line he intended to follow in politics. M. Molé was his kinsman, and no slight admirer of his works. But this proceeding on the part of the Minister ruffled the sensitive pride of Tocqueville. He instantly wrote to M. Molé to decline the support of the Government, and to insist on standing in a position of absolute independence if he were to be elected at all. M. Molé's answer, which has been published, though not written without warmth, is a masterpiece of dignity, good sense, and good breeding. He protested against the supposition that because he had proffered the support of the Government without conditions to a man whom he esteemed, this support was to be considered as an intolerable burden or a humiliating bargain; he observed with truth that isolation is not independence, and that a deputy is more or less engaged to whatever party may return him; lastly, he urged that the ministerial party was not a mere band of dependants, but a body of men acting together from convictions in defence of the parliamentary institutions of the country, a task at no time easy, and certainly rendered more difficult by the opposition and hostility of men of M. de Tocqueville's own character. This correspondence left no unfriendly feeling between these two eminent men; they were both of them consummate gentlemen, and each knew that the other was contending, not for an interest, but for a principle. Men of that stamp are more eager to sacrifice a personal interest than to trade on it.

Two years later, at the general election of 1839, when M. de Tocqueville had made his way in the department, and had become an object of real attachment to his immediate neighbours and of respect to all the country round, he was elected to the Chamber of Deputies by a great majority, and he retained his seat under all circumstances as long as there was a free Parliament in France.

Nevertheless I have adverted to this occurrence because it marks the first important step of M. de Tocqueville in public life by a fixed predetermination to join the Opposition, and to owe nothing at any time to the King's Government. I venture to say that this step on his part, and on the part of several of the able men with whom he acted, was a most unfortunate one for his own public utility, and for the welfare of parliamentary government in France. That form of Government was not so firmly established that it could resist the attacks of those who were in the main sincerely attached to the Constitution, though they disapproved the policy of the Ministry and the Court; and no one repeated more emphatically than M. de Tocqueville his prophetic warnings that it was not this or that Minister, this or that system, but representative government itself which was at stake and in danger. The fixed idea of his life was that the Constitution would be undermined by the democratic passions of the nation, and encroached upon by the insincerity of the Court, until nothing stable would remain, and the overthrow of the Parliamentary system would be followed at no distant time by the despotism of a single ruler. But with a foreknowledge of this danger, which no one else possessed to the same degree, and which as expressed in his earlier writings and speeches looks like a gleam of superhuman intelligence, what political conduct ought he to have pursued? He thought it his duty to throw the weight of his lofty intellect and unblemished character on the side of the Opposition. But what was that Opposition? He himself admits in one of his letters that there never had been a real constituted Opposition in France capable of fighting its way to a majority, and then assuming the direction of affairs. M. Thiers, if he was to be considered its head, was certainly quite as far removed from Tocqueville's standard of political morality as M. Guizot. To thwart the schemes of the court, and once or twice a year to deliver a few set speeches against the policy of a Cabinet, was, after all, a wretched substitute for true political life. He acknowledged himself that he had no party spirit, yet he acted with those to whom party spirit was the sole guide, on the principle, as he himself expressed it, 'On n'a quelque chance de maîtriser les mauvaises passions du peuple, qu'en partageant celles qui sont bonnes.' Under this influence his votes on some of the party divisions of the day were votes which some of his most sincere friends disapproved at the time, and to which they may look back with regret.

Tocqueville was not fitted by nature for opposition; he had none of the passions which belong to it; his speeches were earnest, but not impetuous; his caution and conscientiousness restrained him from extreme steps; and in the tribune of the Chamber he fell far short of the greatest orators of his time. The most useful acts of his parliamentary life were his reports on the questions of negro emancipation in the French colonies, on prison discipline, and on the administration of Algeria, which are masterpieces of their kind, and ought to be republished with his principal speeches. Thus without taking the foremost rank among the politicians of the day, he devoted himself with extreme ardour and industry to the public interests of his country, and his extraordinary sagacity anticipated in the later years of the reign of King Louis Philippe, the ruin of those institutions which he regarded as the sole bulwark of public morality among a democratic people.

At length the storm came. By no other man had it been so clearly foreseen, and for several months before the catastrophe he had carefully abstained from all participation in that mad system of agitation which produced the popular banquets and republican demonstrations of 1847. On the 27th January, 1848, soon after the opening of the last session of the Constitutional Parliament, he rose in the Chamber of Deputies, and said:—

> 'They tell me that there is no danger because there are no disturbances; they say that as there is no visible perturbation on the surface of society, there are no revolutions beneath it. Gentlemen, allow me to say that I think you wrong. Disturbance is not abroad, but it has laid hold of

> men's minds. The working classes are quiet, and are not agitated as they have sometimes been by political passions; but can you not perceive that these passions, which were political, are now social? Can you not see that opinions and ideas are spreading amongst them which tend not only to overthrow this or that law, this or that minister, or even this or that government, but society itself, and to shake the foundations on which it rests? Can you not hear what is daily repeated, that everything which is above their own condition is incapable and unworthy to govern them; that the present division of wealth in the world is unjust; that property rests upon no equitable basis? And are you not aware that when such opinions as these take root, when they are widely diffused, "when they penetrate the masses, they will bring about, sooner or later, I know not when, I know not how, the most tremendous revolutions? Such, Sir, is my conviction; we are slumbering on a volcano. I am certain of it.'

Within four weeks the eruption took place. The King fled. The Republic was proclaimed; and not only the Republic, but all the passions of a socialist revolution were let loose on France.

Then, indeed, neither Tocqueville nor any one of his political friends hesitated as to the part they were called upon to pursue. In the first Revolution, the sanguinary violence of a small faction had prevailed over the great majority of the nation. Under the second Republic, the nation itself, appealed to by universal suffrage, gave an unequivocal answer to the call, and elected an Assembly firmly resolved to defend property and public order. An attempt was made by the Revolutionists to annihilate the Assembly itself; it was saved by a miracle; a few days later the fate of the nation hung on the issue of a battle in the streets of Paris. Thanks to the courage and union of the Assembly, the law triumphed, and the country was saved. In all these events M. de Tocqueville took an active part; and a volume of Memoirs in which he recorded them, for the information of posterity, is complete, and may one day see the light. He had naturally been selected by the constituent body as one of the members of the Committee to frame the new Republican Constitution; and it is a curious example of the difficulty of governing human affairs that a Constitution, now universally acknowledged to be a masterpiece of absurdity, was the work of several men of undoubted intellectual power and political foresight. An attempt was made by Tocqueville to induce his colleagues to adopt the principle of a second Chamber; but this and every other attempt to construct the machinery of a true Republican Government utterly failed. The

Republic was destined to a short-lived existence, between the frenzy of democratic socialism on the one hand, and the violence of that popular reaction which speedily assumed the name of Louis Napoleon Buonaparte. The newly-elected President of the Republic had long appreciated the philosophical insight of M. de Tocqueville into the nature of democratic institutions; and perhaps he inferred that the predictions of a single dominion, with which his books abound, were naturally to be fulfilled by a restoration of the Empire. Soon after his election to the Presidency he invited M. de Tocqueville to dinner, placed him by his side, and paid him marked attentions. On leaving the Élysée Tocqueville said:—'I have been dining with a man who believes in his own hereditary right to the Crown as firmly as Charles X. himself.'

One chance remained to avert the final catastrophe. It was possible that the President might still be content to accept a constitutional position; to govern by responsible Ministers who hoped to effect a revision of the constitution by legal means. At any rate, to abandon or to oppose him was to compel him to resort to an immediate coup d'état. On this principle M. Odilon Barrot and the leading liberals formed an administration on the 2nd June, 1849, in which M. de Tocqueville took the important office of Minister of Foreign Affairs. It would be inappropriate here to enter upon the political transactions in which he was engaged. As he said, on quitting his office four months later:—'I have contributed to maintain order on the 13th June, to preserve the general peace, to improve the relations of France and England. These are recollections which give some value to my passage through affairs. I need hardly say anything to you of the cause which led to the fall of the Cabinet. The President chooses to govern alone, and to have mere agents and creatures in his Ministers. Perhaps he is right. I don't examine that question, but we were not the men to serve him on these terms.'

By a sort of Nemesis the Roman expedition was made the pretext of the downfall of the Cabinet. The President had always disapproved the enterprise, but weary with long negotiations he chose to take the matter into his own hands; his celebrated letter to Edgar Ney was a deathblow to ministerial responsibility in France, and from that moment the violent dissolution of the Assembly and the change of government were only a question of means and of time. Tocqueville retired for some months from the scene, for indeed his frail body, exhausted by the fatigues of office, needed repose. He spent the winter at Sorrento, and there laid the basis of the last of his works, which might be termed the Genesis of the French Revolution, traced by him back to its true source, in the vicious institutions of the 'Ancien Régime.' He already perceived that in the impending contest between the President of the Republic and the Assembly all the chances were in favour of Louis Napoleon. In January 1851 he wrote:—

> 'The general aspect of the time seems to me to be a movement of the nations away from liberty and towards concentration and permanence of power. The circumstance that the most eminent parliamentary chiefs and the best known military commanders are almost all opposed to this movement, does not reassure me; for we live in a democratic age, and a society in which individual men, even the greatest of them, count for very little. To form my opinion, I listen neither to those who exalt nor to those who depreciate the talents of the pretenders. At such times it is not the man we must look at, but that which raises the man and brings him into power. A dwarf on the crest of a huge wave may be washed to the top of a cliff, which a giant could not scale from the sands below.'

Nevertheless, soon afterwards, upon his return to France, M. de Tocqueville drew up the celebrated Report of the Committee on the Revision of the Constitution, which was presented to the National Assembly on the 8th July, 1851. This document is of the highest excellence, and ought to be included in a general edition of his works. He traced in it with masterly precision the fatal situation in which the Constitution had flung the French nation, between two contending powers incapable of union, yet destined both of them to come to an end almost simultaneously, leaving the country without an Assembly and without a government: and he demonstrated that the only possible mode of diverting the impending catastrophe was to alter and amend the organic law of the State. This memorable Report may be regarded as the last public act of his life.

As the crisis approached, in the autumn of 1851, he wrote in increasing perplexity:—

> 'How little we feel ourselves masters of events at such times! There is but one determination that I am always certain to follow, and that is to bring our liberties triumphant through this crisis, or to fell with them. All the rest is secondary; but this is a question of life and death.'

And in common with all that was illustrious in the last free Parliament of France, he did fall. M. de Tocqueville was included in that wholesale act of proscription of the 2nd December, 1851, which, with a sort of insolent derision more odious than the tyranny that prompted it, sent the orators, statesmen, generals, and patriots of France in a felon's cart to the common goal. Their detention lasted not long, but long enough to place their country under the feet of a master, to annihilate the law, to silence the voice of many of them for ever, and to accomplish that revolution which had haunted M. de Tocqueville through life, when a democratic people, weary of anarchy and incapable of self-government, precipitates itself at the feet of despotic authority. The scene itself was described by M. de Tocqueville himself with indignant animation, for it need be now no more a secret that the narrative of the coup d'état published immediately afterwards by the 'Times' newspaper of the 11th December, 1851, was from his pen.[1]

I renounce the painful, the impracticable task of describing the effects of this blow on M. de Tocqueville's mind. It was not the loss of the objects of common ambition, it was not the closing to himself of that career of public utility to which he was passionately attached and devoted; it was the sense of the moral wreck of his country, and of the extinction of the very source of all true public virtue by her own act.

In May 1852, he wrote to M. de Beaumont:—

> 'Work is at present impossible to me. I attribute this painful incapacity to the disturbing conversations one is always having in Paris. If I were in the country I should attribute it to solitude. The truth is, it proceeds from a sickness of the soul, and will not cease till that is better, which can only come with Time, the great healer of sorrow, as everybody knows: we must wait as patiently as we can till its effects are felt. Yet this sorrow, like all true and lawful sorrows, is dear to me as well as poignant. The sight of all that is done, and still more the opinion formed of it, galls every fibre of pride, of rectitude, and of dignity in my frame. I should be grieved to be less sorrowful. On this score, indeed, I have no reason to complain; for, in truth, I am sorrowful to the death. I have reached my present age through many different circumstances, but with one cause, that of regular liberty. Is this cause lost beyond recovery? I feared it was so in 1848; I fear it still more now, though I am not convinced that this country is not destined again to see constitutional institutions. But will it see them last? these or any others? 'Tis sand. It is vain to ask whether it will abide, but what are the winds that will displace it?

'I enclose a copy of the letter addressed to the electors of my department, in which I resign my seat in the Conseil General. I could not take the oath now exacted. This consequence of the 2nd December is perhaps that portion of the event which is personally most painful to myself. I enjoyed in my department a position of unalloyed gratification. It gave me the moral direction of all the chief local affairs, a sort of government of men's minds founded on personal regard, independently of political opinions. This part of my public duties cast a sort of light on my private life, which was very agreeable. But these are very petty miseries.'

The time is not yet come when the burning language in which he denounced the authors of this revolution can with propriety be made public. But the following observations on the probable duration and character of the Imperial power are so just that they may be cited from an unpublished letter:—

> 'Although this government has established itself by one of the greatest crimes recorded in history, nevertheless it will last for some length of time, unless it precipitates itself to destruction. It will last till its excesses, its wars, its corruptions, have effaced in the public mind the dread of socialism; a change requiring time. God grant that in the interval it may not end in a manner almost as prejudicial to us as to itself, in some extravagant foreign enterprise.[2] We know it but too well in France, governments never escape the law of their origin. This government, which comes by the army, which can only last by the army, which traces back its popularity and even its essence to the recollections of military glory, this government will be fatally impelled to seek for aggrandisement of territory and for exclusive influence abroad; in other words, to war. That at last is what I fear, and what all reasonable men dread as I do. War would assuredly be its death, but its death would perhaps cost dear.'
>
> (Letter of 9th January, 1852.)

Henceforth the life of Alexis de Tocqueville was spent in comparative seclusion, and in total estrangement from public affairs. Educated as a French boy, in colleges and towns, he had not acquired in early life any taste for country life or country pursuits. In one of his letters he remarks that from the age of nine to the age of twenty-four he had never spent six weeks in the country at a time; in another letter he expresses his astonishment that people should be able to lead the life of vegetables. But one of the effects of the revolutions to which society in France has been subjected is to teach a wiser lesson. The Revolution of 1789 had forcibly broken the relations formerly existing between the landed proprietors and the peasantry. The Revolutions of 1830 and of 1851, by detaching considerable portions of the upper classes, enjoying the largest amount of landed property and of intellectual cultivation, from the government of the day, have thrown these classes back to their natural position on their own estates. The consequence is that of late years the improvement of agriculture, the restoration of country houses, and a more active participation in rural interests and pursuits, have become engrossing objects of life to the best portion of the French aristocracy. Alexis de Tocqueville applied himself early, and with increasing success, to this laudable and dignified task. He sought in the first place to heal the breach made by the Revolution of 1789 between the cottage and the château, some traces of which were perceptible at his first election in 1837. The simplicity of his manners, the entire absence of any tinge of pride or pretension in his intercourse with persons of all ranks, the genuine interest he felt in their concerns, the patience with which he was ever ready to listen to them, and the readiness with which he placed the stores of his own wisdom and judgment within their reach, inspired the peasantry before long with unfeigned confidence and affection. He practised to the letter, as Father Lacordaire observed, the Divine command, 'Whosoever will be chief among you, let him be your servant.' Speaking of him to a stranger, one of the Norman farmers said, 'The people are very fond of M. de Tocqueville, but it must be confessed he is very grateful for it.' In 1848, on the proclamation of universal suffrage, the whole population of the district voted by acclamation in his favour. While the election was going on, as he leaned exhausted with fatigue against a door-post, one of the peasants, not personally known to him, came up with Norman frankness and said, 'I am surprised, Monsieur de Tocqueville, that you are tired, for did not every one of us bring you here in his pocket?' He was wont to say that in the hearts of these honest fellows the honour and virtue of the French character had taken refuge, that 'Maître Jean' and 'Maître Pierre,' the worthies and notables of the village, were the only titles of dignity which no revolutions could obliterate; and that his peasant neighbours were the only people with whom he cared to converse beyond the circle of his intimate friends. This relish for the homely fare of a rural district was greatly augmented by his inexhaustible sense of the humorous. His biographers appear to have thought it inconsistent with the dignity of a philosophic Academician to admit his love of fun. When a thing presented itself, as it not uncommonly did, to his mind in a droll aspect, his merriment was unquenchable. He was, what is every day becoming more rare, especially in France, a hearty laugher; indeed his laugh, musical and cheerful as his voice, sometimes got the better of him and could not be stopped. It partook of the intensity of all the emotions which alternately swayed his sensitive and delicate nervous organisation. As another instance of the delicacy of his frame, it may be mentioned that he could not support so much as the perfume of a flower, and might literally be said to 'Die of a rose in aromatic pain.'

Thus living in his own ancestral home, without the smallest attempt to humour the democratic passions of his neighbours, he did practically subdue them. He became precisely what he admired in the position of the landed gentlemen of England, independent of the State, independent of the people, but ready and willing to serve the State and to serve the people in all honour. Under these circumstances, he devoted himself to the literary task he had marked out of tracing the Revolution to its true sources: and the originality of his mind can hardly be more demonstrated than by the fact that after all the innumerable commentaries and histories of the French Revolution which have appeared, Alexis de Tocqueville presented to the world an entirely new view of it.

The publication of his last book in 1856 was followed, in 1857, by his last journey to England. The reception he met with here was in fact the last triumph of his life. He was received on all sides with demonstrations of respect and affection; and when the time came for his return to Normandy, the Lords of the Admiralty, hearing that there was no direct steam communication from England to Cherbourg, placed a small vessel at his disposal, which landed him within a mile or two of his own park. At that time nothing appeared to indicate that his life, always precarious, was in any immediate danger. He lived by nervous power, and that seemed unexhausted; indeed, it had repeatedly carried him through dangerous and acute disorders. But in the summer of 1858 a more serious accident showed his lungs to be affected. In the autumn he was ordered to a milder climate than that of his own well-beloved domain. He repaired to Cannes, accompanied by the devoted partner of his life, and by one or two of his nearest relatives and friends. For a time he imagined that the affection of the lungs had been overcome. But in spite of the illusions which attend the closing stages of pulmonary disease, it soon became obvious that life was ebbing away. He received with piety the last sacraments of the Church; for though faith, like every other gift of his nature, had been with him a matter of internal edification rather than of outward display, he had never ceased to entertain the most serious attachment to the Christian religion, and to that Church in which he was born. On the 16th April, 1859, he expired. By his own express desire his mortal remains were interred in the churchyard of Tocqueville, and were attended to the grave by an immense assemblage, not of those who admired him for his genius, but of those who loved him for his goodness; and a plain cross of wood, after the fashion of the country, marks the spot where whatever of him was mortal lies.[3]

DEMOCRACY IN AMERICA

FIRST PART

Introductory Chapter

Amongst the novel objects that attracted my attention during my stay in the United States, nothing struck me more forcibly than the general equality of conditions. I readily discovered the prodigious influence which this primary fact exercises on the whole course of society, by giving a certain direction to public opinion, and a certain tenor to the laws; by imparting new maxims to the governing powers, and peculiar habits to the governed. I speedily perceived that the influence of this fact extends far beyond the political character and the laws of the country, and that it has no less empire over civil society than over the Government; it creates opinions, engenders sentiments, suggests the ordinary practices of life, and modifies whatever it does not produce. The more I advanced in the study of American society, the more I perceived that the equality of conditions is the fundamental fact from which all others seem to be derived, and the central point at which all my observations constantly terminated.

I then turned my thoughts to our own hemisphere, where I imagined that I discerned something analogous to the spectacle which the New World presented to me. I observed that the equality of conditions is daily progressing towards those extreme limits which it seems to have reached in the United States, and that the democracy which governs the American communities appears to be rapidly rising into power in Europe. I hence conceived the idea of the book which is now before the reader.

It is evident to all alike that a great democratic revolution is going on amongst us; but there are two opinions as to its nature and consequences. To some it appears to be a novel accident, which as such may still be checked; to others it seems irresistible, because it is the most uniform, the most ancient, and the most permanent tendency which is to be found in history. Let us recollect the situation of France seven hundred years ago, when the territory was divided amongst a small number of families, who were the owners of the soil and the rulers of the inhabitants; the right of governing descended with the family inheritance from generation to generation; force was the only means by which man could act on man, and landed property was the sole source of power. Soon, however, the political power of the clergy was founded, and began to exert itself: the clergy opened its ranks to all classes, to the poor and the rich, the villein and the lord; equality penetrated into the Government through the Church, and the being who as a serf must have vegetated in perpetual bondage took his place as a priest in the midst of nobles, and not unfrequently above the heads of kings.

The different relations of men became more complicated and more numerous as society gradually became more stable and more civilised. Thence the want of civil laws was felt; and the order of legal functionaries soon rose from the obscurity of the tribunals and their dusty chambers, to appear at the court of the monarch, by the side of the feudal barons in their ermine and their mail. Whilst the kings were ruining themselves by their great enterprises, and the nobles exhausting their resources by private wars, the lower orders were enriching themselves by commerce. The influence of money began to be perceptible in State affairs. The transactions of business opened a new road to power, and the financier rose to a station of political influence in which he was at once flattered and despised. Gradually the spread of mental acquirements, and the increasing taste for literature and art, opened chances of success to talent; science became a means of government, intelligence led to social power, and the man of letters took a part in the affairs of the State. The value attached to the privileges of birth decreased in the exact proportion in which new paths were struck out to advancement. In the eleventh century nobility was beyond all price; in the thirteenth it might be purchased; it was conferred for the first time in 1270; and equality was thus introduced into the Government by the aristocracy itself.

In the course of these seven hundred years it sometimes happened that in order to resist the authority of the Crown, or to diminish the power of their rivals, the nobles granted a certain share of political rights to the people. Or, more frequently, the king permitted the lower orders to enjoy a degree of power, with the intention of repressing the aristocracy. In France the kings have always been the most active and the most constant of levellers. When they were strong and ambitious they spared no pains to raise the people to the level of the nobles; when they were temperate or weak they allowed the people to rise above themselves. Some assisted the democracy by their talents, others by their vices. Louis XI. and Louis XIV. reduced every rank beneath the throne to the same subjection; Louis XV. descended, himself and all his Court, into the dust.

As soon as land was held on any other than a feudal tenure, and personal property began in its turn to confer influence and power, every improvement which was introduced in commerce or manufacture was a fresh element of the equality of conditions. Henceforward every new discovery, every new want which it engendered, and every new desire which craved satisfaction, was a step towards the universal level. The taste for luxury, the love of war, the sway of fashion, and the most superficial as well as the deepest passions of the human heart, cooperated to enrich the poor and to impoverish the rich.

From the time when the exercise of the intellect became the source of strength and of wealth, it is impossible not to consider every addition to science, every fresh truth, and every new idea as a germ of power placed within the reach of the people. Poetry, eloquence, and memory, the grace of wit, the glow of imagination, the depth of thought, and all the gifts which are bestowed by Providence with an equal hand, turned to the advantage of the democracy; and even when they were in the possession of its adversaries they still served its cause by throwing into relief the natural greatness of man; its conquests spread, therefore, with those of civilization and knowledge, and literature became an arsenal where the poorest and the weakest could always find weapons to their hand.

In perusing the pages of our history, we shall scarcely meet with a single great event, in the lapse of seven hundred years, which has not turned to the advantage of equality. The Crusades and the wars of the English decimated the nobles and divided their possessions; the erection of communities introduced an element of democratic liberty into the bosom of feudal monarchy; the invention of fire-arms equalized the villein and the noble on the field of battle; printing opened the same resources to the minds of all classes; the post was organised so as to bring the same information to the door of the poor man's cottage and to the gate of the palace; and Protestantism proclaimed that all men are alike able to find the road to heaven. The discovery of America offered a thousand new paths to fortune, and placed riches and power within the reach of the adventurous and the obscure. If we examine what has happened in France at intervals of fifty years, beginning with the eleventh century, we shall invariably perceive that a twofold revolution has taken place in the state of society. The noble has gone down on the social ladder, and the *roturier* has gone up; the one descends as the other rises. Every half century brings them nearer to each other, and they will very shortly meet.

Nor is this phenomenon at all peculiar to France. Whithersoever we turn our eyes we shall witness the same continual revolution throughout the whole of Christendom. The various occurrences of national existence have everywhere turned to the advantage of democracy; all men have aided it by their exertions: those who have intentionally labored in its cause, and those who have served it unwittingly; those who have fought for it and those who have declared themselves its opponents, have all been driven along in the same track, have all labored to one end, some ignorantly and some unwillingly; all have been blind instruments in the hands of God.

The gradual development of the equality of conditions is therefore a providential fact, and it possesses all the characteristics of a divine decree: it is universal, it is durable, it constantly eludes all human interference, and all events as well as all men contribute to its progress. Would it, then, be wise to imagine that a social impulse which dates from so far back can be checked by the efforts of a generation? Is it credible that the democracy which has annihilated the feudal system and vanquished kings will respect the citizen and the capitalist? Will it stop now that it has grown so strong and its adversaries so weak? None can say which way we are going, for all terms of comparison are wanting: the equality of conditions is more complete in the Christian countries of the present day than it has been at any time or in any part of the world; so that the extent of what already exists prevents us from foreseeing what may be yet to come.

The whole book which is here offered to the public has been written under the impression of a kind of religious dread produced in the author's mind by the contemplation of so irresistible a revolution, which has advanced for centuries in spite of such amazing obstacles, and which is still proceeding in the midst of the ruins it has made. It is not necessary that God himself should speak in order to disclose to us the unquestionable signs of His will; we can discern them in the habitual course of nature, and in the invariable tendency of events: I know, without a special revelation, that the planets move in the orbits traced by the Creator's finger. If the men of our time were led by attentive observation and by sincere reflection to acknowledge that the gradual and progressive development of social equality is at once the past and future of their history, this solitary truth would confer the sacred character of a Divine decree upon the change. To attempt to check democracy would be in that case to resist the will of God; and the nations would then be constrained to make the best of the social lot awarded to them by Providence.

The Christian nations of our age seem to me to present a most alarming spectacle; the impulse which is bearing them along is so strong that it cannot be stopped, but it is not yet so rapid that it cannot be guided: their fate is in their hands; yet a little while and it may be so no longer. The first duty which is at this time imposed upon those who direct our affairs is to educate the democracy; to warm its faith, if that be possible; to purify its morals; to direct its energies; to substitute a knowledge of business for its inexperience, and an acquaintance with its true interests for its blind propensities; to adapt its government to time and place, and to modify it in compliance with the occurrences and the actors of the age. A new science of politics is indispensable to a new world. This, however, is what we think of least; launched in the middle of a rapid stream, we obstinately fix our eyes on the ruins which may still be descried upon the shore we have left, whilst the current sweeps us along, and drives us backwards towards the gulf.

In no country in Europe has the great social revolution which I have been describing made such rapid progress as in France; but it has always been borne on by chance. The heads of the State have never had any forethought for its exigencies, and its victories have been obtained without their consent or without their knowledge. The most powerful, the most intelligent, and the most moral classes of the nation have never attempted to connect themselves with it in order to guide it. The people has consequently been abandoned to its wild propensities, and it has grown up like those outcasts who receive their education in the public streets, and who are unacquainted with aught but the vices and wretchedness of society. The existence of a democracy was seemingly unknown, when on a sudden it took possession of the supreme power. Everything was then submitted to its caprices; it was worshipped as the idol of strength; until, when it was enfeebled by its own excesses, the legislator conceived the rash project of annihilating its power, instead of instructing it and correcting its vices; no attempt was made to fit it to govern, but all were bent on excluding it from the government.

The consequence of this has been that the democratic revolution has been effected only in the material parts of society, without that concomitant change in laws, ideas, customs, and manners which was necessary to render such a revolution beneficial. We have gotten a democracy, but without the conditions which lessen its vices and render its natural advantages more prominent; and although we already perceive the evils it brings, we are ignorant of the benefits it may confer.

While the power of the Crown, supported by the aristocracy, peaceably governed the nations of Europe, society possessed, in the midst of its wretchedness, several different advantages which can now scarcely be appreciated or conceived. The power of a part of his subjects was an insurmountable barrier to the tyranny of the prince; and the monarch, who felt the almost divine character which he enjoyed in the eyes of the multitude, derived a motive for the just use of his power from the respect which he inspired. High as they were placed above the people, the nobles could not but take that calm and benevolent interest in its fate which the shepherd feels towards his flock; and without acknowledging the poor as their equals, they watched over the destiny of those whose welfare Providence had entrusted to their care. The people never having conceived the idea of a social condition different from its own, and entertaining no expectation of ever ranking with its chiefs, received benefits from them without discussing their rights. It grew attached to them when they were clement and just, and it submitted without resistance or servility to their exactions, as to the inevitable visitations of the arm of God. Custom, and the manners of the time, had moreover created a species of law in the midst of violence, and established certain limits to oppression. As the noble never suspected that anyone would attempt to deprive him of the privileges which he believed to be legitimate, and as the serf looked upon his own inferiority as a consequence of the immutable order of nature, it is easy to imagine that a mutual exchange of good-will took place between two classes so differently gifted by fate. Inequality and wretchedness were then to be found in society; but the souls of neither rank of men were degraded. Men are not corrupted by the exercise of power or debased by the habit of obedience, but by the exercise of a power which they believe to be illegal and by obedience to a rule which they consider to be usurped and oppressive. On one side was wealth, strength, and leisure, accompanied by the refinements of luxury, the elegance of taste, the pleasures of wit, and the religion of art. On the other was labor and a rude ignorance; but in the midst of this coarse and ignorant multitude it was not uncommon to meet with energetic passions, generous sentiments, profound religious convictions, and independent virtues. The body of a State thus organized might boast of its stability, its power, and, above all, of its glory.

But the scene is now changed, and gradually the two ranks mingle; the divisions which once severed mankind are lowered, property is divided, power is held in common, the light of intelligence spreads, and the capacities of all classes are equally cultivated; the State becomes democratic, and the empire of democracy is slowly and peaceably introduced into the institutions and the manners of the nation. I can conceive a society in which all men would profess an equal attachment and respect for the laws of which they are the common authors; in which the authority of the State would be respected as necessary, though not as divine; and the loyalty of the subject to its chief magistrate would not be a passion, but a quiet and rational persuasion. Every individual being in the possession of rights which he is sure to retain, a kind of manly reliance and reciprocal courtesy would arise between all classes, alike removed from pride and meanness. The people, well acquainted with its true interests, would allow that in order to profit by the advantages of society it is necessary to satisfy its demands. In this state of things the voluntary association of the citizens might supply the individual exertions of the nobles, and the community would be alike protected from anarchy and from oppression.

I admit that, in a democratic State thus constituted, society will not be stationary; but the impulses of the social body may be regulated and directed forwards; if there be less splendor than in the halls of an aristocracy, the contrast of misery will be less frequent also; the pleasures of enjoyment may be less excessive, but those of comfort will be more general; the sciences may be less perfectly cultivated, but ignorance will be less common; the impetuosity of the feelings will be repressed, and the habits of the nation softened; there will be more vices and fewer crimes. In the absence of enthusiasm and of an ardent faith, great sacrifices may be obtained from the members of a commonwealth by an appeal to their understandings and their experience; each individual will feel the same necessity for uniting with his fellow-citizens to protect his own weakness; and as he knows that if they are to assist he must cooperate, he will readily perceive that his personal interest is identified with the interest of the community. The nation, taken as a whole, will be less brilliant, less glorious, and perhaps less strong; but the majority of the citizens will enjoy a greater degree of prosperity, and the people will remain quiet, not because it despairs of amelioration, but because it is conscious of the advantages of its condition. If all the consequences of this state of things were not good or useful, society would at least have appropriated all such as were useful and good; and having once and for ever renounced the social advantages of aristocracy, mankind would enter into possession of all the benefits which democracy can afford.

But here it may be asked what we have adopted in the place of those institutions, those ideas, and those customs of our forefathers which we have abandoned. The spell of royalty is broken, but it has not been succeeded by the majesty of the laws; the people has learned to despise all authority, but fear now extorts a larger tribute of obedience than that which was formerly paid by reverence and by love.

I perceive that we have destroyed those independent beings which were able to cope with tyranny single-handed; but it is the Government that has inherited the privileges of which families, corporations, and individuals have been deprived; the weakness of the whole community has therefore succeeded that influence of a small body of citizens, which, if it was sometimes oppressive, was often conservative. The division of property has lessened the distance which separated the rich from the poor; but it would seem that the nearer they draw to each other, the greater is their mutual hatred, and the more vehement the envy and the dread with which they resist each other's claims to power; the notion of Right is alike insensible to both classes, and Force affords to both the only argument for the present, and the only guarantee for the future. The poor man retains the prejudices of his forefathers without their faith, and their ignorance without their virtues; he has adopted the doctrine of self-interest as the rule of his actions, without understanding the science which controls it, and his egotism is no less blind than his devotedness was formerly. If society is tranquil, it is not because it relies upon its strength and its well-being, but because it knows its weakness and its infirmities; a single effort may cost it its life; everybody feels the evil, but no one has courage or energy enough to seek the cure; the desires, the regret, the sorrows, and the joys of the time produce nothing that is visible or permanent, like the passions of old men which terminate in impotence.

We have, then, abandoned whatever advantages the old state of things afforded, without receiving any compensation from our present condition; we have destroyed an aristocracy, and we seem inclined to survey its ruins with complacency, and to fix our abode in the midst of them.

The phenomena which the intellectual world presents are not less deplorable. The democracy of France, checked in its course or abandoned to its lawless passions, has overthrown whatever crossed its path, and has shaken all that it has not destroyed. Its empire on society has not been gradually introduced or peaceably established, but it has constantly advanced in the midst of disorder and the agitation of a conflict. In the heat of the struggle each partisan is hurried beyond the limits of his opinions by the opinions and the excesses of his opponents, until he loses sight of the end of his exertions, and holds a language which disguises his real sentiments or secret instincts. Hence arises the strange confusion which we are witnessing. I cannot recall to my mind a passage in history more worthy of sorrow and of pity than the scenes which are happening under our eyes; it is as if the natural bond which unites the opinions of man to his tastes and his actions to his principles was now broken; the sympathy which has always been acknowledged between the feelings and the ideas of mankind appears to be dissolved, and all the laws of moral analogy to be abolished.

Zealous Christians may be found amongst us whose minds are nurtured in the love and knowledge of a future life, and who readily espouse the cause of human liberty as the source of all moral greatness. Christianity, which has declared that all men are equal in the sight of God, will not refuse to acknowledge that all citizens are equal in the eye of the law. But, by a singular concourse of events, religion is entangled in those institutions which democracy assails, and it is not unfrequently brought to reject the equality it loves, and to curse that cause of liberty as a foe which it might hallow by its alliance.

By the side of these religious men I discern others whose looks are turned to the earth more than to Heaven; they are the partisans of liberty, not only as the source of the noblest virtues, but more especially as the root of all solid advantages; and they sincerely desire to extend its sway, and to impart its blessings to mankind. It is natural that they should hasten to invoke the assistance of religion, for they must know that liberty cannot be established without morality, nor morality without faith; but they have seen religion in the ranks of their adversaries, and they inquire no further; some of them attack it openly, and the remainder are afraid to defend it.

In former ages slavery has been advocated by the venal and slavish-minded, whilst the independent and the warm-hearted were struggling without hope to save the liberties of mankind. But men of high and generous characters are now to be met with, whose opinions are at variance with their inclinations, and who praise that servility which they have themselves never known. Others, on the contrary, speak in the name of liberty, as if they were able to feel its sanctity and its majesty, and loudly claim for humanity those rights which they have always disowned. There are virtuous and peaceful individuals whose pure morality, quiet habits, affluence, and talents fit them to be the leaders of the surrounding population; their love of their country is sincere, and they are prepared to make the greatest sacrifices to its welfare, but they confound the abuses of civilization with its benefits, and the idea of evil is inseparable in their minds from that of novelty.

Not far from this class is another party, whose object is to materialize mankind, to hit upon what is expedient without heeding what is just, to acquire knowledge without faith, and prosperity apart from virtue; assuming the title of the champions of modern civilization, and placing themselves in a station which they usurp with insolence, and from which they are driven by their own unworthiness. Where are we then? The religionists are the enemies of liberty, and the friends of liberty attack religion; the high-minded and the noble advocate subjection, and the meanest and most servile minds preach independence; honest and enlightened citizens are opposed to all progress, whilst men without patriotism and without principles are the apostles of civilization and of intelligence. Has such been the fate of the centuries which have preceded our own? and has man always inhabited a world like the present, where nothing is linked together, where virtue is without genius, and genius without honor; where the love of order is confounded with a taste for oppression, and the holy rites of freedom with a contempt of law; where the light thrown by conscience on human actions is dim, and where nothing seems to be any longer forbidden or allowed, honorable or shameful, false or true? I cannot, however, believe that the Creator made man to leave him in an endless struggle with the intellectual miseries which surround us: God destines a calmer and a more certain future to the communities of Europe; I am unacquainted with His designs, but I shall not cease to believe in them because I cannot fathom them, and I had rather mistrust my own capacity than His justice.

There is a country in the world where the great revolution which I am speaking of seems nearly to have reached its natural limits; it has been effected with ease and simplicity, say rather that this country has attained the consequences of the democratic revolution which we are undergoing without having experienced the revolution itself. The emigrants who fixed themselves on the shores of America in the beginning of the seventeenth century severed the democratic principle from all the principles which repressed it in the old communities of Europe, and transplanted it unalloyed to the New World. It has there been allowed to spread in perfect freedom, and to put forth its consequences in the laws by influencing the manners of the country.

It appears to me beyond a doubt that sooner or later we shall arrive, like the Americans, at an almost complete equality of conditions. But I do not conclude from this that we shall ever be necessarily led to draw the same political consequences which the Americans have derived from a similar social organization. I am far from supposing that they have chosen the only form of government which a democracy may adopt; but the identity of the efficient cause of laws and manners in the two countries is sufficient to account for the immense interest we have in becoming acquainted with its effects in each of them.

It is not, then, merely to satisfy a legitimate curiosity that I have examined America; my wish has been to find instruction by which we may ourselves profit. Whoever should imagine that I have intended to write a panegyric will perceive that such was not my design; nor has it been my object to advocate any form of government in particular, for I am of opinion that absolute excellence is rarely to be found in any legislation; I have not even affected to discuss whether the social revolution, which I believe to be irresistible, is advantageous or prejudicial to mankind; I have acknowledged this revolution as a fact already accomplished or on the eve of its accomplishment; and I have selected the nation, from amongst those which have undergone it, in which its development has been the most peaceful and the most complete, in order to discern its natural consequences, and, if it be possible, to distinguish the means by which it may be rendered profitable. I confess that in America I saw more than America; I sought the image of democracy itself, with its inclinations, its character, its prejudices, and its passions, in order to learn what we have to fear or to hope from its progress.

In the first part of this work I have attempted to show the tendency given to the laws by the democracy of America, which is abandoned almost without restraint to its instinctive propensities, and to exhibit the course it prescribes to the Government and the influence it exercises on affairs. I have sought to discover the evils and the advantages which it produces. I have examined the precautions used by the Americans to direct it, as well as those which they have not adopted, and I have undertaken to point out the causes which enable it to govern society. I do not know whether I have succeeded in making known what I saw in America, but I am certain that such has been my sincere desire, and that I have never, knowingly, moulded facts to ideas, instead of ideas to facts.

Whenever a point could be established by the aid of written documents, I have had recourse to the original text, and to the most authentic and approved works. I have cited my authorities in the notes, and any one may refer to them. Whenever an opinion, a political custom, or a remark on the manners of the country was concerned, I endeavored to consult the most enlightened men I met with. If the point in question was important or doubtful, I was not satisfied with one testimony, but I formed my opinion on the evidence of several witnesses. Here the reader must necessarily believe me upon my word. I could frequently have quoted names which are either known to him, or which deserve to be so, in proof of what I advance; but I have carefully abstained from this practice. A stranger frequently hears important truths at the fire-side of his host, which the latter would perhaps conceal from the ear of friendship; he consoles himself with his guest for the silence to which he is restricted, and the shortness of the traveller's stay takes away all fear of his indiscretion. I carefully noted every conversation of this nature as soon as it occurred, but these notes will never leave my writing-case; I had rather injure the success of my statements than add my name to the list of those strangers who repay the generous hospitality they have received by subsequent chagrin and annoyance.

I am aware that, notwithstanding my care, nothing will be easier than to criticise this book, if any one ever chooses to criticise it. Those readers who may examine it closely will discover the fundamental idea which connects the several parts together. But the diversity of the subjects I have had to treat is exceedingly great, and it will not be difficult to oppose an isolated fact to the body of facts which I quote, or an isolated idea to the body of ideas I put forth. I hope to be read in the spirit which has guided my labors, and that my book may be judged by the general impression it leaves, as I have formed my own judgment not on any single reason, but upon the mass of evidence. It must not be forgotten that the author who wishes to be understood is obliged to push all his ideas to their utmost theoretical consequences, and often to the verge of what is false or impracticable; for if it be necessary sometimes to quit the rules of logic in active life, such is not the case in discourse, and a man finds that almost as many difficulties spring from inconsistency of language as usually arise from inconsistency of conduct.

I conclude by pointing out myself what many readers will consider the principal defect of the work. This book is written to favor no particular views, and in composing it I have entertained no designs of serving or attacking any party; I have undertaken not to see differently, but to look further than parties, and whilst they are busied for the morrow I have turned my thoughts to the Future.

Chapter I

Exterior Form of North America

North America divided into two vast regions, one inclining towards the Pole, the other towards the Equator—Valley of the Mississippi—Traces of the Revolutions of the Globe—Shore of the Atlantic Ocean where the English Colonies were founded—Difference in the appearance of North and of South America at the time of their Discovery—Forests of North America—Prairies—Wandering Tribes of Natives—Their outward appearance, manners, and language—Traces of an unknown people.

North America presents in its external form certain general features which it is easy to discriminate at the first glance. A sort of methodical order seems to have regulated the separation of land and water, mountains and valleys. A simple but grand arrangement is discoverable amidst the confusion of objects and the prodigious variety of scenes. This Continent is divided, almost equally, into two vast regions, one of which is bounded on the north by the Arctic Pole, and by the two great Oceans on the east and west. It stretches towards the south, forming a triangle whose irregular sides meet at length below the great lakes of Canada. The second region begins where the other terminates, and includes all the remainder of the continent. The one slopes gently towards the Pole, the other towards the Equator.

The territory comprehended in the first region descends towards the north with so imperceptible a slope that it may almost be said to form a level plain. Within the bounds of this immense tract of country there are neither high mountains nor deep valleys. Streams meander through it irregularly: great rivers mix their currents, separate and meet again, disperse and form vast marshes, losing all trace of their channels in the labyrinth of waters they have themselves created; and thus, at length, after innumerable windings, fall into the Polar Seas. The great lakes which bound this first region are not walled in, like most of those in the Old World, between hills and rocks. Their banks are flat, and rise but a few feet above the level of their waters; each of them thus forming a vast bowl filled to the brim. The slightest change in the structure of the globe would cause their waters to rush either towards the Pole or to the tropical sea.

The second region is more varied on its surface, and better suited for the habitation of man. Two long chains of mountains divide it from one extreme to the other; the Alleghany ridge takes the form of the shores of the Atlantic Ocean; the other is parallel with the Pacific. The space which lies between these two chains of mountains contains 1,341,649 square miles.[1] Its surface is therefore about six times as great as that of France. This vast territory, however, forms a single valley, one side of which descends gradually from the rounded summits of the Alleghanies, while the other rises in an uninterrupted course towards the tops of the Rocky Mountains. At the bottom of the valley flows an immense river, into which the various streams issuing from the mountains fall from all parts. In memory of their native land, the French formerly called this river the St. Louis. The Indians, in their pompous language, have named it the Father of Waters, or the Mississippi.

The Mississippi takes its source above the limit of the two great regions of which I have spoken, not far from the highest point of the table-land where they unite. Near the same spot rises another river,[2] which empties itself into the Polar seas. The course of the Mississippi is at first dubious: it winds several times towards the north, from whence it rose; and at length, after having been delayed in lakes and marshes, it flows slowly onwards to the south. Sometimes quietly gliding along the argillaceous bed which nature has assigned to it, sometimes swollen by storms, the Mississippi waters 2,500 miles in its course.[3] At the distance of 1,364 miles from its mouth this river attains an average depth of fifteen feet; and it is navigated by vessels of 300 tons burden for a course of nearly 500 miles. Fifty-seven large navigable rivers contribute to swell the waters of the Mississippi; amongst others, the Missouri, which traverses a space of 2,500 miles; the Arkansas of 1,300 miles, the Red River 1,000 miles, four whose course is from 800 to 1,000 miles in length, viz., the Illinois, the St. Peter's, the St. Francis, and the Moingona; besides a countless multitude of rivulets which unite from all parts their tributary streams.

The valley which is watered by the Mississippi seems formed to be the bed of this mighty river, which, like a god of antiquity, dispenses both good and evil in its course. On the shores of the stream nature displays an inexhaustible fertility; in proportion as you recede from its banks, the powers of vegetation languish, the soil becomes poor, and the plants that survive have a sickly growth. Nowhere have the great convulsions of the globe left more evident traces than in the valley of the Mississippi; the whole aspect of the country shows the powerful effects of water, both by its fertility and by its barrenness. The waters of the primeval ocean accumulated enormous beds of vegetable mould in the valley, which they levelled as they retired. Upon the right shore of the river are seen immense plains, as smooth as if the husbandman had passed over them with his roller. As you approach the mountains the soil becomes more and more unequal and sterile; the ground is, as it were, pierced in a thousand places by primitive rocks, which appear like the bones of a skeleton whose flesh is partly consumed. The surface of the earth is covered with a granite sand and huge irregular masses of stone, among which a few plants force their growth, and give the appearance of a green field covered with the ruins of a vast edifice. These stones and this sand discover, on examination, a perfect analogy with those which compose the arid and broken summits of the Rocky Mountains. The flood of waters which washed the soil to the bottom of the valley afterwards carried away portions of the rocks themselves; and these, dashed and bruised against the neighboring cliffs, were left scattered like wrecks at their feet.[4] The valley of the Mississippi is, upon the whole, the most magnificent dwelling-place prepared by God for man's abode; and yet it may be said that at present it is but a mighty desert.

On the eastern side of the Alleghanies, between the base of these mountains and the Atlantic Ocean, there lies a long ridge of rocks and sand, which the sea appears to have left behind as it retired. The mean breadth of this territory does not exceed one hundred miles; but it is about nine hundred miles in length. This part of the American continent has a soil which offers every obstacle to the husbandman, and its vegetation is scanty and unvaried.

Upon this inhospitable coast the first united efforts of human industry were made. The tongue of arid land was the cradle of those English colonies which were destined one day to become the United States of America. The centre of power still remains here; whilst in the backwoods the true elements of the great people to whom the future control of the continent belongs are gathering almost in secrecy together.

When the Europeans first landed on the shores of the West Indies, and afterwards on the coast of South America, they thought themselves transported into those fabulous regions of which poets had sung. The sea sparkled with phosphoric light, and the extraordinary transparency of its waters discovered to the view of the navigator all that had hitherto been hidden in the deep abyss.[5] Here and there appeared little islands perfumed with odoriferous plants, and resembling baskets of flowers floating on the tranquil surface of the ocean. Every object which met the sight, in this enchanting region, seemed prepared to satisfy the wants or contribute to the pleasures of man. Almost all the trees were loaded with nourishing fruits, and those which were useless as food delighted the eye by the brilliancy and variety of their colors. In groves of fragrant lemon-trees, wild figs, flowering myrtles, acacias, and oleanders, which were hung with festoons of various climbing plants, covered with flowers, a multitude of birds unknown in Europe displayed their bright plumage, glittering with purple and azure, and mingled their warbling with the harmony of a world teeming with life and motion.[6] Underneath this brilliant exterior death was concealed. But the air of these climates had so enervating an influence that man, absorbed by present enjoyment, was rendered regardless of the future.

North America appeared under a very different aspect; there everything was grave, serious, and solemn: it seemed created to be the domain of intelligence, as the South was that of sensual delight. A turbulent and foggy ocean washed its shores. It was girt round by a belt of granite rocks, or by wide tracts of sand. The foliage of its woods was dark and gloomy, for they were composed of firs, larches, evergreen oaks, wild olive-trees, and laurels. Beyond this outer belt lay the thick shades of the central forest, where the largest trees which are produced in the two hemispheres grow side by side. The plane, the catalpa, the sugar-maple, and the Virginian poplar mingled their branches with those of the oak, the beech, and the lime. In these, as in the forests of the Old World, destruction was perpetually going on. The ruins of vegetation were heaped upon each other; but there was no laboring hand to remove them, and their decay was not rapid enough to make room for the continual work of reproduction. Climbing-plants, grasses, and other herbs forced their way through the mass of dying trees; they crept along their bending trunks, found nourishment in their dusty cavities, and a passage beneath the lifeless bark. Thus decay gave its assistance to life, and their respective productions were mingled together. The depths of these forests were gloomy and obscure, and a thousand rivulets, undirected in their course by human industry, preserved in them a constant moisture. It was rare to meet with flowers, wild fruits, or birds beneath their shades. The fall of a tree overthrown by age, the rushing torrent of a cataract, the lowing of the buffalo, and the howling of the wind were the only sounds which broke the silence of nature.

To the east of the great river, the woods almost disappeared; in their stead were seen prairies of immense extent. Whether Nature in her infinite variety had denied the germs of trees to these fertile plains, or whether they had once been covered with forests, subsequently destroyed by the hand of man, is a question which neither tradition nor scientific research has been able to resolve.

These immense deserts were not, however, devoid of human inhabitants. Some wandering tribes had been for ages scattered among the forest shades or the green pastures of the prairie. From, the mouth of the St. Lawrence to the delta of the Mississippi, and from the Atlantic to the Pacific Ocean, these savages possessed certain points of resemblance which bore witness of their common origin; but at the same time they differed from all other known races of men:[7] they were neither white like the Europeans, nor yellow like most of the Asiatics, nor black like the negroes. Their skin was reddish brown, their hair long and shining, their lips thin, and their cheekbones very prominent. The languages spoken by the North American tribes are various as far as regarded their words, but they were subject to the same grammatical rules. These rules differed in several points from such as had been observed to govern the origin of language. The idiom of the Americans seemed to be the product of new combinations, and bespoke an effort of the understanding of which the Indians of our days would be incapable.[8]

The social state of these tribes differed also in many respects from all that was seen in the Old World. They seemed to have multiplied freely in the midst of their deserts without coming in contact with other races more civilized than their own. Accordingly, they exhibited none of those indistinct, incoherent notions of right and wrong, none of that deep corruption of manners, which is usually joined with ignorance and rudeness among nations which, after advancing to civilization, have relapsed into a state of barbarism. The Indian was indebted to no one but himself; his virtues, his vices, and his prejudices were his own work; he had grown up in the wild independence of his nature.

If, in polished countries, the lowest of the people are rude and uncivil, it is not merely because they are poor and ignorant, but that, being so, they are in daily contact with rich and enlightened men. The sight of their own hard lot and of their weakness, which is daily contrasted with the happiness and power of some of their fellow-creatures, excites in their hearts at the same time the sentiments of anger and of fear: the consciousness of their inferiority and of their dependence irritates while it humiliates them. This state of mind displays itself in their manners and language; they are at once insolent and servile. The truth of this is easily proved by observation; the people are more rude in aristocratic countries than elsewhere, in opulent cities than in rural districts. In those places where the rich and powerful are assembled together the weak and the indigent feel themselves oppressed by their inferior condition. Unable to perceive a single chance of regaining their equality, they give up to despair, and allow themselves to fall below the dignity of human nature.

This unfortunate effect of the disparity of conditions is not observable in savage life: the Indians, although they are ignorant and poor, are equal and free. At the period when Europeans first came among them the natives of North America were ignorant of the value of riches, and indifferent to the enjoyments which civilized man procures to himself by their means. Nevertheless there was nothing coarse in their demeanor; they practised an habitual reserve and a kind of aristocratic politeness. Mild and hospitable when at peace, though merciless in war beyond any known degree of human ferocity, the Indian would expose himself to die of hunger in order to succor the stranger who asked admittance by night at the door of his hut; yet he could tear in pieces with his hands the still quivering limbs of his prisoner. The famous republics of antiquity never gave examples of more unshaken courage, more haughty spirits, or more intractable love of independence than were hidden in former times among the wild forests of the New World.[9] The Europeans produced no great impression when they landed upon the shores of North America; their presence engendered neither envy nor fear. What influence could they possess over such men as we have described? The Indian could live without wants, suffer without complaint, and pour out his death-song at the stake.[10] Like all the other members of the great human family, these savages believed in the existence of a better world, and adored, under different names, God, the creator of the universe. Their notions on the great intellectual truths were in general simple and philosophical.[11]

Although we have here traced the character of a primitive people, yet it cannot be doubted that another people, more civilized and more advanced in all respects, had preceded it in the same regions.

An obscure tradition which prevailed among the Indians to the north of the Atlantic informs us that these very tribes formerly dwelt on the west side of the Mississippi. Along the banks of the Ohio, and throughout the central valley, there are frequently found, at this day, tumuli raised by the hands of men. On exploring these heaps of earth to their centre, it is usual to meet with human bones, strange instruments, arms and utensils of all kinds, made of metal, or destined for purposes unknown to the present race. The Indians of our time are unable to give any information relative to the history of this unknown people. Neither did those who lived three hundred years ago, when America was first discovered, leave any accounts from which even an hypothesis could be formed. Tradition—that perishable, yet ever renewed monument of the pristine world—throws no light upon the subject. It is an undoubted fact, however, that in this part of the globe thousands of our fellow-beings had lived. When they came hither, what was their origin, their destiny, their history, and how they perished, no one can tell. How strange does it appear that nations have existed, and afterwards so completely disappeared from the earth that the remembrance of their very names is effaced; their languages are lost; their glory is vanished like a sound without an echo; though perhaps there is not one which has not left behind it some tomb in memory of its passage! The most durable monument of human labor is that which recalls the wretchedness and nothingness of man.

Although the vast country which we have been describing was inhabited by many indigenous tribes, it may justly be said at the time of its discovery by Europeans to have formed one great desert. The Indians occupied without possessing it. It is by agricultural labor that man appropriates the soil, and the early inhabitants of North America lived by the produce of the chase. Their implacable prejudices, their uncontrolled passions, their vices, and still more perhaps their savage virtues, consigned them to inevitable destruction. The ruin of these nations began from the day when Europeans landed on their shores; it has proceeded ever since, and we are now witnessing the completion of it. They seem to have been placed by Providence amidst the riches of the New World to enjoy them for a season, and then surrender them. Those coasts, so admirably adapted for commerce and industry; those wide and deep rivers; that inexhaustible valley of the Mississippi; the whole continent, in short, seemed prepared to be the abode of a great nation, yet unborn.

In that land the great experiment was to be made, by civilized man, of the attempt to construct society upon a new basis; and it was there, for the first time, that theories hitherto unknown, or deemed impracticable, were to exhibit a spectacle for which the world had not been prepared by the history of the past.

Chapter II

Origin of the Anglo-Americans, and Its Importance in Relation to Their Future Condition

Utility of knowing the origin of nations in order to understand their social condition and their laws—America the only country in which the starting-point of a great people has been clearly observable—In what respects all who emigrated to British America were similar—In what they differed—Remark applicable to all Europeans who established themselves on the shores of the New World—Colonization of Virginia—Colonization of New England—Original character of the first inhabitants of New England—Their arrival—Their first laws—Their social contract—Penal code borrowed from the Hebrew legislation—Religious fervor—Republican spirit—Intimate union of the spirit of religion with the spirit of liberty.

After the birth of a human being his early years are obscurely spent in the toils or pleasures of childhood. As he grows up the world receives him, when his manhood begins, and he enters into contact with his fellows. He is then studied for the first time, and it is imagined that the germ of the vices and the virtues of his maturer years is then formed. This, if I am not mistaken, is a great error. We must begin higher up; we must watch the infant in its mother's arms; we must see the first images which the external world casts upon the dark mirror of his mind; the first occurrences which he witnesses; we must hear the first words which awaken the sleeping powers of thought, and stand by his earliest efforts, if we would understand the prejudices, the habits, and the passions which will rule his life. The entire man is, so to speak, to be seen in the cradle of the child.

The growth of nations presents something analogous to this: they all bear some marks of their origin; and the circumstances which accompanied their birth and contributed to their rise affect the whole term of their being. If we were able to go back to the elements of states, and to examine the oldest monuments of their history, I doubt not that we should discover the primal cause of the prejudices, the habits, the ruling passions, and, in short, of all that constitutes what is called the national character: we should then find the explanation of certain customs which now seem at variance with the prevailing manners; of such laws as conflict with established principles; and of such incoherent opinions as are here and there to be met with in society, like those fragments of broken chains which we sometimes see hanging from the vault of an edifice, and supporting nothing. This might explain the destinies of certain nations, which seem borne on by an unknown force to ends of which they themselves are ignorant. But hitherto facts have been wanting to researches of this kind: the spirit of inquiry has only come upon communities in their latter days; and when they at length contemplated their origin, time had already obscured it, or ignorance and pride adorned it with truth-concealing fables.

America is the only country in which it has been possible to witness the natural and tranquil growth of society, and where the influences exercised on the future condition of states by their origin is clearly distinguishable. At the period when the peoples of Europe landed in the New World their national characteristics were already completely formed; each of them had a physiognomy of its own; and as they had already attained that stage of civilization at which men are led to study themselves, they have transmitted to us a faithful picture of their opinions, their manners, and their laws. The men of the sixteenth century are almost as well known to us as our contemporaries. America, consequently, exhibits in the broad light of day the phenomena which the ignorance or rudeness of earlier ages conceals from our researches. Near enough to the time when the states of America were founded, to be accurately acquainted with their elements, and sufficiently removed from that period to judge of some of their results, the men of our own day seem destined to see further than their predecessors into the series of human events. Providence has given us a torch which our forefathers did not possess, and has allowed us to discern fundamental causes in the history of the world which the obscurity of the past concealed from them. If we carefully examine the social and political state of America, after having studied its history, we shall remain perfectly convinced that not an opinion, not a custom, not a law, I may even say not an event, is upon record which the origin of that people will not explain. The readers of this book will find the germ of all that is to follow in the present chapter, and the key to almost the whole work.

The emigrants who came, at different periods to occupy the territory now covered by the American Union differed from each other in many respects; their aim was not the same, and they governed themselves on different principles. These men had, however, certain features in common, and they were all placed in an analogous situation. The tie of language is perhaps the strongest and the most durable that can unite mankind. All the emigrants spoke the same tongue; they were all offsets from the same people. Born in a country which had been agitated for centuries by the struggles of faction, and in which all parties had been obliged in their turn to place themselves under the protection of the laws, their political education had been perfected in this rude school, and they were more conversant with the notions of right and the principles of true freedom than the greater part of their European contemporaries. At the period of their first emigrations the parish system, that fruitful germ of free institutions, was deeply rooted in the habits of the English; and with it the doctrine of the sovereignty of the people had been introduced into the bosom of the monarchy of the House of Tudor.

The religious quarrels which have agitated the Christian world were then rife. England had plunged into the new order of things with headlong vehemence. The character of its inhabitants, which had always been sedate and reflective, became argumentative and austere. General information had been increased by intellectual debate, and the mind had received a deeper cultivation. Whilst religion was the topic of discussion, the morals of the people were reformed. All these national features are more or less discoverable in the physiognomy of those adventurers who came to seek a new home on the opposite shores of the Atlantic.

Another remark, to which we shall hereafter have occasion to recur, is applicable not only to the English, but to the French, the Spaniards, and all the Europeans who successively established themselves in the New World. All these European colonies contained the elements, if not the development, of a complete democracy. Two causes led to this result. It may safely be advanced, that on leaving the mother-country the emigrants had in general no notion of superiority over one another. The happy and the powerful do not go into exile, and there are no surer guarantees of equality among men than poverty and misfortune. It happened, however, on several occasions, that persons of rank were driven to America by political and religious quarrels. Laws were made to establish a gradation of ranks; but it was soon found that the soil of America was opposed to a territorial aristocracy. To bring that refractory land into cultivation, the constant and interested exertions of the owner himself were necessary; and when the ground was prepared, its produce was found to be insufficient to enrich a master and a farmer at the same time. The land was then naturally broken up into small portions, which the proprietor cultivated for himself. Land is the basis of an aristocracy, which clings to the soil that supports it; for it is not by privileges alone, nor by birth, but by landed property handed down from generation to generation, that an aristocracy is constituted. A nation may present immense fortunes and extreme wretchedness, but unless those fortunes are territorial there is no aristocracy, but simply the class of the rich and that of the poor.

All the British colonies had then a great degree of similarity at the epoch of their settlement. All of them, from their first beginning, seemed destined to witness the growth, not of the aristocratic liberty of their mother-country, but of that freedom of the middle and lower orders of which the history of the world had as yet furnished no complete example.

In this general uniformity several striking differences were however discernible, which it is necessary to point out. Two branches may be distinguished in the Anglo-American family which have hitherto grown up without entirely commingling; the one in the South, the other in the North.

Virginia received the first English colony; the emigrants took possession of it in 1607. The idea that mines of gold and silver are the sources of national wealth was at that time singularly prevalent in Europe; a fatal delusion, which has done more to impoverish the nations which adopted it, and has cost more lives in America, than the united influence of war and bad laws. The men sent to Virginia[1] were seekers of gold, adventurers, without resources and without character, whose turbulent and restless spirit endangered the infant colony,[2] and rendered its progress uncertain. The artizans and agriculturists arrived afterwards; and, although they were a more moral and orderly race of men, they were in nowise above the level of the inferior classes in England.[3] No lofty conceptions, no intellectual system, directed the foundation of these new settlements. The colony was scarcely established when slavery was introduced,[4] and this was the main circumstance which has exercised so prodigious an influence on the character, the laws, and all the future prospects of the South. Slavery, as we shall afterwards show, dishonors labor; it introduces idleness into society, and with idleness, ignorance and pride, luxury and distress. It enervates the powers of the mind, and benumbs the activity of man. The influence of slavery, united to the English character, explains the manners and the social condition of the Southern States.

In the North, the same English foundation was modified by the most opposite shades of character; and here I may be allowed to enter into some details. The two or three main ideas which constitute the basis of the social theory of the United States were first combined in the Northern English colonies, more generally denominated the States of New England.[5] The principles of New England spread at first to the neighboring states; they then passed successively to the more distant ones; and at length they imbued the whole Confederation. They now extend their influence beyond its limits over the whole American world. The civilization of New England has been like a beacon lit upon a hill, which, after it has diffused its warmth around, tinges the distant horizon with its glow.

The foundation of New England was a novel spectacle, and all the circumstances attending it were singular and original. The large majority of colonies have been first inhabited either by men without education and without resources, driven by their poverty and their misconduct from the land which gave them birth, or by speculators and adventurers greedy of gain. Some settlements cannot even boast so honorable an origin; St. Domingo was founded by buccaneers; and the criminal courts of England originally supplied the population of Australia.

The settlers who established themselves on the shores of New England all belonged to the more independent classes of their native country. Their union on the soil of America at once presented the singular phenomenon of a society containing neither lords nor common people, neither rich nor poor. These men possessed, in proportion to their number, a greater mass of intelligence than is to be found in any European nation of our own time. All, without a single exception, had received a good education, and many of them were known in Europe for their talents and their acquirements. The other colonies had been founded by adventurers without family; the emigrants of New England brought with them the best elements of order and morality they landed in the desert accompanied by their wives and children. But what most especially distinguished them was the aim of their undertaking. They had not been obliged by necessity to leave their country; the social position they abandoned was one to be regretted, and their means of subsistence were certain. Nor did they cross the Atlantic to improve their situation or to increase their wealth; the call which summoned them from the comforts of their homes was purely intellectual; and in facing the inevitable sufferings of exile their object was the triumph of an idea.

The emigrants, or, as they deservedly styled themselves, the Pilgrims, belonged to that English sect the austerity of whose principles had acquired for them the name of Puritans. Puritanism was not merely a religious doctrine, but it corresponded in many points with the most absolute democratic and republican theories. It was this tendency which had aroused its most dangerous adversaries. Persecuted by the Government of the mother-country, and disgusted by the habits of a society opposed to the rigor of their own principles, the Puritans went forth to seek some rude and unfrequented part of the world, where they could live according to their own opinions, and worship God in freedom.

A few quotations will throw more light upon the spirit of these pious adventures than all we can say of them. Nathaniel Morton,[6] the historian of the first years of the settlement, thus opens his subject:

'Gentle Reader,—I have for some length of time looked upon it as a duty incumbent, especially on the immediate successors of those that have had so large experience of those many memorable and signal demonstrations of God's goodness, viz. the first beginners of this Plantation in New England, to commit to writing his gracious dispensations on that behalf; having so many inducements thereunto, not onely otherwise but so plentifully in the Sacred Scriptures: that so, what we have seen, and what our fathers have told us (Psalm lxxviii. 3, 4), we may not hide from our children, showing to the generations to come the praises of the Lord; that especially the seed of Abraham his servant, and the children of Jacob his chosen (Psalm cv. 5, 6), may remember his marvellous works in the beginning and progress of the planting of New England, his wonders and the judgments of his mouth; how that God brought a vine into this wilderness; that he cast out the heathen, and planted it; that he made room for it and caused it to take deep root; and it filled the land (Psalm lxxx. 8, 9). And not onely so, but also that he hath guided his people by his strength to his holy habitation and planted them in the mountain of his inheritance in respect of precious Gospel enjoyments: and that as especially God may have the glory of all unto whom it is most due; so also some rays of glory may reach the names of those blessed Saints that were the main instruments and the beginning of this happy enterprise.'

It is impossible to read this opening paragraph without an involuntary feeling of religious awe; it breathes the very savour of Gospel antiquity. The sincerity of the author heightens his power of language. The band which to his eyes was a mere party of adventurers gone forth to seek their fortune beyond seas appears to the reader as the germ of a great nation wafted by Providence to a predestined shore.

The author thus continues his narrative of the departure of the first pilgrims:—

'So they left that goodly and pleasant city of Leyden,[7] which had been their resting-place for above eleven years; but they knew that they were pilgrims and strangers here below, and looked not much on these things, but lifted up their eyes to Heaven, their dearest country, where God hath prepared for them a city (Heb. xi. 16), and therein quieted their spirits. When they came to Delfs-Haven they found the ship and all things ready; and such of their friends as could not come with them followed after them, and sundry came from Amsterdam to see them ship, and to take their leaves of them. One night was spent with little sleep with the most, but with friendly entertainment and Christian discourse, and other real expressions of true Christian love. The next day they went on board, and their friends with them, where truly doleful was the sight of that sad and mournful parting, to hear what sighs and sobs and prayers did sound amongst them; what tears did gush from every eye, and pithy speeches pierced each other's heart, that sundry of the Dutch strangers that stood on the Key as spectators could not refrain from tears. But the tide (which stays for no man) calling them away, that were thus loth to depart, their Reverend Pastor falling down on his knees, and they all with him, with watery cheeks commended them with most fervent prayers unto the Lord and his blessing; and then, with mutual embraces and many tears they took their leaves one of another, which proved to be the last leave to many of them.'

The emigrants were about 150 in number, including the women and the children. Their object was to plant a colony on the shores of the Hudson; but after having been driven about for some time in the Atlantic Ocean, they were forced to land on that arid coast of New England which is now the site of the town of Plymouth. The rock is still shown on which the pilgrims disembarked.[8]

'But before we pass on,' continues our historian, 'let the reader with me make a pause and seriously consider this poor people's present condition, the more to be raised up to admiration of God's goodness towards them in their preservation: for being now passed the vast ocean, and a sea of troubles before them in expectation, they had now no friends to welcome them, no inns to entertain or refresh them, no houses, or much less towns to repair unto to seek for succour: and for the season it was winter, and they that know the winters of the country know them to be sharp and violent, subject to cruel and fierce storms, dangerous to travel to known places, much more to search unknown coasts. Besides, what could they see but a hideous and desolate wilderness, full of wilde beasts, and wilde men? and what multitudes of them there were, they then knew not: for which way soever they turned their eyes (save upward to Heaven) they could have but little solace or content in respect of any outward object; for summer being ended, all things stand in appearance with a weather-beaten face, and the whole country full of woods and thickets, represented a wild and savage hew; if they looked behind them, there was the mighty ocean which they had passed, and was now as a main bar or gulph to separate them from all the civil parts of the world.'

It must not be imagined that the piety of the Puritans was of a merely speculative kind, or that it took no cognizance of the course of worldly affairs. Puritanism, as I have already remarked, was scarcely less a political than a religious doctrine. No sooner had the emigrants landed on the barren coast described by Nathaniel Morton than it was their first care to constitute a society, by passing the following Act:

'In the name of God. Amen. We, whose names are underwritten, the loyal subjects of our dread Sovereign Lord King James, &c., &c., Having undertaken for the glory of God, and advancement of the Christian Faith, and the honour of our King and country, a voyage to plant the first colony in the northern parts of Virginia; Do by these presents solemnly and mutually, in the presence of God and one another, covenant and combine ourselves together into a civil body politick, for our better ordering and preservation, and furtherance of the ends aforesaid: and by virtue hereof do enact, constitute and frame such just and equal laws, ordinances, acts, constitutions, and officers, from time to time, as shall be thought most meet and convenient for the general good of the Colony: unto which we promise all due submission and obedience,' &c.[9]

This happened in 1620, and from that time forwards the emigration went on. The religious and political passions which ravaged the British Empire during the whole reign of Charles I. drove fresh crowds of sectarians every year to the shores of America. In England the stronghold of Puritanism was in the middle classes, and it was from the middle classes that the majority of the emigrants came. The population of New England increased rapidly; and whilst the hierarchy of rank despotically classed the inhabitants of the mother-country, the colony continued to present the novel spectacle of a community homogeneous in all its parts. A democracy, more perfect than any which antiquity had dreamt of, started in full size and panoply from the midst of an ancient feudal society.

The English Government was not dissatisfied with an emigration which removed the elements of fresh discord and of further revolutions. On the contrary, everything was done to encourage it, and great exertions were made to mitigate the hardships of those who sought a shelter from the rigor of their country's laws on the soil of America. It seemed as if New England was a region given up to the dreams of fancy and the unrestrained experiments of innovators.

The English colonies (and this is one of the main causes of their prosperity) have always enjoyed more internal freedom and more political independence than the colonies of other nations; but this principle of liberty was nowhere more extensively applied than in the States of New England.

It was generally allowed at that period that the territories of the New World belonged to that European nation which had been the first to discover them. Nearly the whole coast of North America thus became a British possession towards the end of the sixteenth century. The means used by the English Government to people these new domains were of several kinds; the King sometimes appointed a governor of his own choice, who ruled a portion of the New World in the name and under the immediate orders of the Crown;[10] this is the colonial system adopted by other countries of Europe. Sometimes grants of certain tracts were made by the Crown to an individual or to a company,[11] in which case all the civil and political power fell into the hands of one or more persons, who, under the inspection and control of the Crown, sold the lands and governed the inhabitants. Lastly, a third system consisted in allowing a certain number of emigrants to constitute a political society under the protection of the mother-country, and to govern themselves in whatever was not contrary to her laws. This mode of colonization, so remarkably favorable to liberty, was only adopted in New England.[12]

In 1628[13] a charter of this kind was granted by Charles I. to the emigrants who went to form the colony of Massachusetts. But, in general, charters were not given to the colonies of New England till they had acquired a certain existence. Plymouth, Providence, New Haven, the State of Connecticut, and that of Rhode Island[14] were founded without the cooperation and almost without the knowledge of the mother-country. The new settlers did not derive their incorporation from the seat of the empire, although they did not deny its supremacy; they constituted a society of their own accord, and it was not till thirty or forty years afterwards, under Charles II., that their existence was legally recognized by a royal charter.

This frequently renders it difficult to detect the link which connected the emigrants with the land of their forefathers in studying the earliest historical and legislative records of New England. They exercised the rights of sovereignty; they named their magistrates, concluded peace or declared war, made police regulations, and enacted laws as if their allegiance was due only to God.[15] Nothing can be more curious and, at the same time more instructive, than the legislation of that period; it is there that the solution of the great social problem which the United States now present to the world is to be found.

Amongst these documents we shall notice, as especially characteristic, the code of laws promulgated by the little State of Connecticut in 1650.[16] The legislators of Connecticut[17] begin with the penal laws, and, strange to say, they borrow their provisions from the text of Holy Writ. 'Whosoever shall worship any other God than the Lord,' says the preamble of the Code, 'shall surely be put to death.' This is followed by ten or twelve enactments of the same kind, copied verbatim from the books of Exodus, Leviticus, and Deuteronomy. Blasphemy, sorcery, adultery,[18] and rape were punished with death; an outrage offered by a son to his parents was to be expiated by the same penalty. The legislation of a rude and half-civilized people was thus applied to an enlightened and moral community. The consequence was that the punishment of death was never more frequently prescribed by the statute, and never more rarely enforced towards the guilty.

The chief care of the legislators, in this body of penal laws, was the maintenance of orderly conduct and good morals in the community: they constantly invaded the domain of conscience, and there was scarcely a sin which was not subject to magisterial censure. The reader is aware of the rigor with which these laws punished rape and adultery; intercourse between unmarried persons was likewise severely repressed. The judge was empowered to inflict a pecuniary penalty, a whipping, or marriage[19] on the misdemeanants; and if the records of the old courts of New Haven may be believed, prosecutions of this kind were not unfrequent. We find a sentence bearing date the first of May, 1660, inflicting a fine and reprimand on a young woman who was accused of using improper language, and of allowing herself to be kissed.[20] The Code of 1650 abounds in preventive measures. It punishes idleness and drunkenness with severity.[21] Innkeepers are forbidden to furnish more than a certain quantity of liquor to each consumer; and simple lying, whenever it may be injurious,[22] is checked by a fine or a flogging. In other places, the legislator, entirely forgetting the great principles of religious toleration which he had himself upheld in Europe, renders attendance on divine service compulsory,[23] and goes so far as to visit with severe punishment,[24] and even with death, the Christians who chose to worship God according to a ritual differing from his own.[25] Sometimes indeed the zeal of his enactments induces him to descend to the most frivolous particulars: thus a law is to be found in the same Code which prohibits the use of tobacco.[26] It must not be forgotten that these fantastical and vexatious laws were not imposed by authority, but that they were freely voted by all the persons interested, and that the manners of the community were even more austere and more puritanical than the laws. In 1649 a solemn association was formed in Boston to check the worldly luxury of long hair.[27]

These errors are no doubt discreditable to human reason; they attest the inferiority of our nature, which is incapable of laying firm hold upon what is true and just, and is often reduced to the alternative of two excesses. In strict connection with this penal legislation, which bears such striking marks of a narrow sectarian spirit, and of those religious passions which had been warmed by persecution and were still fermenting among the people, a body of political laws is to be found, which, though written two hundred years ago, is still ahead of the liberties of our age. The general principles which are the groundwork of modern constitutions—principles which were imperfectly known in Europe, and not completely triumphant even in

Great Britain, in the seventeenth century—were all recognised and determined by the laws of New England: the intervention of the people in public affairs, the free voting of taxes, the responsibility of authorities, personal liberty, and trial by jury, were all positively established without discussion. From these fruitful principles consequences have been derived and applications have been made such as no nation in Europe has yet ventured to attempt.

In Connecticut the electoral body consisted, from, its origin, of the whole number of citizens; and this is readily to be understood,[28] when we recollect that this people enjoyed an almost perfect equality of fortune, and a still greater uniformity of opinions.[29] In Connecticut, at this period, all the executive functionaries were elected, including the Governor of the State.[30] The citizens above the age of sixteen were obliged to bear arms; they formed a national militia, which appointed its own officers, and was to hold itself at all times in readiness to march for the defence of the country.[31]

In the laws of Connecticut, as well as in all those of New England, we find the germ and gradual development of that township independence which is the life and mainspring of American liberty at the present day. The political existence of the majority of the nations of Europe commenced in the superior ranks of society, and was gradually and imperfectly communicated to the different members of the social body. In America, on the other hand, it may be said that the township was organized before the county, the county before the State, the State before the Union. In New England townships were completely and definitively constituted as early as 1650. The independence of the township was the nucleus round which the local interests, passions, rights, and duties collected and clung. It gave scope to the activity of a real political life most thoroughly democratic and republican. The colonies still recognized the supremacy of the mother-country; monarchy was still the law of the State; but the republic was already established in every township. The towns named their own magistrates of every kind, rated themselves, and levied their own taxes.[32] In the parish of New England the law of represervation was not adopted, but the affairs of the community were discussed, as at Athens, in the market-place, by a general assembly of the citizens.

In studying the laws which were promulgated at this first era of the American republics, it is impossible not to be struck by the remarkable acquaintance with the science of government and the advanced theory of legislation which they display. The ideas there formed of the duties of society towards its members are evidently much loftier and more comprehensive than those of the European legislators at that time: obligations were there imposed which were elsewhere slighted. In the States of New

England, from the first, the condition of the poor was provided for;[33] strict measures were taken for the maintenance of roads, and surveyors were appointed to attend to them;[34] registers were established in every parish, in which the results of public deliberations, and the births, deaths, and marriages of the citizens were entered;[35] clerks were directed to keep these registers;[36] officers were charged with the administration of vacant inheritances, and with the arbitration of litigated landmarks; and many others were created whose chief functions were the maintenance of public order in the community.[37] The law enters into a thousand useful provisions for a number of social wants which are at present very inadequately felt in France.

But it is by the attention it pays to Public Education that the original character of American civilization is at once placed in the clearest light. 'It being,' says the law, 'one chief project of Satan to keep men from the knowledge of the Scripture by persuading from the use of tongues, to the end that learning may not be buried in the graves of our forefathers, in church and commonwealth, the Lord assisting our endeavors. . . .'[38] Here follow clauses establishing schools in every township, and obliging the inhabitants, under pain of heavy fines, to support them. Schools of a superior kind were founded in the same manner in the more populous districts. The municipal authorities were bound to enforce the sending of children to school by their parents; they were empowered to inflict fines upon all who refused compliance; and in case of continued resistance society assumed the place of the parent, took possession of the child, and deprived the father of those natural rights which he used to so bad a purpose. The reader will undoubtedly have remarked the preamble of these enactments: in America religion is the road to knowledge, and the observance of the divine laws leads man to civil freedom.

If, after having cast a rapid glance over the state of American society in 1650, we turn to the condition of Europe, and more especially to that of the Continent, at the same period, we cannot fail to be struck with astonishment. On the continent of Europe, at the beginning of the seventeenth century, absolute monarchy had everywhere triumphed over the ruins of the oligarchical and feudal liberties of the Middle Ages. Never were the notions of right more completely confounded than in the midst of the splendour and literature of Europe; never was there less political activity among the people; never were the principles of true freedom less widely circulated; and at that very time those principles, which were scorned or unknown by the nations of Europe, were proclaimed in the deserts of the New World, and were accepted as the future creed of a great people. The boldest theories of the human reason were put into practice by a community so humble that not a statesman condescended to attend to it; and a legislation without a precedent was produced offhand by the imagination of the citizens. In the bosom of this obscure democracy, which had as yet brought forth neither generals, nor philosophers, nor authors, a man might stand up in the face of a free people and pronounce the following fine definition of liberty.[39]

'Nor would I have you to mistake in the point of your own liberty. There is a liberty of a corrupt nature which is effected both by men and beasts to do what they list, and this liberty is inconsistent with authority, impatient of all restraint; by this liberty *"sumus omnes deteriores:"* 'tis the grand enemy of truth and peace, and all the ordinances of God are bent against it. But there is a civil, a moral, a federal liberty which is the proper end and object of authority; it is a liberty for that only which is just and good: for this liberty you are to stand with the hazard of your very lives and whatsoever crosses it is not authority, but a distemper thereof. This liberty is maintained in a way of subjection to authority; and the authority set over you will, in all administrations for your good, be quietly submitted unto by all but such as have a disposition to shake off the yoke and lose their true liberty, by their murmuring at the honor and power of authority.'

The remarks I have made will suffice to display the character of Anglo-American civilization in its true light. It is the result (and this should be constantly present to the mind) of two distinct elements, which in other places have been in frequent hostility, but which in America have been admirably incorporated and combined with one another. I allude to the spirit of religion and the spirit of Liberty.

The settlers of New England were at the same time ardent sectarians and daring innovators. Narrow as the limits of some of their religious opinions were, they were entirely free from political prejudices. Hence arose two tendencies, distinct but not opposite, which are constantly discernible in the manners as well as in the laws of the country.

It might be imagined that men who sacrificed their friends, their family, and their native land to a religious conviction were absorbed in the pursuit of the intellectual advantages which they purchased at so dear a rate. The energy, however, with which they strove for the acquirement of wealth, moral enjoyment, and the comforts as well as liberties of the world, is scarcely inferior to that with which they devoted themselves to Heaven.

Political principles and all human laws and institutions were moulded and altered at their pleasure; the barriers of the society in which they were born were broken down before them; the old principles which had governed the world for ages were no more; a path without a turn and a field without an horizon were opened to the exploring and ardent curiosity of man: but at the limits of the political world he checks his researches, he discreetly lays aside the use of his most formidable faculties, he no longer consents to doubt or to innovate, but carefully abstaining from raising the curtain of the sanctuary, he yields with submissive respect to truths which he will not discuss. Thus, in the moral world everything is classed, adapted, decided, and foreseen; in the political world everything is agitated, uncertain, and disputed: in the one is a passive, though a voluntary, obedience; in the other an independence scornful of experience and jealous of authority.

These two tendencies, apparently so discrepant, are far from conflicting; they advance together, and mutually support each other. Religion perceives that civil liberty affords a noble exercise to the faculties of man, and that the political world is a field prepared by the Creator for the efforts of the intelligence. Contented with the freedom and the power which it enjoys in its own sphere, and with the place which it occupies, the empire of religion is never more surely established than when it reigns in the hearts of men unsupported by aught beside its native strength. Religion is no less the companion of liberty in all its battles and its triumphs; the cradle of its infancy, and the divine source of its claims. The safeguard of morality is religion, and morality is the best security of law and the surest pledge of freedom.[40]

Reasons of Certain Anomalies Which the Laws and Customs of the Anglo-Americans Present

Remains of aristocratic institutions in the midst of a complete democracy—Why?—Distinction carefully to be drawn between what is of Puritanical and what is of English origin.

The reader is cautioned not to draw too general or too absolute an inference from what has been said. The social condition, the religion, and the manners of the first emigrants undoubtedly exercised an immense influence on the destiny of their new country. Nevertheless they were not in a situation to found a state of things solely dependent on themselves: no man can entirely shake off the influence of the past, and the settlers, intentionally or involuntarily, mingled habits and notions derived from their education and from the traditions of their country with those habits and notions which were exclusively their own. To form a judgment on the Anglo-Americans of the present day it is therefore necessary to distinguish what is of Puritanical and what is of English origin.

Laws and customs are frequently to be met with in the United States which contrast strongly with all that surrounds them. These laws seem to be drawn up in a spirit contrary to the prevailing tenor of the American legislation; and these customs are no less opposed to the tone of society. If the English colonies had been founded in an age of darkness, or if their origin was already lost in the lapse of years, the problem would be insoluble.

I shall quote a single example to illustrate what I advance. The civil and criminal procedure of the Americans has only two means of action committal and bail. The first measure taken by the magistrate is to exact security from the defendant, or, in case of refusal, to incarcerate him: the ground of the accusation and the importance of the charges against him are then discussed. It is evident that a legislation of this kind is hostile to the poor man, and favourable only to the rich. The poor man has not always a security to produce, even in a civil cause; and if he is obliged to wait for justice in prison, he is speedily reduced to distress. The wealthy individual, on the contrary, always escapes imprisonment in civil causes; nay, more, he may readily elude the punishment which awaits him for a delinquency by breaking his bail. So that all the penalties of the law are, for him, reducible to fines.[41] Nothing can be more aristocratic than this system of legislation. Yet in America it is the poor who make the law, and they usually reserve the greatest social advantages to themselves. The explanation of the phenomenon is to be found in England; the laws of which I speak are English,[42] and the Americans have retained them, however repugnant they may be to the tenor of their legislation and the mass of their ideas. Next to its habits, the thing which a nation is least apt to change is its civil legislation. Civil laws are only familiarly known to legal men, whose direct interest it is to maintain them as they are, whether good or bad, simply because they themselves are conversant with them. The body of the nation is scarcely acquainted with them; it merely perceives their action in particular cases; but it has some difficulty in seizing their tendency, and obeys them without premeditation. I have quoted one instance where it would have been easy to adduce a great number of others. The surface of American society is, if I may use the expression, covered with a layer of democracy, from beneath which the old aristocratic colors sometimes peep.

CHAPTER III

Social Condition of the Anglo-Americans

A social condition is commonly the result of circumstances, sometimes of laws, oftener still of these two causes united; but wherever it exists, it may justly be considered as the source of almost all the laws, the usages, and the ideas which regulate the conduct of nations; whatever it does not produce it modifies. It is therefore necessary, if we would become acquainted with the legislation and the manners of a nation, to begin by the study of its social condition.

The Striking Characteristic of the Social Condition of the Anglo-Americans Is Its Essential Democracy

The first emigrants of New England—Their equality—
Aristocratic laws introduced in the South—Period of the Revolution—
Change in the law of descent. Effects produced by this change—
Democracy carried to its utmost limits in the new States
of the West—Equality of education.

Many important observations suggest themselves upon the social condition of the Anglo-Americans, but there is one which takes precedence of all the rest. The social condition of the Americans is eminently democratic; this was its character at the foundation of the Colonies, and is still more strongly marked at the present day. I have stated in the preceding chapter that great equality existed among the emigrants who settled on the shores of New England. The germ of aristocracy was never planted in that part of the Union. The only influence which obtained there was that of intellect; the people were used to reverence certain names as the emblems of knowledge and virtue. Some of their fellow-citizens acquired a power over the rest which might truly have been called aristocratic, if it had been capable of transmission from father to son.

This was the state of things to the east of the Hudson: to the south-west of that river, and in the direction of the Floridas, the case was different. In most of the States situated to the south-west of the Hudson some great English proprietors had settled, who had imported with them aristocratic principles and the English law of descent. I have explained the reasons why it was impossible ever to establish a powerful aristocracy in America; these reasons existed with less force to the south-west of the Hudson. In the South, one man, aided by slaves, could cultivate a great extent of country: it was therefore common to see rich landed proprietors. But their influence was not altogether aristocratic as that term is understood in Europe, since they possessed no privileges; and the cultivation of their estates being carried on by slaves, they had no tenants depending on them, and consequently no patronage. Still, the great proprietors south of the Hudson constituted a superior class, having ideas and tastes of its own, and forming the centre of political action. This kind of aristocracy sympathized with the body of the people, whose passions and interests it easily embraced; but it was too weak and too short-lived to excite either love or hatred for itself. This was the class which headed the insurrection in the South, and furnished the best leaders of the American revolution.

At the period of which we are now speaking society was shaken to its centre: the people, in whose name the struggle had taken place, conceived the desire of exercising the authority which it had acquired; its democratic tendencies were awakened; and having thrown off the yoke of the mother-country, it aspired to independence of every kind. The influence of individuals gradually ceased to be felt, and custom and law united together to produce the same result.

But the law of descent was the last step to equality. I am surprised that ancient and modern jurists have not attributed to this law a greater influence on human affairs.[1] It is true that these laws belong to civil affairs; but they ought nevertheless to be placed at the head of all political institutions; for, whilst political laws are only the symbol of a nation's condition, they exercise an incredible influence upon its social state. They have, moreover, a sure and uniform manner of operating upon society, affecting, as it were, generations yet unborn.

Through their means man acquires a kind of preternatural power over the future lot of his fellow-creatures. When the legislator has regulated the law of inheritance, he may rest from his labor. The machine once put in motion will go on for ages, and advance, as if self-guided, towards a given point. When framed in a particular manner, this law unites, draws together, and vests property and power in a few hands: its tendency is clearly aristocratic. On opposite principles its action is still more rapid; it divides, distributes, and disperses both property and power. Alarmed by the rapidity of its progress, those who despair of arresting its motion endeavor to obstruct it by difficulties and impediments; they vainly seek to counteract its effect by contrary efforts; but it gradually reduces or destroys every obstacle, until by its incessant activity the bulwarks of the influence of wealth are ground down to the fine and shifting sand which is the basis of democracy. When the law of inheritance permits, still more when it decrees, the equal division of a father's property amongst all his children, its effects are of two kinds: it is important to distinguish them from each other, although they tend to the same end.

In virtue of the law of partible inheritance, the death of every proprietor brings about a kind of revolution in property; not only do his possessions change hands, but their very nature is altered, since they are parcelled into shares, which become smaller and smaller at each division. This is the direct and, as it were, the physical effect of the law. It follows, then, that in countries where equality of inheritance is established by law, property, and especially landed property, must have a tendency to perpetual diminution. The effects, however, of such legislation would only be perceptible after a lapse of time, if the law was abandoned to its own working; for supposing the family to consist of two children (and in a country peopled as France is the average number is not above three), these children, sharing amongst them the fortune of both parents, would not be poorer than their father or mother.

But the law of equal division exercises its influence not merely upon the property itself, but it affects the minds of the heirs, and brings their passions into play. These indirect consequences tend powerfully to the destruction of large fortunes, and especially of large domains. Among nations whose law of descent is founded upon the right of primogeniture landed estates often pass from generation to generation without undergoing division, the consequence of which is that family feeling is to a certain degree incorporated with the estate. The family represents the estate, the estate the family; whose name, together with its origin, its glory, its power, and its virtues, is thus perpetuated in an imperishable memorial of the past and a sure pledge of the future.

When the equal partition of property is established by law, the intimate connection is destroyed between family feeling and the preservation of the paternal estate; the property ceases to represent the family; for as it must inevitably be divided after one or two generations, it has evidently a constant tendency to diminish, and must in the end be completely dispersed. The sons of the great landed proprietor, if they are few in number, or if fortune befriends them, may indeed entertain the hope of being as wealthy as their father, but not that of possessing the same property as he did; the riches must necessarily be composed of elements different from his.

Now, from the moment that you divest the landowner of that interest in the preservation of his estate which he derives from association, from tradition, and from family pride, you may be certain that sooner or later he will dispose of it; for there is a strong pecuniary interest in favour of selling, as floating capital produces higher interest than real property, and is more readily available to gratify the passions of the moment.

Great landed estates which have once been divided never come together again; for the small proprietor draws from his land a better revenue, in proportion, than the large owner does from his, and of course he sells it at a higher rate.[2] The calculations of gain, therefore, which decide the rich man to sell his domain will still more powerfully influence him against buying small estates to unite them into a large one.

What is called family pride is often founded upon an illusion of self-love. A man wishes to perpetuate and immortalize himself, as it were, in his great-grandchildren. Where the *esprit de famille* ceases to act individual selfishness comes into play. When the idea of family becomes vague, indeterminate, and uncertain, a man thinks of his present convenience; he provides for the establishment of his succeeding generation, and no more. Either a man gives up the idea of perpetuating his family, or at any rate he seeks to accomplish it by other means than that of a landed estate. Thus not only does the law of partible inheritance render it difficult for families to preserve their ancestral domains entire, but it deprives them of the inclination to attempt it, and compels them in some measure to cooperate with the law in their own extinction.

The law of equal distribution proceeds by two methods: by acting upon things, it acts upon persons; by influencing persons, it affects things. By these means the law succeeds in striking at the root of landed property, and dispersing rapidly both families and fortunes.[3]

Most certainly it is not for us Frenchmen of the nineteenth century, who daily witness the political and social changes which the law of partition is bringing to pass, to question its influence. It is perpetually conspicuous in our country, overthrowing the walls of our dwellings and removing the landmarks of our fields. But although it has produced great effects in France, much still remains for it to do. Our recollections, opinions, and habits present powerful obstacles to its progress.

In the United States it has nearly completed its work of destruction, and there we can best study its results. The English laws concerning the transmission of property were abolished in almost all the States at the time of the Revolution. The law of entail was so modified as not to interrupt the free circulation of property.[4] The first generation having passed away, estates began to be parcelled out, and the change became more and more rapid with the progress of time. At this moment, after a lapse of a little more than sixty years, the aspect of society is totally altered; the families of the great landed proprietors are almost all commingled with the general mass. In the State of New York, which formerly contained many of these, there are but two who still keep their heads above the stream, and they must shortly disappear. The sons of these opulent citizens are become merchants, lawyers, or physicians. Most of them have lapsed into obscurity. The last trace of hereditary ranks and distinctions is destroyed the law of partition has reduced all to one level.

I do not mean that there is any deficiency of wealthy individuals in the United States; I know of no country, indeed, where the love of money has taken stronger hold on the affections of men, and where the profounder contempt is expressed for the theory of the permanent equality of property. But wealth circulates with inconceivable rapidity, and experience shows that it is rare to find two succeeding generations in the full enjoyment of it.

This picture, which may perhaps be thought to be overcharged, still gives a very imperfect idea of what is taking place in the new States of the West and South-west. At the end of the last century a few bold adventurers began to penetrate into the valleys of the Mississippi, and the mass of the population very soon began to move in that direction: communities unheard of till then were seen to emerge from the wilds: States whose names were not in existence a few years before claimed their place in the American Union; and in the Western settlements we may behold democracy arrived at its utmost extreme. In these States, founded off-hand, and, as it were, by chance, the inhabitants are but of yesterday. Scarcely known to one another, the nearest neighbours are ignorant of each other's history. In this part of the American continent, therefore, the population has not experienced the influence of great names and great wealth, nor even that of the natural aristocracy of knowledge and virtue. None are there to wield that respectable power which men willingly grant to the remembrance of a life spent in doing good before their eyes. The new states of the West are already inhabited, but society has no existence among them.[5]

It is not only the fortunes of men which are equal in America; even their requirements partake in some degree of the same uniformity. I do not believe that there is a country in the world where, in proportion to the population, there are so few uninstructed and at the same time so few learned individuals. Primary instruction is within the reach of everybody; superior instruction is scarcely to be obtained by any.

This is not surprising; it is in fact the necessary consequence of what we have advanced above. Almost all the Americans are in easy circumstances, and can therefore obtain the first elements of human knowledge.

In America there are comparatively few who are rich enough to live without a profession. Every profession requires an apprenticeship, which limits the time of instruction to the early years of life. At fifteen they enter upon their calling, and thus their education ends at the age when ours begins. Whatever is done afterwards is with a view to some special and lucrative object; a science is taken up as a matter of business, and the only branch of it which is attended to is such as admits of an immediate practical application. In America most of the rich men were formerly poor; most of those who now enjoy leisure were absorbed in business during their youth; the consequence of which is, that when they might have had a taste for study they had no time for it, and when time is at their disposal they have no longer the inclination.

There is no class, then, in America, in which the taste for intellectual pleasures is transmitted with hereditary fortune and leisure, and by which the labours of the intellect are held in honour. Accordingly there is an equal want of the desire and the power of application to these objects.

A middle standard is fixed in America for human knowledge. All approach as near to it as they can; some as they rise, others as they descend. Of course, an immense multitude of persons are to be found who entertain the same number of ideas on religion, history, science, political economy, legislation, and government. The gifts of intellect proceed directly from God, and man cannot prevent their unequal distribution. But in consequence of the state of things which we have here represented it happens that, although the capacities of men are widely different, as the Creator has doubtless intended they should be, they are submitted to the same method of treatment.

In America the aristocratic element has always been feeble from its birth; and if at the present day it is not actually destroyed, it is at any rate so completely disabled that we can scarcely assign to it any degree of influence in the course of affairs. The democratic principle, on the contrary, has gained so much strength by time, by events, and by legislation, as to have become not only predominant but all-powerful. There is no family or corporate authority, and it is rare to find even the influence of individual character enjoy any durability.

America, then, exhibits in her social state a most extraordinary phenomenon. Men are there seen on a greater equality in point of fortune and intellect, or, in other words, more equal in their strength, than in any other country of the world, or in any age of which history has preserved the remembrance.

Political Consequences of the Social Condition of the Anglo-Americans

The political consequences of such a social condition as this are easily deducible. It is impossible to believe that equality will not eventually find its way into the political world as it does everywhere else. To conceive of men remaining forever unequal upon one single point, yet equal on all others, is impossible; they must come in the end to be equal upon all. Now I know of only two methods of establishing equality in the political world; every citizen must be put in possession of his rights, or rights must be granted to no one. For nations which are arrived at the same stage of social existence as the Anglo-Americans, it is therefore very difficult to discover a medium between the sovereignty of all and the absolute power of one man: and it would be vain to deny that the social condition which I have been describing is equally liable to each of these consequences.

There is, in fact, a manly and lawful passion for equality which excites men to wish all to be powerful and honoured. This passion tends to elevate the humble to the rank of the great; but there exists also in the human heart a depraved taste for equality, which impels the weak to attempt to lower the powerful to their own level, and reduces men to prefer equality in slavery to inequality with freedom. Not that those nations whose social condition is democratic naturally despise liberty; on the contrary, they have an instinctive love of it. But liberty is not the chief and constant object of their desires; equality is their idol: they make rapid and sudden efforts to obtain liberty, and if they miss their aim resign themselves to their disappointment; but nothing can satisfy them except equality, and rather than lose it they resolve to perish.

On the other hand, in a State where the citizens are nearly on an equality, it becomes difficult for them to preserve their independence against the aggressions of power. No one among them being strong enough to engage in the struggle with advantage, nothing but a general combination can protect their liberty. And such a union is not always to be found.

From the same social position, then, nations may derive one or the other of two great political results; these results are extremely different from each other, but they may both proceed from the same cause.

The Anglo-Americans are the first nations who, having been exposed to this formidable alternative, have been happy enough to escape the dominion of absolute power. They have been allowed by their circumstances, their origin, their intelligence, and especially by their moral feeling, to establish and maintain the sovereignty of the people.

Chapter IV

The Principle of the Sovereignty of the People in America

It predominates over the whole of society in America—Application made of this principle by the Americans even before their Revolution—Development given to it by that Revolution—Gradual and irresistible extension of the elective qualification.

Whenever the political laws of the United States are to be discussed, it is with the doctrine of the sovereignty of the people that we must begin. The principle of the sovereignty of the people, which is to be found, more or less, at the bottom of almost all human institutions, generally remains concealed from view. It is obeyed without being recognised, or if for a moment it be brought to light, it is hastily cast back into the gloom of the sanctuary. 'The will of the nation' is one of those expressions which have been most profusely abused by the wily and the despotic of every age. To the eyes of some it has been represented by the venal suffrages of a few of the satellites of power; to others by the votes of a timid or an interested minority; and some have even discovered it in the silence of a people, on the supposition that the fact of submission established the right of command.

In America the principle of the sovereignty of the people is not either barren or concealed, as it is with some other nations; it is recognised by the customs and proclaimed by the laws; it spreads freely, and arrives without impediment at its most remote consequences. If there be a country in the world where the doctrine of the sovereignty of the people can be fairly appreciated, where it can be studied in its application to the affairs of society, and where its dangers and its advantages may be foreseen, that country is assuredly America.

I have already observed that, from their origin, the sovereignty of the people was the fundamental principle of the greater number of British colonies in America. It was far, however, from then exercising as much influence on the government of society as it now does. Two obstacles, the one external, the other internal, checked its invasive progress. It could not ostensibly disclose itself in the laws of colonies which were still constrained to obey the mother-country: it was therefore obliged to spread secretly, and to gain ground in the provincial assemblies, and especially in the townships.

American society was not yet prepared to adopt it with all its consequences. The intelligence of New England, and the wealth of the country to the south of the Hudson (as I have shown in the preceding chapter), long exercised a sort of aristocratic influence, which tended to retain the exercise of social authority in the hands of a few. The public functionaries were not universally elected, and the citizens were not all of them electors. The electoral franchise was everywhere placed within certain limits, and made dependent on a certain qualification, which was exceedingly low in the North and more considerable in the South.

The American revolution broke out, and the doctrine of the sovereignty of the people, which had been nurtured in the townships and municipalities, took possession of the State: every class was enlisted in its cause; battles were fought, and victories obtained for it, until it became the law of laws.

A no less rapid change was effected in the interior of society, where the law of descent completed the abolition of local influences.

At the very time when this consequence of the laws and of the revolution was apparent to every eye, victory was irrevocably pronounced in favor of the democratic cause. All power was, in fact, in its hands, and resistance was no longer possible. The higher orders submitted without a murmur and without a struggle to an evil which was thenceforth inevitable. The ordinary fate of falling powers awaited them; each of their several members followed his own interests; and as it was impossible to wring the power from the hands of a people which they did not detest sufficiently to brave, their only aim was to secure its good-will at any price. The most democratic laws were consequently voted by the very men whose interests they impaired; and thus, although the higher classes did not excite the passions of the people against their order, they accelerated the triumph of the new state of things; so that by a singular change the democratic impulse was found to be most irresistible in the very States where the aristocracy had the firmest hold. The State of Maryland, which had been founded by men of rank, was the first to proclaim universal suffrage, and to introduce the most democratic forms into the conduct of its government.

When a nation modifies the elective qualification, it may easily be foreseen that sooner or later that qualification will be entirely abolished. There is no more invariable rule in the history of society: the further electoral rights are extended, the greater is the need of extending them; for after each concession the strength of the democracy increases, and its demands increase with its strength. The ambition of those who are below the appointed rate is irritated in exact proportion to the great number of those who are above it. The exception at last becomes the rule, concession follows concession, and no stop can be made short of universal suffrage.

At the present day the principle of the sovereignty of the people has acquired, in the United States, all the practical development which the imagination can conceive. It is unencumbered by those fictions which have been thrown over it in other countries, and it appears in every possible form according to the exigency of the occasion. Sometimes the laws are made by the people in a body, as at Athens; and sometimes its representatives, chosen by universal suffrage, transact business in its name, and almost under its immediate control.

In some countries a power exists which, though it is in a degree foreign to the social body, directs it, and forces it to pursue a certain track. In others the ruling force is divided, being partly within and partly without the ranks of the people. But nothing of the kind is to be seen in the United States; there society governs itself for itself. All power centres in its bosom; and scarcely an individual is to be met with who would venture to conceive, or, still less, to express, the idea of seeking it elsewhere. The nation participates in the making of its laws by the choice of its legislators, and in the execution of them by the choice of the agents of the executive government; it may almost be said to govern itself, so feeble and so restricted is the share left to the administration, so little do the authorities forget their popular origin and the power from which they emanate.[1]

Chapter V

Necessity of Examining the Condition of the States before That of the Union at Large

It is proposed to examine in the following chapter what is the form of government established in America on the principle of the sovereignty of the people; what are its resources, its hindrances, its advantages, and its dangers. The first difficulty which presents itself arises from the complex nature of the constitution of the United States, which consists of two distinct social structures, connected and, as it were, encased one within the other; two governments, completely separate and almost independent, the one fulfilling the ordinary duties and responding to the daily and indefinite calls of a community, the other circumscribed within certain limits, and only exercising an exceptional authority over the general interests of the country. In short, there are twenty-four small sovereign nations, whose agglomeration constitutes the body of the Union. To examine the Union before we have studied the States would be to adopt a method filled with obstacles. The form of the Federal Government of the United States was the last which was adopted; and it is in fact nothing more than a modification or a summary of those republican principles which were current in the whole community before it existed, and independently of its existence. Moreover, the Federal Government is, as I have just observed, the exception; the Government of the States is the rule. The author who should attempt to exhibit the picture as a whole before he had explained its details would necessarily fall into obscurity and repetition.

The great political principles which govern American society at this day undoubtedly took their origin and their growth in the State. It is therefore necessary to become acquainted with the State in order to possess a clue to the remainder. The States which at present compose the American Union all present the same features, as far as regards the external aspect of their institutions. Their political or administrative existence is centred in three focuses of action, which may not inaptly be compared to the different nervous centres which convey motion to the human body. The township is the lowest in order, then the county, and lastly the State; and I propose to devote the following chapter to the examination of these three divisions.

The American System of Townships and Municipal Bodies

Why the Author begins the examination of the political institutions with the township—Its existence in all nations. Difficulty of establishing and preserving municipal independence—Its importance—Why the Author has selected the township system of New England as the main topic of his discussion.

It is not undesignedly that I begin this subject with the Township. The village or township is the only association which is so perfectly natural that wherever a number of men are collected it seems to constitute itself.

The town, or tithing, as the smallest division of a community, must necessarily exist in all nations, whatever their laws and customs may be: if man makes monarchies and establishes republics, the first association of mankind seems constituted by the hand of God. But although the existence of the township is coeval with that of man, its liberties are not the less rarely respected and easily destroyed. A nation is always able to establish great political assemblies, because it habitually contains a certain number of individuals fitted by their talents, if not by their habits, for the direction of affairs. The township is, on the contrary, composed of coarser materials, which are less easily fashioned by the legislator. The difficulties which attend the consolidation of its independence rather augment than diminish with the increasing enlightenment of the people. A highly civilized community spurns the attempts of a local independence, is disgusted at its numerous blunders, and is apt to despair of success before the experiment is completed. Again, no immunities are so ill protected from the encroachments of the supreme power as those of municipal bodies in general: they are unable to struggle, single-handed, against a strong or an enterprising government, and they cannot defend their cause with success unless it be identified with the customs of the nation and supported by public opinion. Thus until the independence of townships is amalgamated with the manners of a people it is easily destroyed, and it is only after a long existence in the laws that it can be thus amalgamated. Municipal freedom is not the fruit of human device; it is rarely created; but it is, as it were, secretly and spontaneously engendered in the midst of a semi-barbarous state of society. The constant action of the laws and the national habits, peculiar circumstances, and above all time, may consolidate it; but there is certainly no nation on the continent of Europe which has experienced its advantages. Nevertheless local assemblies of citizens constitute the strength of free nations. Town-meetings are to liberty what primary schools are to science; they bring it within the people's reach, they teach men how to use and how to enjoy it. A nation may establish a system of free government, but without the spirit of municipal institutions it cannot have the spirit of liberty. The transient passions and the interests of an hour, or the chance of circumstances, may have created the external forms of independence; but the despotic tendency which has been repelled will, sooner or later, inevitably reappear on the surface.

In order to explain to the reader the general principles on which the political organization of the counties and townships of the United States rests, I have thought it expedient to choose one of the States of New England as an example, to examine the mechanism of its constitution, and then to cast a general glance over the country. The township and the county are not organized in the same manner in every part of the Union; it is, however, easy to perceive that the same principles have guided the formation of both of them throughout the Union. I am inclined to believe that these principles have been carried further in New England than elsewhere, and consequently that they offer greater facilities to the observations of a stranger. The institutions of New England form a complete and regular whole; they have received the sanction of time, they have the support of the laws, and the still stronger support of the manners of the community, over which they exercise the most prodigious influence; they consequently deserve our attention on every account.

Limits of the Township

The township of New England is a division which stands between the *commune* and the *canton* of France, and which corresponds in general to the English tithing, or town. Its average population is from two to three thousand;[1] so that, on the one hand, the interests of its inhabitants are not likely to conflict, and, on the other, men capable of conducting its affairs are always to be found among its citizens.

Authorities of the Township in New England

The people the source of all power here as elsewhere—Manages its own affairs—No corporation—The greater part of the authority vested in the

hands of the Selectmen—How the Selectmen act—
Town-meeting—Enumeration of the public officers of the
township—Obligatory and remunerated functions.

In the township, as well as everywhere else, the people is the only source of power; but in no stage of government does the body of citizens exercise a more immediate influence. In America the people is a master whose exigencies demand obedience to the utmost limits of possibility.

In New England the majority acts by representatives in the conduct of the public business of the State; but if such an arrangement be necessary in general affairs, in the townships, where the legislative and administrative action of the government is in more immediate contact with the subject, the system of representation is not adopted. There is no corporation; but the body of electors, after having designated its magistrates, directs them in everything that exceeds the simple and ordinary executive business of the State.[2]

This state of things is so contrary to our ideas, and so different from our customs, that it is necessary for me to adduce some examples to explain it thoroughly.

The public duties in the township are extremely numerous and minutely divided, as we shall see further on; but the larger proportion of administrative power is vested in the hands of a small number of individuals, called 'the Selectmen.'[3] The general laws of the State impose a certain number of obligations on the selectmen, which they may fulfil without the authorization of the body they represent, but which they can only neglect on their own responsibility. The law of the State obliges them, for instance, to draw up the list of electors in their townships; and if they omit this part of their functions, they are guilty of a misdemeanour. In all the affairs, however, which are determined by the town-meeting, the selectmen are the organs of the popular mandate, as in France the Maire executes the decree of the municipal council. They usually act upon their own responsibility, and merely put in practice principles which have been previously recognized by the majority. But if any change is to be introduced in the existing state of things, or if they wish to undertake any new enterprise, they are obliged to refer to the source of their power. If, for instance, a school is to be established, the selectmen convoke the whole body of the electors on a certain day at an appointed place; they explain the urgency of the case; they give their opinion on the means of satisfying it, on the probable expense, and the site which seems to be most favourable. The meeting is consulted on these several points; it adopts the principle, marks out the site, votes the rate, and confides the execution of its resolution to the selectmen.

The selectmen have alone the right of calling a town-meeting, but they may be requested to do so: if ten citizens are desirous of submitting a new project to the assent of the township, they may demand a general convocation of the inhabitants; the selectmen are obliged to comply, but they have only the right of presiding at the meeting.[4]

The selectmen are elected every year in the month of April or of May. The town-meeting chooses at the same time a number of other municipal magistrates, who are entrusted with important administrative functions. The assessors rate the township; the collectors receive the rate. A constable is appointed to keep the peace, to watch the streets, and to forward the execution of the laws; the town-clerk records all the town votes, orders, grants, births, deaths, and marriages; the treasurer keeps the funds; the overseer of the poor performs the difficult task of superintending the action of the poor-laws; committee-men are appointed to attend to the schools and to public instruction; and the road-surveyors, who take care of the greater and lesser thoroughfares of the township, complete the list of the principal functionaries. They are, however, still further subdivided; and amongst the municipal officers are to be found parish commissioners, who audit the expenses of public worship; different classes of inspectors, some of whom are to direct the citizens in case of fire; tithing-men, listers, haywards, chimney-viewers, fence-viewers to maintain the bounds of property, timber-measurers, and sealers of weights and measures.[5]

There are nineteen principal officers in a township. Every inhabitant is constrained, on the pain of being fined, to undertake these different functions; which, however, are almost all paid, in order that the poorer citizens may be able to give up their time without loss. In general the American system is not to grant a fixed salary to its functionaries. Every service has its price, and they are remunerated in proportion to what they have done.

Existence of the Township

Every one the best judge of his own interest—Corollary of the principle of the sovereignty of the people—Application of those doctrines in the townships of America—The township of New England is sovereign in all that concerns itself alone: subject to the State in all other matters—Bond of the township and the State—
In France the Government lends its agent to the *Commune*—
In America the reverse occurs.

I have already observed that the principle of the sovereignty of the people governs the whole political system of the Anglo-Americans. Every page of this book will afford new instances of the same doctrine. In the nations by which the sovereignty of the people is recognized every individual possesses an equal share of power, and participates alike in the government of the State. Every individual is, therefore, supposed to be as well informed, as virtuous, and as strong as any of his fellow-citizens. He obeys the government, not because he is inferior to the authorities which conduct it, or that he is less capable than his neighbor of governing himself, but because he acknowledges the utility of an association with his fellow-men, and because he knows that no such association can exist without a regulating force. If he be a subject in all that concerns the mutual relations of citizens, he is free and responsible to God alone for all that concerns himself. Hence arises the maxim that every one is the best and the sole judge of his own private interest, and that society has no right to control a man's actions, unless they are prejudicial to the common weal, or unless the common weal demands his cooperation. This doctrine is universally admitted in the United States. I shall hereafter examine the general influence which it exercises on the ordinary actions of life; I am now speaking of the nature of municipal bodies.

The township, taken as a whole, and in relation to the government of the country, may be looked upon as an individual to whom the theory I have just alluded to is applied. Municipal independence is therefore a natural consequence of the principle of the sovereignty of the people in the United States: all the American republics recognize it more or less; but circumstances have peculiarly favoured its growth in New England.

In this part of the Union the impulsion of political activity was given in the townships; and it may almost be said that each of them originally formed an independent nation. When the Kings of England asserted their supremacy, they were contented to assume the central power of the State. The townships of New England remained as they were before; and although they are now subject to the State, they were at first scarcely dependent upon it. It is important to remember that they have not been invested with privileges, but that they have, on the contrary, forfeited a portion of their independence to the State. The townships are only subordinate to the State in those interests which I shall term *social,* as they are common to all the citizens. They are independent in all that concerns themselves; and amongst the inhabitants of New England I believe that not a man is to be found who would acknowledge that the State has any right to interfere in their local interests. The towns of New England buy and sell, sue or are sued, augment or diminish their rates, without the slightest opposition on the part of the administrative authority of the State.

They are bound, however, to comply with the demands of the community. If the State is in need of money, a town can neither give nor withhold the supplies. If the State projects a road, the township cannot refuse to let it cross its territory; if a police regulation is made by the State, it must be enforced by the town. A uniform system of instruction is organized all over the country, and every town is bound to establish the schools which the law ordains. In speaking of the administration of the United States I shall have occasion to point out the means by which the townships are compelled to obey in these different cases: I here merely show the existence of the obligation. Strict as this obligation is, the government of the State imposes it in principle only, and in its performance the township resumes all its independent rights. Thus, taxes are voted by the State, but they are levied and collected by the township; the existence of a school is obligatory, but the township builds, pays, and superintends it. In France the State-collector receives the local imposts; in America the town-collector receives the taxes of the State. Thus the French Government lends its agents to the commune; in America the township is the agent of the Government. This fact alone shows the extent of the differences which exist between the two nations.

Public Spirit of the Townships of New England

How the township of New England wins the affections of its inhabitants—Difficulty of creating local public spirit in Europe—The rights and duties of the American township favourable to it—Characteristics of home in the United States—Manifestations of public spirit in New England—Its happy effects.

In America, not only do municipal bodies exist, but they are kept alive and supported by public spirit. The township of New England possesses two advantages which infallibly secure the attentive interest of mankind, namely, independence and authority. Its sphere is indeed small and limited, but within that sphere its action is unrestrained; and its independence gives to it a real importance which its extent and population may not always ensure.

It is to be remembered that the affections of men generally lie on the side of authority. Patriotism is not durable in a conquered nation. The New Englander is attached to his township, not only because he was born in it, but because it constitutes a social body of which he is a member, and whose government claims and deserves the exercise of his sagacity. In Europe the absence of local public spirit is a frequent subject of regret to those who are in power; every one agrees that there is no surer guarantee of order and tranquillity, and yet nothing is more difficult to create. If the municipal bodies were made powerful and independent, the authorities of the nation might be disunited and the peace of the country endangered. Yet, without power and independence, a town may contain good subjects, but it can have no active citizens. Another important fact is that the township of New England is so constituted as to excite the warmest of human affections, without arousing the ambitious passions of the heart of man. The officers of the country are not elected, and their authority is very limited. Even the State is only a second-rate community, whose tranquil and obscure administration offers no inducement sufficient to draw men away from the circle of their interests into the turmoil of public affairs. The federal government confers power and honour on the men who conduct it; but these individuals can never be very numerous. The high station of the Presidency can only be reached at an advanced period of life, and the other federal functionaries are generally men who have been favoured by fortune, or distinguished in some other career. Such cannot be the permanent aim of the ambitious. But the township serves as a centre for the desire of public esteem, the want of exciting interests, and the taste for authority and popularity, in the midst of the ordinary relations of life; and the passions which commonly embroil society change their character when they find a vent so near the domestic hearth and the family circle.

In the American States power has been disseminated with admirable skill for the purpose of interesting the greatest possible number of persons in the common weal. Independently of the electors who are from time to time called into action, the body politic is divided into innumerable functionaries and officers, who all, in their several spheres, represent the same powerful whole in whose name they act. The local administration thus affords an unfailing source of profit and interest to a vast number of individuals.

The American system, which divides the local authority among so many citizens, does not scruple to multiply the functions of the town officers. For in the United States it is believed, and with truth, that patriotism is a kind of devotion which is strengthened by ritual observance. In this manner the activity of the township is continually perceptible; it is daily manifested in the fulfilment of a duty or the exercise of a right, and a constant though gentle motion is thus kept up in society which animates without disturbing it.

The American attaches himself to his home as the mountaineer clings to his hills, because the characteristic features of his country are there more distinctly marked than elsewhere. The existence of the townships of New England is in general a happy one. Their government is suited to their tastes, and chosen by themselves. In the midst of the profound peace and general comfort which reign in America the commotions of municipal discord are unfrequent. The conduct of local business is easy. The political education of the people has long been complete; say rather that it was complete when the people first set foot upon the soil. In New England no tradition exists of a distinction of ranks; no portion of the community is tempted to oppress the remainder; and the abuses which may injure isolated individuals are forgotten in the general contentment which prevails. If the government is defective (and it would no doubt be easy to point out its deficiencies), the fact that it really emanates from those it governs, and that it acts, either ill or well, casts the protecting spell of a parental pride over its faults. No term of comparison disturbs the satisfaction of the citizen: England formerly governed the mass of the colonies, but the people was always sovereign in the township where its rule is not only an ancient but a primitive state.

The native of New England is attached to his township because it is independent and free: his cooperation in its affairs ensures his attachment to its interest; the well-being it affords him secures his affection; and its welfare is the aim of his ambition and of his future exertions: he takes a part in every occurrence in the place; he practises the art of government in the small sphere within his reach; he accustoms himself to those forms which can alone ensure the steady progress of liberty; he imbibes their spirit; he acquires a taste for order, comprehends the union or the balance of powers, and collects clear practical notions on the nature of his duties and the extent of his rights.

The Counties of New England

The division of the counties in America has considerable analogy with that of the arrondissements of France. The limits of the counties are arbitrarily laid down, and the various districts which they contain have no necessary connection, no common tradition or natural sympathy; their object is simply to facilitate the administration of justice.

The extent of the township was too small to contain a system of judicial institutions; each county has, however, a court of justice,[6] a sheriff to execute its decrees, and a prison for criminals. There are certain wants which are felt alike by all the townships of a county; it is therefore natural that they should be satisfied by a central authority. In the State of Massachusetts this authority is vested in the hands of several magistrates, who are appointed by the Governor of the State, with the advice[7] of his council.[8] The officers of the county have only a limited and occasional authority, which is applicable to certain predetermined cases. The State and the townships possess all the power requisite to conduct public business. The budget of the county is drawn up by its officers, and is voted by the legislature, but there is no assembly which directly or indirectly represents the county. It has, therefore, properly speaking, no political existence.

A twofold tendency may be discerned in the American constitutions, which impels the legislator to centralize the legislative and to disperse the executive power. The township of New England has in itself an indestructible element of independence; and this distinct existence could only be fictitiously introduced into the county, where its utility has not been felt. But all the townships united have but one representation, which is the State, the centre of the national authority: beyond the action of the township and that of the nation, nothing can be said to exist but the influence of individual exertion.

Administration in New England

Administration not perceived in America—Why?—The Europeans believe that liberty is promoted by depriving the social authority of some of its rights; the Americans, by dividing its exercise—Almost all the administration confined to the township, and divided amongst the town-officers—No trace of an administrative body to be perceived, either in the township or above it—The reason of this—How it happens that the administration of the State is uniform—Who is empowered to enforce the obedience of the township and the county to the law—The introduction of judicial power into the administration—Consequence of the extension of the elective principle to all functionaries—The Justice of the Peace in New England—By whom appointed—County officer: ensures the administration of the townships—Court of Sessions—Its action—Right of inspection and indictment disseminated like the other administrative functions—Informers encouraged by the division of fines.

Nothing is more striking to an European traveller in the United States than the absence of what we term the Government, or the Administration. Written laws exist in America, and one sees that they are daily executed; but although everything is in motion, the hand which gives the impulse to the social machine can nowhere be discovered. Nevertheless, as all peoples are obliged to have recourse to certain grammatical forms, which are the foundation of human language, in order to express their thoughts; so all communities are obliged to secure their existence by submitting to a certain dose of authority, without which they fall a prey to anarchy. This authority may be distributed in several ways, but it must always exist somewhere.

There are two methods of diminishing the force of authority in a nation: The first is to weaken the supreme power in its very principle, by forbidding or preventing society from acting in its own defence under certain circumstances. To weaken authority in this manner is what is generally termed in Europe to lay the foundations of freedom. The second manner of diminishing the influence of authority does not consist in stripping society of any of its rights, nor in paralysing its efforts, but in distributing the exercise of its privileges in various hands, and in multiplying functionaries, to each of whom the degree of power necessary for him to perform his duty is entrusted. There may be nations whom this distribution of social powers might lead to anarchy; but in itself it is not anarchical. The action of authority is indeed thus rendered less irresistible and less perilous, but it is not totally suppressed.

The revolution of the United States was the result of a mature and dignified taste for freedom, and not of a vague or ill-defined craving for independence. It contracted no alliance with the turbulent passions of anarchy; but its course was marked, on the contrary, by an attachment to whatever was lawful and orderly.

It was never assumed in the United States that the citizen of a free country has a right to do whatever he pleases; on the contrary, social obligations were there imposed upon him more various than anywhere else. No idea was ever entertained of attacking the principles or of contesting the rights of society; but the exercise of its authority was divided, to the end that the office might be powerful and the officer insignificant, and that the community should be at once regulated and free. In no country in the world does the law hold so absolute a language as in America, and in no country is the right of applying it vested in so many hands. The administrative power in the United States presents nothing either central or hierarchical in its constitution, which accounts for its passing unperceived. The power exists, but its representative is not to be perceived.

We have already seen that the independent townships of New England protect their own private interests; and the municipal magistrates are the persons to whom the execution of the laws of the State is most frequently entrusted.[9] Besides the general laws, the State sometimes passes general police regulations; but more commonly the townships and town-officers, conjointly with justices of the peace, regulate the minor details of social life, according to the necessities of the different localities, and promulgate such enactments as concern the health of the community, and the peace as well as morality of the citizens.[10] Lastly, these municipal magistrates provide, of their own accord and without any delegated powers, for those unforeseen emergencies which frequently occur in society.[11]

It results from what we have said that in the State of Massachusetts the administrative authority is almost entirely restricted to the township,[12] but that it is distributed among a great number of individuals. In the French commune there is properly but one official functionary, namely, the Maire; and in New England we have seen that there are nineteen. These nineteen functionaries do not in general depend upon one another. The law carefully prescribes a circle of action to each of these magistrates; and within that circle they have an entire right to perform their functions independently of any other authority. Above the township scarcely any trace of a series of official dignitaries is to be found. It sometimes happens that the county officers alter a decision of the townships or town magistrates,[13] but in general the authorities of the county have no right to interfere with the authorities of the township,[14] except in such matters as concern the county.

The magistrates of the township, as well as those of the county, are bound to communicate their acts to the central government in a very small number of predetermined cases.[15] But the central government is not represented by an individual whose business it is to publish police regulations and ordonnances enforcing the execution of the laws; to keep up a regular communication with the officers of the township and the county; to inspect their conduct, to direct their actions, or to reprimand their faults. There is no point which serves as a centre to the radii of the administration.

What, then, is the uniform plan on which the government is conducted, and how is the compliance of the counties and their magistrates or the townships and their officers enforced? In the States of New England the legislative authority embraces more subjects than it does in France; the legislator penetrates to the very core of the administration; the law descends to the most minute details; the same enactment prescribes the principle and the method of its application, and thus imposes a multitude of strict and rigorously defined obligations on the secondary functionaries of the State. The consequence of this is that if all the secondary functionaries of the administration conform to the law, society in all its branches proceeds with the greatest uniformity: the difficulty remains of compelling the secondary functionaries of the administration to conform to the law. It may be affirmed that, in general, society has only two methods of enforcing the execution of the laws at its disposal: a discretionary power may be entrusted to a superior functionary of directing all the others, and of cashiering them in case of disobedience; or the courts of justice may be authorized to inflict judicial penalties on the offender: but these two methods are not always available.

The right of directing a civil officer presupposes that of cashiering him if he does not obey orders, and of rewarding him by promotion if he fulfils his duties with propriety. But an elected magistrate can neither be cashiered nor promoted. All elective functions are inalienable until their term is expired. In fact, the elected magistrate has nothing either to expect or to fear from his constituents; and when all public offices are filled by ballot there can be no series of official dignities, because the double right of commanding and of enforcing obedience can never be vested in the same individual, and because the power of issuing an order can never be joined to that of inflicting a punishment or bestowing a reward.

The communities therefore in which the secondary functionaries of the government are elected are perforce obliged to make great use of judicial penalties as a means of administration. This is not evident at first sight; for those in power are apt to look upon the institution of elective functionaries as one concession, and the subjection of the elected magistrate to the judges of the land as another. They are equally averse to both these innovations; and as they are more pressingly solicited to grant the former than the latter, they accede to the election of the magistrate, and leave him independent of the judicial power. Nevertheless, the second of these measures is the only thing that can possibly counterbalance the first; and it will be found that an elective authority which is not subject to judicial power will, sooner or later, either elude all control or be destroyed. The courts of justice are the only possible medium between the central power and the administrative bodies; they alone can compel the elected functionary to obey, without violating the rights of the elector. The extension of judicial power in the political world ought therefore to be in the exact ratio of the extension of elective offices: if these two institutions do not go hand in hand, the State must fall into anarchy or into subjection.

It has always been remarked that habits of legal business do not render men apt to the exercise of administrative authority. The Americans have borrowed from the English, their fathers, the idea of an institution which is unknown upon the continent of Europe: I allude to that of the Justices of the Peace. The Justice of the Peace is a sort of *mezzo termine* between the magistrate and the man of the world, between the civil officer and the judge. A justice of the peace is a well-informed citizen, though he is not necessarily versed in the knowledge of the laws. His office simply obliges him to execute the police regulations of society; a task in which good sense and integrity are of more avail than legal science. The justice introduces into the administration a certain taste for established forms and publicity, which renders him a most unserviceable instrument of despotism; and, on the other hand, he is not blinded by those superstitions which render legal officers unfit members of a government. The Americans have adopted the system of the English justices of the peace, but they have deprived it of that aristocratic character which is discernible in the mother-country. The Governor of Massachusetts[16] appoints a certain number of justices of the peace in every county, whose functions last seven years.[17] He further designates three individuals from amongst the whole body of justices who form in each county what is called the Court of Sessions. The justices take a personal share in public business; they are sometimes entrusted with administrative functions in conjunction with elected officers,[18] they sometimes constitute a tribunal, before which the magistrates summarily prosecute a refractory citizen, or the citizens inform against the abuses of the magistrate. But it is in the Court of Sessions that they exercise their most important functions. This court meets twice a year in the county town; in Massachusetts it is empowered to enforce the obedience of the greater number[19] of public officers.[20] It must be observed, that in the State of Massachusetts the Court of Sessions is at the same time an administrative body, properly so called, and a political tribunal. It has been asserted that the county is a purely administrative division. The Court of Sessions presides over that small number of affairs which, as they concern several townships, or all the townships of the county in common, cannot be entrusted to any one of them in particular.[21] In all that concerns county business the duties of the Court of Sessions are purely administrative; and if in its investigations it occasionally borrows the forms of judicial procedure, it is only with a view to its own information,[22] or as a guarantee to the community over which it presides. But when the administration of the township is brought before it, it always acts as a judicial body, and in some few cases as an official assembly.

The first difficulty is to procure the obedience of an authority as entirely independent of the general laws of the State as the township is. We have stated that assessors are annually named by the town-meetings to levy the taxes. If a township attempts to evade the payment of the taxes by neglecting to name its assessors, the Court of Sessions condemns it to a heavy penalty.[23] The fine is levied on each of the inhabitants; and the sheriff of the county, who is the officer of justice, executes the mandate. Thus it is that in the United States the authority of the Government is mysteriously concealed under the forms of a judicial sentence; and its influence is at the same time fortified by that irresistible power with which men have invested the formalities of law.

These proceedings are easy to follow and to understand. The demands made upon a township are in general plain and accurately defined; they consist in a simple fact without any complication, or in a principle without its application in detail.[24] But the difficulty increases when it is not the obedience of the township, but that of the town officers which is to be enforced. All the reprehensible actions of which a public functionary may be guilty are reducible to the following heads:

He may execute the law without energy or zeal;

He may neglect to execute the law;

He may do what the law enjoins him not to do.

The last two violations of duty can alone come under the cognizance of a tribunal; a positive and appreciable fact is the indispensable foundation of an action at law. Thus, if the selectmen omit to fulfil the legal formalities usual at town-elections, they may be condemned to pay a fine;[25] but when the public officer performs his duty without ability, and when he obeys the letter of the law without zeal or energy, he is at least beyond the reach of judicial interference. The Court of Sessions, even when it is invested with its official powers, is in this case unable to compel him to a more satisfactory obedience. The fear of removal is the only check to these quasi-offences; and as the Court of Sessions does not originate the town authorities, it cannot remove functionaries whom it does not appoint. Moreover, a perpetual investigation would be necessary to convict the officer of negligence or lukewarmness; and the Court of Sessions sits but twice a year and then only judges such offences as are brought before its notice. The only security of that active and enlightened obedience which a court of justice cannot impose upon public officers lies in the possibility of their arbitrary removal. In France this security is sought for in powers exercised by the heads of the administration; in America it is sought for in the principle of election.

Thus, to recapitulate in a few words what I have been showing: If a public officer in New England commits a crime in the exercise of his functions, the ordinary courts of justice are always called upon to pass sentence upon him. If he commits a fault in his official capacity, a purely administrative tribunal is empowered to punish him; and, if the affair is important or urgent, the judge supplies the omission of the functionary.[26] Lastly, if the same individual is guilty of one of those intangible offences of which human justice has no cognizance, he annually appears before a tribunal from which there is no appeal, which can at once reduce him to insignificance and deprive him of his charge. This system undoubtedly possesses great advantages, but its execution is attended with a practical difficulty which it is important to point out.

I have already observed that the administrative tribunal, which is called the Court of Sessions, has no right of inspection over the town-officers. It can only interfere when the conduct of a magistrate is specially brought under its notice; and this is the delicate part of the system. The Americans of New England are unacquainted with the office of public prosecutor in the Court of Sessions,[27] and it may readily be perceived that it could not have been established without difficulty. If an accusing magistrate had merely been appointed in the chief town of each county, and if he had been unassisted by agents in the townships, he would not have been better acquainted with what was going on in the county than the members of the Court of Sessions. But to appoint agents in each township would have been to centre in his person the most formidable of powers, that of a judicial administration. Moreover, laws are the children of habit, and nothing of the kind exists in the legislation of England. The Americans have therefore divided the offices of inspection and of prosecution, as well as all the other functions of the administration. Grand-jurors are bound by the law to apprise the court to which they belong of all the misdemeanours which may have been committed in their county.[28] There are certain great offences which are officially prosecuted by the States;[29] but more frequently the task of punishing delinquents devolves upon the fiscal officer, whose province it is to receive the fine: thus the treasurer of the township is charged with the prosecution of such administrative offences as fall under his notice. But a more special appeal is made by American legislation to the private interest of the citizen;[30] and this great principle is constantly to be met with in studying the laws of the United States. American legislators are more apt to give men credit for intelligence than for honesty, and they rely not a little on personal cupidity for the execution of the laws. When an individual is really and sensibly injured by an administrative abuse, it is natural that his personal interest should induce him to prosecute. But if a legal formality be required, which, however advantageous to the community, is of small importance to individuals, plaintiffs may be less easily found; and thus, by a tacit agreement, the laws may fall into disuse. Reduced by their system to this extremity, the Americans are obliged to encourage informers by bestowing on them a portion of the penalty in certain cases,[31] and to insure the execution of the laws by the dangerous expedient of degrading the morals of the people. The only administrative authority above the county magistrates is, properly speaking, that of the Government.

General Remarks on the Administration of the United States

Differences of the States of the Union in their system of administration—Activity and perfection of the local authorities decrease towards the South—Power of the magistrate increases; that of the elector diminishes—Administration passes from the township to the country—States of New York, Ohio, Pennsylvania—Principles of administration applicable to the whole Union—Election of public officers, and inalienability of their functions—Absence of gradation of ranks—Introduction of judicial resources into the administration.

I have already premised that, after having examined the constitution of the township and the county of New England in detail, I should take a general view of the remainder of the Union. Townships and a local activity exist in every State; but in no part of the confederation is a township to be met with precisely similar to those of New England. The more we descend towards the South, the less active does the business of the township or parish become; the number of magistrates, of functions, and of rights decreases; the population exercises a less immediate influence on affairs; town meetings are less frequent, and the subjects of debate less numerous. The power of the elected magistrate is augmented and that of the elector diminished, whilst the public spirit of the local communities is less awakened and less influential.[32] These differences may be perceived to a certain extent in the State of New York; they are very sensible in Pennsylvania; but they become less striking as we advance to the north-west. The majority of the emigrants who settle in the north-western States are natives of New England, and they carry the habits of their mother-country with them into that which they adopt. A township in Ohio is by no means dissimilar from a township in Massachusetts.

We have seen that in Massachusetts the mainspring of public administration lies in the township. It forms the common centre of the interests and affections of the citizens. But this ceases to be the case as we descend to States in which knowledge is less generally diffused, and where the township consequently offers fewer guarantees of a wise and active administration. As we leave New England, therefore, we find that the importance of the town is gradually transferred to the county, which becomes the centre of administration, and the intermediate power between the Government and the citizen. In Massachusetts the business of the county is conducted by the Court of Sessions, which is composed of a quorum named by the Governor and his council; but the county has no representative assembly, and its expenditure is voted by the national legislature. In the great State of New York, on the contrary, and in those of Ohio and Pennsylvania, the inhabitants of each county choose a certain number of representatives, who constitute the assembly of the county.[33] The county assembly has the right of taxing the inhabitants to a certain extent; and in this respect it enjoys the privileges of a real legislative body: at the same time it exercises an executive power in the county, frequently directs the administration of the townships, and restricts their authority within much narrower bounds than in Massachusetts.

Such are the principal differences which the systems of county and town administration present in the Federal States. Were it my intention to examine the provisions of American law minutely, I should have to point out still further differences in the executive details of the several communities. But what I have already said may suffice to show the general principles on which the administration of the United States rests. These principles are differently applied; their consequences are more or less numerous in various localities; but they are always substantially the same. The laws differ, and their outward features change, but their character does not vary. If the township and the county are not everywhere constituted in the same manner, it is at least true that in the United States the county and the township are always based upon the same principle, namely, that everyone is the best judge of what concerns himself alone, and the most proper person to supply his private wants. The township and the county are therefore bound to take care of their special interests: the State governs, but it does not interfere with their administration. Exceptions to this rule may be met with, but not a contrary principle.

The first consequence of this doctrine has been to cause all the magistrates to be chosen either by or at least from amongst the citizens. As the officers are everywhere elected or appointed for a certain period, it has been impossible to establish the rules of a dependent series of authorities; there are almost as many independent functionaries as there are functions, and the executive power is disseminated in a multitude of hands. Hence arose the indispensable necessity of introducing the control of the courts of justice over the administration, and the system of pecuniary penalties, by which the secondary bodies and their representatives are constrained to obey the laws. This system obtains from one end of the Union to the other. The power of punishing the misconduct of public officers, or of performing the part of the executive in urgent cases, has not, however, been bestowed on the same judges in all the States. The Anglo-Americans derived the institution of justices of the peace from a common source; but although it exists in all the States, it is not always turned to the same use. The justices of the peace everywhere participate in the administration of the townships and the counties,[34] either as public officers or as the judges of public misdemeanours, but in most of the States the more important classes of public offences come under the cognizance of the ordinary tribunals.

The election of public officers, or the inalienability of their functions, the absence of a gradation of powers, and the introduction of a judicial control over the secondary branches of the administration, are the universal characteristics of the American system from Maine to the Floridas. In some States (and that of New York has advanced most in this direction) traces of a centralized administration begin to be discernible. In the State of New York the officers of the central government exercise, in certain cases, a sort of inspection or control over the secondary bodies.[35]

At other times they constitute a court of appeal for the decision of affairs.[36] In the State of New York judicial penalties are less used than in other parts as a means of administration, and the right of prosecuting the offences of public officers is vested in fewer hands.[37] The same tendency is faintly observable in some other States;[38] but in general the prominent feature of the administration in the United States is its excessive local independence.

Of the State

I have described the townships and the administration; it now remains for me to speak of the State and the Government. This is ground I may pass over rapidly, without fear of being misunderstood; for all I have to say is to be found in written forms of the various constitutions, which are easily to be procured. These constitutions rest upon a simple and rational theory; their forms have been adopted by all constitutional nations, and are become familiar to us. In this place, therefore, it is only necessary for me to give a short analysis; I shall endeavour afterwards to pass judgment upon what I now describe.

Legislative Power of the State

Division of the Legislative Body into two Houses—Senate—
House of Representatives—Different functions of these two Bodies.

The legislative power of the State is vested in two assemblies, the first of which generally bears the name of the Senate. The Senate is commonly a legislative body; but it sometimes becomes an executive and judicial one. It takes a part in the government in several ways, according to the constitution of the different States;[39] but it is in the nomination of public functionaries that it most commonly assumes an executive power. It partakes of judicial power in the trial of certain political offences, and sometimes also in the decision of certain civil cases.[40] The number of its members is always small. The other branch of the legislature, which is usually called the House of Representatives, has no share whatever in the administration, and only takes a part in the judicial power inasmuch as it impeaches public functionaries before the Senate. The members of the two Houses are nearly everywhere subject to the same conditions of election. They are chosen in the same manner, and by the same citizens. The only difference which exists between them is, that the term for which the Senate is chosen is in general longer than that of the House of Representatives. The latter seldom remain in office longer than a year; the former usually sit two or three years. By granting to the senators the privilege of being chosen for several years, and being renewed seriatim, the law takes care to preserve in the legislative body a nucleus of men already accustomed to public business, and capable of exercising a salutary influence upon the junior members.

The Americans, plainly, did not desire, by this separation of the legislative body into two branches, to make one house hereditary and the other elective; one aristocratic and the other democratic. It was not their object to create in the one a bulwark to power, whilst the other represented the interests and passions of the people. The only advantages which result from the present constitution of the United States are the division of the legislative power and the consequent check upon political assemblies; with the creation of a tribunal of appeal for the revision of the laws.

Time and experience, however, have convinced the Americans that if these are its only advantages, the division of the legislative power is still a principle of the greatest necessity. Pennsylvania was the only one of the United States which at first attempted to establish a single House of Assembly, and Franklin himself was so far carried away by the necessary consequences of the principle of the sovereignty of the people as to have concurred in the measure; but the Pennsylvanians were soon obliged to change the law, and to create two Houses. Thus the principle of the division of the legislative power was finally established, and its necessity may henceforward be regarded as a demonstrated truth. This theory, which was nearly unknown to the republics of antiquity—which was introduced into the world almost by accident, like so many other great truths—and misunderstood by several modern nations, is at length become an axiom in the political science of the present age.

The Executive Power of the State

Office of Governor in an American State—The place he occupies in relation to the Legislature—His rights and his duties—His dependence on the people.

The executive power of the State may with truth be said to be *represented* by the Governor, although he enjoys but a portion of its rights. The supreme magistrate, under the title of Governor, is the official moderator and counsellor of the legislature. He is armed with a veto or suspensive power, which allows him to stop, or at least to retard, its movements at pleasure. He lays the wants of the country before the legislative body, and points out the means which he thinks may be usefully employed in providing for them; he is the natural executor of its decrees in all the undertakings which interest the nation at large.[41] In the absence of the legislature, the Governor is bound to take all necessary steps to guard the State against violent shocks and unforeseen dangers. The whole military power of the State is at the disposal of the Governor. He is the commander of the militia, and head of the armed force. When the authority, which is by general consent awarded to the laws, is disregarded, the Governor puts himself at the head of the armed force of the State, to quell resistance, and to restore order. Lastly, the Governor takes no share in the administration of townships and counties, except it be indirectly in the nomination of Justices of the Peace, which nomination he has not the power to cancel.[42] The Governor is an elected magistrate, and is generally chosen for one or two years only; so that he always continues to be strictly dependent upon the majority who returned him.

Political Effects of the System of Local Administration in the United States

Necessary distinction between the general centralization of Government and the centralization of the local administration—Local administration not centralized in the United States: great general centralization of the Government—Some bad consequences resulting to the United States from the local administration—Administrative advantages attending this order of things—The power which conducts the Government is less regular, less enlightened, less learned, but much greater than in Europe—Political advantages of this order of things—In the United States the interests of the country are everywhere kept in view—Support given to the Government by the community—Provincial institutions more necessary in proportion as the social condition becomes more democratic—Reason of this.

Centralization is become a word of general and daily use, without any precise meaning being attached to it. Nevertheless, there exist two distinct kinds of centralization, which it is necessary to discriminate with accuracy. Certain interests are common to all parts of a nation, such as the enactment of its general laws and the maintenance of its foreign relations. Other interests are peculiar to certain parts of the nation; such, for instance, as the business of different townships. When the power which directs the general interests is centred in one place, or vested in the same persons, it constitutes a central government. In like manner the power of directing partial or local interests, when brought together into one place, constitutes what may be termed a central administration.

Upon some points these two kinds of centralization coalesce; but by classifying the objects which fall more particularly within the province of each of them, they may easily be distinguished. It is evident that a central government acquires immense power when united to administrative centralization. Thus combined, it accustoms men to set their own will habitually and completely aside; to submit, not only for once, or upon one point, but in every respect, and at all times. Not only, therefore, does this union of power subdue them compulsorily, but it affects them in the ordinary habits of life, and influences each individual, first separately and then collectively.

These two kinds of centralization mutually assist and attract each other; but they must not be supposed to be inseparable. It is impossible to imagine a more completely central government than that which existed in France under Louis XIV.; when the same individual was the author and the interpreter of the laws, and the representative of France at home and abroad, he was justified in asserting that the State was identified with his person. Nevertheless, the administration was much less centralized under Louis XIV. than it is at the present day.

In England the centralization of the government is carried to great perfection; the State has the compact vigour of a man, and by the sole act of its will it puts immense engines in motion, and wields or collects the efforts of its authority. Indeed, I cannot conceive that a nation can enjoy a secure or prosperous existence without a powerful centralization of government. But I am of opinion that a central administration enervates the nations in which it exists by incessantly diminishing their public spirit. If such an administration succeeds in condensing at a given moment, on a given point, all the disposable resources of a people, it impairs at least the renewal of those resources. It may ensure a victory in the hour of strife, but it gradually relaxes the sinews of strength. It may contribute admirably to the transient greatness of a man, but it cannot ensure the durable prosperity of a nation.

If we pay proper attention, we shall find that whenever it is said that a State cannot act because it has no central point, it is the centralization of the government in which it is deficient. It is frequently asserted, and we are prepared to assent to the proposition, that the German empire was never able to bring all its powers into action. But the reason was, that the State was never able to enforce obedience to its general laws, because the several members of that great body always claimed the right, or found the means, of refusing their cooperation to the representatives of the common authority, even in the affairs which concerned the mass of the people; in other words, because there was no centralization of government. The same remark is applicable to the Middle Ages; the cause of all the confusion of feudal society was that the control, not only of local but of general interests, was divided amongst a thousand hands, and broken up in a thousand different ways; the absence of a central government prevented the nations of Europe from advancing with energy in any straightforward course.

We have shown that in the United States no central administration and no dependent series of public functionaries exist. Local authority has been carried to lengths which no European nation could endure without great inconvenience, and which has even produced some disadvantageous consequences in America. But in the United States the centralization of the Government is complete; and it would be easy to prove that the national power is more compact than it has ever been in the old nations of Europe. Not only is there but one legislative body in each State; not only does there exist but one source of political authority; but district assemblies and county courts have not in general been multiplied, lest they should be tempted to exceed their administrative duties, and interfere with the Government. In America the legislature of each State is supreme; nothing can impede its authority; neither privileges, nor local immunities, nor personal influence, nor even the empire of reason, since it represents that majority which claims to be the sole organ of reason. Its own determination is, therefore, the only limit to this action. In juxtaposition to it, and under its immediate control, is the representative of the executive power, whose duty it is to constrain the refractory to submit by superior force. The only symptom of weakness lies in certain details of the action of the Government. The American republics have no standing armies to intimidate a discontented minority; but as no minority has as yet been reduced to declare open war, the necessity of an army has not been felt.[43] The State usually employs the officers of the township or the county to deal with the citizens. Thus, for instance, in New England, the assessor fixes the rate of taxes; the collector receives them; the town-treasurer transmits the amount to the public treasury; and the disputes which may arise are brought before the ordinary courts of justice. This method of collecting taxes is slow as well as inconvenient, and it would prove a perpetual hindrance to a Government whose pecuniary demands were large. It is desirable that, in whatever materially affects its existence, the Government should be served by officers of its own, appointed by itself, removable at pleasure, and accustomed to rapid methods of proceeding. But it will always be easy for the central government, organized as it is in America, to introduce new and more efficacious modes of action, proportioned to its wants.

The absence of a central government will not, then, as has often been asserted, prove the destruction of the republics of the New World; far from supposing that the American governments are not sufficiently centralized, I shall prove hereafter that they are too much so. The legislative bodies daily encroach upon the authority of the Government, and their tendency, like that of the French Convention, is to appropriate it entirely to themselves. Under these circumstances the social power is constantly changing hands, because it is subordinate to the power of the people, which is too apt to forget the maxims of wisdom and of foresight in the consciousness of its strength: hence arises its danger; and thus its vigour, and not its impotence, will probably be the cause of its ultimate destruction.

The system of local administration produces several different effects in America. The Americans seem to me to have outstepped the limits of sound policy in isolating the administration of the Government; for order, even in second-rate affairs, is a matter of national importance.[44] As the State has no administrative functionaries of its own, stationed on different points of its territory, to whom it can give a common impulse, the consequence is that it rarely attempts to issue any general police regulations. The want of these regulations is severely felt, and is frequently observed by Europeans. The appearance of disorder which prevails on the surface leads him at first to imagine that society is in a state of anarchy; nor does he perceive his mistake till he has gone deeper into the subject. Certain undertakings are of importance to the whole State; but they cannot be put in execution, because there is no national administration to direct them. Abandoned to the exertions of the towns or counties, under the care of elected or temporary agents, they lead to no result, or at least to no durable benefit.

The partisans of centralization in Europe are wont to maintain that the Government directs the affairs of each locality better than the citizens could do it for themselves; this may be true when the central power is enlightened, and when the local districts are ignorant; when it is as alert as they are slow; when it is accustomed to act, and they to obey. Indeed, it is evident that this double tendency must augment with the increase of centralization, and that the readiness of the one and the incapacity of the others must become more and more prominent. But I deny that such is the case when the people is as enlightened, as awake to its interests, and as accustomed to reflect on them, as the Americans are. I am persuaded, on the contrary, that in this case the collective strength of the citizens will always conduce more efficaciously to the public welfare than the authority of the Government. It is difficult to point out with certainty the means of arousing a sleeping population, and of giving it passions and knowledge which it does not possess; it is, I am well aware, an arduous task to persuade men to busy themselves about their own affairs; and it would frequently be easier to interest them in the punctilios of court etiquette than in the repairs of their common dwelling. But whenever a central administration affects to supersede the persons most interested, I am inclined to suppose that it is either misled or desirous to mislead. However enlightened and however skilful a central power may be, it cannot of itself embrace all the details of the existence of a great nation. Such vigilance exceeds the powers of man. And when it attempts to create and set in motion so many complicated springs, it must submit to a very imperfect result, or consume itself in bootless efforts.

Centralization succeeds more easily, indeed, in subjecting the external actions of men to a certain uniformity, which at least commands our regard, independently of the objects to which it is applied, like those devotees who worship the statue and forget the deity it represents. Centralization imparts without difficulty an admirable regularity to the routine of business; provides for the details of the social police with sagacity; represses the smallest disorder and the most petty misdemeanours; maintains society in a *status quo* alike secure from improvement and decline; and perpetuates a drowsy precision in the conduct of affairs, which is hailed by the heads of the administration as a sign of perfect order and public tranquillity:[45] in short, it excels more in prevention than in action. Its force deserts it when society is to be disturbed or accelerated in its course; and if once the cooperation of private citizens is necessary to the furtherance of its measures, the secret of its impotence is disclosed. Even whilst it invokes their assistance, it is on the condition that they shall act exactly as much as the Government chooses, and exactly in the manner it appoints. They are to take charge of the details, without aspiring to guide the system; they are to work in a dark and subordinate sphere, and only to judge the acts in which they have themselves co-operated by their results. These, however, are not conditions on which the alliance of the human will is to be obtained; its carriage must be free and its actions responsible, or (such is the constitution of man) the citizen had rather remain a passive spectator than a dependent actor in schemes with which he is unacquainted.

It is undeniable that the want of those uniform regulations which control the conduct of every inhabitant of France is not unfrequently felt in the United States. Gross instances of social indifference and neglect are to be met with, and from time to time disgraceful blemishes are seen in complete contrast with the surrounding civilization. Useful undertakings which cannot succeed without perpetual attention and rigorous exactitude are very frequently abandoned in the end; for in America, as well as in other countries, the people is subject to sudden impulses and momentary exertions. The European who is accustomed to find a functionary always at hand to interfere with all he undertakes has some difficulty in accustoming himself to the complex mechanism of the administration of the townships. In general it may be affirmed that the lesser details of the police, which render life easy and comfortable, are neglected in America; but that the essential guarantees of man in society are as strong there as elsewhere. In America the power which conducts the Government is far less regular, less enlightened, and less learned, but an hundredfold more authoritative than in Europe. In no country in the world do the citizens make such exertions for the common weal; and I am acquainted with no people which has established schools as numerous and as efficacious, places of public worship better suited to the wants of the inhabitants, or roads kept in better repair. Uniformity or permanence of design, the minute arrangement of details,[46] and the perfection of an ingenious administration, must not be sought for in the United States; but it will be easy to find, on the other hand, the symptoms of a power which, if it is somewhat barbarous, is at least robust; and of an existence which is checkered with accidents indeed, but cheered at the same time by animation and effort.

Granting for an instant that the villages and counties of the United States would be more usefully governed by a remote authority which they had never seen than by functionaries taken from the midst of them admitting, for the sake of argument, that the country would be more secure, and the resources of society better employed, if the whole administration centred in a single arm—still the *political* advantages which the Americans derive from their system would induce me to prefer it to the contrary plan. It profits me but little, after all, that a vigilant authority should protect the tranquillity of my pleasures and constantly avert all dangers from my path, without my care or my concern, if this same authority is the absolute mistress of my liberty and of my life, and if it so monopolizes all the energy of existence that when it languishes everything languishes around it, that when it sleeps everything must sleep, that when it dies the State itself must perish.

In certain countries of Europe the natives consider themselves as a kind of settlers, indifferent to the fate of the spot upon which they live. The greatest changes are effected without their concurrence and (unless chance may have apprised them of the event) without their knowledge; nay more, the citizen is unconcerned as to the condition of his village, the police of his street, the repairs of the church or of the parsonage; for he looks upon all these things as unconnected with himself, and as the property of a powerful stranger whom he calls the Government. He has only a life-interest in these possessions, and he entertains no notions of ownership or of improvement. This want of interest in his own affairs goes so far that, if his own safety or that of his children is endangered, instead of trying to avert the peril, he will fold his arms, and wait till the nation comes to his assistance. This same individual, who has so completely sacrificed his own free will, has no natural propensity to obedience; he cowers, it is true, before the pettiest officer; but he braves the law with the spirit of a conquered foe as soon as its superior force is removed: his oscillations between servitude and license are perpetual. When a nation has arrived at this state it must either change its customs and its laws or perish: the source of public virtue is dry, and, though it may contain subjects, the race of citizens is extinct. Such communities are a natural prey to foreign conquests, and if they do not disappear from the scene of life, it is because they are surrounded by other nations similar or inferior to themselves: it is because the instinctive feeling of their country's claims still exists in their hearts; and because an involuntary pride in the name it bears, or a vague reminiscence of its bygone fame, suffices to give them the impulse of self-preservation.

Nor can the prodigious exertions made by tribes in the defence of a country to which they did not belong be adduced in favour of such a system; for it will be found that in these cases their main incitement was religion. The permanence, the glory, or the prosperity of the nation were become parts of their faith, and in defending the country they inhabited they defended that Holy City of which they were all citizens. The Turkish tribes have never taken an active share in the conduct of the affairs of society, but they accomplished stupendous enterprises as long as the victories of the Sultan were the triumphs of the Mohammedan faith. In the present age they are in rapid decay, because their religion is departing, and despotism only remains. Montesquieu, who attributed to absolute power an authority peculiar to itself, did it, as I conceive, an undeserved honour; for despotism, taken by itself, can produce no durable results. On close inspection we shall find that religion, and not fear, has ever been the cause of the long-lived prosperity of an absolute government. Whatever exertions may be made, no true power can be founded among men which does not depend upon the free union of their inclinations; and patriotism and religion are the only two motives in the world which can permanently direct the whole of a body politic to one end.

Laws cannot succeed in rekindling the ardour of an extinguished faith, but men may be interested in the fate of their country by the laws. By this influence the vague impulse of patriotism, which never abandons the human heart, may be directed and revived; and if it be connected with the thoughts, the passions, and the daily habits of life, it may be consolidated into a durable and rational sentiment. Let it not be said that the time for the experiment is already past; for the old age of nations is not like the old age of men, and every fresh generation is a new people ready for the care of the legislator.

It is not the *administrative* but the *political* effects of the local system that I most admire in America. In the United States the interests of the country are everywhere kept in view; they are an object of solicitude to the people of the whole Union, and every citizen is as warmly attached to them as if they were his own. He takes pride in the glory of his nation; he boasts of its success, to which he conceives himself to have contributed, and he rejoices in the general prosperity by which he profits. The feeling he entertains towards the State is analogous to that which unites him to his family, and it is by a kind of egotism that he interests himself in the welfare of his country.

The European generally submits to a public officer because he represents a superior force; but to an American he represents a right. In America it may be said that no one renders obedience to man, but to justice and to law. If the opinion which the citizen entertains of himself is exaggerated, it is at least salutary; he unhesitatingly confides in his own powers, which appear to him to be all-sufficient. When a private individual meditates an undertaking, however directly connected it may be with the welfare of society, he never thinks of soliciting the cooperation of the Government, but he publishes his plan, offers to execute it himself, courts the assistance of other individuals, and struggles manfully against all obstacles. Undoubtedly he is often less successful than the State might have been in his position; but in the end the sum of these private undertakings far exceeds all that the Government could have done.

As the administrative authority is within the reach of the citizens, whom it in some degree represents, it excites neither their jealousy nor their hatred; as its resources are limited, every one feels that he must not rely solely on its assistance. Thus, when the administration thinks fit to interfere, it is not abandoned to itself as in Europe; the duties of the private citizens are not supposed to have lapsed because the State assists in their fulfilment, but every one is ready, on the contrary, to guide and to support it. This action of individual exertions, joined to that of the public authorities, frequently performs what the most energetic central administration would be unable to execute. It would be easy to adduce several facts in proof of what I advance, but I had rather give only one, with which I am more thoroughly acquainted.[47] In America the means which the authorities have at their disposal for the discovery of crimes and the arrest of criminals are few. The State police does not exist, and passports are unknown. The criminal police of the United States cannot be compared to that of France; the magistrates and public prosecutors are not numerous, and the examinations of prisoners are rapid and oral. Nevertheless in no country does crime more rarely elude punishment. The reason is, that every one conceives himself to be interested in furnishing evidence of the act committed, and in stopping the delinquent. During my stay in the United States I witnessed the spontaneous formation of committees for the pursuit and prosecution of a man who had committed a great crime in a certain county. In Europe a criminal is an unhappy being who is struggling for his life against the ministers of justice, whilst the population is merely a spectator of the conflict; in America he is looked upon as an enemy of the human race, and the whole of mankind is against him.

I believe that provincial institutions are useful to all nations, but nowhere do they appear to me to be more indispensable than amongst a democratic people. In an aristocracy order can always be maintained in the midst of liberty, and as the rulers have a great deal to lose order is to them a first-rate consideration. In like manner an aristocracy protects the people from the excesses of despotism, because it always possesses an organised power ready to resist a despot. But a democracy without provincial institutions has no security against these evils. How can a populace, unaccustomed to freedom in small concerns, learn to use it temperately in great affairs? What resistance can be offered to tyranny in a country where every private individual is impotent, and where the citizens are united by no common tie? Those who dread the licence of the mob, and those who fear the rule of absolute power, ought alike to desire the progressive growth of provincial liberties.

On the other hand, I am convinced that democratic nations are most exposed to fall beneath the yoke of a central administration, for several reasons, amongst which is the following. The constant tendency of these nations is to concentrate all the strength of the Government in the hands of the only power which directly represents the people, because beyond the people nothing is to be perceived but a mass of equal individuals confounded together. But when the same power is already in possession of all the attributes of the Government, it can scarcely refrain from penetrating into the details of the administration, and an opportunity of doing so is sure to present itself in the end, as was the case in France. In the French Revolution there were two impulses in opposite directions, which must never be confounded—the one was favourable to liberty, the other to despotism. Under the ancient monarchy the King was the sole author of the laws, and below the power of the sovereign certain vestiges of provincial institutions, half destroyed, were still distinguishable. These provincial institutions were incoherent, ill compacted, and frequently absurd; in the hands of the aristocracy they had sometimes been converted into instruments of oppression. The Revolution declared itself the enemy of royalty and of provincial institutions at the same time; it confounded all that had preceded it—despotic power and the checks to its abuses in indiscriminate hatred, and its tendency was at once to overthrow and to centralize. This double character of the French Revolution is a fact which has been adroitly handled by the friends of absolute power. Can they be accused of labouring in the cause of despotism when they are defending that central administration which was one of the great innovations of the Revolution?[48] In this manner popularity may be conciliated with hostility to the rights of the people, and the secret slave of tyranny may be the professed admirer of freedom.

I have visited the two nations in which the system of provincial liberty has been most perfectly established, and I have listened to the opinions of different parties in those countries. In America I met with men who secretly aspired to destroy the democratic institutions of the Union; in England I found others who attacked the aristocracy openly, but I know of no one who does not regard provincial independence as a great benefit. In both countries I have heard a thousand different causes assigned for the evils of the State, but the local system was never mentioned amongst them. I have heard citizens attribute the power and prosperity of their country to a multitude of reasons, but they *all* placed the advantages of local institutions in the foremost rank. Am I to suppose that when men who are naturally so divided on religious opinions and on political theories agree on one point (and that one of which they have daily experience), they are all in error? The only nations which deny the utility of provincial liberties are those which have fewest of them; in other words, those who are unacquainted with the institution are the only persons who pass a censure upon it.

Chapter VI

Judicial Power in the United States, and Its Influence on Political Society

The Anglo-Americans have retained the characteristics of judicial power which are common to all nations—They have, however, made it a powerful political organ—How—In what the judicial system of the Anglo-Americans differs from that of all other nations—Why the American judges have the right of declaring the laws to be unconstitutional—How they use this right—Precautions taken by the legislator to prevent its abuse.

I have thought it essential to devote a separate chapter to the judicial authorities of the United States, lest their great political importance should be lessened in the reader's eyes by a merely incidental mention of them. Confederations have existed in other countries beside America, and republics have not been established upon the shores of the New World alone; the representative system of government has been adopted in several States of Europe, but I am not aware that any nation of the globe has hitherto organised a judicial power on the principle now adopted by the Americans. The judicial organisation of the United States is the institution which a stranger has the greatest difficulty in understanding. He hears the authority of a judge invoked in the political occurrences of every day, and he naturally concludes that in the United States the judges are important political functionaries; nevertheless, when he examines the nature of the tribunals, they offer nothing which is contrary to the usual habits and privileges of those bodies, and the magistrates seem to him to interfere in public affairs of chance, but by a chance which recurs every day.

When the Parliament of Paris remonstrated, or refused to enregister an edict, or when it summoned a functionary accused of malversation to its bar, its political influence as a judicial body was clearly visible; but nothing of the kind is to be seen in the United States. The Americans have retained all the ordinary characteristics of judicial authority, and have carefully restricted its action to the ordinary circle of its functions.

The first characteristic of judicial power in all nations is the duty of arbitration. But rights must be contested in order to warrant the interference of a tribunal; and an action must be brought to obtain the decision of a judge. As long, therefore, as the law is uncontested, the judicial authority is not called upon to discuss it, and it may exist without being perceived. When a judge in a given case attacks a law relating to that case, he extends the circle of his customary duties, without however stepping beyond it; since he is in some measure obliged to decide upon the law in order to decide the case. But if he pronounces upon a law without resting upon a case, he clearly steps beyond his sphere, and invades that of the legislative authority.

The second characteristic of judicial power is that it pronounces on special cases, and not upon general principles. If a judge in deciding a particular point destroys a general principle, by passing a judgment which tends to reject all the inferences from that principle, and consequently to annul it, he remains within the ordinary limits of his functions. But if he directly attacks a general principle without having a particular case in view, he leaves the circle in which all nations have agreed to confine his authority, he assumes a more important, and perhaps a more useful, influence than that of the magistrate, but he ceases to be a representative of the judicial power.

The third characteristic of the judicial power is its inability to act unless it is appealed to, or until it has taken cognizance of an affair. This characteristic is less general than the other two; but, notwithstanding the exceptions, I think it may be regarded as essential. The judicial power is by its nature devoid of action; it must be put in motion in order to produce a result. When it is called upon to repress a crime, it punishes the criminal; when a wrong is to be redressed, it is ready to redress it; when an act requires interpretation, it is prepared to interpret it; but it does not pursue criminals, hunt out wrongs, or examine into evidence of its own accord. A judicial functionary who should open proceedings, and usurp the censorship of the laws, would in some measure do violence to the passive nature of his authority.

The Americans have retained these three distinguishing characteristics of the judicial power; an American judge can only pronounce a decision when litigation has arisen, he is only conversant with special cases, and he cannot act until the cause has been duly brought before the court. His position is therefore perfectly similar to that of the magistrate of other nations; and he is nevertheless invested with immense political power. If the sphere of his authority and his means of action are the same as those of other judges, it may be asked whence he derives a power which they do not possess. The cause of this difference lies in the simple fact that the Americans have acknowledged the right of the judges to found their decisions on the constitution rather than on the laws. In other words, they have left them at liberty not to apply such laws as may appear to them to be unconstitutional.

I am aware that a similar right has been claimed—but claimed in vain—by courts of justice in other countries; but in America it is recognised by all authorities; and not a party, nor so much as an individual, is found to contest it. This fact can only be explained by the principles of the American constitution. In France the constitution is (or at least is supposed to be) immutable; and the received theory is that no power has the right of changing any part of it.[1] In England the Parliament has an acknowledged right to modify the constitution; as, therefore, the constitution may undergo perpetual changes, it does not in reality exist; the Parliament is at once a legislative and a constituent assembly.[2] The political theories of America are more simple and more rational. An American constitution is not supposed to be immutable as in France, nor is it susceptible of modification by the ordinary powers of society as in England. It constitutes a detached whole, which, as it represents the determination of the whole people, is no less binding on the legislator than on the private citizen, but which may be altered by the will of the people in predetermined cases, according to established rules. In America the constitution may therefore vary, but as long as it exists it is the origin of all authority, and the sole vehicle of the predominating force.[3]

It is easy to perceive in what manner these differences must act upon the position and the rights of the judicial bodies in the three countries I have cited. If in France the tribunals were authorised to disobey the laws on the ground of their being opposed to the constitution, the supreme power would in fact be placed in their hands, since they alone would have the right of interpreting a constitution, the clauses of which can be modified by no authority. They would therefore take the place of the nation, and exercise as absolute a sway over society as the inherent weakness of judicial power would allow them to do. Undoubtedly, as the French judges are incompetent to declare a law to be unconstitutional, the power of changing the constitution is indirectly given to the legislative body, since no legal barrier would oppose the alterations which it might prescribe. But it is better to grant the power of changing the constitution of the people to men who represent (however imperfectly) the will of the people, than to men who represent no one but themselves.

It would be still more unreasonable to invest the English judges with the right of resisting the decisions of the legislative body, since the Parliament which makes the laws also makes the constitution; and consequently a law emanating from the three powers of the State can in no case be unconstitutional. But neither of these remarks is applicable to America.

In the United States the constitution governs the legislator as much as the private citizen; as it is the first of laws it cannot be modified by a law, and it is therefore just that the tribunals should obey the constitution in preference to any law. This condition is essential to the power of the judicature, for to select that legal obligation by which he is most strictly bound is the natural right of every magistrate.

In France the constitution is also the first of laws, and the judges have the same right to take it as the ground of their decisions, but were they to exercise this right they must perforce encroach on rights more sacred than their own, namely, on those of society, in whose name they are acting. In this case the State-motive clearly prevails over the motives of an individual. In America, where the nation can always reduce its magistrates to obedience by changing its constitution, no danger of this kind is to be feared. Upon this point, therefore, the political and the logical reasons agree, and the people as well as the judges preserve their privileges.

Whenever a law which the judge holds to be unconstitutional is argued in a tribunal of the United States he may refuse to admit it as a rule; this power is the only one which is peculiar to the American magistrate, but it gives rise to immense political influence. Few laws can escape the searching analysis of the judicial power for any length of time, for there are few which are not prejudicial to some private interest or other, and none which may not be brought before a court of justice by the choice of parties, or by the necessity of the case. But from the time that a judge has refused to apply any given law in a case, that law loses a portion of its moral cogency. The persons to whose interests it is prejudicial learn that means exist of evading its authority, and similar suits are multiplied, until it becomes powerless. One of two alternatives must then be resorted to: the people must alter the constitution, or the legislature must repeal the law. The political power which the Americans have intrusted to their courts of justice is therefore immense, but the evils of this power are considerably diminished by the obligation which has been imposed of attacking the laws through the courts of justice alone. If the judge had been empowered to contest the laws on the ground of theoretical generalities, if he had been enabled to open an attack or to pass a censure on the legislator, he would have played a prominent part in the political sphere; and as the champion or the antagonist of a party, he would have arrayed the hostile passions of the nation in the conflict. But when a judge contests a law applied to some particular case in an obscure proceeding, the importance of his attack is concealed from the public gaze, his decision bears upon the interest of an individual, and if the law is slighted it is only collaterally. Moreover, although it is censured, it is not abolished; its moral force may be diminished, but its cogency is by no means suspended, and its final destruction can only be accomplished by the reiterated attacks of judicial functionaries. It will readily be understood that by connecting the censorship of the laws with the private interests of members of the community, and by intimately uniting the prosecution of the law with the prosecution of an individual, legislation is protected from wanton assailants, and from the daily aggressions of party-spirit. The errors of the legislator are exposed whenever their evil consequences are most felt, and it is always a positive and appreciable fact which serves as the basis of a prosecution.

I am inclined to believe this practice of the American courts to be at once the most favourable to liberty as well as to public order. If the judge could only attack the legislator openly and directly, he would sometimes be afraid to oppose any resistance to his will; and at other moments party-spirit might encourage him to brave it at every turn. The laws would consequently be attacked when the power from which they emanate is weak, and obeyed when it is strong. That is to say, when it would be useful to respect them they would be contested, and when it would be easy to convert them into an instrument of oppression they would be respected. But the American judge is brought into the political arena independently of his own will. He only judges the law because he is obliged to judge a case. The political question which he is called upon to resolve is connected with the interest of the suitors, and he cannot refuse to decide it without abdicating the duties of his post. He performs his functions as a citizen by fulfilling the precise duties which belong to his profession as a magistrate. It is true that upon this system the judicial censorship which is exercised by the courts of justice over the legislation cannot extend to all laws indiscriminately, inasmuch as some of them can never give rise to that exact species of contestation which is termed a lawsuit; and even when such a contestation is possible, it may happen that no one cares to bring it before a court of justice. The Americans have often felt this disadvantage, but they have left the remedy incomplete, lest they should give it an efficacy which might in some cases prove dangerous. Within these limits the power vested in the American courts of justice of pronouncing a statute to be unconstitutional forms one of the most powerful barriers which has ever been devised against the tyranny of political assemblies.

Other Powers Granted to American Judges

In the United States all the citizens have the right of indicting public functionaries before the ordinary tribunals—How they use this right Art—75 of the French Constitution of the An VIII—The Americans and the English cannot understand the purport of this clause.

It is perfectly natural that in a free country like America all the citizens should have the right of indicting public functionaries before the ordinary tribunals, and that all the judges should have the power of punishing public offences. The right granted to the courts of justice of judging the agents of the executive government, when they have violated the laws, is so natural a one that it cannot be looked upon as an extraordinary privilege. Nor do the springs of government appear to me to be weakened in the United States by the custom which renders all public officers responsible to the judges of the land. The Americans seem, on the contrary, to have increased by this means that respect which is due to the authorities, and at the same time to have rendered those who are in power more scrupulous of offending public opinion. I was struck by the small number of political trials which occur in the United States, but I had no difficulty in accounting for this circumstance. A lawsuit, of whatever nature it may be, is always a difficult and expensive undertaking. It is easy to attack a public man in a journal, but the motives which can warrant an action at law must be serious. A solid ground of complaint must therefore exist to induce an individual to prosecute a public officer, and public officers are careful not to furnish these grounds of complaint when they are afraid of being prosecuted.

This does not depend upon the republican form of American institutions, for the same facts present themselves in England. These two nations do not regard the impeachment of the principal officers of State as a sufficient guarantee of their independence. But they hold that the right of minor prosecutions, which are within the reach of the whole community, is a better pledge of freedom than those great judicial actions which are rarely employed until it is too late.

In the Middle Ages, when it was very difficult to overtake offenders, the judges inflicted the most dreadful tortures on the few who were arrested, which by no means diminished the number of crimes. It has since been discovered that when justice is more certain and more mild, it is at the same time more efficacious. The English and the Americans hold that tyranny and oppression are to be treated like any other crime, by lessening the penalty and facilitating conviction.

In the year VIII of the French Republic a constitution was drawn up in which the following clause was introduced: 'Art. 75. All the agents of the government below the rank of ministers can only be prosecuted for offences relating to their several functions by virtue of a decree of the Conseil d'Etat; in which the case the prosecution takes place before the ordinary tribunals.' This clause survived the 'Constitution de l'An VIII,' and it is still maintained in spite of the just complaints of the nation. I have always found the utmost difficulty in explaining its meaning to Englishmen or Americans. They were at once led to conclude that the Conseil d'Etat in France was a great tribunal, established in the centre of the kingdom, which exercised a preliminary and somewhat tyrannical jurisdiction in all political causes. But when I told them that the Conseil d'Etat was not a judicial body, in the common sense of the term, but an administrative council composed of men dependent on the Crown, so that the king, after having ordered one of his servants, called a Prefect, to commit an injustice, has the power of commanding another of his servants, called a Councillor of State, to prevent the former from being punished; when I demonstrated to them that the citizen who has been injured by the order of the sovereign is obliged to solicit from the sovereign permission to obtain redress, they refused to credit so flagrant an abuse, and were tempted to accuse me of falsehood or of ignorance. It frequently happened before the Revolution that a Parliament issued a warrant against a public officer who had committed an offence, and sometimes the proceedings were stopped by the authority of the Crown, which enforced compliance with its absolute and despotic will. It is painful to perceive how much lower we are sunk than our forefathers, since we allow things to pass under the color of justice and the sanction of the law which violence alone could impose upon them.

Chapter VII

Political Jurisdiction in the United States

Definition of political jurisdiction—What is understood by political jurisdiction in France, in England, and in the United States—In America the political judge can only pass sentence on public officers—He more frequently passes a sentence of removal from office than a penalty—Political jurisdiction as it exists in the United States is, notwithstanding its mildness, and perhaps in consequence of that mildness, a most powerful instrument in the hands of the majority.

I understand, by political jurisdiction, that temporary right of pronouncing a legal decision with which a political body may be invested.

In absolute governments no utility can accrue from the introduction of extraordinary forms of procedure; the prince in whose name an offender is prosecuted is as much the sovereign of the courts of justice as of everything else, and the idea which is entertained of his power is of itself a sufficient security. The only thing he has to fear is, that the external formalities of justice should be neglected, and that his authority should be dishonoured from a wish to render it more absolute. But in most free countries, in which the majority can never exercise the same influence upon the tribunals as an absolute monarch, the judicial power has occasionally been vested for a time in the representatives of the nation. It has been thought better to introduce a temporary confusion between the functions of the different authorities than to violate the necessary principle of the unity of government.

England, France, and the United States have established this political jurisdiction by law; and it is curious to examine the different adaptations which these three great nations have made of the principle. In England and in France the House of Lords and the Chambre des Paris[1] constitute the highest criminal court of their respective nations, and although they do not habitually try all political offences, they are competent to try them all. Another political body enjoys the right of impeachment before the House of Lords: the only difference which exists between the two countries in this respect is, that in England the Commons may impeach whomsoever they please before the Lords, whilst in France the Deputies can only employ this mode of prosecution against the ministers of the Crown.

In both countries the Upper House may make use of all the existing penal laws of the nation to punish the delinquents.

In the United States, as well as in Europe, one branch of the legislature is authorised to impeach and another to judge: the House of Representatives arraigns the offender, and the Senate awards his sentence. But the Senate can only try such persons as are brought before it by the House of Representatives, and those persons must belong to the class of public functionaries. Thus the jurisdiction of the Senate is less extensive than that of the Peers of France, whilst the right of impeachment by the Representatives is more general than that of the Deputies. But the great difference which exists between Europe and America is, that in Europe political tribunals are empowered to inflict all the dispositions of the penal code, while in America, when they have deprived the offender of his official rank, and have declared him incapable of filling any political office for the future, their jurisdiction terminates and that of the ordinary tribunals begins.

Suppose, for instance, that the President of the United States has committed the crime of high treason; the House of Representatives impeaches him, and the Senate degrades him; he must then be tried by a jury, which alone can deprive him of his liberty or his life. This accurately illustrates the subject we are treating. The political jurisdiction which is established by the laws of Europe is intended to try great offenders, whatever may be their birth, their rank, or their powers in the State; and to this end all the privileges of the courts of justice are temporarily extended to a great political assembly. The legislator is then transformed into the magistrate; he is called upon to admit, to distinguish, and to punish the offence; and as he exercises all the authority of a judge, the law restricts him to the observance of all the duties of that high office, and of all the formalities of justice. When a public functionary is impeached before an English or a French political tribunal, and is found guilty, the sentence deprives him *ipso facto* of his functions, and it may pronounce him to be incapable of resuming them or any others for the future. But in this case the political interdict is a consequence of the sentence, and not the sentence itself. In Europe the sentence of a political tribunal is to be regarded as a judicial verdict rather than as an administrative measure. In the United States the contrary takes place; and although the decision of the Senate is judicial in its form, since the Senators are obliged to comply with the practices and formalities of a court of justice; although it is judicial in respect to the motives on which it is founded, since the Senate is in general obliged to take an offence at common law as the basis of its sentence; nevertheless the object of the proceeding is purely administrative. If it had been the intention of the American legislator to invest a political body with great judicial authority, its action would not have been limited to the circle of public functionaries, since the most dangerous enemies of the State may be in the possession of no functions at all; and this is especially true in republics, where party influence is the first of authorities, and where the strength of many a leader is increased by his exercising no legal power.

If it had been the intention of the American legislator to give society the means of repressing State offences by exemplary punishment, according to the practice of ordinary justice, the resources of the penal code would all have been placed at the disposal of the political tribunals. But the weapon with which they are entrusted is an imperfect one, and it can never reach the most dangerous offenders, since men who aim at the entire subversion of the laws are not likely to murmur at a political interdict.

The main object of the political jurisdiction which obtains in the United States is, therefore, to deprive the ill-disposed citizen of an authority which he has used amiss, and to prevent him from ever acquiring it again. This is evidently an administrative measure sanctioned by the formalities of a judicial decision. In this matter the Americans have created a mixed system; they have surrounded the act which removes a public functionary with the securities of a political trial; and they have deprived all political condemnations of their severest penalties. Every link of the system may easily be traced from this point; we at once perceive why the American constitutions subject all the civil functionaries to the jurisdiction of the Senate, whilst the military, whose crimes are nevertheless more formidable, are exempted from that tribunal. In the civil service none of the American functionaries can be said to be removable; the places which some of them occupy are inalienable, and the others are chosen for a term which cannot be shortened. It is therefore necessary to try them all in order to deprive them of their authority. But military officers are dependent on the chief magistrate of the State, who is himself a civil functionary, and the decision which condemns him is a blow upon them all.

If we now compare the American and the European systems, we shall meet with differences no less striking in the different effects which each of them produces or may produce. In France and in England the jurisdiction of political bodies is looked upon as an extraordinary resource, which is only to be employed in order to rescue society from unwonted dangers. It is not to be denied that these tribunals, as they are constituted in Europe, are apt to violate the conservative principle of the balance of power in the State, and to threaten incessantly the lives and liberties of the subject. The same political jurisdiction in the United States is only indirectly hostile to the balance of power; it cannot menace the lives of the citizens, and it does not hover, as in Europe, over the heads of the community, since those only who have submitted to its authority on accepting office are exposed to the severity of its investigations. It is at the same time less formidable and less efficacious; indeed, it has not been considered by the legislators of the United States as a remedy for the more violent evils of society, but as an ordinary means of conducting the government. In this respect it probably exercises more real influence on the social body in America than in Europe. We must not be misled by the apparent mildness of the American legislation in all that relates to political jurisdiction. It is to be observed, in the first place, that in the United States the tribunal which passes sentence is composed of the same elements, and subject to the same influences, as the body which impeaches the offender, and that this uniformity gives an almost irresistible impulse to the vindictive passions of parties. If political judges in the United States cannot inflict such heavy penalties as those of Europe, there is the less chance of their acquitting a prisoner; and the conviction, if it is less formidable, is more certain. The principal object of the political tribunals of Europe is to punish the offender; the purpose of those in America is to deprive him of his authority. A political condemnation in the United States may, therefore, be looked upon as a preventive measure; and there is no reason for restricting the judges to the exact definitions of criminal law. Nothing can be more alarming than the excessive latitude with which political offences are described in the laws of America. Article II. Section iv. of the Constitution of the United States runs thus:—'The President, Vice-President, and all civil officers of the United States shall be removed from office on impeachment for, and conviction of, treason, bribery, *or other high crimes and misdemeanors.*' Many of the Constitutions of the States are even less explicit. "Public officers," says the Constitution of Massachusetts,[2] "shall be impeached for misconduct or maladministration;" the Constitution of Virginia declares that all the civil officers who shall have offended against the State, by maladministration, corruption, or other high crimes, may be impeached by the House of Delegates; in some constitutions no offences are specified, in order to subject the public functionaries to an unlimited responsibility.[3] But I will venture to affirm that it is precisely their mildness which renders the American laws most formidable in this respect. We have shown that in Europe the removal of a functionary and his political interdiction are the consequences of the penalty he is to undergo, and that in America they constitute the penalty itself. The consequence is that in Europe political tribunals are invested with rights which they are afraid to use, and that the fear of punishing too much hinders them from punishing at all. But in America no one hesitates to inflict a penalty from which humanity does not recoil. To condemn a political opponent to death, in order to deprive him of his power, is to commit what all the world would execrate as a horrible assassination; but to declare that opponent unworthy to exercise that authority, to deprive him of it, and to leave him uninjured in life and limb, may be judged to be the fair issue of the struggle. But this sentence, which it is so easy to pronounce, is not the less fatally severe to the majority of those upon whom it is inflicted. Great criminals may undoubtedly brave its intangible rigour, but ordinary offenders will dread it as a condemnation which destroys their position in the world, casts a blight upon their honour, and condemns them to a shameful inactivity worse than death. The influence exercised in the United States upon the progress of society by the jurisdiction of political bodies may not appear to be formidable, but it is only the more immense. It does not directly coerce the subject, but it renders the majority more absolute over those in power; it does not confer an unbounded authority on the legislator which can be exerted at some momentous crisis, but it establishes a temperate and regular influence, which is at all times available. If the power is decreased, it can, on the other hand, be more conveniently employed and more easily abused. By preventing political tribunals from inflicting judicial punishments the Americans seem to have eluded the worst consequences of legislative tyranny, rather than tyranny itself; and I am not sure that political jurisdiction, as it is constituted in the United States, is not the most formidable weapon which has ever been placed in the rude grasp of a popular majority. When the American republics begin to degenerate it will be easy to verify the truth of this observation, by remarking whether the number of political impeachments augments.[4]

Chapter VIII

The Federal Constitution

I have hitherto considered each State as a separate whole, and I have explained the different springs which the people sets in motion, and the different means of action which it employs. But all the States which I have considered as independent are forced to submit, in certain cases, to the supreme authority of the Union. The time is now come for me to examine separately the supremacy with which the Union has been invested, and to cast a rapid glance over the Federal Constitution.

History of the Federal Constitution

Origin of the first Union—Its weakness Congress—appeals to the constituent authority—Interval of two years between this appeal and the promulgation of the new Constitution.

The thirteen colonies which simultaneously threw off the yoke of England towards the end of the last century professed, as I have already observed, the same religion, the same language, the same customs, and almost the same laws; they were struggling against a common enemy; and these reasons were sufficiently strong to unite them one to another, and to consolidate them into one nation. But as each of them had enjoyed a separate existence and a government within its own control, the peculiar interests and customs which resulted from this system were opposed to a compact and intimate union which would have absorbed the individual importance of each in the general importance of all. Hence arose two opposite tendencies, the one prompting the Anglo-Americans to unite, the other to divide their strength. As long as the war with the mother-country lasted the principle of union was kept alive by necessity; and although the laws which constituted it were defective, the common tie subsisted in spite of their imperfections.[1] But no sooner was peace concluded than the faults of the legislation became manifest, and the State seemed to be suddenly dissolved. Each colony became an independent republic, and assumed an absolute sovereignty. The federal government, condemned to impotence by its constitution, and no longer sustained by the presence of a common danger, witnessed the outrages offered to its flag by the great nations of Europe, whilst it was scarcely able to maintain its ground against the Indian tribes, and to pay the interest of the debt which had been contracted during the war of independence. It was already on the verge of destruction, when it officially proclaimed its inability to conduct the government, and appealed to the constituent authority of the nation.[2] If America ever approached (for however brief a time) that lofty pinnacle of glory to which the fancy of its inhabitants is wont to point, it was at the solemn moment at which the power of the nation abdicated, as it were, the empire of the land. All ages have furnished the spectacle of a people struggling with energy to win its independence; and the efforts of the Americans in throwing off the English yoke have been considerably exaggerated. Separated from their enemies by three thousand miles of ocean, and backed by a powerful ally, the success of the United States may be more justly attributed to their geographical position than to the valour of their armies or the patriotism of their citizens. It would be ridiculous to compare the American war to the wars of the French Revolution, or the efforts of the Americans to those of the French when they were attacked by the whole of Europe, without credit and without allies, yet capable of opposing a twentieth part of their population to the world, and of bearing the torch of revolution beyond their frontiers whilst they stifled its devouring flame within the bosom of their country. But it is a novelty in the history of society to see a great people turn a calm and scrutinising eye upon itself, when apprised by the legislature that the wheels of government are stopped; to see it carefully examine the extent of the evil, and patiently wait for two whole years until a remedy was discovered, which it voluntarily adopted without having wrung a tear or a drop of blood from mankind. At the time when the inadequacy of the first constitution was discovered America possessed the double advantage of that calm which had succeeded the effervescence of the revolution, and of those great men who had led the revolution to a successful issue. The assembly which accepted the task of composing the second constitution was small;[3] but George Washington was its President, and it contained the choicest talents and the noblest hearts which had ever appeared in the New World. This national commission, after long and mature deliberation, offered to the acceptance of the people the body of general laws which still rules the Union. All the States adopted it successively.[4] The new Federal Government commenced its functions in 1789, after an interregnum of two years. The Revolution of America terminated when that of France began.

Summary of the Federal Constitution

Division of authority between the Federal Government and the States—The Government of the States is the rule, the Federal Government the exception.

The first question which awaited the Americans was intricate, and by no means easy of solution: the object was so to divide the authority of the different States which composed the Union that each of them should continue to govern itself in all that concerned its internal prosperity, whilst the entire nation, represented by the Union, should continue to form a compact body, and to provide for the general exigencies of the people. It was as impossible to determine beforehand, with any degree of accuracy, the share of authority which each of two governments was to enjoy, as to foresee all the incidents in the existence of a nation.

The obligations and the claims of the Federal Government were simple and easily definable, because the Union had been formed with the express purpose of meeting the general exigencies of the people; but the claims and obligations of the States were, on the other hand, complicated and various, because those Governments had penetrated into all the details of social life. The attributes of the Federal Government were therefore carefully enumerated and all that was not included amongst them was declared to constitute a part of the privileges of the several Governments of the States. Thus the government of the States remained the rule, and that of the Confederation became the exception.[5]

But as it was foreseen that, in practice, questions might arise as to the exact limits of this exceptional authority, and that it would be dangerous to submit these questions to the decision of the ordinary courts of justice, established in the States by the States themselves, a high Federal court was created,[6] which was destined, amongst other functions, to maintain the balance of power which had been established by the Constitution between the two rival Governments.[7]

Prerogative of the Federal Government

Power of declaring war, making peace, and levying general taxes vested in the Federal Government—What part of the internal policy of the country it may direct—The Government of the Union in some respects more central than the King's Government in the old French monarchy.

The external relations of a people may be compared to those of private individuals, and they cannot be advantageously maintained without the agency of a single head of a Government. The exclusive right of making peace and war, of concluding treaties of commerce, of raising armies, and equipping fleets, was granted to the Union.[8] The necessity of a national Government was less imperiously felt in the conduct of the internal policy of society; but there are certain general interests which can only be attended to with advantage by a general authority. The Union was invested with the power of controlling the monetary system, of directing the post office, and of opening the great roads which were to establish a communication between the different parts of the country.[9] The independence of the Government of each State was formally recognized in its sphere; nevertheless, the Federal Government was authorized to interfere in the internal affairs of the States[10] in a few predetermined cases, in which an indiscreet abuse of their independence might compromise the security of the Union at large. Thus, whilst the power of modifying and changing their legislation at pleasure was preserved in all the republics, they were forbidden to enact *ex post facto* laws, or to create a class of nobles in their community.[11] Lastly, as it was necessary that the Federal Government should be able to fulfil its engagements, it was endowed with an unlimited power of levying taxes.[12]

In examining the balance of power as established by the Federal Constitution; in remarking on the one hand the portion of sovereignty which has been reserved to the several States, and on the other the share of power which the Union has assumed, it is evident that the Federal legislators entertained the clearest and most accurate notions on the nature of the centralization of government. The United States form not only a republic, but a confederation; nevertheless the authority of the nation is more central than it was in several of the monarchies of Europe when the American Constitution was formed. Take, for instance, the two following examples.

Thirteen supreme courts of justice existed in France, which, generally speaking, had the right of interpreting the law without appeal; and those provinces which were styled *pays d'Etats* were authorized to refuse their assent to an impost which had been levied by the sovereign who represented the nation. In the Union there is but one tribunal to interpret, as there is one legislature to make the laws; and an impost voted by the representatives of the nation is binding upon all the citizens. In these two essential points, therefore, the Union exercises more central authority than the French monarchy possessed, although the Union is only an assemblage of confederate republics.

In Spain certain provinces had the right of establishing a system of custom-house duties peculiar to themselves, although that privilege belongs, by its very nature, to the national sovereignty. In America the Congress alone has the right of regulating the commercial relations of the States. The government of the Confederation is therefore more centralized in this respect than the kingdom of Spain. It is true that the power of the Crown in France or in Spain was always able to obtain by force whatever the Constitution of the country denied, and that the ultimate result was consequently the same; but I am here discussing the theory of the Constitution.

Federal Powers

After having settled the limits within which the Federal Government was to act, the next point was to determine the powers which it was to exert.

Legislative Powers[13]

Division of the Legislative Body into two branches—Difference in the manner of forming the two Houses—The principle of the independence of the States predominates in the formation of the Senate—The principle of the sovereignty of the nation in the composition of the House of Representatives—Singular effects of the fact that a Constitution can only be logical in the early stages of a nation.

The plan which had been laid down beforehand for the Constitutions of the several States was followed, in many points, in the organization of the powers of the Union. The Federal legislature of the Union was composed of a Senate and a House of Representatives. A spirit of conciliation prescribed the observance of distinct principles in the formation of these two assemblies. I have already shown that two contrary interests were opposed to each other in the establishment of the Federal Constitution. These two interests had given rise to two opinions. It was the wish of one party to convert the Union into a league of independent States, or a sort of congress, at which the representatives of the several peoples would meet to discuss certain points of their common interests. The other party desired to unite the inhabitants of the American colonies into one sole nation, and to establish a Government which should act as the sole representative of the nation, as far as the limited sphere of its authority would permit. The practical consequences of these two theories were exceedingly different.

The question was, whether a league was to be established instead of a national Government; whether the majority of the States, instead of the majority of the inhabitants of the Union, was to give the law: for every State, the small as well as the great, would then remain in the full enjoyment of its independence, and enter the Union upon a footing of perfect equality. If, however, the inhabitants of the United States were to be considered as belonging to one and the same nation, it would be just that the majority of the citizens of the Union should prescribe the law. Of course the lesser States could not subscribe to the application of this doctrine without, in fact, abdicating their existence in relation to the sovereignty of the Confederation; since they would have passed from the condition of a coequal and colegislative authority to that of an insignificant fraction of a great people. But if the former system would have invested them with an excessive authority, the latter would have annulled their influence altogether. Under these circumstances the result was, that the strict rules of logic were evaded, as is usually the case when interests are opposed to arguments. A middle course was hit upon by the legislators, which brought together by force two systems theoretically irreconcilable.

The principle of the independence of the States prevailed in the formation of the Senate, and that of the sovereignty of the nation predominated in the composition of the House of Representatives. It was decided that each State should send two senators to Congress, and a number of representatives proportioned to its population.[14] It results from this arrangement that the State of New York has at the present day forty representatives and only two senators; the State of Delaware has two senators and only one representative; the State of Delaware is therefore equal to the State of New York in the Senate, whilst the latter has forty times the influence of the former in the House of Representatives. Thus, if the minority of the nation preponderates in the Senate, it may paralyse the decisions of the majority represented in the other House, which is contrary to the spirit of constitutional government.

These facts show how rare and how difficult it is rationally and logically to combine all the several parts of legislation. In the course of time different interests arise, and different principles are sanctioned by the same people; and when a general constitution is to be established, these interests and principles are so many natural obstacles to the rigorous application of any political system, with all its consequences. The early stages of national existence are the only periods at which it is possible to maintain the complete logic of legislation; and when we perceive a nation in the enjoyment of this advantage, before we hasten to conclude that it is wise, we should do well to remember that it is young. When the Federal Constitution was formed, the interests of independence for the separate States, and the interest of union for the whole people, were the only two conflicting interests which existed amongst the Anglo-Americans, and a compromise was necessarily made between them.

It is, however, just to acknowledge that this part of the Constitution has not hitherto produced those evils which might have been feared. All the States are young and contiguous; their customs, their ideas, and their exigencies are not dissimilar; and the differences which result from their size or inferiority do not suffice to set their interests at variance. The small States have consequently never been induced to league themselves together in the Senate to oppose the designs of the larger ones; and indeed there is so irresistible an authority in the legitimate expression of the will of a people that the Senate could offer but a feeble opposition to the vote of the majority of the House of Representatives.

It must not be forgotten, on the other hand, that it was not in the power of the American legislators to reduce to a single nation the people for whom they were making laws. The object of the Federal Constitution was not to destroy the independence of the States, but to restrain it. By acknowledging the real authority of these secondary communities (and it was impossible to deprive them of it), they disavowed beforehand the habitual use of constraint in enforcing the decisions of the majority. Upon this principle the introduction of the influence of the States into the mechanism of the Federal Government was by no means to be wondered at, since it only attested the existence of an acknowledged power, which was to be humored and not forcibly checked.

A Further Difference between the Senate and the House of Representatives

The Senate named by the provincial legislators, the Representatives by the people—Double election of the former; single election of the latter—Term of the different offices—Peculiar functions of each House.

The Senate not only differs from the other House in the principle which it represents, but also in the mode of its election, in the term for which it is chosen, and in the nature of its functions. The House of Representatives is named by the people, the Senate by the legislators of each State; the former is directly elected, the latter is elected by an elected body; the term for which the representatives are chosen is only two years, that of the senators is six. The functions of the House of Representatives are purely legislative, and the only share it takes in the judicial power is in the impeachment of public officers. The Senate cooperates in the work of legislation, and tries those political offences which the House of Representatives submits to its decision. It also acts as the great executive council of the nation; the treaties which are concluded by the President must be ratified by the Senate, and the appointments he may make must be definitely approved by the same body.[15]

The Executive Power[16]

Dependence of the President—He is elective and responsible—He is free to act in his own sphere under the inspection, but not under the direction, of the Senate—His salary fixed at his entry into office—Suspensive veto.

The American legislators undertook a difficult task in attempting to create an executive power dependent on the majority of the people, and nevertheless sufficiently strong to act without restraint in its own sphere. It was indispensable to the maintenance of the republican form of government that the representative of the executive power should be subject to the will of the nation.

The President is an elective magistrate. His honour, his property, his liberty, and his life are the securities which the people has for the temperate use of his power. But in the exercise of his authority he cannot be said to be perfectly independent; the Senate takes cognizance of his relations with foreign powers, and of the distribution of public appointments, so that he can neither be bribed nor can he employ the means of corruption. The legislators of the Union acknowledged that the executive power would be incompetent to fulfil its task with dignity and utility, unless it enjoyed a greater degree of stability and of strength than had been granted to it in the separate States.

The President is chosen for four years, and he may be reelected; so that the chances of a prolonged administration may inspire him with hopeful undertakings for the public good, and with the means of carrying them into execution. The President was made the sole representative of the executive power of the Union, and care was taken not to render his decisions subordinate to the vote of a council a dangerous measure, which tends at the same time to clog the action of the Government and to diminish its responsibility. The Senate has the right of annulling certain acts of the President; but it cannot compel him to take any steps, nor does it participate in the exercise of the executive power.

The action of the legislature on the executive power may be direct; and we have just shown that the Americans carefully obviated this influence; but it may, on the other hand, be indirect. Public assemblies which have the power of depriving an officer of state of his salary encroach upon his independence; and as they are free to make the laws, it is to be feared lest they should gradually appropriate to themselves a portion of that authority which the Constitution had vested in his hands. This dependence of the executive power is one of the defects inherent in republican constitutions. The Americans have not been able to counteract the tendency which legislative assemblies have to get possession of the government, but they have rendered this propensity less irresistible. The salary of the President is fixed, at the time of his entering upon office, for the whole period of his magistracy. The President is, moreover, provided with a suspensive veto, which allows him to oppose the passing of such laws as might destroy the portion of independence which the Constitution awards him. The struggle between the President and the legislature must always be an unequal one, since the latter is certain of bearing down all resistance by persevering in its plans; but the suspensive veto forces it at least to reconsider the matter, and, if the motion be persisted in, it must then be backed by a majority of two-thirds of the whole house. The veto is, in fact, a sort of appeal to the people. The executive power, which, without this security, might have been secretly oppressed, adopts this means of pleading its cause and stating its motives. But if the legislature is certain of overpowering all resistance by persevering in its plans, I reply, that in the constitutions of all nations, of whatever kind they may be, a certain point exists at which the legislator is obliged to have recourse to the good sense and the virtue of his fellow-citizens. This point is more prominent and more discoverable in republics, whilst it is more remote and more carefully concealed in monarchies, but it always exists somewhere. There is no country in the world in which everything can be provided for by the laws, or in which political institutions can prove a substitute for common sense and public morality.

Differences between the Position of the President of the United States and That of a Constitutional King of France

Executive power in the Northern States as limited and as partial as the supremacy which it represents. Executive power in France as universal as the supremacy it represents—The King a branch of the legislature—The President the mere executor of the law—Other differences resulting from the duration of the two powers—The President checked in the exercise of the executive authority—The King independent in its exercise—Notwithstanding these discrepancies France is more akin to a republic than the Union to a monarchy—Comparison of the number of public officers depending upon the executive power in the two countries.

The executive power has so important an influence on the destinies of nations that I am inclined to pause for an instant at this portion of my subject, in order more clearly to explain the part it sustains in America. In order to form an accurate idea of the position of the President of the United States, it may not be irrelevant to compare it to that of one of the constitutional kings of Europe. In this comparison I shall pay but little attention to the external signs of power, which are more apt to deceive the eye of the observer than to guide his researches. When a monarchy is being gradually transformed into a republic, the executive power retains the titles, the honours, the etiquette, and even the funds of royalty long after its authority has disappeared. The English, after having cut off the head of one king and expelled another from his throne, were accustomed to accost the successor of those princes upon their knees. On the other hand, when a republic falls under the sway of a single individual, the demeanour of the sovereign is simple and unpretending, as if his authority was not yet paramount. When the emperors exercised an unlimited control over the fortunes and the lives of their fellow-citizens, it was customary to call them Caesar in conversation, and they were in the habit of supping without formality at their friends' houses. It is therefore necessary to look below the surface.

The sovereignty of the United States is shared between the Union and the States, whilst in France it is undivided and compact: hence arises the first and the most notable difference which exists between the President of the United States and the King of France. In the United States the executive power is as limited and partial as the sovereignty of the Union in whose name it acts; in France it is as universal as the authority of the State. The Americans have a federal and the French a national Government.

This cause of inferiority results from the nature of things, but it is not the only one; the second in importance is as follows: Sovereignty may be defined to be the right of making laws: in France, the King really exercises a portion of the sovereign power, since the laws have no weight till he has given his assent to them; he is, moreover, the executor of all they ordain. The President is also the executor of the laws, but he does not really cooperate in their formation, since the refusal of his assent does not annul them. He is therefore merely to be considered as the agent of the sovereign power. But not only does the King of France exercise a portion of the sovereign power, he also contributes to the nomination of the legislature, which exercises the other portion. He has the privilege of appointing the members of one chamber, and of dissolving the other at his pleasure; whereas the President of the United States has no share in the formation of the legislative body, and cannot dissolve any part of it. The King has the same right of bringing forward measures as the Chambers; a right which the President does not possess. The King is represented in each assembly by his ministers, who explain his intentions, support his opinions, and maintain the principles of the Government. The President and his ministers are alike excluded from Congress; so that his influence and his opinions can only penetrate indirectly into that great body. The King of France is therefore on an equal footing with the legislature, which can no more act without him than he can without it. The President exercises an authority inferior to, and depending upon, that of the legislature.

Even in the exercise of the executive power, properly so called the point upon which his position seems to be most analogous to that of the King of France the President labours under several causes of inferiority. The authority of the King, in France, has, in the first place, the advantage of duration over that of the President, and durability is one of the chief elements of strength; nothing is either loved or feared but what is likely to endure. The President of the United States is a magistrate elected for four years; the King, in France, is an hereditary sovereign. In the exercise of the executive power the President of the United States is constantly subject to a jealous scrutiny. He may make, but he cannot conclude, a treaty; he may designate, but he cannot appoint, a public officer.[17] The King of France is absolute within the limits of his authority. The President of the United States is responsible for his actions; but the person of the King is declared inviolable by the French Charter.[18]

Nevertheless, the supremacy of public opinion is no less above the head of the one than of the other. This power is less definite, less evident, and less sanctioned by the laws in France than in America, but in fact it exists. In America, it acts by elections and decrees; in France it proceeds by revolutions; but notwithstanding the different constitutions of these two countries, public opinion is the predominant authority in both of them. The fundamental principle of legislation—a principle essentially republican—is the same in both countries, although its consequences may be different, and its results more or less extensive. Whence I am led to conclude that France with its King is nearer akin to a republic than the Union with its President is to a monarchy.

In what I have been saying I have only touched upon the main points of distinction; and if I could have entered into details, the contrast would have been rendered still more striking.

I have remarked that the authority of the President in the United States is only exercised within the limits of a partial sovereignty, whilst that of the King in France is undivided. I might have gone on to show that the power of the King's government in France exceeds its natural limits, however extensive they may be, and penetrates in a thousand different ways into the administration of private interests. Amongst the examples of this influence may be quoted that which results from the great number of public functionaries, who all derive their appointments from the Government. This number now exceeds all previous limits; it amounts to 138,000[19] nominations, each of which may be considered as an element of power. The President of the United States has not the exclusive right of making any public appointments, and their whole number scarcely exceeds 12,000.[20]

Accidental Causes Which May Increase the Influence of the Executive Government

External security of the Union—Army of six thousand men—Few ships—The President has no opportunity of exercising his great prerogatives—In the prerogatives he exercises he is weak.

If the executive government is feebler in America than in France, the cause is more attributable to the circumstances than to the laws of the country.

It is chiefly in its foreign relations that the executive power of a nation is called upon to exert its skill and its vigour. If the existence of the Union were perpetually threatened, and if its chief interests were in daily connection with those of other powerful nations, the executive government would assume an increased importance in proportion to the measures expected of it, and those which it would carry into effect. The President of the United States is the commander-in-chief of the army, but of an army composed of only six thousand men; he commands the fleet, but the fleet reckons but few sail; he conducts the foreign relations of the Union, but the United States are a nation without neighbours. Separated from the rest of the world by the ocean, and too weak as yet to aim at the dominion of the seas, they have no enemies, and their interests rarely come into contact with those of any other nation of the globe.

The practical part of a Government must not be judged by the theory of its constitution. The President of the United States is in the possession of almost royal prerogatives, which he has no opportunity of exercising; and those privileges which he can at present use are very circumscribed. The laws allow him to possess a degree of influence which circumstances do not permit him to employ.

On the other hand, the great strength of the royal prerogative in France arises from circumstances far more than from the laws. There the executive government is constantly struggling against prodigious obstacles, and exerting all its energies to repress them; so that it increases by the extent of its achievements, and by the importance of the events it controls, without modifying its constitution. If the laws had made it as feeble and as circumscribed as it is in the Union, its influence would very soon become still more preponderant.

Why the President of the United States Does Not Require the Majority of the Two Houses in Order to Carry on the Government

It is an established axiom in Europe that a constitutional King cannot persevere in a system of government which is opposed by the two other branches of the legislature. But several Presidents of the United States have been known to lose the majority in the legislative body without being obliged to abandon the supreme power, and without inflicting a serious evil upon society. I have heard this fact quoted as an instance of the independence and the power of the executive government in America: a moment's reflection will convince us, on the contrary, that it is a proof of its extreme weakness.

A King in Europe requires the support of the legislature to enable him to perform the duties imposed upon him by the Constitution, because those duties are enormous. A constitutional King in Europe is not merely the executor of the law but the execution of its provisions devolves so completely upon him that he has the power of paralysing its influence if it opposes his designs. He requires the assistance of the legislative assemblies to make the law, but those assemblies stand in need of his aid to execute it: these two authorities cannot subsist without each other, and the mechanism of government is stopped as soon as they are at variance.

In America the President cannot prevent any law from being passed, nor can he evade the obligation of enforcing it. His sincere and zealous cooperation is no doubt useful, but it is not indispensable, in the carrying on of public affairs. All his important acts are directly or indirectly submitted to the legislature, and of his own free authority he can do but little. It is therefore his weakness, and not his power, which enables him to remain in opposition to Congress. In Europe, harmony must reign between the Crown and the other branches of the legislature, because a collision between them may prove serious; in America, this harmony is not indispensable, because such a collision is impossible.

Election of the President

Dangers of the elective system increase in proportion to the extent of the prerogative—This system possible in America because no powerful executive authority is required—What circumstances are favourable to the elective system—Why the election of the President does not cause a deviation from the principles of the Government—Influence of the election of the President on secondary functionaries.

The dangers of the system of election applied to the head of the executive government of a great people have been sufficiently exemplified by experience and by history, and the remarks I am about to make refer to America alone. These dangers may be more or less formidable in proportion to the place which the executive power occupies, and to the importance it possesses in the State; and they may vary according to the mode of election and the circumstances in which the electors are placed. The most weighty argument against the election of a chief magistrate is, that it offers so splendid a lure to private ambition, and is so apt to inflame men in the pursuit of power, that when legitimate means are wanting force may not unfrequently seize what right denied.

It is clear that the greater the privileges of the executive authority are, the greater is the temptation; the more the ambition of the candidates is excited, the more warmly are their interests espoused by a throng of partisans who hope to share the power when their patron has won the prize. The dangers of the elective system increase, therefore, in the exact ratio of the influence exercised by the executive power in the affairs of State. The revolutions of Poland were not solely attributable to the elective system in general, but to the fact that the elected monarch was the sovereign of a powerful kingdom. Before we can discuss the absolute advantages of the elective system we must make preliminary inquiries as to whether the geographical position, the laws, the habits, the manners, and the opinions of the people amongst whom it is to be introduced will admit of the establishment of a weak and dependent executive government; for to attempt to render the representative of the State a powerful sovereign, and at the same time elective, is, in my opinion, to entertain two incompatible designs. To reduce hereditary royalty to the condition of an elective authority, the only means that I am acquainted with are to circumscribe its sphere of action beforehand, gradually to diminish its prerogatives, and to accustom the people to live without its protection. Nothing, however, is further from the designs of the republicans of Europe than this course: as many of them owe their hatred of tyranny to the sufferings which they have personally undergone, it is oppression, and not the extent of the executive power, which excites their hostility, and they attack the former without perceiving how nearly it is connected with the latter.

Hitherto no citizen has shown any disposition to expose his honour and his life in order to become the President of the United States; because the power of that office is temporary, limited, and subordinate. The prize of fortune must be great to encourage adventurers in so desperate a game. No candidate has as yet been able to arouse the dangerous enthusiasm or the passionate sympathies of the people in his favour, for the very simple reason that when he is at the head of the Government he has but little power, but little wealth, and but little glory to share amongst his friends; and his influence in the State is too small for the success or the ruin of a faction to depend upon the elevation of an individual to power.

The great advantage of hereditary monarchies is, that as the private interest of a family is always intimately connected with the interests of the State, the executive government is never suspended for a single instant; and if the affairs of a monarchy are not better conducted than those of a republic, at least there is always some one to conduct them, well or ill, according to his capacity. In elective States, on the contrary, the wheels of government cease to act, as it were, of their own accord at the approach of an election, and even for some time previous to that event. The laws may indeed accelerate the operation of the election, which may be conducted with such simplicity and rapidity that the seat of power will never be left vacant; but, notwithstanding these precautions, a break necessarily occurs in the minds of the people.

At the approach of an election the head of the executive government is wholly occupied by the coming struggle; his future plans are doubtful; he can undertake nothing new, and he will only prosecute with indifference those designs which another will perhaps terminate. 'I am so near the time of my retirement from office,' said President Jefferson on the 21st of January, 1809 (six weeks before the election), 'that I feel no passion, I take no part, I express no sentiment. It appears to me just to leave to my successor the commencement of those measures which he will have to prosecute, and for which he will be responsible.'

On the other hand, the eyes of the nation are centred on a single point; all are watching the gradual birth of so important an event. The wider the influence of the executive power extends, the greater and the more necessary is its constant action, the more fatal is the term of suspense; and a nation which is accustomed to the government, or, still more, one used to the administrative protection of a powerful executive authority, would be infallibly convulsed by an election of this kind. In the United States the action of the Government may be slackened with impunity, because it is always weak and circumscribed.[21]

One of the principal vices of the elective system is that it always introduces a certain degree of instability into the internal and external policy of the State. But this disadvantage is less sensibly felt if the share of power vested in the elected magistrate is small. In Rome the principles of the Government underwent no variation, although the Consuls were changed every year, because the Senate, which was an hereditary assembly, possessed the directing authority. If the elective system were adopted in Europe, the condition of most of the monarchical States would be changed at every new election. In America the President exercises a certain influence on State affairs, but he does not conduct them; the preponderating power is vested in the representatives of the whole nation. The political maxims of the country depend therefore on the mass of the people, not on the President alone; and consequently in America the elective system has no very prejudicial influence on the fixed principles of the Government. But the want of fixed principles is an evil so inherent in the elective system that it is still extremely perceptible in the narrow sphere to which the authority of the President extends.

The Americans have admitted that the head of the executive power, who has to bear the whole responsibility of the duties he is called upon to fulfil, ought to be empowered to choose his own agents, and to remove them at pleasure: the legislative bodies watch the conduct of the President more than they direct it. The consequence of this arrangement is, that at every new election the fate of all the Federal public officers is in suspense. Mr. Quincy Adams, on his entry into office, discharged the majority of the individuals who had been appointed by his predecessor: and I am not aware that General Jackson allowed a single removable functionary employed in the Federal service to retain his place beyond the first year which succeeded his election. It is sometimes made a subject of complaint that in the constitutional monarchies of Europe the fate of the humbler servants of an Administration depends upon that of the Ministers. But in elective Governments this evil is far greater. In a constitutional monarchy successive ministries are rapidly formed; but as the principal representative of the executive power does not change, the spirit of innovation is kept within bounds; the changes which take place are in the details rather than in the principles of the administrative system; but to substitute one system for another, as is done in America every four years, by law, is to cause a sort of revolution. As to the misfortunes which may fall upon individuals in consequence of this state of things, it must be allowed that the uncertain situation of the public officers is less fraught with evil consequences in America than elsewhere. It is so easy to acquire an independent position in the United States that the public officer who loses his place may be deprived of the comforts of life, but not of the means of subsistence.

I remarked at the beginning of this chapter that the dangers of the elective system applied to the head of the State are augmented or decreased by the peculiar circumstances of the people which adopts it. However the functions of the executive power may be restricted, it must always exercise a great influence upon the foreign policy of the country, for a negotiation cannot be opened or successfully carried on otherwise than by a single agent. The more precarious and the more perilous the position of a people becomes, the more absolute is the want of a fixed and consistent external policy, and the more dangerous does the elective system of the Chief Magistrate become. The policy of the Americans in relation to the whole world is exceedingly simple; for it may almost be said that no country stands in need of them, nor do they require the cooperation of any other people. Their independence is never threatened. In their present condition, therefore, the functions of the executive power are no less limited by circumstances than by the laws; and the President may frequently change his line of policy without involving the State in difficulty or destruction.

Whatever the prerogatives of the executive power may be, the period which immediately precedes an election and the moment of its duration must always be considered as a national crisis, which is perilous in proportion to the internal embarrassments and the external dangers of the country. Few of the nations of Europe could escape the calamities of anarchy or of conquest every time they might have to elect a new sovereign. In America society is so constituted that it can stand without assistance upon its own basis; nothing is to be feared from the pressure of external dangers, and the election of the President is a cause of agitation, but not of ruin.

Mode of Election

Skill of the American legislators shown in the mode of election adopted by them—Creation of a special electoral body—Separate votes of these electors—Case in which the House of Representatives is called upon to choose the President—Results of the twelve elections which have taken place since the Constitution has been established.

Besides the dangers which are inherent in the system, many other difficulties may arise from the mode of election, which may be obviated by the precaution of the legislator. When a people met in arms on some public spot to choose its head, it was exposed to all the chances of civil war resulting from so martial a mode of proceeding, besides the dangers of the elective system in itself. The Polish laws, which subjected the election of the sovereign to the veto of a single individual, suggested the murder of that individual or prepared the way to anarchy.

In the examination of the institutions and the political as well as social condition of the United States, we are struck by the admirable harmony of the gifts of fortune and the efforts of man. The nation possessed two of the main causes of internal peace; it was a new country, but it was inhabited by a people grown old in the exercise of freedom. America had no hostile neighbours to dread; and the American legislators, profiting by these favourable circumstances, created a weak and subordinate executive power which could without danger be made elective.

It then only remained for them to choose the least dangerous of the various modes of election; and the rules which they laid down upon this point admirably correspond to the securities which the physical and political constitution of the country already afforded. Their object was to find the mode of election which would best express the choice of the people with the least possible excitement and suspense. It was admitted in the first place that the simple majority should be decisive; but the difficulty was to obtain this majority without an interval of delay which it was most important to avoid. It rarely happens that an individual can at once collect the majority of the suffrages of a great people; and this difficulty is enhanced in a republic of confederate States, where local influences are apt to preponderate. The means by which it was proposed to obviate this second obstacle was to delegate the electoral powers of the nation to a body of representatives. This mode of election rendered a majority more probable; for the fewer the electors are, the greater is the chance of their coming to a final decision. It also offered an additional probability of a judicious choice. It then remained to be decided whether this right of election was to be entrusted to a legislative body, the habitual representative assembly of the nation, or whether an electoral assembly should be formed for the express purpose of proceeding to the nomination of a President. The Americans chose the latter alternative, from a belief that the individuals who were returned to make the laws were incompetent to represent the wishes of the nation in the election of its chief magistrate; and that, as they are chosen for more than a year, the constituency they represent might have changed its opinion in that time. It was thought that if the legislature was empowered to elect the head of the executive power, its members would, for some time before the election, be exposed to the manœuvres of corruption and the tricks of intrigue; whereas the special electors would, like a jury, remain mixed up with the crowd till the day of action, when they would appear for the sole purpose of giving their votes.

It was therefore established that every State should name a certain number of electors,[22] who in their turn should elect the President; and as it had been observed that the assemblies to which the choice of a chief magistrate had been entrusted in elective countries inevitably became the centres of passion and of cabal; that they sometimes usurped an authority which did not belong to them; and that their proceedings, or the uncertainty which resulted from them, were sometimes prolonged so much as to endanger the welfare of the State, it was determined that the electors should all vote upon the same day, without being convoked to the same place.[23] This double election rendered a majority probable, though not certain; for it was possible that as many differences might exist between the electors as between their constituents. In this case it was necessary to have recourse to one of three measures; either to appoint new electors, or to consult a second time those already appointed, or to defer the election to another authority. The first two of these alternatives, independently of the uncertainty of their results, were likely to delay the final decision, and to perpetuate an agitation which must always be accompanied with danger. The third expedient was therefore adopted, and it was agreed that the votes should be transmitted sealed to the President of the Senate, and that they should be opened and counted in the presence of the Senate and the House of Representatives. If none of the candidates has a majority, the House of Representatives then proceeds immediately to elect a President, but with the condition that it must fix upon one of the three candidates who have the highest numbers.[24]

Thus it is only in case of an event which cannot often happen, and which can never be foreseen, that the election is entrusted to the ordinary representatives of the nation; and even then they are obliged to choose a citizen who has already been designated by a powerful minority of the special electors. It is by this happy expedient that the respect which is due to the popular voice is combined with the utmost celerity of execution and those precautions which the peace of the country demands. But the decision of the question by the House of Representatives does not necessarily offer an immediate solution of the difficulty, for the majority of that assembly may still be doubtful, and in this case the Constitution prescribes no remedy. Nevertheless, by restricting the number of candidates to three, and by referring the matter to the judgment of an enlightened public body, it has smoothed all the obstacles[25] which are not inherent in the elective system.

In the forty-four years which have elapsed since the promulgation of the Federal Constitution the United States have twelve times chosen a President. Ten of these elections took place simultaneously by the votes of the special electors in the different States. The House of Representatives has only twice exercised its conditional privilege of deciding in cases of uncertainty; the first time was at the election of Mr. Jefferson in 1801; the second was in 1825, when Mr. Quincy Adams was named.[26]

Crisis of the Election

The Election may be considered as a national crisis—Why?—Passions of the people—Anxiety of the President—Calm which succeeds the agitation of the election.

I have shown what the circumstances are which favoured the adoption of the elective system in the United States, and what precautions were taken by the legislators to obviate its dangers. The Americans are habitually accustomed to all kinds of elections, and they know by experience the utmost degree of excitement which is compatible with security. The vast extent of the country and the dissemination of the inhabitants render a collision between parties less probable and less dangerous there than elsewhere. The political circumstances under which the elections have hitherto been carried on have presented no real embarrassments to the nation.

Nevertheless, the epoch of the election of a President of the United States may be considered as a crisis in the affairs of the nation. The influence which he exercises on public business is no doubt feeble and indirect; but the choice of the President, which is of small importance to each individual citizen, concerns the citizens collectively; and however trifling an interest may be, it assumes a great degree of importance as soon as it becomes general. The President possesses but few means of rewarding his supporters in comparison to the kings of Europe, but the places which are at his disposal are sufficiently numerous to interest, directly or indirectly, several thousand electors in his success. Political parties in the United States are led to rally round an individual, in order to acquire a more tangible shape in the eyes of the crowd, and the name of the candidate for the Presidency is put forward as the symbol and personification of their theories. For these reasons parties are strongly interested in gaining the election, not so much with a view to the triumph of their principles under the auspices of the President-elect as to show by the majority which returned him, the strength of the supporters of those principles.

For a long while before the appointed time is at hand the election becomes the most important and the all-engrossing topic of discussion. The ardour of faction is redoubled; and all the artificial passions which the imagination can create in the bosom of a happy and peaceful land are agitated and brought to light. The President, on the other hand, is absorbed by the cares of self-defence. He no longer governs for the interest of the State, but for that of his reelection; he does homage to the majority, and instead of checking its passions, as his duty commands him to do, he frequently courts its worst caprices. As the election draws near, the activity of intrigue and the agitation of the populace increase; the citizens are divided into hostile camps, each of which assumes the name of its favourite candidate; the whole nation glows with feverish excitement; the election is the daily theme of the public papers, the subject of private conversation, the end of every thought and every action, the sole interest of the present. As soon as the choice is determined, this ardour is dispelled; and as a calmer season returns, the current of the State, which had nearly broken its banks, sinks to its usual level:[27] but who can refrain from astonishment at the causes of the storm?

Re-Election of the President

When the head of the executive power is reeligible, it is the State which is the source of intrigue and corruption—The desire of being reelected the chief aim of a President of the United States. Disadvantage of the system peculiar to America—The natural evil of democracy is that it subordinates all authority to the slightest desires of the majority—The reelection of the President encourages this evil.

It may be asked whether the legislators of the United States did right or wrong in allowing the reelection of the President. It seems at first sight contrary to all reason to prevent the head of the executive power from being elected a second time. The influence which the talents and the character of a single individual may exercise upon the fate of a whole people, in critical circumstances or arduous times, is well known: a law preventing the reelection of the chief magistrate would deprive the citizens of the surest pledge of the prosperity and the security of the commonwealth; and, by a singular inconsistency, a man would be excluded from the government at the very time when he had shown his ability in conducting its affairs.

But if these arguments are strong, perhaps still more powerful reasons may be advanced against them. Intrigue and corruption are the natural defects of elective government; but when the head of the State can be reelected these evils rise to a great height, and compromise the very existence of the country. When a simple candidate seeks to rise by intrigue, his manœuvres must necessarily be limited to a narrow sphere; but when the chief magistrate enters the lists, he borrows the strength of the government for his own purposes. In the former case the feeble resources of an individual are in action; in the latter, the State itself, with all its immense influence, is busied in the work of corruption and cabal. The private citizen, who employs the most immoral practices to acquire power, can only act in a manner indirectly prejudicial to the public prosperity. But if the representative of the executive descends into the combat, the cares of government dwindle into second-rate importance, and the success of his election is his first concern. All laws and all the negotiations he undertakes are to him nothing more than electioneering schemes; places become the reward of services rendered, not to the nation, but to its chief; and the influence of the government, if not injurious to the country, is at least no longer beneficial to the community for which it was created.

It is impossible to consider the ordinary course of affairs in the United States without perceiving that the desire of being reelected is the chief aim of the President; that his whole administration, and even his most indifferent measures, tend to this object; and that, as the crisis approaches, his personal interest takes the place of his interest in the public good. The principle of reeligibility renders the corrupt influence of elective government still more extensive and pernicious.

In America it exercises a peculiarly fatal influence on the sources of national existence. Every Government seems to be afflicted by some evil which is inherent in its nature, and the genius of the legislator is shown in eluding its attacks. A State may survive the influence of a host of bad laws, and the mischief they cause is frequently exaggerated; but a law which encourages the growth of the canker within must prove fatal in the end, although its bad consequences may not be immediately perceived.

The principle of destruction in absolute monarchies lies in the excessive and unreasonable extension of the prerogative of the crown; and a measure tending to remove the constitutional provisions which counterbalance this influence would be radically bad, even if its immediate consequences were unattended with evil. By a parity of reasoning, in countries governed by a democracy, where the people is perpetually drawing all authority to itself, the laws which increase or accelerate its action are the direct assailants of the very principle of the government.

The greatest proof of the ability of the American legislators is, that they clearly discerned this truth, and that they had the courage to act up to it. They conceived that a certain authority above the body of the people was necessary, which should enjoy a degree of independence, without, however, being entirely beyond the popular control; an authority which would be forced to comply with the *permanent* determinations of the majority, but which would be able to resist its caprices, and to refuse its most dangerous demands. To this end they centred the whole executive power of the nation in a single arm; they granted extensive prerogatives to the President, and they armed him with the veto to resist the encroachments of the legislature.

But by introducing the principle of reelection they partly destroyed their work; and they rendered the President but little inclined to exert the great power they had vested in his hands. If ineligible a second time, the President would be far from independent of the people, for his responsibility would not be lessened; but the favour of the people would not be so necessary to him as to induce him to court it by humouring its desires. If reeligible (and this is more especially true at the present day, when political morality is relaxed, and when great men are rare), the President of the United States becomes an easy tool in the hands of the majority. He adopts its likings and its animosities, he hastens to anticipate its wishes, he forestalls its complaints, he yields to its idlest cravings, and instead of guiding it, as the legislature intended that he should do, he is ever ready to follow its bidding. Thus, in order not to deprive the State of the talents of an individual, those talents have been rendered almost useless; and to reserve an expedient for extraordinary perils, the country has been exposed to daily dangers.

Federal Courts[28]

Political importance of the judiciary in the United States—Difficulty of treating this subject—Utility of judicial power in confederations—What tribunals could be introduced into the Union—Necessity of establishing federal courts of justice—Organization of the national judiciary—The Supreme Court—In what it differs from all known tribunals.

I have inquired into the legislative and executive power of the Union, and the judicial power now remains to be examined; but in this place I cannot conceal my fears from the reader. Their judicial institutions exercise a great influence on the condition of the Anglo-Americans, and they occupy a prominent place amongst what are probably called political institutions: in this respect they are peculiarly deserving of our attention. But I am at a loss to explain the political action of the American tribunals without entering into some technical details of their constitution and their forms of proceeding; and I know not how to descend to these minutiæ without wearying the curiosity of the reader by the natural aridity of the subject, or without risking to fall into obscurity through a desire to be succinct. I can scarcely hope to escape these various evils; for if I appear too lengthy to a man of the world, a lawyer may on the other hand complain of my brevity. But these are the natural disadvantages of my subject, and more especially of the point which I am about to discuss.

The great difficulty was, not to devise the Constitution to the Federal Government, but to find out a method of enforcing its laws. Governments have in general but two means of overcoming the opposition of the people they govern, viz., the physical force which is at their own disposal, and the moral force which they derive from the decisions of the courts of justice.

A government which should have no other means of exacting obedience than open war must be very near its ruin, for one of two alternatives would then probably occur: if its authority was small and its character temperate, it would not resort to violence till the last extremity, and it would connive at a number of partial acts of insubordination, in which case the State would gradually fall into anarchy; if it was enterprising and powerful, it would perpetually have recourse to its physical strength, and would speedily degenerate into a military despotism. So that its activity would not be less prejudicial to the community than its inaction.

The great end of justice is to substitute the notion of right for that of violence, and to place a legal barrier between the power of the government and the use of physical force. The authority which is awarded to the intervention of a court of justice by the general opinion of mankind is so surprisingly great that it clings to the mere formalities of justice, and gives a bodily influence to the shadow of the law. The moral force which courts of justice possess renders the introduction of physical force exceedingly rare, and is very frequently substituted for it; but if the latter proves to be indispensable, its power is doubled by the association of the idea of law.

A Federal Government stands in greater need of the support of judicial institutions than any other, because it is naturally weak and exposed to formidable opposition.[29] If it were always obliged to resort to violence in the first instance, it could not fulfil its task. The Union, therefore, required a national judiciary to enforce the obedience of the citizens to the laws, and to repeal the attacks which might be directed against them. The question then remained as to what tribunals were to exercise these privileges; were they to be entrusted to the courts of justice which were already organized in every State? or was it necessary to create federal courts? It may easily be proved that the Union could not adapt the judicial power of the States to its wants. The separation of the judiciary from the administrative power of the State no doubt affects the security of every citizen and the liberty of all. But it is no less important to the existence of the nation that these several powers should have the same origin, should follow the same principles, and act in the same sphere; in a word, that they should be correlative and homogeneous. No one,

I presume, ever suggested the advantage of trying offences committed in France by a foreign court of justice, in order to secure the impartiality of the judges. The Americans form one people in relation to their Federal Government; but in the bosom of this people divers political bodies have been allowed to subsist which are dependent on the national Government in a few points, and independent in all the rest; which have all a distinct origin, maxims peculiar to themselves, and special means of carrying on their affairs. To entrust the execution of the laws of the Union to tribunals instituted by these political bodies would be to allow foreign judges to preside over the nation. Nay, more; not only is each State foreign to the Union at large, but it is in perpetual opposition to the common interests, since whatever authority the Union loses turns to the advantage of the States. Thus to enforce the laws of the Union by means of the tribunals of the States would be to allow not only foreign but partial judges to preside over the nation.

But the number, still more than the mere character, of the tribunals of the States rendered them unfit for the service of the nation. When the Federal Constitution was formed there were already thirteen courts of justice in the United States which decided causes without appeal. That number is now increased to twenty-four. To suppose that a State can subsist when its fundamental laws may be subjected to four-and-twenty different interpretations at the same time is to advance a proposition alike contrary to reason and to experience.

The American legislators therefore agreed to create a federal judiciary power to apply the laws of the Union, and to determine certain questions affecting general interests, which were carefully determined beforehand. The entire judicial power of the Union was centred in one tribunal, which was denominated the Supreme Court of the United States. But, to facilitate the expedition of business, inferior courts were appended to it, which were empowered to decide causes of small importance without appeal, and with appeal causes of more magnitude. The members of the Supreme Court are named neither by the people nor the legislature, but by the President of the United States, acting with the advice of the Senate. In order to render them independent of the other authorities, their office was made inalienable; and it was determined that their salary, when once fixed, should not be altered by the legislature.[30] It was easy to proclaim the principle of a Federal judiciary, but difficulties multiplied when the extent of its jurisdiction was to be determined.

Means of Determining the Jurisdiction of the Federal Courts

Difficulty of determining the jurisdiction of separate courts of justice in confederations—The courts of the Union obtained the right of fixing their own jurisdiction—In what respect this rule attacks the portion of sovereignty reserved to the several States—The sovereignty of these States restricted by the laws, and the interpretation of the laws—Consequently, the danger of the several States is more apparent than real.

As the Constitution of the United States recognized two distinct powers in presence of each other, represented in a judicial point of view by two distinct classes of courts of justice, the utmost care which could be taken in defining their separate jurisdictions would have been insufficient to prevent frequent collisions between those tribunals. The question then arose to whom the right of deciding the competency of each court was to be referred.

In nations which constitute a single body politic, when a question is debated between two courts relating to their mutual jurisdiction, a third tribunal is generally within reach to decide the difference; and this is effected without difficulty, because in these nations the questions of judicial competency have no connection with the privileges of the national supremacy. But it was impossible to create an arbiter between a superior court of the Union and the superior court of a separate State which would not belong to one of these two classes. It was, therefore, necessary to allow one of these courts to judge its own cause, and to take or to retain cognizance of the point which was contested. To grant this privilege to the different courts of the States would have been to destroy the sovereignty of the Union *de facto* after having established it *dejure;* for the interpretation of the Constitution would soon have restored that portion of independence to the States of which the terms of that act deprived them. The object of the creation of a Federal tribunal was to prevent the courts of the States from deciding questions affecting the national interests in their own department, and so to form a uniform body of jurisprudence for the interpretation of the laws of the Union. This end would not have been accomplished if the courts of the several States had been competent to decide upon cases in their separate capacities from which they were obliged to abstain as Federal tribunals. The Supreme Court of the United States was therefore invested with the right of determining all questions of jurisdiction.[31]

This was a severe blow upon the independence of the States, which was thus restricted not only by the laws, but by the interpretation of them; by one limit which was known, and by another which was dubious; by a rule which was certain, and a rule which was arbitrary. It is true the Constitution had laid down the precise limits of the Federal supremacy, but whenever this supremacy is contested by one of the States, a Federal tribunal decides the question. Nevertheless, the dangers with which the independence of the States was threatened by this mode of proceeding are less serious than they appeared to be. We shall see hereafter that in America the real strength of the country is vested in the provincial far more than in the Federal Government. The Federal judges are conscious of the relative weakness of the power in whose name they act, and they are more inclined to abandon a right of jurisdiction in cases where it is justly their own than to assert a privilege to which they have no legal claim.

Different Cases of Jurisdiction

The matter and the party are the first conditions of the Federal jurisdiction—Suits in which ambassadors are engaged—Suits of the Union—Of a separate State—By whom tried—Causes resulting from the laws of the Union—Why judged by the Federal tribunals—Causes relating to the performance of contracts tried by the Federal courts—Consequence of this arrangement.

After having appointed the means of fixing the competency of the Federal courts, the legislators of the Union defined the cases which should come within their jurisdiction. It was established, on the one hand, that certain parties must always be brought before the Federal courts, without any regard to the special nature of the cause; and, on the other, that certain causes must always be brought before the same courts, without any regard to the quality of the parties in the suit. These distinctions were therefore admitted to be the basis of the Federal jurisdiction.

Ambassadors are the representatives of nations in a state of amity with the Union, and whatever concerns these personages concerns in some degree the whole Union. When an ambassador is a party in a suit, that suit affects the welfare of the nation, and a Federal tribunal is naturally called upon to decide it.

The Union itself may be invoked in legal proceedings, and in this case it would be alike contrary to the customs of all nations and to common sense to appeal to a tribunal representing any other sovereignty than its own; the Federal courts, therefore, take cognizance of these affairs.

When two parties belonging to two different States are engaged in a suit, the case cannot with propriety be brought before a court of either State. The surest expedient is to select a tribunal like that of the Union, which can excite the suspicions of neither party, and which offers the most natural as well as the most certain remedy.

When the two parties are not private individuals, but States, an important political consideration is added to the same motive of equity. The quality of the parties in this case gives a national importance to all their disputes; and the most trifling litigation of the States may be said to involve the peace of the whole Union.[32]

The nature of the cause frequently prescribes the rule of competency. Thus all the questions which concern maritime commerce evidently fall under the cognizance of the Federal tribunals.[33] Almost all these questions are connected with the interpretation of the law of nations, and in this respect they essentially interest the Union in relation to foreign powers. Moreover, as the sea is not included within the limits of any peculiar jurisdiction, the national courts can only hear causes which originate in maritime affairs.

The Constitution comprises under one head almost all the cases which by their very nature come within the limits of the Federal courts. The rule which it lays down is simple, but pregnant with an entire system of ideas, and with a vast multitude of facts. It declares that the judicial power of the Supreme Court shall extend to all cases in law and equity *arising under the laws of the United States.*

Two examples will put the intention of the legislator in the clearest light:

The Constitution prohibits the States from making laws on the value and circulation of money: If, notwithstanding this prohibition, a State passes a law of this kind, with which the interested parties refuse to comply because it is contrary to the Constitution, the case must come before a Federal court, because it arises under the laws of the United States. Again, if difficulties arise in the levying of import duties which have been voted by Congress, the Federal court must decide the case, because it arises under the interpretation of a law of the United States.

This rule is in perfect accordance with the fundamental principles of the Federal Constitution. The Union, as it was established in 1789, possesses, it is true, a limited supremacy; but it was intended that within its limits it should form one and the same people.[34] Within those limits the Union is sovereign. When this point is established and admitted, the inference is easy; for if it be acknowledged that the United States constitute one and the same people within the bounds prescribed by their Constitution, it is impossible to refuse them the rights which belong to other nations. But it has been allowed, from the origin of society, that every nation has the right of deciding by its own courts those questions which concern the execution of its own laws. To this it is answered that the Union is in so singular a position that in relation to some matters it constitutes a people, and that in relation to all the rest it is a nonentity. But the inference to be drawn is, that in the laws relating to these matters the Union possesses all the rights of absolute sovereignty. The difficulty is to know what these matters are;

and when once it is resolved (and we have shown how it was resolved, in speaking of the means of determining the jurisdiction of the Federal courts) no further doubt can arise; for as soon as it is established that a suit is Federal—that is to say, that it belongs to the share of sovereignty reserved by the Constitution to the Union—the natural consequence is that it should come within the jurisdiction of a Federal court.

Whenever the laws of the United States are attacked, or whenever they are resorted to in self-defence, the Federal courts must be appealed to. Thus the jurisdiction of the tribunals of the Union extends and narrows its limits exactly in the same ratio as the sovereignty of the Union augments or decreases. We have shown that the principal aim of the legislators of 1789 was to divide the sovereign authority into two parts. In the one they placed the control of all the general interests of the Union, in the other the control of the special interests of its component States. Their chief solicitude was to arm the Federal Government with sufficient power to enable it to resist, within its sphere, the encroachments of the several States. As for these communities, the principle of independence within certain limits of their own was adopted in their behalf; and they were concealed from the inspection, and protected from the control, of the central Government. In speaking of the division of authority, I observed that this latter principle had not always been held sacred, since the States are prevented from passing certain laws which apparently belong to their own particular sphere of interest. When a State of the Union passes a law of this kind, the citizens who are injured by its execution can appeal to the Federal courts.

Thus the jurisdiction of the Federal courts extends not only to all the cases which arise under the laws of the Union, but also to those which arise under laws made by the several States in opposition to the Constitution. The States are prohibited from making *ex post facto* laws in criminal cases, and any person condemned by virtue of a law of this kind can appeal to the judicial power of the Union. The States are likewise prohibited from making laws which may have a tendency to impair the obligations of contracts.[35] If a citizen thinks that an obligation of this kind is impaired by a law passed in his State, he may refuse to obey it, and may appeal to the Federal courts.[36]

This provision appears to me to be the most serious attack upon the independence of the States. The rights awarded to the Federal Government for purposes of obvious national importance are definite and easily comprehensible; but those with which this last clause invests it are not either clearly appreciable or accurately defined. For there are vast numbers of political laws which influence the existence of obligations of contracts, which may thus furnish an easy pretext for the aggressions of the central authority.

Procedure of the Federal Courts

Natural weakness of the judiciary power in confederations—Legislators ought to strive as much as possible to bring private individuals, and not States, before the Federal Courts—How the Americans have succeeded in this—
Direct prosecution of private individuals in the Federal Courts—
Indirect prosecution of the States which violate the laws of the Union—The decrees of the Supreme Court enervate but do not destroy the provincial laws.

I have shown what the privileges of the Federal courts are, and it is no less important to point out the manner in which they are exercised. The irresistible authority of justice in countries in which the sovereignty is undivided is derived from the fact that the tribunals of those countries represent the entire nation at issue with the individual against whom their decree is directed, and the idea of power is thus introduced to corroborate the idea of right. But this is not always the case in countries in which the sovereignty is divided; in them the judicial power is more frequently opposed to a fraction of the nation than to an isolated individual, and its moral authority and physical strength are consequently diminished. In Federal States the power of the judge is naturally decreased, and that of the justiciable parties is augmented. The aim of the legislator in confederate States ought therefore to be to render the position of the courts of justice analogous to that which they occupy in countries where the sovereignty is undivided; in other words, his efforts ought constantly to tend to maintain the judicial power of the confederation as the representative of the nation, and the justiciable party as the representative of an individual interest.

Every Government, whatever may be its constitution, requires the means of constraining its subjects to discharge their obligations, and of protecting its privileges from their assaults. As far as the direct action of the Government on the community is concerned, the Constitution of the United States contrived, by a master-stroke of policy, that the Federal courts, acting in the name of the laws, should only take cognizance of parties in an individual capacity. For, as it had been declared that the Union consisted of one and the same people within the limits laid down by the Constitution, the inference was that the Government created by this Constitution, and acting within these limits, was invested with all the privileges of a national Government, one of the principals of which is the right of transmitting its injunctions directly to the private citizen. When, for instance, the Union votes an impost, it does not apply to the States for the levying of it, but to every American citizen in proportion to his assessment. The Supreme Court, which is empowered to enforce the execution of this law of the Union, exerts its influence not upon a refractory State, but upon the private taxpayer; and, like the judicial power of other nations, it is opposed to the person of an individual. It is to be observed that the Union chose its own antagonist; and as that antagonist is feeble, he is naturally worsted.

But the difficulty increases when the proceedings are not brought forward *by* but *against* the Union. The Constitution recognises the legislative power of the States; and a law so enacted may impair the privileges of the Union, in which case a collision in unavoidable between that body and the State which has passed the law: and it only remains to select the least dangerous remedy, which is very clearly deducible from the general principles I have before established.[37]

It may be conceived that, in the case under consideration, the Union might have sued the State before a Federal court, which would have annulled the act, and by this means it would have adopted a natural course of proceeding; but the judicial power would have been placed in open hostility to the State, and it was desirable to avoid this predicament as much as possible. The Americans hold that it is nearly impossible that a new law should not impair the interests of some private individual by its provisions: these private interests are assumed by the American legislators as the ground of attack against such measures as may be prejudicial to the Union, and it is to these cases that the protection of the Supreme Court is extended.

Suppose a State vends a certain portion of its territory to a company, and that a year afterwards it passes a law by which the territory is otherwise disposed of, and that clause of the Constitution which prohibits laws impairing the obligation of contracts violated. When the purchaser under the second act appears to take possession, the possessor under the first act brings his action before the tribunals of the Union, and causes the title of the claimant to be pronounced null and void.[38] Thus, in point of fact, the judicial power of the Union is contesting the claims of the sovereignty of a State; but it only acts indirectly and upon a special application of detail: it attacks the law in its consequences, not in its principle, and it rather weakens than destroys it.

The last hypothesis that remained was that each State formed a corporation enjoying a separate existence and distinct civil rights, and that it could therefore sue or be sued before a tribunal. Thus a State could bring an action against another State. In this instance the Union was not called upon to contest a provincial law, but to try a suit in which a State was a party. This suit was perfectly similar to any other cause, except that the quality of the parties was different; and here the danger pointed out at the beginning of this chapter exists with less chance of being avoided.

The inherent disadvantage of the very essence of Federal constitutions is that they engender parties in the bosom of the nation which present powerful obstacles to the free course of justice.

High Rank of the Supreme Court amongst the Great Powers of State

No nation ever constituted so great a judicial power as the Americans—Extent of its prerogative—Its political influence—The tranquillity and the very existence of the Union depend on the discretion of the seven Federal Judges.

When we have successively examined in detail the organization of the Supreme Court, and the entire prerogatives which it exercises, we shall readily admit that a more imposing judicial power was never constituted by any people. The Supreme Court is placed at the head of all known tribunals, both by the nature of its rights and the class of justiciable parties which it controls.

In all the civilized countries of Europe the Government has always shown the greatest repugnance to allow the cases to which it was itself a party to be decided by the ordinary course of justice. This repugnance naturally attains its utmost height in an absolute Government; and, on the other hand, the privileges of the courts of justice are extended with the increasing liberties of the people: but no European nation has at present held that all judicial controversies, without regard to their origin, can be decided by the judges of common law.

In America this theory has been actually put in practice, and the Supreme Court of the United States is the sole tribunal of the nation. Its power extends to all the cases arising under laws and treaties made by the executive and legislative authorities, to all cases of admiralty and maritime jurisdiction, and in general to all points which affect the law of nations. It may even be affirmed that, although its constitution is essentially judicial, its prerogatives are almost entirely political. Its sole object is to enforce the execution of the laws of the Union; and the Union only regulates the relations of the Government with the citizens, and of the nation with Foreign Powers: the relations of citizens amongst themselves are almost exclusively regulated by the sovereignty of the States.

A second and still greater cause of the preponderance of this court may be adduced. In the nations of Europe the courts of justice are only called upon to try the controversies of private individuals; but the Supreme Court of the United States summons sovereign powers to its bar. When the clerk of the court advances on the steps of the tribunal, and simply says, 'The State of New York *versus* the State of Ohio,' it is impossible not to feel that the Court which he addresses is no ordinary body; and when it is recollected that one of these parties represents one million, and the other two millions of men, one is struck by the responsibility of the seven judges whose decision is about to satisfy or to disappoint so large a number of their fellow-citizens.

The peace, the prosperity, and the very existence of the Union are vested in the hands of the seven judges. Without their active cooperation the Constitution would be a dead letter: the Executive appeals to them for assistance against the encroachments of the legislative powers; the Legislature demands their protection from the designs of the Executive; they defend the Union from the disobedience of the States, the States from the exaggerated claims of the Union, the public interest against the interests of private citizens, and the conservative spirit of order against the fleeting innovations of democracy. Their power is enormous, but it is clothed in the authority of public opinion. They are the all-powerful guardians of a people which respects law, but they would be impotent against popular neglect or popular contempt. The force of public opinion is the most intractable of agents, because its exact limits cannot be defined; and it is not less dangerous to exceed than to remain below the boundary prescribed.

The Federal judges must not only be good citizens, and men possessed of that information and integrity which are indispensable to magistrates, but they must be statesmen-politicians, not unread in the signs of the times, not afraid to brave the obstacles which can be subdued, nor slow to turn aside such encroaching elements as may threaten the supremacy of the Union and the obedience which is due to the laws.

The President, who exercises a limited power, may err without causing great mischief in the State. Congress may decide amiss without destroying the Union, because the electoral body in which Congress originates may cause it to retract its decision by changing its members. But if the Supreme Court is ever composed of imprudent men or bad citizens, the Union may be plunged into anarchy or civil war.

The real cause of this danger, however, does not he in the constitution of the tribunal, but in the very nature of Federal Governments. We have observed that in confederate peoples it is especially necessary to consolidate the judicial authority, because in no other nations do those independent persons who are able to cope with the social body exist in greater power or in a better condition to resist the physical strength of the Government. But the more a power requires to be strengthened, the more extensive and independent it must be made; and the dangers which its abuse may create are heightened by its independence and its strength. The source of the evil is not, therefore, in the constitution of the power, but in the constitution of those States which render its existence necessary.

In What Respects the Federal Constitution Is Superior to That of the States

In what respects the Constitution, of the Union can be compared to that of the States—Superiority of the Constitution of the Union attributable to the wisdom of the Federal legislators—Legislature of the Union less dependent on the people than that of the States—Executive power more independent in its sphere—
Judicial power less subjected to the inclinations of the majority—Practical consequence of these facts—The dangers inherent in a democratic government eluded by the Federal legislators, and increased by the legislators of the States.

The Federal Constitution differs essentially from that of the States in the ends which it is intended to accomplish, but in the means by which these ends are promoted a greater analogy exists between them. The objects of the Governments are different, but their forms are the same; and in this special point of view there is some advantage in comparing them together.

I am of opinion that the Federal Constitution is superior to all the Constitutions of the States, for several reasons.

The present Constitution of the Union was formed at a later period than those of the majority of the States, and it may have derived some ameliorations from past experience. But we shall be led to acknowledge that this is only a secondary cause of its superiority, when we recollect that eleven new States[39] have been added to the American Confederation since the promulgation of the Federal Constitution, and that these new republics have always rather exaggerated than avoided the defects which existed in the former Constitutions.

The chief cause of the superiority of the Federal Constitution lay in the character of the legislators who composed it. At the time when it was formed the dangers of the Confederation were imminent, and its ruin seemed inevitable. In this extremity the people chose the men who most deserved the esteem, rather than those who had gained the affections, of the country. I have already observed that distinguished as almost all the legislators of the Union were for their intelligence, they were still more so for their patriotism. They had all been nurtured at a time when the spirit of liberty was braced by a continual struggle against a powerful and predominant authority. When the contest was terminated, whilst the excited passions of the populace persisted in warring with dangers which had ceased to threaten them, these men stopped short in their career; they cast a calmer and more penetrating look upon the country which was now their own; they perceived that the war of independence was definitely ended, and that the only dangers which America had to fear were those which might result from the abuse of the freedom she had won. They had the courage to say what they believed to be true, because they were animated by a warm and sincere love of liberty; and they ventured to propose restrictions, because they were resolutely opposed to destruction.[40]

The greater number of the Constitutions of the States assign one year for the duration of the House of Representatives, and two years for that of the Senate; so that members of the legislative body are constantly and narrowly tied down by the slightest desires of their constituents. The legislators of the Union were of opinion that this excessive dependence of the Legislature tended to alter the nature of the main consequences of the representative system, since it vested the source, not only of authority, but of government, in the people. They increased the length of the time for which the representatives were returned, in order to give them freer scope for the exercise of their own judgment.

The Federal Constitution, as well as the Constitutions of the different States, divided the legislative body into two branches. But in the States these two branches were composed of the same elements, and elected in the same manner. The consequence was that the passions and inclinations of the populace were as rapidly and as energetically represented in one chamber as in the other, and that laws were made with all the characteristics of violence and precipitation. By the Federal Constitution the two houses originate in like manner in the choice of the people; but the conditions of eligibility and the mode of election were changed, to the end that, if, as is the case in certain nations, one branch of the Legislature represents the same interests as the other, it may at least represent a superior degree of intelligence and discretion. A mature age was made one of the conditions of the senatorial dignity, and the Upper House was chosen by an elected assembly of a limited number of members.

To concentrate the whole social force in the hands of the legislative body is the natural tendency of democracies; for as this is the power which emanates the most directly from the people, it is made to participate most fully in the preponderating authority of the multitude, and it is naturally led to monopolize every species of influence. This concentration is at once prejudicial to a well-conducted administration, and favourable to the despotism of the majority. The legislators of the States frequently yielded to these democratic propensities, which were invariably and courageously resisted by the founders of the Union.

In the States the executive power is vested in the hands of a magistrate, who is apparently placed upon a level with the Legislature, but who is in reality nothing more than the blind agent and the passive instrument of its decisions. He can derive no influence from the duration of his functions, which terminate with the revolving year, or from the exercise of prerogatives which can scarcely be said to exist. The Legislature can condemn him to inaction by intrusting the execution of the laws to special committees of its own members, and can annul his temporary dignity by depriving him of his salary. The Federal Constitution vests all the privileges and all the responsibility of the executive power in a single individual. The duration of the Presidency is fixed at four years; the salary of the individual who fills that office cannot be altered during the term of his functions; he is protected by a body of official dependents, and armed with a suspensive veto. In short, every effort was made to confer a strong and independent position upon the executive authority within the limits which had been prescribed to it.

In the Constitutions of all the States the judicial power is that which remains the most independent of the legislative authority; nevertheless, in all the States the Legislature has reserved to itself the right of regulating the emoluments of the judges, a practice which necessarily subjects these magistrates to its immediate influence. In some States the judges are only temporarily appointed, which deprives them of a great portion of their power and their freedom. In others the legislative and judicial powers are entirely confounded; thus the Senate of New York, for instance, constitutes in certain cases the superior court of the State. The Federal Constitution, on the other hand, carefully separates the judicial authority from all external influences; and it provides for the independence of the judges, by declaring that their salary shall not be altered, and that their functions shall be inalienable.

The practical consequences of these different systems may easily be perceived. An attentive observer will soon remark that the business of the Union is incomparably better conducted than that of any individual State. The conduct of the Federal Government is more fair and more temperate than that of the States, its designs are more fraught with wisdom, its projects are more durable and more skilfully combined, its measures are put into execution with more vigour and consistency.

I recapitulate the substance of this chapter in a few words: The existence of democracies is threatened by two dangers, viz., the complete subjection of the legislative body to the caprices of the electoral body, and the concentration of all the powers of the Government in the legislative authority. The growth of these evils has been encouraged by the policy of the legislators of the States, but it has been resisted by the legislators of the Union by every means which lay within their control.

Characteristics Which Distinguish the Federal Constitution of the United States of America from All Other Federal Constitutions

American Union appears to resemble all other confederations—Nevertheless its effects are different—Reason of this—Distinctions between the Union and all other confederations—The American Government not a federal but an imperfect national Government.

The United States of America do not afford either the first or the only instance of confederate States, several of which have existed in modern Europe, without adverting to those of antiquity. Switzerland, the Germanic Empire, and the Republic of the United Provinces either have been or still are confederations. In studying the constitutions of these different countries, the politician is surprised to observe that the powers with which they invested the Federal Government are nearly identical with the privileges awarded by the American Constitution to the Government of the United States. They confer upon the central power the same rights of making peace and war, of raising money and troops, and of providing for the general exigencies and the common interests of the nation. Nevertheless the Federal Government of these different peoples has always been as remarkable for its weakness and inefficiency as that of the Union is for its vigorous and enterprising spirit. Again, the first American Confederation perished through the excessive weakness of its Government; and this weak Government was, notwithstanding, in possession of rights even more extensive than those of the Federal Government of the present day. But the more recent Constitution of the United States contains certain principles which exercise a most important influence, although they do not at once strike the observer.

This Constitution, which may at first sight be confounded with the federal constitutions which preceded it, rests upon a novel theory, which may be considered as a great invention in modern political science. In all the confederations which had been formed before the American Constitution of 1789 the allied States agreed to obey the injunctions of a Federal Government; but they reserved to themselves the right of ordaining and enforcing the execution of the laws of the Union. The American States which combined in 1789 agreed that the Federal Government should not only dictate the laws, but that it should execute it own enactments. In both cases the right is the same, but the exercise of the right is different; and this alteration produced the most momentous consequences.

In all the confederations which had been formed before the American Union the Federal Government demanded its supplies at the hands of the separate Governments; and if the measure it prescribed was onerous to any one of those bodies means were found to evade its claims: if the State was powerful, it had recourse to arms; if it was weak, it connived at the resistance which the law of the Union, its sovereign, met with, and resorted to inaction under the plea of inability. Under these circumstances one of the two alternatives has invariably occurred; either the most preponderant of the allied peoples has assumed the privileges of the Federal authority and ruled all the States in its name,[41] or the Federal Government has been abandoned by its natural supporters, anarchy has arisen between the confederates, and the Union has lost all powers of action.[42]

In America the subjects of the Union are not States, but private citizens: the national Government levies a tax, not upon the State of Massachusetts, but upon each inhabitant of Massachusetts. All former confederate governments presided over communities, but that of the Union rules individuals; its force is not borrowed, but self-derived; and it is served by its own civil and military officers, by its own army, and its own courts of justice. It cannot be doubted that the spirit of the nation, the passions of the multitude, and the provincial prejudices of each State tend singularly to diminish the authority of a Federal authority thus constituted, and to facilitate the means of resistance to its mandates; but the comparative weakness of a restricted sovereignty is an evil inherent in the Federal system. In America, each State has fewer opportunities of resistance and fewer temptations to non-compliance; nor can such a design be put in execution (if indeed it be entertained) without an open violation of the laws of the Union, a direct interruption of the ordinary course of justice, and a bold declaration of revolt; in a word, without taking a decisive step which men hesitate to adopt.

In all former confederations the privileges of the Union furnished more elements of discord than of power, since they multiplied the claims of the nation without augmenting the means of enforcing them: and in accordance with this fact it may be remarked that the real weakness of federal governments has almost always been in the exact ratio of their nominal power. Such is not the case in the American Union, in which, as in ordinary governments, the Federal Government has the means of enforcing all it is empowered to demand.

The human understanding more easily invents new things than new words, and we are thence constrained to employ a multitude of improper and inadequate expressions. When several nations form a permanent league and establish a supreme authority, which, although it has not the same influence over the members of the community as a national government, acts upon each of the Confederate States in a body, this Government, which is so essentially different from all others, is denominated a Federal one. Another form of society is afterwards discovered, in which several peoples are fused into one and the same nation with regard to certain common interests, although they remain distinct, or at least only confederate, with regard to all their other concerns. In this case the central power acts directly upon those whom it governs, whom it rules, and whom it judges, in the same manner, as, but in a more limited circle than, a national government. Here the term Federal Government is clearly no longer applicable to a state of things which must be styled an incomplete national Government: a form of government has been found out which is neither exactly national nor federal; but no further progress has been made, and the new word which will one day designate this novel invention does not yet exist.

The absence of this new species of confederation has been the cause which has brought all Unions to civil war, to subjection, or to a stagnant apathy, and the peoples which formed these leagues have been either too dull to discern, or too pusillanimous to apply this great remedy. The American Confederation perished by the same defects.

But the confederate States of America had been long accustomed to form a portion of one empire before they had won their independence; they had not contracted the habit of governing themselves, and their national prejudices had not taken deep root in their minds. Superior to the rest of the world in political knowledge, and sharing that knowledge equally amongst themselves, they were little agitated by the passions which generally oppose the extension of federal authority in a nation, and those passions were checked by the wisdom of the chief citizens. The Americans applied the remedy with prudent firmness as soon as they were conscious of the evil; they amended their laws, and they saved their country.

Advantages of the Federal System in General, and Its Special Utility in America

Happiness and freedom of small nations—Power of great nations—Great empires favourable to the growth of civilization—Strength often the first element of national prosperity—Aim of the Federal system to unite the twofold advantages resulting from a small and from a large territory—Advantages derived by the United States from this system—

The law adapts itself to the exigencies of the population; population does not conform to the exigencies of the law—Activity, amelioration, love and enjoyment of freedom in the American communities—Public spirit of the Union the abstract of provincial patriotism—Principles and things circulate freely over the territory of the United States—The Union is happy and free as a little nation, and respected as a great empire.

In small nations the scrutiny of society penetrates into every part, and the spirit of improvement enters into the most trifling details; as the ambition of the people is necessarily checked by its weakness, all the efforts and resources of the citizens are turned to the internal benefit of the community, and are not likely to evaporate in the fleeting breath of glory. The desires of every individual are limited, because extraordinary faculties are rarely to be met with. The gifts of an equal fortune render the various conditions of life uniform, and the manners of the inhabitants are orderly and simple. Thus, if one estimate the gradations of popular morality and enlightenment, we shall generally find that in small nations there are more persons in easy circumstances, a more numerous population, and a more tranquil state of society, than in great empires.

When tyranny is established in the bosom of a small nation, it is more galling than elsewhere, because, as it acts within a narrow circle, every point of that circle is subject to its direct influence. It supplies the place of those great designs which it cannot entertain by a violent or an exasperating interference in a multitude of minute details; and it leaves the political world, to which it properly belongs, to meddle with the arrangements of domestic life. Tastes as well as actions are to be regulated at its pleasure; and the families of the citizens as well as the affairs of the State are to be governed by its decisions. This invasion of rights occurs, however, but seldom, and freedom is in truth the natural state of small communities. The temptations which the Government offers to ambition are too weak, and the resources of private individuals are too slender, for the sovereign power easily to fall within the grasp of a single citizen; and should such an event have occurred, the subjects of the State can without difficulty overthrow the tyrant and his oppression by a simultaneous effort.

Small nations have therefore ever been the cradle of political liberty; and the fact that many of them have lost their immunities by extending their dominion shows that the freedom they enjoyed was more a consequence of the inferior size than of the character of the people.

The history of the world affords no instance of a great nation retaining the form of republican government for a long series of years,[43] and this has led to the conclusion that such a state of things is impracticable. For my own part, I cannot but censure the imprudence of attempting to limit the possible and to judge the future on the part of a being who is hourly deceived by the most palpable realities of life, and who is constantly taken by surprise in the circumstances with which he is most familiar. But it may be advanced with confidence that the existence of a great republic will always be exposed to far greater perils than that of a small one.

All the passions which are most fatal to republican institutions spread with an increasing territory, whilst the virtues which maintain their dignity do not augment in the same proportion. The ambition of the citizens increases with the power of the State; the strength of parties with the importance of the ends they have in view; but that devotion to the common weal which is the surest check on destructive passions is not stronger in a large than in a small republic. It might, indeed, be proved without difficulty that it is less powerful and less sincere. The arrogance of wealth and the dejection of wretchedness, capital cities of unwonted extent, a lax morality, a vulgar egotism, and a great confusion of interests, are the dangers which almost invariably arise from the magnitude of States. But several of these evils are scarcely prejudicial to a monarchy, and some of them contribute to maintain its existence. In monarchical States the strength of the government is its own; it may use, but it does not depend on, the community, and the authority of the prince is proportioned to the prosperity of the nation; but the only security which a republican government possesses against these evils lies in the support of the majority. This support is not, however, proportionably greater in a large republic than it is in a small one; and thus, whilst the means of attack perpetually increase both in number and in influence, the power of resistance remains the same, or it may rather be said to diminish, since the propensities and interests of the people are diversified by the increase of the population, and the difficulty of forming a compact majority is constantly augmented. It has been observed, moreover, that the intensity of human passions is heightened, not only by the importance of the end which they propose to attain, but by the multitude of individuals who are animated by them at the same time. Every one has had occasion to remark that his emotions in the midst of a sympathising crowd are far greater than those which he would have felt in solitude. In great republics the impetus of political passion is irresistible, not only because it aims at gigantic purposes, but because it is felt and shared by millions of men at the same time.

It may therefore be asserted as a general proposition that nothing is more opposed to the well-being and the freedom of man than vast empires. Nevertheless it is important to acknowledge the peculiar advantages of great States. For the very reason which renders the desire of power more intense in these communities than amongst ordinary men, the love of glory is also more prominent in the hearts of a class of citizens, who regard the applause of a great people as a reward worthy of their exertions, and an elevating encouragement to man. If we would learn why it is that great nations contribute more powerfully to the spread of human improvement than small States, we shall discover an adequate cause in the rapid and energetic circulation of ideas, and in those great cities which are the intellectual centres where all the rays of human genius are reflected and combined. To this it may be added that most important discoveries demand a display of national power which the Government of a small State is unable to make; in great nations the Government entertains a greater number of general notions, and is more completely disengaged from the routine of precedent and the egotism of local prejudice; its designs are conceived with more talent, and executed with more boldness.

In time of peace the well-being of small nations is undoubtedly more general and more complete, but they are apt to suffer more acutely from the calamities of war than those great empires whose distant frontiers may for ages avert the presence of the danger from the mass of the people, which is therefore more frequently afflicted than ruined by the evil.

But in this matter, as in many others, the argument derived from the necessity of the case predominates over all others. If none but small nations existed, I do not doubt that mankind would be more happy and more free; but the existence of great nations is unavoidable.

This consideration introduces the element of physical strength as a condition of national prosperity. It profits a people but little to be affluent and free if it is perpetually exposed to be pillaged or subjugated; the number of its manufactures and the extent of its commerce are of small advantage if another nation has the empire of the seas and gives the law in all the markets of the globe. Small nations are often impoverished, not because they are small, but because they are weak; the great empires prosper less because they are great than because they are strong. Physical strength is therefore one of the first conditions of the happiness and even of the existence of nations. Hence it occurs that, unless very peculiar circumstances intervene, small nations are always united to large empires in the end, either by force or by their own consent: yet I am unacquainted with a more deplorable spectacle than that of a people unable either to defend or to maintain its independence.

The Federal system was created with the intention of combining the different advantages which result from the greater and the lesser extent of nations; and a single glance over the United States of America suffices to discover the advantages which they have derived from its adoption.

In great centralized nations the legislator is obliged to impart a character of uniformity to the laws which does not always suit the diversity of customs and of districts; as he takes no cognizance of special cases, he can only proceed upon general principles; and the population is obliged to conform to the exigencies of the legislation, since the legislation cannot adapt itself to the exigencies and the customs of the population, which is the cause of endless trouble and misery. This disadvantage does not exist in confederations. Congress regulates the principal measures of the national Government, and all the details of the administration are reserved to the provincial legislatures. It is impossible to imagine how much this division of sovereignty contributes to the well-being of each of the States which compose the Union. In these small communities, which are never agitated by the desire of aggrandizement or the cares of self-defence, all public authority and private energy is employed in internal amelioration. The central government of each State, which is in immediate juxtaposition to the citizens, is daily apprised of the wants which arise in society; and new projects are proposed every year, which are discussed either at town meetings or by the legislature of the State, and which are transmitted by the press to stimulate the zeal and to excite the interest of the citizens. This spirit of amelioration is constantly alive in the American republics, without compromising their tranquillity; the ambition of power yields to the less refined and less dangerous love of comfort. It is generally believed in America that the existence and the permanence of the republican form of government in the New World depend upon the existence and the permanence of the Federal system; and it is not unusual to attribute a large share of the misfortunes which have befallen the new States of South America to the injudicious erection of great republics, instead of a divided and confederate sovereignty.

It is incontestably true that the love and the habits of republican government in the United States were engendered in the townships and in the provincial assemblies. In a small State, like that of Connecticut for instance, where cutting a canal or laying down a road is a momentous political question, where the State has no army to pay and no wars to carry on, and where much wealth and much honour cannot be bestowed upon the chief citizens, no form of government can be more natural or more appropriate than that of a republic. But it is this same republican spirit, it is these manners and customs of a free people, which are engendered and nurtured in the different States, to be afterwards applied to the country at large. The public spirit of the Union is, so to speak, nothing more than an abstract of the patriotic zeal of the provinces. Every citizen of the United States transfuses his attachment to his little republic in the common store of American patriotism. In defending the Union he defends the increasing prosperity of his own district, the right of conducting its affairs, and the hope of causing measures of improvement to be adopted which may be favourable to his own interest; and these are motives which are wont to stir men more readily than the general interests of the country and the glory of the nation.

On the other hand, if the temper and the manners of the inhabitants especially fitted them to promote the welfare of a great republic, the Federal system smoothed the obstacles which they might have encountered. The confederation of all the American States presents none of the ordinary disadvantages resulting from great agglomerations of men. The Union is a great republic in extent, but the paucity of objects for which its Government provides assimilates it to a small State. Its acts are important, but they are rare. As the sovereignty of the Union is limited and incomplete, its exercise is not incompatible with liberty; for it does not excite those insatiable desires of fame and power which have proved so fatal to great republics. As there is no common centre to the country, vast capital cities, colossal wealth, abject poverty, and sudden revolutions are alike unknown; and political passion, instead of spreading over the land like a torrent of desolation, spends its strength against the interests and the individual passions of every State.

Nevertheless, all commodities and ideas circulate throughout the Union as freely as in a country inhabited by one people. Nothing checks the spirit of enterprise. Government avails itself of the assistance of all who have talents or knowledge to serve it. Within the frontiers of the Union the profoundest peace prevails, as within the heart of some great empire; abroad, it ranks with the most powerful nations of the earth; two thousand miles of coast are open to the commerce of the world; and as it possesses the keys of the globe, its flag is respected in the most remote seas. The Union is as happy and as free as a small people, and as glorious and as strong as a great nation.

Why the Federal System Is Not Adapted to All Peoples, and How the Anglo-Americans Were Enabled to Adopt It

Every Federal system contains defects which baffle the efforts of the legislator—The Federal system is complex—It demands a daily exercise of discretion on the part of the citizens—Practical knowledge of government common amongst the Americans—Relative weakness of the Government of the Union, another defect inherent in the Federal system—The Americans have diminished without remedying it—The sovereignty of the separate States apparently weaker, but really stronger, than that of the Union—Why?—Natural causes of union must exist between confederate peoples besides the laws—What these causes are amongst the Anglo-Americans—Maine and Georgia, separated by a distance of a thousand miles, more naturally united than Normandy and Brittany—War, the main peril of confederations—This proved even by the example of the United States—The Union has no great wars to fear—Why?—Dangers to which Europeans would be exposed if they adopted the Federal system of the Americans.

When a legislator succeeds, after persevering efforts, in exercising an indirect influence upon the destiny of nations, his genius is lauded by mankind, whilst, in point of fact, the geographical position of the country which he is unable to change, a social condition which arose without his cooperation, manners and opinions which he cannot trace to their source, and an origin with which he is unacquainted, exercise so irresistible an influence over the courses of society that he is himself borne away by the current, after an ineffectual resistance. Like the navigator, he may direct the vessel which bears him along, but he can neither change its structure, nor raise the winds, nor lull the waters which swell beneath him.

I have shown the advantages which the Americans derive from their Federal system; it remains for me to point out the circumstances which rendered that system practicable, as its benefits are not to be enjoyed by all nations. The incidental defects of the Federal system which originate in the laws may be corrected by the skill of the legislator, but there are further evils inherent in the system which cannot be counteracted by the peoples which adopt it. These nations must therefore find the strength necessary to support the natural imperfections of their Government.

The most prominent evil of all Federal systems is the very complex nature of the means they employ. Two sovereignties are necessarily in presence of each other. The legislator may simplify and equalize the action of these two sovereignties, by limiting each of them to a sphere of authority accurately defined; but he cannot combine them into one, or prevent them from coming into collision at certain points. The Federal system therefore rests upon a theory which is necessarily complicated, and which demands the daily exercise of a considerable share of discretion on the part of those it governs.

A proposition must be plain to be adopted by the understanding of a people. A false notion which is clear and precise will always meet with a greater number of adherents in the world than a true principle which is obscure or involved. Hence it arises that parties, which are like small communities in the heart of the nation, invariably adopt some principle or some name as a symbol, which very inadequately represents the end they have in view and the means which are at their disposal, but without which they could neither act nor subsist. The Governments which are founded upon a single principle or a single feeling which is easily defined are perhaps not the best, but they are unquestionably the strongest and the most durable in the world.

In examining the Constitution of the United States, which is the most perfect Federal Constitution that ever existed, one is startled, on the other hand, at the variety of information and the excellence of discretion which it presupposes in the people whom it is meant to govern. The Government of the Union depends entirely upon legal fictions; the Union is an ideal nation which only exists in the mind, and whose limits and extent can only be discerned by the understanding.

When once the general theory is comprehended, numberless difficulties remain to be solved in its application; for the sovereignty of the Union is so involved in that of the States that it is impossible to distinguish its boundaries at the first glance. The whole structure of the Government is artificial and conventional; and it would be ill adapted to a people which has not been long accustomed to conduct its own affairs, or to one in which the science of politics has not descended to the humblest classes of society. I have never been more struck by the good sense and the practical judgment of the Americans than in the ingenious devices by which they elude the numberless difficulties resulting from their Federal Constitution. I scarcely ever met with a plain American citizen who could not distinguish, with surprising facility, the obligations created by the laws of Congress from those created by the laws of his own State; and who, after having discriminated between the matters which come under the cognizance of the Union and those which the local legislature is competent to regulate, could not point out the exact limit of the several jurisdictions of the Federal Courts and the tribunals of the State.

The Constitution of the United States is like those exquisite productions of human industry which ensure wealth and renown to their inventors, but which are profitless in any other hands. This truth is exemplified by the condition of Mexico at the present time. The Mexicans were desirous of establishing a Federal system, and they took the Federal Constitution of their neighbours, the Anglo-Americans, as their model, and copied it with considerable accuracy.[44] But although they had borrowed the letter of the law, they were unable to create or to introduce the spirit and the sense which give it life. They were involved in ceaseless embarrassments between the mechanism of their double Government; the sovereignty of the States and that of the Union perpetually exceeded their respective privileges, and entered into collision; and to the present day Mexico is alternately the victim of anarchy and the slave of military despotism.

The second and the most fatal of all the defects I have alluded to, and that which I believe to be inherent in the Federal system, is the relative weakness of the Government of the Union. The principle upon which all confederations rest is that of a divided sovereignty. The legislator may render this partition less perceptible, he may even conceal it for a time from the public eye, but he cannot prevent it from existing, and a divided sovereignty must always be less powerful than an entire supremacy. The reader has seen in the remarks I have made on the Constitution of the United States that the Americans have displayed singular ingenuity in combining the restriction of the power of the Union within the narrow limits of a Federal Government with the semblance and, to a certain extent, with the force of a national Government. By this means the legislators of the Union have succeeded in diminishing, though not in counteracting the natural danger of confederations.

It has been remarked that the American Government does not apply itself to the States, but that it immediately transmits its injunctions to the citizens, and compels them as isolated individuals to comply with its demands. But if the Federal law were to clash with the interests and the prejudices of a State, it might be feared that all the citizens of that State would conceive themselves to be interested in the cause of a single individual who should refuse to obey. If all the citizens of the State were aggrieved at the same time and in the same manner by the authority of the Union, the Federal Government would vainly attempt to subdue them individually; they would instinctively unite in a common defence, and they would derive a ready-prepared organization from the share of sovereignty which the institution of their State allows them to enjoy. Fiction would give way to reality, and an organized portion of the territory might then contest the central authority.[45] The same observation holds good with regard to the Federal jurisdiction. If the courts of the Union violated an important law of a State in a private case, the real, if not the apparent, contest would arise between the aggrieved State represented by a citizen and the Union represented by its courts of justice.[46]

He would have but a partial knowledge of the world who should imagine that it is possible, by the aid of legal fictions, to prevent men from finding out and employing those means of gratifying their passions which have been left open to them; and it may be doubted whether the American legislators, when they rendered a collision between the two sovereigns less probable, destroyed the cause of such a misfortune.

But it may even be affirmed that they were unable to ensure the preponderance of the Federal element in a case of this kind. The Union is possessed of money and of troops, but the affections and the prejudices of the people are in the bosom of the States. The sovereignty of the Union is an abstract being, which is connected with but few external objects; the sovereignty of the States is hourly perceptible, easily understood, constantly active; and if the former is of recent creation, the latter is coeval with the people itself. The sovereignty of the Union is factitious, that of the States is natural, and derives its existence from its own simple influence, like the authority of a parent. The supreme power of the nation only affects a few of the chief interests of society; it represents an immense but remote country, and claims a feeling of patriotism which is vague and ill defined; but the authority of the States controls every individual citizen at every hour and in all circumstances; it protects his property, his freedom, and his life; and when we recollect the traditions, the customs, the prejudices of local and familiar attachment with which it is connected, we cannot doubt of the superiority of a power which is interwoven with every circumstance that renders the love of one's native country instinctive in the human heart.

Since legislators are unable to obviate such dangerous collisions as occur between the two sovereignties which coexist in the Federal system, their first object must be, not only to dissuade the confederate States from warfare, but to encourage such institutions as may promote the maintenance of peace. Hence it results that the Federal compact cannot be lasting unless there exists in the communities which are leagued together a certain number of inducements to union which render their common dependence agreeable, and the task of the Government light, and that system cannot succeed without the presence of favourable circumstances added to the influence of good laws. All the peoples which have ever formed a confederation have been held together by a certain number of common interests, which served as the intellectual ties of association.

But the sentiments and the principles of man must be taken into consideration as well as his immediate interests. A certain uniformity of civilization is not less necessary to the durability of a confederation than a uniformity of interests in the States which compose it. In Switzerland the difference which exists between the Canton of Uri and the Canton of Vaud is equal to that between the fifteenth and the nineteenth centuries; and, properly speaking, Switzerland has never possessed a Federal Government. The union between these two cantons only subsists upon the map, and their discrepancies would soon be perceived if an attempt were made by a central authority to prescribe the same laws to the whole territory.

One of the circumstances which most powerfully contribute to support the Federal Government in America is that the States have not only similar interests, a common origin, and a common tongue, but that they are also arrived at the same stage of civilization; which almost always renders a union feasible. I do not know of any European nation, how small soever it may be, which does not present less uniformity in its different provinces than the American people, which occupies a territory as extensive as one-half of Europe. The distance from the State of Maine to that of Georgia is reckoned at about one thousand miles; but the difference between the civilization of Maine and that of Georgia is slighter than the difference between the habits of Normandy and those of Brittany. Maine and Georgia, which are placed at the opposite extremities of a great empire, are consequently in the natural possession of more real inducements to form a confederation than Normandy and Brittany, which are only separated by a bridge.

The geographical position of the country contributed to increase the facilities which the American legislators derived from the manners and customs of the inhabitants; and it is to this circumstance that the adoption and the maintenance of the Federal system are mainly attributable.

The most important occurrence which can mark the annals of a people is the breaking out of a war. In war a people struggles with the energy of a single man against foreign nations in the defence of its very existence. The skill of a government, the good sense of the community, and the natural fondness which men entertain for their country, may suffice to maintain peace in the interior of a district, and to favour its internal prosperity; but a nation can only carry on a great war at the cost of more numerous and more painful sacrifices; and to suppose that a great number of men will of their own accord comply with these exigencies of the State is to betray an ignorance of mankind. All the peoples which have been obliged to sustain a long and serious warfare have consequently been led to augment the power of their Government. Those which have not succeeded in this attempt have been subjugated. A long war almost always places nations in the wretched alternative of being abandoned to ruin by defeat or to despotism by success. War therefore renders the symptoms of the weakness of a Government most palpable and most alarming; and I have shown that the inherent defeat of Federal Governments is that of being weak.

The Federal system is not only deficient in every kind of centralized administration, but the central government itself is imperfectly organized, which is invariably an influential cause of inferiority when the nation is opposed to other countries which are themselves governed by a single authority. In the Federal Constitution of the United States, by which the central Government possesses more real force, this evil is still extremely sensible. An example will illustrate the case to the reader.

The Constitution confers upon Congress the right of 'calling forth militia to execute the laws of the Union, suppress insurrections, and repel invasions;' and another article declares that the President of the United States is the commander-in-chief of the militia. In the war of 1812 the President ordered the militia of the Northern States to march to the frontiers; but Connecticut and Massachusetts, whose interests were impaired by the war, refused to obey the command. They argued that the Constitution authorizes the Federal Government to call forth the militia in case of insurrection or invasion, but that in the present instance there was neither invasion nor insurrection. They added, that the same Constitution which conferred upon the Union the right of calling forth the militia reserved to the States that of naming the officers; and that consequently (as they understood the clause) no officer of the Union had any right to command the militia, even during war, except the President in person; and in this case they were ordered to join an army commanded by another individual. These absurd and pernicious doctrines received the sanction not only of the Governors and the Legislative bodies, but also of the courts of justice in both States; and the Federal Government was constrained to raise elsewhere the troops which it required.[47]

The only safeguard which the American Union, with all the relative perfection of its laws, possesses against the dissolution which would be produced by a great war, lies in its probable exemption from that calamity. Placed in the centre of an immense continent, which offers a boundless field for human industry, the Union is almost as much insulated from the world as if its frontiers were girt by the ocean. Canada contains only a million of inhabitants, and its population is divided into two inimical nations. The rigour of the climate limits the extension of its territory, and shuts up its ports during the six months of winter. From Canada to the Gulf of Mexico a few savage tribes are to be met with, which retire, perishing in their retreat, before six thousand soldiers. To the South, the Union has a point of contact with the empire of Mexico; and it is thence that serious hostilities may one day be expected to arise. But for a long while to come the uncivilized state of the Mexican community, the depravity of its morals, and its extreme poverty, will prevent that country from ranking high amongst nations.[48] As for the Powers of Europe, they are too distant to be formidable.[49]

The great advantage of the United States does not, then, consist in a Federal Constitution which allows them to carry on great wars, but in a geographical position which renders such enterprises extremely improbable.

No one can be more inclined than I am myself to appreciate the advantages of the Federal system, which I hold to be one of the combinations most favourable to the prosperity and freedom of man. I envy the lot of those nations which have been enabled to adopt it; but I cannot believe that any confederate peoples could maintain a long or an equal contest with a nation of similar strength in which the Government should be centralized. A people which should divide its sovereignty into fractional powers, in the presence of the great military monarchies of Europe, would, in my opinion, by that very act, abdicate its power, and perhaps its existence and its name. But such is the admirable position of the New World that man has no other enemy than himself; and that, in order to be happy and to be free, it suffices to seek the gifts of prosperity and the knowledge of freedom.

Chapter IX

Why the People May Strictly Be Said to Govern in the United States

I have hitherto examined the institutions of the United States; I have passed their legislation in review, and I have depicted the present characteristics of political society in that country. But a sovereign power exists above these institutions and beyond these characteristic features which may destroy or modify them at its pleasure—I mean that of the people. It remains to be shown in what manner this power, which regulates the laws, acts: its propensities and its passions remain to be pointed out, as well as the secret springs which retard, accelerate, or direct its irresistible course; and the effects of its unbounded authority, with the destiny which is probably reserved for it.

Why the People May Strictly Be Said to Govern in the United States

In America the people appoints the legislative and the executive power, and furnishes the jurors who punish all offences against the laws. The American institutions are democratic, not only in their principle but in all their consequences; and the people elects its representatives *directly,* and for the most part *annually,* in order to ensure their dependence. The people is therefore the real directing power; and although the form of government is representative, it is evident that the opinions, the prejudices, the interests, and even the passions of the community are hindered by no durable obstacles from exercising a perpetual influence on society. In the United States the majority governs in the name of the people, as is the case in all the countries in which the people is supreme. The majority is principally composed of peaceable citizens who, either by inclination or by interest, are sincerely desirous of the welfare of their country. But they are surrounded by the incessant agitation of parties, which attempt to gain their cooperation and to avail themselves of their support.

Chapter X

Parties in the United States

Great distinction to be made between parties—Parties which are to each other as rival nations—Parties properly so called—Difference between great and small parties—Epochs which produce them—Their characteristics—America has had great parties—They are extinct—Federalists Republicans—Defeat of the Federalists—Difficulty of creating parties in the United States—What is done with this intention—Aristocratic or democratic character to be met with in all parties—Struggle of General Jackson against the Bank.

A great distinction must be made between parties. Some countries are so large that the different populations which inhabit them have contradictory interests, although they are the subjects of the same Government, and they may thence be in a perpetual state of opposition. In this case the different fractions of the people may more properly be considered as distinct nations than as mere parties; and if a civil war breaks out, the struggle is carried on by rival peoples rather than by factions in the State.

But when the citizens entertain different opinions upon subjects which affect the whole country alike, such, for instance, as the principles upon which the government is to be conducted, then distinctions arise which may correctly be styled parties. Parties are a necessary evil in free governments; but they have not at all times the same character and the same propensities.

At certain periods a nation may be oppressed by such insupportable evils as to conceive the design of effecting a total change in its political constitution; at other times the mischief lies still deeper, and the existence of society itself is endangered. Such are the times of great revolutions and of great parties. But between these epochs of misery and of confusion there are periods during which human society seems to rest, and mankind to make a pause. This pause is, indeed, only apparent, for time does not stop its course for nations any more than for men; they are all advancing towards a goal with which they are unacquainted; and we only imagine them to be stationary when their progress escapes our observation, as men who are going at a foot-pace seem to be standing still to those who run.

But however this may be, there are certain epochs at which the changes that take place in the social and political constitution of nations are so slow and so insensible that men imagine their present condition to be a final state; and the human mind, believing itself to be firmly based upon certain foundations, does not extend its researches beyond the horizon which it descries. These are the times of small parties and of intrigue.

The political parties which I style great are those which cling to principles more than to their consequences; to general, and not to especial cases; to ideas, and not to men. These parties are usually distinguished by a nobler character, by more generous passions, more genuine convictions, and a more bold and open conduct than the others. In them private interest, which always plays the chief part in political passions, is more studiously veiled under the pretext of the public good; and it may even be sometimes concealed from the eyes of the very persons whom it excites and impels.

Minor parties are, on the other hand, generally deficient in political faith. As they are not sustained or dignified by a lofty purpose, they ostensibly display the egotism of their character in their actions. They glow with a factitious zeal; their language is vehement, but their conduct is timid and irresolute. The means they employ are as wretched as the end at which they aim. Hence it arises that when a calm state of things succeeds a violent revolution, the leaders of society seem suddenly to disappear, and the powers of the human mind to lie concealed. Society is convulsed by great parties, by minor ones it is agitated; it is torn by the former, by the latter it is degraded; and if these sometimes save it by a salutary perturbation, those invariably disturb it to no good end.

America has already lost the great parties which once divided the nation; and if her happiness is considerably increased, her morality has suffered by their extinction. When the War of Independence was terminated, and the foundations of the new Government were to be laid down, the nation was divided between two opinions—two opinions which are as old as the world, and which are perpetually to be met with under all the forms and all the names which have ever obtained in free communities—the one tending to limit, the other to extend indefinitely, the power of the people. The conflict of these two opinions never assumed that degree of violence in America which it has frequently displayed elsewhere. Both parties of the Americans were, in fact, agreed upon the most essential points; and neither of them had to destroy a traditionary constitution, or to overthrow the structure of society, in order to ensure its own triumph. In neither of them, consequently, were a great number of private interests affected by success or by defeat; but moral principles of a high order, such as the love of equality and of independence, were concerned in the struggle, and they sufficed to kindle violent passions.

The party which desired to limit the power of the people endeavoured to apply its doctrines more especially to the Constitution of the Union, whence it derived its name of *Federal.* The other party, which affected to be more exclusively attached to the cause of liberty, took that of *Republican.* America is a land of democracy, and the Federalists were always in a minority; but they reckoned on their side almost all the great men who had been called forth by the War of Independence, and their moral influence was very considerable. Their cause was, moreover, favoured by circumstances. The ruin of the Confederation had impressed the people with a dread of anarchy, and the Federalists did not fail to profit by this transient disposition of the multitude. For ten or twelve years they were at the head of affairs, and they were able to apply some, though not all, of their principles; for the hostile current was becoming from day to day too violent to be checked or stemmed. In 1801 the Republicans got possession of the Government; Thomas Jefferson was named President; and he increased the influence of their party by the weight of his celebrity, the greatness of his talents, and the immense extent of his popularity.

The means by which the Federalists had maintained their position were artificial, and their resources were temporary; it was by the virtues or the talents of their leaders that they had risen to power. When the Republicans attained to that lofty station, their opponents were overwhelmed by utter defeat. An immense majority declared itself against the retiring party, and the Federalists found themselves in so small a minority that they at once despaired of their future success. From that moment the Republican or Democratic party[1] has proceeded from conquest to conquest, until it has acquired absolute supremacy in the country. The Federalists, perceiving that they were vanquished without resource, and isolated in the midst of the nation, fell into two divisions, of which one joined the victorious Republicans, and the other abandoned its rallying-point and its name. Many years have already elapsed since they ceased to exist as a party.

The accession of the Federalists to power was, in my opinion, one of the most fortunate incidents which accompanied the formation of the great American Union; they resisted the inevitable propensities of their age and of the country. But whether their theories were good or bad, they had the effect of being inapplicable, as a system, to the society which they professed to govern, and that which occurred under the auspices of Jefferson must therefore have taken place sooner or later. But their Government gave the new republic time to acquire a certain stability, and afterwards to support the rapid growth of the very doctrines which they had combated. A considerable number of their principles were in point of fact embodied in the political creed of their opponents; and the Federal Constitution which subsists at the present day is a lasting monument of their patriotism and their wisdom.

Great political parties are not, then, to be met with in the United States at the present time. Parties, indeed, may be found which threaten the future tranquillity of the Union; but there are none which seem to contest the present form of Government or the present course of society. The parties by which the Union is menaced do not rest upon abstract principles, but upon temporal interests. These interests, disseminated in the provinces of so vast an empire, may be said to constitute rival nations rather than parties. Thus, upon a recent occasion, the North contended for the system of commercial prohibition, and the South took up arms in favour of free trade, simply because the North is a manufacturing and the South an agricultural district; and that the restrictive system which was profitable to the one was prejudicial to the other.[2]

In the absence of great parties, the United States abound with lesser controversies; and public opinion is divided into a thousand minute shades of difference upon questions of very little moment. The pains which are taken to create parties are inconceivable, and at the present day it is no easy task. In the United States there is no religious animosity, because all religion is respected, and no sect is predominant; there is no jealousy of rank, because the people is everything, and none can contest its authority; lastly, there is no public indigence to supply the means of agitation, because the physical position of the country opens so wide a field to industry that man is able to accomplish the most surprising undertakings with his own native resources. Nevertheless, ambitious men are interested in the creation of parties, since it is difficult to eject a person from authority upon the mere ground that his place is coveted by others. The skill of the actors in the political world lies therefore in the art of creating parties. A political aspirant in the United States begins by discriminating his own interest, and by calculating upon those interests which may be collected around and amalgamated with it; he then contrives to discover some doctrine or some principle which may suit the purposes of this new association, and which he adopts in order to bring forward his party and to secure his popularity; just as the *imprimatur* of a King was in former days incorporated with the volume which it authorised, but to which it nowise belonged. When these preliminaries are terminated, the new party is ushered into the political world.

All the domestic controversies of the Americans at first appear to a stranger to be so incomprehensible and so puerile that he is at a loss whether to pity a people which takes such arrant trifles in good earnest, or to envy the happiness which enables it to discuss them. But when he comes to study the secret propensities which govern the factions of America, he easily perceives that the greater part of them are more or less connected with one or the other of those two divisions which have always existed in free communities. The deeper we penetrate into the working of these parties, the more do we perceive that the object of the one is to limit, and that of the other to extend, the popular authority. I do not assert that the ostensible end, or even that the secret aim, of American parties is to promote the rule of aristocracy or democracy in the country; but I affirm that aristocratic or democratic passions may easily be detected at the bottom of all parties, and that, although they escape a superficial observation, they are the main point and the very soul of every faction in the United States.

To quote a recent example. When the President attacked the Bank, the country was excited and parties were formed; the well-informed classes rallied round the Bank, the common people round the President. But it must not be imagined that the people had formed a rational opinion upon a question which offers so many difficulties to the most experienced statesmen. The Bank is a great establishment which enjoys an independent existence, and the people, accustomed to make and unmake whatsoever it pleases, is startled to meet with this obstacle to its authority. In the midst of the perpetual fluctuation of society the community is irritated by so permanent an institution, and is led to attack it in order to see whether it can be shaken and controlled, like all the other institutions of the country.

Remains of the Aristocratic Party in the United States

Secret opposition of wealthy individuals to democracy—
Their retirement—Their taste for exclusive pleasures and for luxury at home—Their simplicity abroad—Their affected condescension towards the people.

It sometimes happens in a people amongst which various opinions prevail that the balance of the several parties is lost, and one of them obtains an irresistible preponderance, overpowers all obstacles, harasses its opponents, and appropriates all the resources of society to its own purposes. The vanquished citizens despair of success and they conceal their dissatisfaction in silence and in general apathy. The nation seems to be governed by a single principle, and the prevailing party assumes the credit of having restored peace and unanimity to the country. But this apparent unanimity is merely a cloak to alarming dissensions and perpetual opposition.

This is precisely what occurred in America; when the democratic party got the upper hand, it took exclusive possession of the conduct of affairs, and from, that time the laws and the customs of society have been adapted to its caprices. At the present day the more affluent classes of society are so entirely removed from the direction of political affairs in the United States that wealth, far from conferring a right to the exercise of power, is rather an obstacle than a means of attaining to it. The wealthy members of the community abandon the lists, through unwillingness to contend, and frequently to contend in vain, against the poorest classes of their fellow citizens. They concentrate all their enjoyments in the privacy of their homes, where they occupy a rank which cannot be assumed in public; and they constitute a private society in the State, which has its own tastes and its own pleasures. They submit to this state of things as an irremediable evil, but they are careful not to show that they are galled by its continuance; it is even not uncommon to hear them laud the delights of a republican government, and the advantages of democratic institutions when they are in public. Next to hating their enemies, men are most inclined to flatter them.

Mark, for instance, that opulent citizen, who is as anxious as a Jew of the Middle Ages to conceal his wealth. His dress is plain, his demeanour unassuming; but the interior of his dwelling glitters with luxury, and none but a few chosen guests whom he haughtily styles his equals are allowed to penetrate into this sanctuary. No European noble is more exclusive in his pleasures, or more jealous of the smallest advantages which his privileged station confers upon him. But the very same individual crosses the city to reach a dark counting-house in the centre of traffic, where every one may accost him who pleases. If he meets his cobbler upon the way, they stop and converse; the two citizens discuss the affairs of the State in which they have an equal interest, and they shake hands before they part.

But beneath this artificial enthusiasm, and these obsequious attentions to the preponderating power, it is easy to perceive that the wealthy members of the community entertain a hearty distaste to the democratic institutions of their country. The populace is at once the object of their scorn and of their fears. If the maladministration of the democracy ever brings about a revolutionary crisis, and if monarchical institutions ever become practicable in the United States, the truth of what I advance will become obvious.

The two chief weapons which parties use in order to ensure success are the *public press* and the formation of *associations.*

Chapter XI

Liberty of the Press in the United States

Difficulty of restraining the liberty of the press—Particular reasons which some nations have to cherish this liberty—The liberty of the press a necessary consequence of the sovereignty of the people as it is understood in America—Violent language of the periodical press in the United States—Propensities of the periodical press—Illustrated by the United States—Opinion of the Americans upon the repression of the abuse of the liberty of the press by judicial prosecutions—Reasons for which the press is less powerful in America than in France.

The influence of the liberty of the press does not affect political opinions alone, but it extends to all the opinions of men, and it modifies customs as well as laws. In another part of this work I shall attempt to determine the degree of influence which the liberty of the press has exercised upon civil society in the United States, and to point out the direction which it has given to the ideas, as well as the tone which it has imparted to the character and the feelings, of the Anglo-Americans, but at present I purpose simply to examine the effects produced by the liberty of the press in the political world.

I confess that I do not entertain that firm and complete attachment to the liberty of the press which things that are supremely good in their very nature are wont to excite in the mind; and I approve of it more from a recollection of the evils it prevents than from a consideration of the advantages it ensures.

If any one could point out an intermediate and yet a tenable position between the complete independence and the entire subjection of the public expression of opinion, I should perhaps be inclined to adopt it; but the difficulty is to discover this position. If it is your intention to correct the abuses of unlicensed printing and to restore the use of orderly language, you may in the first instance try the offender by a jury; but if the jury acquits him, the opinion which was that of a single individual becomes the opinion of the country at large. Too much and too little has therefore hitherto been done. If you proceed, you must bring the delinquent before a court of permanent judges. But even here the cause must be heard before it can be decided; and the very principles which no book would have ventured to avow are blazoned forth in the pleadings, and what was obscurely hinted at in a single composition is then repeated in a multitude of other publications. The language in which a thought is embodied is the mere carcass of the thought, and not the idea itself; tribunals may condemn the form, but the sense and spirit of the work is too subtle for their authority. Too much has still been done to recede, too little to attain your end; you must therefore proceed. If you establish a censorship of the press, the tongue of the public speaker will still make itself heard, and you have only increased the mischief. The powers of thought do not rely, like the powers of physical strength, upon the number of their mechanical agents, nor can a host of authors be reckoned like the troops which compose an army; on the contrary, the authority of a principle is often increased by the smallness of the number of men by whom it is expressed. The words of a strongminded man, which penetrate amidst the passions of a listening assembly, have more power than the vociferations of a thousand orators; and if it be allowed to speak freely in any public place, the consequence is the same as if free speaking was allowed in every village. The liberty of discourse must therefore be destroyed as well as the liberty of the press; this is the necessary term of your efforts; but if your object was to repress the abuses of liberty, they have brought you to the feet of a despot. You have been led from the extreme of independence to the extreme of subjection without meeting with a single tenable position for shelter or repose.

There are certain nations which have peculiar reasons for cherishing the liberty of the press, independently of the general motives which I have just pointed out. For in certain countries which profess to enjoy the privileges of freedom every individual agent of the Government may violate the laws with impunity, since those whom he oppresses cannot prosecute him before the courts of justice. In this case the liberty of the press is not merely a guarantee, but it is the only guarantee, of their liberty and their security which the citizens possess. If the rulers of these nations propose to abolish the independence of the press, the people would be justified in saying: Give us the right of prosecuting your offences before the ordinary tribunals, and perhaps we may then waive our right of appeal to the tribunal of public opinion.

But in the countries in which the doctrine of the sovereignty of the people ostensibly prevails, the censorship of the press is not only dangerous, but it is absurd. When the right of every citizen to cooperate in the government of society is acknowledged, every citizen must be presumed to possess the power of discriminating between the different opinions of his contemporaries, and of appreciating the different facts from which inferences may be drawn. The sovereignty of the people and the liberty of the press may therefore be looked upon as correlative institutions; just as the censorship of the press and universal suffrage are two things which are irreconcilably opposed, and which cannot long be retained among the institutions of the same people. Not a single individual of the twelve millions who inhabit the territory of the United States has as yet dared to propose any restrictions to the liberty of the press. The first newspaper over which I cast my eyes, upon my arrival in America, contained the following article:

> In all this affair the language of Jackson has been that of a heartless despot, solely occupied with the preservation of his own authority. Ambition is his crime, and it will be his punishment too: intrigue is his native element, and intrigue will confound his tricks, and will deprive him of his power: he governs by means of corruption, and his immoral practices will redound to his shame and confusion. His conduct in the political arena has been that of a shameless and lawless gamester. He succeeded at the time, but the hour of retribution approaches, and he will be obliged to disgorge his winnings, to throw aside his false dice, and to end his days in some retirement, where he may curse his madness at his leisure; for repentance is a virtue with which his heart is likely to remain forever unacquainted.

It is not uncommonly imagined in France that the virulence of the press originates in the uncertain social condition, in the political excitement, and the general sense of consequent evil which prevail in that country; and it is therefore supposed that as soon as society has resumed a certain degree of composure the press will abandon its present vehemence. I am inclined to think that the above causes explain the reason of the extraordinary ascendency it has acquired over the nation, but that they do not exercise much influence upon the tone of its language. The periodical press appears to me to be actuated by passions and propensities independent of the circumstances in which it is placed, and the present position of America corroborates this opinion.

America is perhaps, at this moment, the country of the whole world which contains the fewest germs of revolution; but the press is not less destructive in its principles than in France, and it displays the same violence without the same reasons for indignation. In America, as in France, it constitutes a singular power, so strangely composed of mingled good and evil that it is at the same time indispensable to the existence of freedom, and nearly incompatible with the maintenance of public order. Its power is certainly much greater in France than in the United States; though nothing is more rare in the latter country than to hear of a prosecution having been instituted against it. The reason of this is perfectly simple: the Americans, having once admitted the doctrine of the sovereignty of the people, apply it with perfect consistency. It was never their intention to found a permanent state of things with elements which undergo daily modifications; and there is consequently nothing criminal in an attack upon the existing laws, provided it be not attended with a violent infraction of them. They are moreover of opinion that courts of justice are unable to check the abuses of the press; and that as the subtilty of human language perpetually eludes the severity of judicial analysis, offences of this nature are apt to escape the hand which attempts to apprehend them. They hold that to act with efficacy upon the press it would be necessary to find a tribunal, not only devoted to the existing order of things, but capable of surmounting the influence of public opinion; a tribunal which should conduct its proceedings without publicity, which should pronounce its decrees without assigning its motives, and punish the intentions even more than the language of an author. Whosoever should have the power of creating and maintaining a tribunal of this kind would waste his time in prosecuting the liberty of the press; for he would be the supreme master of the whole community, and he would be as free to rid himself of the authors as of their writings. In this question, therefore, there is no medium between servitude and extreme license; in order to enjoy the inestimable benefits which the liberty of the press ensures, it is necessary to submit to the inevitable evils which it engenders. To expect to acquire the former and to escape the latter is to cherish one of those illusions which commonly mislead nations in their times of sickness, when, tired with faction and exhausted by effort, they attempt to combine hostile opinions and contrary principles upon the same soil.

The small influence of the American journals is attributable to several reasons, amongst which are the following:

The liberty of writing, like all other liberty, is most formidable when it is a novelty; for a people which has never been accustomed to cooperate in the conduct of State affairs places implicit confidence in the first tribune who arouses its attention. The

Anglo-Americans have enjoyed this liberty ever since the foundation of the settlements; moreover, the press cannot create human passions by its own power, however skilfully it may kindle them where they exist. In America politics are discussed with animation and a varied activity, but they rarely touch those deep passions which are excited whenever the positive interest of a part of the community is impaired: but in the United States the interests of the community are in a most prosperous condition. A single glance upon a French and an American newspaper is sufficient to show the difference which exists between the two nations on this head. In France the space allotted to commercial advertisements is very limited, and the intelligence is not considerable, but the most essential part of the journal is that which contains the discussion of the politics of the day. In America three-quarters of the enormous sheet which is set before the reader are filled with advertisements, and the remainder is frequently occupied by political intelligence or trivial anecdotes: it is only from time to time that one finds a corner devoted to passionate discussions like those with which the journalists of France are wont to indulge their readers.

It has been demonstrated by observation, and discovered by the innate sagacity of the pettiest as well as the greatest of despots, that the influence of a power is increased in proportion as its direction is rendered more central. In France the press combines a twofold centralization; almost all its power is centred in the same spot, and vested in the same hands, for its organs are far from numerous. The influence of a public press thus constituted, upon a sceptical nation, must be unbounded. It is an enemy with which a Government may sign an occasional truce, but which it is difficult to resist for any length of time.

Neither of these kinds of centralization exists in America. The United States have no metropolis; the intelligence as well as the power of the country are dispersed abroad, and instead of radiating from a point, they cross each other in every direction; the Americans have established no central control over the expression of opinion, any more than over the conduct of business. These are circumstances which do not depend on human foresight; but it is owing to the laws of the Union that there are no licenses to be granted to printers, no securities demanded from editors as in France, and no stamp duty as in France and formerly in England. The consequence of this is that nothing is easier than to set up a newspaper, and a small number of readers suffices to defray the expenses of the editor.

The number of periodical and occasional publications which appears in the United States actually surpasses belief. The most enlightened Americans attribute the subordinate influence of the press to this excessive dissemination; and it is adopted as an axiom of political science in that country that the only way to neutralize the effect of public journals is to multiply them indefinitely. I cannot conceive that a truth which is so self-evident should not already have been more generally admitted in Europe; it is comprehensible that the persons who hope to bring about revolutions by means of the press should be desirous of confining its action to a few powerful organs, but it is perfectly incredible that the partisans of the existing state of things, and the natural supporters of the law, should attempt to diminish the influence of the press by concentrating its authority. The Governments of Europe seem to treat the press with the courtesy of the knights of old; they are anxious to furnish it with the same central power which they have found to be so trusty a weapon, in order to enhance the glory of their resistance to its attacks.

In America there is scarcely a hamlet which has not its own newspaper. It may readily be imagined that neither discipline nor unity of design can be communicated to so multifarious a host, and each one is consequently led to fight under his own standard. All the political journals of the United States are indeed arrayed on the side of the administration or against it; but they attack and defend in a thousand different ways. They cannot succeed in forming those great currents of opinion which overwhelm the most solid obstacles. This division of the influence of the press produces a variety of other consequences which are scarcely less remarkable. The facility with which journals can be established induces a multitude of individuals to take a part in them; but as the extent of competition precludes the possibility of considerable profit, the most distinguished classes of society are rarely led to engage in these undertakings. But such is the number of the public prints that, even if they were a source of wealth, writers of ability could not be found to direct them all. The journalists of the United States are usually placed in a very humble position, with a scanty education and a vulgar turn of mind. The will of the majority is the most general of laws, and it establishes certain habits which form the characteristics of each peculiar class of society; thus it dictates the etiquette practised at courts and the etiquette of the bar. The characteristics of the French journalist consist in a violent, but frequently an eloquent and lofty, manner of discussing the politics of the day; and the exceptions to this habitual practice are only occasional. The characteristics of the American journalist consist in an open and coarse appeal to the passions of the populace; and he habitually abandons the principles of political science to assail the characters of individuals, to track them into private life, and disclose all their weaknesses and errors.

Nothing can be more deplorable than this abuse of the powers of thought; I shall have occasion to point out hereafter the influence of the newspapers upon the taste and the morality of the American people, but my present subject exclusively concerns the political world. It cannot be denied that the effects of this extreme license of the press tend indirectly to the maintenance of public order. The individuals who are already in the possession of a high station in the esteem of their fellow-citizens are afraid to write in the newspapers, and they are thus deprived of the most powerful instrument which they can use to excite the passions of the multitude to their own advantage.[1]

The personal opinions of the editors have no kind of weight in the eyes of the public: the only use of a journal is, that it imparts the knowledge of certain facts, and it is only by altering or distorting those facts that a journalist can contribute to the support of his own views.

But although the press is limited to these resources, its influence in America is immense. It is the power which impels the circulation of political life through all the districts of that vast territory. Its eye is constantly open to detect the secret springs of political designs, and to summon the leaders of all parties to the bar of public opinion. It rallies the interests of the community round certain principles, and it draws up the creed which factions adopt; for it affords a means of intercourse between parties which hear, and which address each other without ever having been in immediate contact. When a great number of the organs of the press adopt the same line of conduct, their influence becomes irresistible; and public opinion, when it is perpetually assailed from the same side, eventually yields to the attack. In the United States each separate journal exercises but little authority, but the power of the periodical press is only second to that of the people.[2]

The opinions established in the United States under the empire of the liberty of the press are frequently more firmly rooted than those which are formed elsewhere under the sanction of a censor.

In the United States the democracy perpetually raises fresh individuals to the conduct of public affairs; and the measures of the administration are consequently seldom regulated by the strict rules of consistency or of order. But the general principles of the Government are more stable, and the opinions most prevalent in society are generally more durable than in many other countries. When once the Americans have taken up an idea, whether it be well or ill-founded, nothing is more difficult than to eradicate it from their minds. The same tenacity of opinion has been observed in England, where, for the last century, greater freedom of conscience and more invincible prejudices have existed than in all the other countries of Europe. I attribute this consequence to a cause which may at first sight appear to have a very opposite tendency, namely, to the liberty of the press. The nations amongst which this liberty exists are as apt to cling to their opinions from pride as from conviction. They cherish them because they hold them to be just, and because they exercised their own free-will in choosing them; and they maintain them not only because they are true, but because they are their own. Several other reasons conduce to the same end.

It was remarked by a man of genius that 'ignorance lies at the two ends of knowledge.' Perhaps it would have been more correct to have said, that absolute convictions are to be met with at the two extremities, and that doubt lies in the middle; for the human intellect may be considered in three distinct states, which frequently succeed one another. A man believes implicitly, because he adopts a proposition without inquiry. He doubts as soon as he is assailed by the objections which his inquiries may have aroused. But he frequently succeeds in satisfying these doubts, and then he begins to believe afresh: he no longer lays hold on a truth in its most shadowy and uncertain form, but he sees it clearly before him, and he advances onwards by the light it gives him.[3]

When the liberty of the press acts upon men who are in the first of these three states, it does not immediately disturb their habit of believing implicitly without investigation, but it constantly modifies the objects of their intuitive convictions. The human mind continues to discern but one point upon the whole intellectual horizon, and that point is in continual motion. Such are the symptoms of sudden revolutions, and of the misfortunes which are sure to befall those generations which abruptly adopt the unconditional freedom of the press.

The circle of novel ideas is, however, soon terminated; the touch of experience is upon them, and the doubt and mistrust which their uncertainty produces become universal. We may rest assured that the majority of mankind will either believe they know not wherefore, or will not know what to believe. Few are the beings who can ever hope to attain to that state of rational and independent conviction which true knowledge can beget in defiance of the attacks of doubt.

It has been remarked that in times of great religious fervour men sometimes change their religious opinions; whereas in times of general scepticism everyone clings to his own persuasion. The same thing takes place in politics under the liberty of the press. In countries where all the theories of social science have been contested in their turn, the citizens who have adopted one of them stick to it, not so much because they are assured of its excellence, as because they are not convinced of the superiority of any other. In the present age men are not very ready to die in defence of their opinions, but they are rarely inclined to change them; and there are fewer martyrs as well as fewer apostates.

Another still more valid reason may yet be adduced: when no abstract opinions are looked upon as certain, men cling to the mere propensities and external interests of their position, which are naturally more tangible and more permanent than any opinions in the world.

It is not a question of easy solution whether aristocracy or democracy is most fit to govern a country. But it is certain that democracy annoys one part of the community, and that aristocracy oppresses another part. When the question is reduced to the simple expression of the struggle between poverty and wealth, the tendency of each side of the dispute becomes perfectly evident without further controversy.

Chapter XII

Political Associations in the United States

Daily use which the Anglo-Americans make of the right of association—Three kinds of political associations—In what manner the Americans apply the representative system to associations—Dangers resulting to the State—Great Convention of 1831 relative to the Tariff—Legislative character of this Convention—Why the unlimited exercise of the right of association is less dangerous in the United States than elsewhere—Why it may be looked upon as necessary—Utility of associations in a democratic people.

In no country in the world has the principle of association been more successfully used, or more unsparingly applied to a multitude of different objects, than in America. Besides the permanent associations which are established by law under the names of townships, cities, and counties, a vast number of others are formed and maintained by the agency of private individuals.

The citizen of the United States is taught from his earliest infancy to rely upon his own exertions in order to resist the evils and the difficulties of life; he looks upon social authority with an eye of mistrust and anxiety, and he only claims its assistance when he is quite unable to shift without it. This habit may even be traced in the schools of the rising generation, where the children in their games are wont to submit to rules which they have themselves established, and to punish misdemeanours which they have themselves defined. The same spirit pervades every act of social life. If a stoppage occurs in a thoroughfare, and the circulation of the public is hindered, the neighbours immediately constitute a deliberative body; and this extemporaneous assembly gives rise to an executive power which remedies the inconvenience before anybody has thought of recurring to an authority superior to that of the persons immediately concerned. If the public pleasures are concerned, an association is formed to provide for the splendour and the regularity of the entertainment. Societies are formed to resist enemies which are exclusively of a moral nature, and to diminish the vice of intemperance: in the United States associations are established to promote public order, commerce, industry, morality, and religion; for there is no end which the human will, seconded by the collective exertions of individuals, despairs of attaining.

I shall hereafter have occasion to show the effects of association upon the course of society, and I must confine myself for the present to the political world. When once the right of association is recognised, the citizens may employ it in several different ways.

An association consists simply in the public assent which a number of individuals give to certain doctrines, and in the engagement which they contract to promote the spread of those doctrines by their exertions. The right of association with these views is very analogous to the liberty of unlicensed writing; but societies thus formed possess more authority than the press. When an opinion is represented by a society, it necessarily assumes a more exact and explicit form. It numbers its partisans, and compromises their welfare in its cause: they, on the other hand, become acquainted with each other, and their zeal is increased by their number. An association unites the efforts of minds which have a tendency to diverge in one single channel, and urges them vigorously towards one single end which it points out.

The second degree in the right of association is the power of meeting. When an association is allowed to establish centres of action at certain important points in the country, its activity is increased and its influence extended. Men have the opportunity of seeing each other; means of execution are more readily combined, and opinions are maintained with a degree of warmth and energy which written language cannot approach.

Lastly, in the exercise of the right of political association, there is a third degree: the partisans of an opinion may unite in electoral bodies, and choose delegates to represent them in a central assembly. This is, properly speaking, the application of the representative system to a party.

Thus, in the first instance, a society is formed between individuals professing the same opinion, and the tie which keeps it together is of a purely intellectual nature; in the second case, small assemblies are formed which only represent a fraction of the party. Lastly, in the third case, they constitute a separate nation in the midst of the nation, a government within the Government. Their delegates, like the real delegates of the majority, represent the entire collective force of their party; and they enjoy a certain degree of that national dignity and great influence which belong to the chosen representatives of the people. It is true that they have not the right of making the laws, but they have the power of attacking those which are in being, and of drawing up beforehand those which they may afterwards cause to be adopted.

If, in a people which is imperfectly accustomed to the exercise of freedom, or which is exposed to violent political passions, a deliberating minority, which confines itself to the contemplation of future laws, be placed in juxtaposition to the legislative majority, I cannot but believe that public tranquillity incurs very great risks in that nation. There is doubtless a very wide difference between proving that one law is in itself better than another and proving that the former ought to be substituted for the latter. But the imagination of the populace is very apt to overlook this difference, which is so apparent to the minds of thinking men. It sometimes happens that a nation is divided into two nearly equal parties, each of which affects to represent the majority. If, in immediate contiguity to the directing power, another power be established, which exercises almost as much moral authority as the former, it is not to be believed that it will long be content to speak without acting; or that it will always be restrained by the abstract consideration of the nature of associations which are meant to direct but not to enforce opinions, to suggest but not to make the laws.

The more we consider the independence of the press in its principal consequences, the more are we convinced that it is the chief and, so to speak, the constitutive element of freedom in the modern world. A nation which is determined to remain free is therefore right in demanding the unrestrained exercise of this independence. But the *unrestrained* liberty of political association cannot be entirely assimilated to the liberty of the press. The one is at the same time less necessary and more dangerous than the other. A nation may confine it within certain limits without forfeiting any part of its self-control; and it may sometimes be obliged to do so in order to maintain its own authority.

In America the liberty of association for political purposes is unbounded. An example will show in the clearest light to what an extent this privilege is tolerated.

The question of the tariff, or of free trade, produced a great manifestation of party feeling in America; the tariff was not only a subject of debate as a matter of opinion, but it exercised a favourable or a prejudicial influence upon several very powerful interests of the States. The North attributed a great portion of its prosperity, and the South all its sufferings, to this system; insomuch that for a long time the tariff was the sole source of the political animosities which agitated the Union.

In 1831, when the dispute was raging with the utmost virulence, a private citizen of Massachusetts proposed to all the enemies of the tariff, by means of the public prints, to send delegates to Philadelphia in order to consult together upon the means which were most fitted to promote freedom of trade. This proposal circulated in a few days from Maine to New Orleans by the power of the printing-press: the opponents of the tariff adopted it with enthusiasm; meetings were formed on all sides, and delegates were named. The majority of these individuals were well known, and some of them had earned a considerable degree of celebrity. South Carolina alone, which afterwards took up arms in the same cause, sent sixty-three delegates. On the 1st October, 1831, this assembly, which according to the American custom had taken the name of a Convention, met at Philadelphia; it consisted of more than two hundred members. Its debates were public, and they at once assumed a legislative character; the extent of the powers of Congress, the theories of free trade, and the different clauses of the tariff, were discussed in turn. At the end often days' deliberation the Convention broke up, after having published an address to the American people, in which it declared:

I. That Congress had not the right of making a tariff, and that the existing tariff was unconstitutional;

II. That the prohibition of free trade was prejudicial to the interests of all nations, and to that of the American people in particular.

It must be acknowledged that the unrestrained liberty of political association has not hitherto produced, in the United States, those fatal consequences which might perhaps be expected from it elsewhere. The right of association was imported from England, and it has always existed in America; so that the exercise of this privilege is now amalgamated with the manners and customs of the people. At the present time the liberty of association is become a necessary guarantee against the tyranny of the majority. In the United States, as soon as a party is become preponderant, all public authority passes under its control; its private supporters occupy all the places, and have all the force of the administration at their disposal. As the most distinguished partisans of the other side of the question are unable to surmount the obstacles which exclude them from power, they require some means of establishing themselves upon their own basis, and of opposing the moral authority of the minority to the physical power which domineers over it. Thus a dangerous expedient is used to obviate a still more formidable danger.

The omnipotence of the majority appears to me to present such extreme perils to the American Republics that the dangerous measure which is used to repress it seems to be more advantageous than prejudicial. And here I am about to advance a proposition which may remind the reader of what I said before in speaking of municipal freedom: There are no countries in which associations are more needed, to prevent the despotism of faction or the arbitrary power of a prince, than those which are democratically constituted. In aristocratic nations the body of the nobles and the more opulent part of the community are in themselves natural associations, which act as checks upon the abuses of power. In countries in which these associations do not exist, if private individuals are unable to create an artificial and a temporary substitute for them, I can imagine no permanent protection against the most galling tyranny; and a great people may be oppressed by a small faction, or by a single individual, with impunity.

The meeting of a great political Convention (for there are Conventions of all kinds), which may frequently become a necessary measure, is always a serious occurrence, even in America, and one which is never looked forward to, by the judicious friends of the country, without alarm. This was very perceptible in the Convention of 1831, at which the exertions of all the most distinguished members of the Assembly tended to moderate its language, and to restrain the subjects which it treated within certain limits. It is probable, in fact, that the Convention of 1831 exercised a very great influence upon the minds of the malcontents, and prepared them for the open revolt against the commercial laws of the Union which took place in 1832.

It cannot be denied that the unrestrained liberty of association for political purposes is the privilege which a people is longest in learning how to exercise. If it does not throw the nation into anarchy, it perpetually augments the chances of that calamity. On one point, however, this perilous liberty offers a security against dangers of another kind; in countries where associations are free, secret societies are unknown. In America there are numerous factions, but no conspiracies.

Different ways in which the right of association is understood in Europe and in the United States—Different use which is made of it.

The most natural privilege of man, next to the right of acting for himself, is that of combining his exertions with those of his fellow-creatures, and of acting in common with them. I am therefore led to conclude that the right of association is almost as inalienable as the right of personal liberty. No legislator can attack it without impairing the very foundations of society. Nevertheless, if the liberty of association is a fruitful source of advantages and prosperity to some nations, it may be perverted or carried to excess by others, and the element of life may be changed into an element of destruction. A comparison of the different methods which associations pursue in those countries in which they are managed with discretion, as well as in those where liberty degenerates into license, may perhaps be thought useful both to governments and to parties.

The greater part of Europeans look upon an association as a weapon which is to be hastily fashioned, and immediately tried in the conflict. A society is formed for discussion, but the idea of impending action prevails in the minds of those who constitute it: it is, in fact, an army; and the time given to parley serves to reckon up the strength and to animate the courage of the host, after which they direct their march against the enemy. Resources which lie within the bounds of the law may suggest themselves to the persons who compose it as means, but never as the only means, of success.

Such, however, is not the manner in which the right of association is understood in the United States. In America the citizens who form the minority associate, in order, in the first place, to show their numerical strength, and so to diminish the moral authority of the majority; and, in the second place, to stimulate competition, and to discover those arguments which are most fitted to act upon the majority; for they always entertain hopes of drawing over their opponents to their own side, and of afterwards disposing of the supreme power in their name. Political associations in the United States are therefore peaceable in their intentions, and strictly legal in the means which they employ; and they assert with perfect truth that they only aim at success by lawful expedients.

The difference which exists between the Americans and ourselves depends on several causes. In Europe there are numerous parties so diametrically opposed to the majority that they can never hope to acquire its support, and at the same time they think that they are sufficiently strong in themselves to struggle and to defend their cause. When a party of this kind forms an association, its object is, not to conquer, but to fight. In America the individuals who hold opinions very much opposed to those of the majority are no sort of impediment to its power, and all other parties hope to win it over to their own principles in the end. The exercise of the right of association becomes dangerous in proportion to the impossibility which excludes great parties from acquiring the majority. In a country like the United States, in which the differences of opinion are mere differences of hue, the right of association may remain unrestrained without evil consequences. The inexperience of many of the European nations in the enjoyment of liberty leads them only to look upon the liberty of association as a right of attacking the Government. The first notion which presents itself to a party, as well as to an individual, when it has acquired a consciousness of its own strength, is that of violence: the notion of persuasion arises at a later period and is only derived from experience. The English, who are divided into parties which differ most essentially from each other, rarely abuse the right of association, because they have long been accustomed to exercise it. In France the passion for war is so intense that there is no undertaking so mad, or so injurious to the welfare of the State, that a man does not consider himself honoured in defending it, at the risk of his life.

But perhaps the most powerful of the causes which tend to mitigate the excesses of political association in the United States is Universal Suffrage. In countries in which universal suffrage exists the majority is never doubtful, because neither party can pretend to represent that portion of the community which has not voted. The associations which are formed are aware, as well as the nation at large, that they do not represent the majority: this is, indeed, a condition inseparable from their existence; for if they did represent the preponderating power, they would change the law instead of soliciting its reform. The consequence of this is that the moral influence of the Government which they attack is very much increased, and their own power is very much enfeebled.

In Europe there are few associations which do not affect to represent the majority, or which do not believe that they represent it. This conviction or this pretension tends to augment their force amazingly, and contributes no less to legalize their measures. Violence may seem to be excusable in defence of the cause of oppressed right. Thus it is, in the vast labyrinth of human laws, that extreme liberty sometimes corrects the abuses of license, and that extreme democracy obviates the dangers of democratic government. In Europe, associations consider themselves, in some degree, as the legislative and executive councils of the people, which is unable to speak for itself. In America, where they only represent a minority of the nation, they argue and they petition.

The means which the associations of Europe employ are in accordance with the end which they propose to obtain. As the principal aim of these bodies is to act, and not to debate, to fight rather than to persuade, they are naturally led to adopt a form of organization which differs from the ordinary customs of civil bodies, and which assumes the habits and the maxims of military life. They centralize the direction of their resources as much as possible, and they intrust the power of the whole party to a very small number of leaders.

The members of these associations respond to a watchword, like soldiers on duty; they profess the doctrine of passive obedience; say rather, that in uniting together they at once abjure the exercise of their own judgment and free will; and the tyrannical control which these societies exercise is often far more insupportable than the authority possessed over society by the Government which they attack. Their moral force is much diminished by these excesses, and they lose the powerful interest which is always excited by a struggle between oppressors and the oppressed. The man who in given cases consents to obey his fellows with servility, and who submits his activity and even his opinions to their control, can have no claim to rank as a free citizen.

The Americans have also established certain forms of government which are applied to their associations, but these are invariably borrowed from the forms of the civil administration. The independence of each individual is formally recognised; the tendency of the members of the association points, as it does in the body of the community, towards the same end, but they are not obliged to follow the same track. No one abjures the exercise of his reason and his free will; but every one exerts that reason and that will for the benefit of a common undertaking.

Chapter XIII

Government of the Democracy in America

I am well aware of the difficulties which attend this part of my subject, but although every expression which I am about to make use of may clash, upon some one point, with the feelings of the different parties which divide my country, I shall speak my opinion with the most perfect openness.

In Europe we are at a loss how to judge the true character and the more permanent propensities of democracy, because in Europe two conflicting principles exist, and we do not know what to attribute to the principles themselves, and what to refer to the passions which they bring into collision. Such, however, is not the case in America; there the people reigns without any obstacle, and it has no perils to dread and no injuries to avenge. In America, democracy is swayed by its own free propensities; its course is natural and its activity is unrestrained; the United States consequently afford the most favourable opportunity of studying its real character. And to no people can this inquiry be more vitally interesting than to the French nation, which is blindly driven onwards by a daily and irresistible impulse towards a state of things which may prove either despotic or republican, but which will assuredly be democratic.

Universal Suffrage

I have already observed that universal suffrage has been adopted in all the States of the Union; it consequently occurs amongst different populations which occupy very different positions in the scale of society. I have had opportunities of observing its effects in different localities, and amongst races of men who are nearly strangers to each other by their language, their religion, and their manner of life; in Louisiana as well as in New England, in Georgia and in Canada. I have remarked that Universal Suffrage is far from producing in America either all the good or all the evil consequences which are assigned to it in Europe, and that its effects differ very widely from those which are usually attributed to it.

Choice of the People, and Instinctive Preferences of the American Democracy

In the United States the most able men are rarely placed at the head of affairs—Reason of this peculiarity—The envy which prevails in the lower orders of France against the higher classes is not a French, but a purely democratic sentiment—For what reason the most distinguished men in America frequently seclude themselves from public affairs.

Many people in Europe are apt to believe without saying it, or to say without believing it, that one of the great advantages of universal suffrage is, that it entrusts the direction of public affairs to men who are worthy of the public confidence. They admit that the people is unable to govern for itself, but they aver that it is always sincerely disposed to promote the welfare of the State, and that it instinctively designates those persons who are animated by the same good wishes, and who are the most fit to wield the supreme authority. I confess that the observations I made in America by no means coincide with these opinions. On my arrival in the United States I was surprised to find so much distinguished talent among the subjects, and so little among the heads of the Government. It is a well-authenticated fact, that at the present day the most able men in the United States are very rarely placed at the head of affairs; and it must be acknowledged that such has been the result in proportion as democracy has outstepped all its former limits. The race of American statesmen has evidently dwindled most remarkably in the course of the last fifty years.

Several causes may be assigned to this phenomenon. It is impossible, notwithstanding the most strenuous exertions, to raise the intelligence of the people above a certain level. Whatever may be the facilities of acquiring information, whatever may be the profusion of easy methods and of cheap science, the human mind can never be instructed and educated without devoting a considerable space of time to those objects.

The greater or the lesser possibility of subsisting without labor is therefore the necessary boundary of intellectual improvement. This boundary is more remote in some countries and more restricted in others; but it must exist somewhere as long as the people is constrained to work in order to procure the means of physical subsistence, that is to say, as long as it retains its popular character. It is therefore quite as difficult to imagine a State in which all the citizens should be very well informed as a State in which they should all be wealthy; these two difficulties may be looked upon as correlative. It may very readily be admitted that the mass of the citizens are sincerely disposed to promote the welfare of their country; nay more, it may even be allowed that the lower classes are less apt to be swayed by considerations of personal interest than the higher orders: but it is always more or less impossible for them to discern the best means of attaining the end which they desire with sincerity. Long and patient observation, joined to a multitude of different notions, is required to form a just estimate of the character of a single individual; and can it be supposed that the vulgar have the power of succeeding in an inquiry which misleads the penetration of genius itself? The people has neither the time nor the means which are essential to the prosecution of an investigation of this kind: its conclusions are hastily formed from a superficial inspection of the more prominent features of a question. Hence it often assents to the clamour of a mountebank who knows the secret of stimulating its tastes, while its truest friends frequently fail in their exertions.

Moreover, the democracy is not only deficient in that soundness of judgment which is necessary to select men really deserving of its confidence, but it has neither the desire nor the inclination to find them out. It cannot be denied that democratic institutions have a very strong tendency to promote the feeling of envy in the human heart; not so much because they afford to every one the means of rising to the level of any of his fellow-citizens, as because those means perpetually disappoint the persons who employ them. Democratic institutions awaken and foster a passion for equality which they can never entirely satisfy. This complete equality eludes the grasp of the people at the very moment at which it thinks to hold it fast, and 'flies,' as Pascal says, 'with eternal flight;' the people is excited in the pursuit of an advantage, which is more precious because it is not sufficiently remote to be unknown, or sufficiently near to be enjoyed. The lower orders are agitated by the chance of success, they are irritated by its uncertainty; and they pass from the enthusiasm of pursuit to the exhaustion of ill-success, and lastly to the acrimony of disappointment. Whatever transcends their own limits appears to be an obstacle to their desires, and there is no kind of superiority, however legitimate it may be, which is not irksome in their sight.

It has been supposed that the secret instinct which leads the lower orders to remove their superiors as much as possible from the direction of public affairs is peculiar to France. This, however, is an error; the propensity to which I allude is not inherent in any particular nation, but in democratic institutions in general; and although it may have been heightened by peculiar political circumstances, it owes its origin to a higher cause.

In the United States the people is not disposed to hate the superior classes of society; but it is not very favourably inclined towards them, and it carefully excludes them from the exercise of authority. It does not entertain any dread of distinguished talents, but it is rarely captivated by them; and it awards its approbation very sparingly to such as have risen without the popular support.

Whilst the natural propensities of democracy induce the people to reject the most distinguished citizens as its rulers, these individuals are no less apt to retire from a political career in which it is almost impossible to retain their independence, or to advance without degrading themselves. This opinion has been very candidly set forth by Chancellor Kent, who says, in speaking with great eulogiums of that part of the Constitution which empowers the Executive to nominate the judges: 'It is indeed probable that the men who are best fitted to discharge the duties of this high office would have too much reserve in their manners, and too much austerity in their principles, for them to be returned by the majority at an election where universal suffrage is adopted.' Such were the opinions which were printed without contradiction in America in the year 1830!

I hold it to be sufficiently demonstrated that universal suffrage is by no means a guarantee of the wisdom of the popular choice; and that, whatever its advantages may be, this is not one of them.

Causes Which May Partly Correct These Tendencies of the Democracy

Contrary effects produced on peoples as well as on individuals by great dangers—Why so many distinguished men stood at the head of affairs in America fifty years ago—Influence which the intelligence and the manners of the people exercise upon its choice. Example of New—England—States of the Southwest—Influence of certain laws upon the choice of the people—Election by an elected body—Its effects upon the composition of the Senate.

When a State is threatened by serious dangers, the people frequently succeeds in selecting the citizens who are the most able to save it. It has been observed that man rarely retains his customary level in presence of very critical circumstances; he rises above or he sinks below his usual condition, and the same thing occurs in nations at large. Extreme perils sometimes quench the energy of a people instead of stimulating it; they excite without directing its passions, and instead of clearing they confuse its powers of perception. The Jews deluged the smoking ruins of their temple with the carnage of the remnant of their host. But it is more common, both in the case of nations and in that of individuals, to find extraordinary virtues arising from the very imminence of the danger. Great characters are then thrown into relief, as edifices which are concealed by the gloom of night are illuminated by the glare of a conflagration. At those dangerous times genius no longer abstains from presenting itself in the arena; and the people, alarmed by the perils of its situation, buries its envious passions in a short oblivion. Great names may then be drawn from the balloting-box.

I have already observed that the American statesmen of the present day are very inferior to those who stood at the head of affairs fifty years ago. This is as much a consequence of the circumstances as of the laws of the country. When America was struggling in the high cause of independence to throw off the yoke of another country, and when it was about to usher a new nation into the world, the spirits of its inhabitants were roused to the height which their great efforts required. In this general excitement the most distinguished men were ready to forestall the wants of the community, and the people clung to them for support, and placed them at its head. But events of this magnitude are rare, and it is from an inspection of the ordinary course of affairs that our judgment must be formed.

If passing occurrences sometimes act as checks upon the passions of democracy, the intelligence and the manners of the community exercise an influence which is not less powerful and far more permanent. This is extremely perceptible in the United States.

In New England the education and the liberties of the communities were engendered by the moral and religious principles of their founders. Where society has acquired a sufficient degree of stability to enable it to hold certain maxims and to retain fixed habits, the lower orders are accustomed to respect intellectual superiority and to submit to it without complaint, although they set at naught all those privileges which wealth and birth have introduced among mankind. The democracy in New England consequently makes a more judicious choice than it does elsewhere.

But as we descend towards the South, to those States in which the constitution of society is more modern and less strong, where instruction is less general, and where the principles of morality, of religion, and of liberty are less happily combined, we perceive that the talents and the virtues of those who are in authority become more and more rare.

Lastly, when we arrive at the new South-western States, in which the constitution of society dates but from yesterday, and presents an agglomeration of adventurers and speculators, we are amazed at the persons who are invested with public authority, and we are led to ask by what force, independent of the legislation and of the men who direct it, the State can be protected, and society be made to flourish.

There are certain laws of a democratic nature which contribute, nevertheless, to correct, in some measure, the dangerous tendencies of democracy. On entering the House of Representatives of Washington one is struck by the vulgar demeanour of that great assembly. The eye frequently does not discover a man of celebrity within its walls. Its members are almost all obscure individuals whose names present no associations to the mind: they are mostly village lawyers, men in trade, or even persons belonging to the lower classes of society. In a country in which education is very general, it is said that the representatives of the people do not always know how to write correctly.

At a few yards' distance from this spot is the door of the Senate, which contains within a small space a large proportion of the celebrated men of America. Scarcely an individual is to be perceived in it who does not recall the idea of an active and illustrious career: the Senate is composed of eloquent advocates, distinguished generals, wise magistrates, and statesmen of note, whose language would at all times do honour to the most remarkable parliamentary debates of Europe.

What then is the cause of this strange contrast, and why are the most able citizens to be found in one assembly rather than in the other? Why is the former body remarkable for its vulgarity and its poverty of talent, whilst the latter seems to enjoy a monopoly of intelligence and of sound judgment? Both of these assemblies emanate from the people; both of them are chosen by universal suffrage; and no voice has hitherto been heard to assert in America that the Senate is hostile to the interests of the people. From what cause, then, does so startling a difference arise? The only reason which appears to me adequately to account for it is, that the House of Representatives is elected by the populace directly, and that the Senate is elected by elected bodies. The whole body of the citizens names the legislature of each State, and the Federal Constitution converts these legislatures into so many electoral bodies, which return the members of the Senate. The senators are elected by an indirect application of universal suffrage; for the legislatures which name them are not aristocratic or privileged bodies which exercise the electoral franchise in their own right; but they are chosen by the totality of the citizens; they are generally elected every year, and new members may constantly be chosen who will employ their electoral rights in conformity with the wishes of the public. But this transmission of the popular authority through an assembly of chosen men operates an important change in it, by refining its discretion and improving the forms which it adopts. Men who are chosen in this manner accurately represent the majority of the nation which governs them; but they represent the elevated thoughts which are current in the community, the propensities which prompt its nobler actions, rather than the petty passions which disturb or the vices which disgrace it.

The time may be already anticipated at which the American Republics will be obliged to introduce the plan of election by an elected body more frequently into their system of representation, or they will incur no small risk of perishing miserably amongst the shoals of democracy.

And here I have no scruple in confessing that I look upon this peculiar system of election as the only means of bringing the exercise of political power to the level of all classes of the people. Those thinkers who regard this institution as the exclusive weapon of a party, and those who fear, on the other hand, to make use of it, seem to me to fall into as great an error in the one case as in the other.

Influence Which the American Democracy Has Exercised on the Laws Relating to Elections

When elections are rare, they expose the State to a violent crisis—When they are frequent, they keep up a degree of feverish excitement—The Americans have preferred the second of these two evils—Mutability of the laws—Opinions of Hamilton and Jefferson on this subject.

When elections recur at long intervals the State is exposed to violent agitation every time they take place. Parties exert themselves to the utmost in order to gain a prize which is so rarely within their reach; and as the evil is almost irremediable for the candidates who fail, the consequences of their disappointed ambition may prove most disastrous; if, on the other hand, the legal struggle can be repeated within a short space of time, the defeated parties take patience. When elections occur frequently, their recurrence keeps society in a perpetual state of feverish excitement, and imparts a continual instability to public affairs.

Thus, on the one hand the State is exposed to the perils of a revolution, on the other to perpetual mutability; the former system threatens the very existence of the Government, the latter is an obstacle to all steady and consistent policy. The

Americans have preferred the second of these evils to the first; but they were led to this conclusion by their instinct much more than by their reason; for a taste for variety is one of the characteristic passions of democracy. An extraordinary mutability has, by this means, been introduced into their legislation. Many of the Americans consider the instability of their laws as a necessary consequence of a system whose general results are beneficial. But no one in the United States affects to deny the fact of this instability, or to contend that it is not a great evil.

Hamilton, after having demonstrated the utility of a power which might prevent, or which might at least impede, the promulgation of bad laws, adds: 'It might perhaps be said that the power of preventing bad laws includes that of preventing good ones, and may be used to the one purpose as well as to the other. But this objection will have little weight with those who can properly estimate the mischiefs of that inconstancy and mutability in the laws which form the greatest blemish in the character and genius of our governments.' (Federalist, No. 73.) And again in No. 62 of the same work he observes: 'The facility and excess of law-making seem to be the diseases to which our governments are most liable. . . . The mischievous effects of the mutability in the public councils arising from a rapid succession of new members would fill a volume: every new election in the States is found to change one-half of the representatives. From this change of men must proceed a change of opinions and of measures, which forfeits the respect and confidence of other nations, poisons the blessings of liberty itself, and diminishes the attachment and reverence of the people toward a political system which betrays so many marks of infirmity.'

Jefferson himself, the greatest Democrat whom the democracy of America has yet produced, pointed out the same evils. 'The instability of our laws,' said he in a letter to Madison, 'is really a very serious inconvenience. I think that we ought to have obviated it by deciding that a whole year should always be allowed to elapse between the bringing in of a bill and the final passing of it. It should afterward be discussed and put to the vote without the possibility of making any alteration in it; and if the circumstances of the case required a more speedy decision, the question should not be decided by a simple majority, but by a majority of at least two-thirds of both houses.'

Public Offivers under the Control of the Democracy in America

Simple exterior of the American public officers—
No official costume—All public officers are remunerated—
Political consequences of this system—No public career
exists in America—Result of this.

Public officers in the United States are commingled with the crowd of citizens; they have neither palaces, nor guards, nor ceremonial costumes. This simple exterior of the persons in authority is connected not only with the peculiarities of the American character, but with the fundamental principles of that society. In the estimation of the democracy a government is not a benefit, but a necessary evil. A certain degree of power must be granted to public officers, for they would be of no use without it. But the ostensible semblance of authority is by no means indispensable to the conduct of affairs, and it is needlessly offensive to the susceptibility of the public. The public officers themselves are well aware that they only enjoy the superiority over their fellow-citizens which they derive from their authority upon condition of putting themselves on a level with the whole community by their manners. A public officer in the United States is uniformly civil, accessible to all the world, attentive to all requests, and obliging in his replies. I was pleased by these characteristics of a democratic government; and I was struck by the manly independence of the citizens, who respect the office more than the officer, and who are less attached to the emblems of authority than to the man who bears them.

I am inclined to believe that the influence which costumes really exercise, in an age like that in which we live, has been a good deal exaggerated. I never perceived that a public officer in America was the less respected whilst he was in the discharge of his duties because his own merit was set off by no adventitious signs. On the other hand, it is very doubtful whether a peculiar dress contributes to the respect which public characters ought to have for their own position, at least when they are not otherwise inclined to respect it. When a magistrate (and in France such instances are not rare) indulges his trivial wit at the expense of the prisoner, or derides the predicament in which a culprit is placed, it would be well to deprive him of his robes of office, to see whether he would recall some portion of the natural dignity of mankind when he is reduced to the apparel of a private citizen.

A democracy may, however, allow a certain show of magisterial pomp, and clothe its officers in silks and gold, without seriously compromising its principles. Privileges of this kind are transitory; they belong to the place, and are distinct from the individual: but if public officers are not uniformly remunerated by the State, the public charges must be entrusted to men of opulence and independence, who constitute the basis of an aristocracy; and if the people still retains its right of election, that election can only be made from a certain class of citizens. When a democratic republic renders offices which had formerly been remunerated gratuitous, it may safely be believed that the State is advancing to monarchical institutions; and when a monarchy begins to remunerate such officers as had hitherto been unpaid, it is a sure sign that it is approaching toward a despotic or a republican form of government. The substitution of paid for unpaid functionaries is of itself, in my opinion, sufficient to constitute a serious revolution.

I look upon the entire absence of gratuitous functionaries in America as one of the most prominent signs of the absolute dominion which democracy exercises in that country. All public services, of whatsoever nature they may be, are paid; so that every one has not merely the right, but also the means of performing them. Although, in democratic States, all the citizens are qualified to occupy stations in the Government, all are not tempted to try for them. The number and the capacities of the candidates are more apt to restrict the choice of electors than the conditions of the candidateship.

In nations in which the principle of election extends to every place in the State no political career can, properly speaking, be said to exist. Men are promoted as if by chance to the rank which they enjoy, and they are by no means sure of retaining it. The consequence is that in tranquil times public functions offer but few lures to ambition. In the United States the persons who engage in the perplexities of political life are individuals of very moderate pretensions. The pursuit of wealth generally diverts men of great talents and of great passions from the pursuit of power, and it very frequently happens that a man does not undertake to direct the fortune of the State until he has discovered his incompetence to conduct his own affairs. The vast number of very ordinary men who occupy public stations is quite as attributable to these causes as to the bad choice of the democracy. In the United States, I am not sure that the people would return the men of superior abilities who might solicit its support, but it is certain that men of this description do not come forward.

Arbitrary Power of Magistrates[1] under the Rule of the American Democracy

For what reason the arbitrary power of Magistrates is greater in absolute monarchies and in democratic republics than it is in limited monarchies—Arbitrary power of the Magistrates in New England.

In two different kinds of government the magistrates exercise a considerable degree of arbitrary power; namely, under the absolute government of a single individual, and under that of a democracy. This identical result proceeds from causes which are nearly analogous.

In despotic States the fortune of no citizen is secure; and public officers are not more safe than private individuals. The sovereign, who has under his control the lives, the property, and sometimes the honour of the men whom he employs, does not scruple to allow them a great latitude of action, because he is convinced that they will not use it to his prejudice. In despotic States the sovereign is so attached to the exercise of his power, that he dislikes the constraint even of his own regulations; and he is well pleased that his agents should follow a somewhat fortuitous line of conduct, provided he be certain that their actions will never counteract his desires.

In democracies, as the majority has every year the right of depriving the officers whom it has appointed of their power, it has no reason to fear any abuse of their authority. As the people is always able to signify its wishes to those who conduct the Government, it prefers leaving them to make their own exertions to prescribing an invariable rule of conduct which would at once fetter their activity and the popular authority.

It may even be observed, on attentive consideration, that under the rule of a democracy the arbitrary power of the magistrate must be still greater than in despotic States. In the latter the sovereign has the power of punishing all the faults with which he becomes acquainted, but it would be vain for him to hope to become acquainted with all those which are committed. In the former the sovereign power is not only supreme, but it is universally present. The American functionaries are, in point of fact, much more independent in the sphere of action which the law traces out for them than any public officer in Europe. Very frequently the object which they are to accomplish is simply pointed out to them, and the choice of the means is left to their own discretion.

In New England, for instance, the selectmen of each township are bound to draw up the list of persons who are to serve on the jury; the only rule which is laid down to guide them in their choice is that they are to select citizens possessing the elective franchise and enjoying a fair reputation.[2] In France the lives and liberties of the subjects would be thought to be in danger if a public officer of any kind was entrusted with so formidable a right. In New England the same magistrates are empowered to post the names of habitual drunkards in public-houses, and to prohibit the inhabitants of a town from supplying them with liquor.[3] A censorial power of this excessive kind would be revolting to the population of the most absolute monarchies; here, however, it is submitted to without difficulty.

Nowhere has so much been left by the law to the arbitrary determination of the magistrate as in democratic republics, because this arbitrary power is unattended by any alarming consequences. It may even be asserted that the freedom of the magistrate increases as the elective franchise is extended, and as the duration of the time of office is shortened. Hence arises the great difficulty which attends the conversion of a democratic republic into a monarchy. The magistrate ceases to be elective, but he retains the rights and the habits of an elected officer, which lead directly to despotism.

It is only in limited monarchies that the law, which prescribes the sphere in which public officers are to act, superintends all their measures. The cause of this may be easily detected. In limited monarchies the power is divided between the King and the people, both of whom are interested in the stability of the magistrate. The King does not venture to place the public officers under the control of the people, lest they should be tempted to betray his interests; on the other hand, the people fears lest the magistrates should serve to oppress the liberties of the country, if they were entirely dependent upon the Crown; they cannot therefore be said to depend on either one or the other. The same cause which induces the king and the people to render public officers independent suggests the necessity of such securities as may prevent their independence from encroaching upon the authority of the former and the liberties of the latter. They consequently agree as to the necessity of restricting the functionary to a line of conduct laid down beforehand, and they are interested in confining him by certain regulations which he cannot evade.

Instability of the Administration in the United States

In America the public acts of a community frequently leave fewer traces than the occurrences of a family—Newspapers the only historical remains—Instability of the administration prejudicial to the art of government.

The authority which public men possess in America is so brief, and they are so soon commingled with the ever-changing population of the country, that the acts of a community frequently leave fewer traces than the occurrences of a private family. The public administration is, so to speak, oral and traditionary. But little is committed to writing, and that little is wafted away forever, like the leaves of the Sibyl, by the smallest breeze.

The only historical remains in the United States are the newspapers; but if a number be wanting, the chain of time is broken, and the present is severed from the past. I am convinced that in fifty years it will be more difficult to collect authentic documents concerning the social condition of the Americans at the present day than it is to find remains of the administration of France during the Middle Ages; and if the United States were ever invaded by barbarians, it would be necessary to have recourse to the history of other nations in order to learn anything of the people which now inhabits them.

The instability of the administration has penetrated into the habits of the people: it even appears to suit the general taste, and no one cares for what occurred before his time. No methodical system is pursued; no archives are formed; and no documents are brought together when it would be very easy to do so. Where they exist, little store is set upon them; and I have amongst my papers several original public documents which were given to me in answer to some of my inquiries. In America society seems to live from hand to mouth, like an army in the field. Nevertheless, the art of administration may undoubtedly be ranked as a science, and no sciences can be improved if the discoveries and observations of successive generations are not connected together in the order in which they occur. One man, in the short space of his life remarks a fact; another conceives an idea; the former invents a means of execution, the latter reduces a truth to a fixed proposition; and mankind gathers the fruits of individual experience upon its way and gradually forms the sciences. But the persons who conduct the administration in America can seldom afford any instruction to each other; and when they assume the direction of society, they simply possess those attainments which are most widely disseminated in the community, and no experience peculiar to themselves. Democracy, carried to its furthest limits, is therefore prejudicial to the art of government; and for this reason it is better adapted to a people already versed in the conduct of an administration than to a nation which is uninitiated in public affairs.

This remark, indeed, is not exclusively applicable to the science of administration. Although a democratic government is founded upon a very simple and natural principle, it always presupposes the existence of a high degree of culture and enlightenment in society.[4] At the first glance it may be imagined to belong to the earliest ages of the world; but maturer observation will convince us that it could only come last in the succession of human history.

Charges Levied by the State under the Rule of the American Democracy

In all communities citizens divisible into three classes—
Habits of each of these classes in the direction of public finances—
Why public expenditure must tend to increase when the people

governs—What renders the extravagance of a democracy less to be feared in America—Public expenditure under a democracy.

Before we can affirm whether a democratic form of government is economical or not, we must establish a suitable standard of comparison. The question would be one of easy solution if we were to attempt to draw a parallel between a democratic republic and an absolute monarchy. The public expenditure would be found to be more considerable under the former than under the latter; such is the case with all free States compared to those which are not so. It is certain that despotism ruins individuals by preventing them from producing wealth, much more than by depriving them of the wealth they have produced; it dries up the source of riches, whilst it usually respects acquired property. Freedom, on the contrary, engenders far more benefits than it destroys; and the nations which are favoured by free institutions invariably find that their resources increase even more rapidly than their taxes.

My present object is to compare free nations to each other, and to point out the influence of democracy upon the finances of a State.

Communities, as well as organic bodies, are subject to certain fixed rules in their formation which they cannot evade. They are composed of certain elements which are common to them at all times and under all circumstances. The people may always be mentally divided into three distinct classes. The first of these classes consists of the wealthy; the second, of those who are in easy circumstances; and the third is composed of those who have little or no property, and who subsist more especially by the work which they perform for the two superior orders. The proportion of the individuals who are included in these three divisions may vary according to the condition of society, but the divisions themselves can never be obliterated.

It is evident that each of these classes will exercise an influence peculiar to its own propensities upon the administration of the finances of the State. If the first of the three exclusively possesses the legislative power, it is probable that it will not be sparing of the public funds, because the taxes which are levied on a large fortune only tend to diminish the sum of superfluous enjoyment, and are, in point of fact, but little felt. If the second class has the power of making the laws, it will certainly not be lavish of taxes, because nothing is so onerous as a large impost which is levied upon a small income. The government of the middle classes appears to me to be the most economical, though perhaps not the most enlightened, and certainly not the most generous, of free governments.

But let us now suppose that the legislative authority is vested in the lowest orders: there are two striking reasons which show that the tendency of the expenditure will be to increase, not to diminish. As the great majority of those who create the laws are possessed of no property upon which taxes can be imposed, all the money which is spent for the community appears to be spent to their advantage, at no cost of their own; and those who are possessed of some little property readily find means of regulating the taxes so that they are burdensome to the wealthy and profitable to the poor, although the rich are unable to take the same advantage when they are in possession of the Government.

In countries in which the poor[5] should be exclusively invested with the power of making the laws no great economy of public expenditure ought to be expected: that expenditure will always be considerable; either because the taxes do not weigh upon those who levy them, or because they are levied in such a manner as not to weigh upon those classes. In other words, the government of the democracy is the only one under which the power which lays on taxes escapes the payment of them.

It may be objected (but the argument has no real weight) that the true interest of the people is indissolubly connected with that of the wealthier portion of the community, since it cannot but suffer by the severe measures to which it resorts. But is it not the true interest of kings to render their subjects happy, and the true interest of nobles to admit recruits into their order on suitable grounds? If remote advantages had power to prevail over the passions and the exigencies of the moment, no such thing as a tyrannical sovereign or an exclusive aristocracy could ever exist.

Again, it may be objected that the poor are never invested with the sole power of making the laws; but I reply, that wherever universal suffrage has been established the majority of the community unquestionably exercises the legislative authority; and if it be proved that the poor always constitute the majority, it may be added, with perfect truth, that in the countries in which they possess the elective franchise they possess the sole power of making laws. But it is certain that in all the nations of the world the greater number has always consisted of those persons who hold no property, or of those whose property is insufficient to exempt them from the necessity of working in order to procure an easy subsistence. Universal suffrage does therefore, in point of fact, invest the poor with the government of society.

The disastrous influence which popular authority may sometimes exercise upon the finances of a State was very clearly seen in some of the democratic republics of antiquity, in which the public treasure was exhausted in order to relieve indigent citizens, or to supply the games and theatrical amusements of the populace. It is true that the representative system was then very imperfectly known, and that, at the present time, the influence of popular passion is less felt in the conduct of public affairs; but it may be believed that the delegate will in the end conform to the principles of his constituents, and favour their propensities as much as their interests.

The extravagance of democracy is, however, less to be dreaded in proportion as the people acquires a share of property, because on the one hand the contributions of the rich are then less needed, and, on the other, it is more difficult to lay on taxes which do not affect the interests of the lower classes. On this account universal suffrage would be less dangerous in France than in England, because in the latter country the property on which taxes may be levied is vested in fewer hands. America, where the great majority of the citizens possess some fortune, is in a still more favourable position than France.

There are still further causes which may increase the sum of public expenditure in democratic countries. When the aristocracy governs, the individuals who conduct the affairs of State are exempted by their own station in society from every kind of privation; they are contented with their position; power and renown are the objects for which they strive; and, as they are placed far above the obscurer throng of citizens, they do not always distinctly perceive how the well-being of the mass of the people ought to redound to their own honour. They are not indeed callous to the sufferings of the poor, but they cannot feel those miseries as acutely as if they were themselves partakers of them. Provided that the people appear to submit to its lot, the rulers are satisfied, and they demand nothing further from the Government. An aristocracy is more intent upon the means of maintaining its influence than upon the means of improving its condition.

When, on the contrary, the people is invested with the supreme authority, the perpetual sense of their own miseries impels the rulers of society to seek for perpetual ameliorations. A thousand different objects are subjected to improvement; the most trivial details are sought out as susceptible of amendment; and those changes which are accompanied with considerable expense are more especially advocated, since the object is to render the condition of the poor more tolerable, who cannot pay for themselves.

Moreover, all democratic communities are agitated by an ill-defined excitement and by a kind of feverish impatience, that engender a multitude of innovations, almost all of which are attended with expense.

In monarchies and aristocracies the natural taste which the rulers have for power and for renown is stimulated by the promptings of ambition, and they are frequently incited by these temptations to very costly undertakings. In democracies, where the rulers labour under privations, they can only be courted by such means as improve their well-being, and these improvements cannot take place without a sacrifice of money. When a people begins to reflect upon its situation, it discovers a multitude of wants to which it had not before been subject, and to satisfy these exigencies recourse must be had to the coffers of the State. Hence it arises that the public charges increase in proportion as civilization spreads, and that imposts are augmented as knowledge pervades the community.

The last cause which frequently renders a democratic government dearer than any other is, that a democracy does not always succeed in moderating its expenditure, because it does not understand the art of being economical. As the designs which it entertains are frequently changed, and the agents of those designs are still more frequently removed, its undertakings are often ill conducted or left unfinished: in the former case the State spends sums out of all proportion to the end which it proposes to accomplish; in the second, the expense itself is unprofitable.[6]

Tendencies of the American Democracy as Regards the Salaries of Public Officers

In the democracies those who establish high salaries have no chance of profiting by them—Tendency of the American democracy to increase the salaries of subordinate officers and to lower those of the more important functionaries—Reason of this—Comparative statement of the salaries of public officers in the United States and in France.

There is a powerful reason which usually induces democracies to economize upon the salaries of public officers. As the number of citizens who dispense the remuneration is extremely large in democratic countries, so the number of persons who can hope to be benefited by the receipt of it is comparatively small. In aristocratic countries, on the contrary, the individuals who fix high salaries have almost always a vague hope of profiting by them. These appointments may be looked upon as a capital which they create for their own use, or at least as a resource for their children.

It must, however, be allowed that a democratic State is most parsimonious towards its principal agents. In America the secondary officers are much better paid, and the dignitaries of the administration much worse, than they are elsewhere.

These opposite effects result from the same cause; the people fixes the salaries of the public officers in both cases; and the scale of remuneration is determined by the consideration of its own wants. It is held to be fair that the servants of the public should be placed in the same easy circumstances as the public itself;[7] but when the question turns upon the salaries of the great officers of State, this rule fails, and chance alone can guide the popular decision. The poor have no adequate conception of the wants which the higher classes of society may feel. The sum which is scanty to the rich appears enormous to the poor man whose wants do not extend beyond the necessaries of life; and in his estimation the Governor of a State, with his two or three hundred a year, is a very fortunate and enviable being.[8] If you undertake to convince him that the representative of a great people ought to be able to maintain some show of splendour in the eyes of foreign nations, he will perhaps assent to your meaning; but when he reflects on his own humble dwelling, and on the hard-earned produce of his wearisome toil, he remembers all that he could do with a salary which you say is insufficient, and he is startled or almost frightened at the sight of such uncommon wealth. Besides, the secondary public officer is almost on a level with the people, whilst the others are raised above it. The former may therefore excite his interest, but the latter begins to arouse his envy.

This is very clearly seen in the United States, where the salaries seem to decrease as the authority of those who receive them augments.[9]

Under the rule of an aristocracy it frequently happens, on the contrary, that whilst the high officers are receiving munificent salaries, the inferior ones have not more than enough to procure the necessaries of life. The reason of this fact is easily discoverable from causes very analogous to those to which I have just alluded. If a democracy is unable to conceive the pleasures of the rich or to witness them without envy, an aristocracy is slow to understand, or, to speak more correctly, is unacquainted with, the privations of the poor. The poor man is not (if we use the term aright) the fellow of the rich one; but he is a being of another species. An aristocracy is therefore apt to care but little for the fate of its subordinate agents; and their salaries are only raised when they refuse to perform their service for too scanty a remuneration.

It is the parsimonious conduct of democracy towards its principal officers which has countenanced a supposition of far more economical propensities than any which it really possesses. It is true that it scarcely allows the means of honourable subsistence to the individuals who conduct its affairs; but enormous sums are lavished to meet the exigencies or to facilitate the enjoyments of the people.[10] The money raised by taxation may be better employed, but it is not saved. In general, democracy gives largely to the community, and very sparingly to those who govern it. The reverse is the case in aristocratic countries, where the money of the State is expended to the profit of the persons who are at the head of affairs.

Difficulty of Distinguishing the Causes Which Contribute to the Economy of the American Government

We are liable to frequent errors in the research of those facts which exercise a serious influence upon the fate of mankind, since nothing is more difficult than to appreciate their real value. One people is naturally inconsistent and enthusiastic; another is sober and calculating; and these characteristics originate in their physical constitution or in remote causes with which we are unacquainted.

These are nations which are fond of parade and the bustle of festivity, and which do not regret the costly gaieties of an hour. Others, on the contrary, are attached to more retiring pleasures, and seem almost ashamed of appearing to be pleased. In some countries the highest value is set upon the beauty of public edifices; in others the productions of art are treated with indifference, and everything which is unproductive is looked down upon with contempt. In some renown, in others money, is the ruling passion.

Independently of the laws, all these causes concur to exercise a very powerful influence upon the conduct of the finances of the State. If the Americans never spend the money of the people in galas, it is not only because the imposition of taxes is under the control of the people, but because the people takes no delight in public rejoicings. If they repudiate all ornament from their architecture, and set no store on any but the more practical and homely advantages, it is not only because they live under democratic institutions, but because they are a commercial nation. The habits of private life are continued in public; and we ought carefully to distinguish that economy which depends upon their institutions from that which is the natural result of their manners and customs.

Whether the Expenditure of the United States Can Be Compared to That of France

Two points to be established in order to estimate the extent of the public charges, viz., the national wealth and the rate of taxation—The wealth and the charges of France not accurately known. Why the wealth and charges of the Union cannot be accurately known—Researches of the author with a view to discover the amount of taxation of Pennsylvania—General symptoms which may serve to indicate the amount of the public charges in a given nation—Result of this investigation for the Union.

Many attempts have recently been made in France to compare the public expenditure of that country with the expenditure of the United States; all these attempts have, however, been unattended by success, and a few words will suffice to show that they could not have had a satisfactory result.

In order to estimate the amount of the public charges of a people two preliminaries are indispensable: it is necessary, in the first place, to know the wealth of that people; and in the second, to learn what portion of that wealth is devoted to the expenditure of the State. To show the amount of taxation without showing the resources which are destined to meet the demand, is to undertake a futile labour; for it is not the expenditure, but the relation of the expenditure to the revenue, which it is desirable to know.

The same rate of taxation which may easily be supported by a wealthy contributor will reduce a poor one to extreme misery. The wealth of nations is composed of several distinct elements, of which population is the first, real property the second, and personal property the third. The first of these three elements may be discovered without difficulty. Amongst civilized nations it is easy to obtain an accurate census of the inhabitants; but the two others cannot be determined with so much facility. It is difficult to take an exact account of all the lands in a country which are under cultivation, with their natural or their acquired value; and it is still more impossible to estimate the entire personal property which is at the disposal of a nation, and which eludes the strictest analysis by the diversity and the number of shapes under which it may occur. And, indeed, we find that the most ancient civilized nations of Europe, including even those in which the administration is most central, have not succeeded, as yet, in determining the exact condition of their wealth.

In America the attempt has never been made; for how would such an investigation be possible in a country where society has not yet settled into habits of regularity and tranquillity; where the national Government is not assisted by a multiple of agents whose exertions it can command and direct to one sole end; and where statistics are not studied, because no one is able to collect the necessary documents, or to find time to peruse them? Thus the primary elements of the calculations which have been made in France cannot be obtained in the Union; the relative wealth of the two countries is unknown; the property of the former is not accurately determined, and no means exist of computing that of the latter. I consent, therefore, for the sake of the discussion, to abandon this necessary term of the comparison, and I confine myself to a computation of the actual amount of taxation, without investigating the relation which subsists between the taxation and the revenue. But the reader will perceive that my task has not been facilitated by the limits which I here lay down for my researches.

It cannot be doubted that the central administration of France, assisted by all the public officers who are at its disposal, might determine with exactitude the amount of the direct and indirect taxes levied upon the citizens. But this investigation, which no private individual can undertake, has not hitherto been completed by the French Government, or, at least, its results have not been made public. We are acquainted with the sum total of the charges of the State; we know the amount of the departmental expenditure; but the expenses of the communal divisions have not been computed, and the amount of the public expenses of France is consequently unknown.

If we now turn to America, we shall perceive that the difficulties are multiplied and enhanced. The Union publishes an exact return of the amount of its expenditure; the budgets of the four and twenty States furnish similar returns of their revenues; but the expenses incident to the affairs of the counties and the townships are unknown.[11]

The authority of the Federal government cannot oblige the provincial governments to throw any light upon this point; and even if these governments were inclined to afford their simultaneous cooperation, it may be doubted whether they possess the means of procuring a satisfactory answer. Independently of the natural difficulties of the task, the political organization of the country would act as a hindrance to the success of their efforts. The county and town magistrates are not appointed by the authorities of the State, and they are not subjected to their control. It is therefore very allowable to suppose that, if the State was desirous of obtaining the returns which we require, its design would be counteracted by the neglect of those subordinate officers whom it would be obliged to employ.[12] It is, in point of fact, useless to inquire what the Americans might do to forward this inquiry, since it is certain that they have hitherto done nothing at all. There does not exist a single individual at the present day, in America or in Europe, who can inform us what each citizen of the Union annually contributes to the public charges of the nation.[13]

This, however, is not the last of the difficulties which prevent us from comparing the expenditure of the Union with that of France. The French Government contracts certain obligations which do not exist in America, and *vice versa*. The French Government pays the clergy; in America the voluntary principle prevails. In America there is a legal provision for the poor; in France they are abandoned to the charity of the public. The French public officers are paid by a fixed salary; in America they are allowed certain perquisites. In France contributions in kind take place on very few roads; in America upon almost all the thoroughfares: in the former country the roads are free to all travellers; in the latter turnpikes abound. All these differences in the manner in which contributions are levied in the two countries enhance the difficulty of comparing their expenditure; for there are certain expenses which the citizens would not be subject to, or which would at any rate be much less considerable, if the State did not take upon itself to act in the name of the public.

Hence we must conclude that it is no less difficult to compare the social expenditure than it is to estimate the relative wealth of France and America. I will even add that it would be dangerous to attempt this comparison; for when statistics are not based upon computations which are strictly accurate, they mislead instead of guiding aright. The mind is easily imposed upon by the false affectation of exactness, which prevails even in the mis-statements of science, and it adopts with confidence errors which are dressed in the forms of mathematical truth.

We abandon, therefore, our numerical investigation, with the hope of meeting with data of another kind. In the absence of positive documents, we may form an opinion as to the proportion which the taxation of a people bears to its real prosperity, by observing whether its external appearance is flourishing; whether, after having discharged the calls of the State, the poor man retains the means of subsistence, and the rich the means of enjoyment; and whether both classes are contented with their position, seeking, however, to ameliorate it by perpetual exertions, so that industry is never in want of capital, nor capital unemployed by industry. The observer who draws his inferences from these signs will, undoubtedly, be led to the conclusion that the American of the United States contributes a much smaller portion of his income to the State than the citizen of France. Nor, indeed, can the result be otherwise.

A portion of the French debt is the consequence of two successive invasions; and the Union has no similar calamity to fear. A nation placed upon the continent of Europe is obliged to maintain a large standing army; the isolated position of the Union enables it to have only 6,000 soldiers. The French have a fleet of 300 sail; the Americans have 52 vessels.[14] How, then, can the inhabitants of the Union be called upon to contribute as largely as the inhabitants of France? No parallel can be drawn between the finances of two countries so differently situated.

It is by examining what actually takes place in the Union, and not by comparing the Union with France, that we may discover whether the American Government is really economical. On casting my eyes over the different republics which form the confederation, I perceive that their Governments lack perseverance in their undertakings, and that they exercise no steady control over the men whom they employ.

Whence I naturally infer that they must often spend the money of the people to no purpose, or consume more of it than is really necessary to their undertakings. Great efforts are made, in accordance with the democratic origin of society, to satisfy the exigencies of the lower orders, to open the career of power to their endeavours, and to diffuse knowledge and comfort amongst them. The poor are maintained, immense sums are annually devoted to public instruction, all services whatsoever are remunerated, and the most subordinate agents are liberally paid. If this kind of government appears to me to be useful and rational, I am nevertheless constrained to admit that it is expensive.

Wherever the poor direct public affairs and dispose of the national resources, it appears certain that, as they profit by the expenditure of the State, they are apt to augment that expenditure.

I conclude, therefore, without having recourse to inaccurate computations, and without hazarding a comparison which might prove incorrect, that the democratic government of the Americans is not a cheap government, as is sometimes asserted; and I have no hesitation in predicting that, if the people of the United States is ever involved in serious difficulties, its taxation will speedily be increased to the rate of that which prevails in the greater part of the aristocracies and the monarchies of Europe.[15]

Corruption and Vices of the Rulers in a Democracy, and Consequent Effects upon Public Morality

In aristocracies rulers sometimes endeavour to corrupt the people—In democracies rulers frequently show themselves to be corrupt—In the former their vices are directly prejudicial to the morality of the people—In the latter their indirect influence is still more pernicious.

A distinction must be made, when the aristocratic and the democratic principles mutually inveigh against each other, as tending to facilitate corruption. In aristocratic governments the individuals who are placed at the head of affairs are rich men, who are solely desirous of power. In democracies statesmen are poor, and they have their fortunes to make. The consequence is that in aristocratic States the rulers are rarely accessible to corruption, and have very little craving for money; whilst the reverse is the case in democratic nations.

But in aristocracies, as those who are desirous of arriving at the head of affairs are possessed of considerable wealth, and as the number of persons by whose assistance they may rise is comparatively small, the government is, if I may use the expression, put up to a sort of auction. In democracies, on the contrary, those who are covetous of power are very seldom wealthy, and the number of citizens who confer that power is extremely great. Perhaps in democracies the number of men who might be bought is by no means smaller, but buyers are rarely to be met with; and, besides, it would be necessary to buy so many persons at once that the attempt is rendered nugatory.

Many of the men who have been in the administration in France during the last forty years have been accused of making their fortunes at the expense of the State or of its allies; a reproach which was rarely addressed to the public characters of the ancient monarchy. But in France the practice of bribing electors is almost unknown, whilst it is notoriously and publicly carried on in England. In the United States I never heard a man accused of spending his wealth in corrupting the populace; but I have often heard the probity of public officers questioned; still more frequently have I heard their success attributed to low intrigues and immoral practices.

If, then, the men who conduct the government of an aristocracy sometimes endeavour to corrupt the people, the heads of a democracy are themselves corrupt. In the former case the morality of the people is directly assailed; in the latter an indirect influence is exercised upon the people which is still more to be dreaded.

As the rulers of democratic nations are almost always exposed to the suspicion of dishonourable conduct, they in some measure lend the authority of the Government to the base practices of which they are accused. They thus afford an example which must prove discouraging to the struggles of virtuous independence, and must foster the secret calculations of a vicious ambition. If it be asserted that evil passions are displayed in all ranks of society, that they ascend the throne by hereditary right, and that despicable characters are to be met with at the head of aristocratic nations as well as in the sphere of a democracy, this objection has but little weight in my estimation. The corruption of men who have casually risen to power has a coarse and vulgar infection in it which renders it contagious to the multitude. On the contrary, there is a kind of aristocratic refinement and an air of grandeur in the depravity of the great, which frequently prevent it from spreading abroad.

The people can never penetrate into the perplexing labyrinth of court intrigue, and it will always have difficulty in detecting the turpitude which lurks under elegant manners, refined tastes, and graceful language. But to pillage the public purse, and to vend the favours of the State, are arts which the meanest villain may comprehend, and hope to practice in his turn.

In reality it is far less prejudicial to witness the immorality of the great than to witness that immorality which leads to greatness. In a democracy private citizens see a man of their own rank in life, who rises from that obscure position, and who becomes possessed of riches and of power in a few years; the spectacle excites their surprise and their envy, and they are led to inquire how the person who was yesterday their equal is to-day their ruler. To attribute his rise to his talents or his virtues is unpleasant; for it is tacitly to acknowledge that they are themselves less virtuous and less talented than he was. They are therefore led (and not unfrequently their conjecture is a correct one) to impute his success mainly to some one of his defects; and an odious mixture is thus formed of the ideas of turpitude and power, unworthiness and success, utility and dishonour.

Efforts of Which a Democracy Is Capable

The Union has only had one struggle hitherto for its existence—Enthusiasm, at the commencement of the war—Indifference towards its close—Difficulty of establishing military conscription or impressment of seamen in America—Why a democratic people is less capable of sustained effort than another.

I here warn the reader that I speak of a government which implicitly follows the real desires of a people, and not of a government which simply commands in its name. Nothing is so irresistible as a tyrannical power commanding in the name of the people, because, whilst it exercises that moral influence which belongs to the decision of the majority, it acts at the same time with the promptitude and the tenacity of a single man.

It is difficult to say what degree of exertion a democratic government may be capable of making a crisis in the history of the nation. But no great democratic republic has hitherto existed in the world. To style the oligarchy which ruled over France in 1793 by that name would be to offer an insult to the republican form of government. The United States afford the first example of the kind.

The American Union has now subsisted for half a century, in the course of which time its existence has only once been attacked, namely, during the War of Independence. At the commencement of that long war, various occurrences took place which betokened an extraordinary zeal for the service of the country.[16] But as the contest was prolonged, symptoms of private egotism began to show themselves. No money was poured into the public treasury; few recruits could be raised to join the army; the people wished to acquire independence, but was very ill-disposed to undergo the privations by which alone it could be obtained. 'Tax laws,' says

Hamilton in the 'Federalist' (No. 12), 'have in vain been multiplied; new methods to enforce the collection have in vain been tried; the public expectation has been uniformly disappointed and the treasuries of the States have remained empty. The popular system of administration inherent in the nature of popular government, coinciding with the real scarcity of money incident to a languid and mutilated state of trade, has hitherto defeated every experiment for extensive collections, and has at length taught the different legislatures the folly of attempting them.'

The United States have not had any serious war to carry on ever since that period. In order, therefore, to appreciate the sacrifices which democratic nations may impose upon themselves, we must wait until the American people is obliged to put half its entire income at the disposal of the Government, as was done by the English; or until it sends forth a twentieth part of its population to the field of battle, as was done by France.[17]

In America the use of conscription is unknown, and men are induced to enlist by bounties. The notions and habits of the people of the United States are so opposed to compulsory enlistment that I do not imagine it can ever be sanctioned by the laws. What is termed the conscription in France is assuredly the heaviest tax upon the population of that country; yet how could a great continental war be carried on without it? The Americans have not adopted the British impressment of seamen, and they have nothing which corresponds to the French system of maritime conscription; the navy, as well as the merchant service, is supplied by voluntary service. But it is not easy to conceive how a people can sustain a great maritime war without having recourse to one or the other of these two systems. Indeed, the Union, which has fought with some honour upon the seas, has never possessed a very numerous fleet, and the equipment of the small number of American vessels has always been excessively expensive.

I have heard American statesmen confess that the Union will have great difficulty in maintaining its rank on the seas without adopting the system of impressment or of maritime conscription; but the difficulty is to induce the people, which exercises the supreme authority, to submit to impressment or any compulsory system.

It is incontestable that in times of danger a free people displays far more energy than one which is not so. But I incline to believe that this is more especially the case in those free nations in which the democratic element preponderates. Democracy appears to me to be much better adapted for the peaceful conduct of society, or for an occasional effort of remarkable vigour, than for the hardy and prolonged endurance of the storms which beset the political existence of nations. The reason is very evident; it is enthusiasm which prompts men to expose themselves to dangers and privations, but they will not support them long without reflection. There is more calculation, even in the impulses of bravery, than is generally attributed to them; and although the first efforts are suggested by passion, perseverance is maintained by a distinct regard of the purpose in view. A portion of what we value is exposed, in order to save the remainder.

But it is this distinct perception of the future, founded upon a sound judgment and an enlightened experience, which is most frequently wanting in democracies. The populace is more apt to feel than to reason; and if its present sufferings are great, it is to be feared that the still greater sufferings attendant upon defeat will be forgotten.

Another cause tends to render the efforts of a democratic government less persevering than those of an aristocracy. Not only are the lower classes less awakened than the higher orders to the good or evil chances of the future, but they are liable to suffer far more acutely from present privations. The noble exposes his life, indeed, but the chance of glory is equal to the chance of harm. If he sacrifices a large portion of his income to the State, he deprives himself for a time of the pleasures of affluence; but to the poor man death is embellished by no pomp or renown, and the imposts which are irksome to the rich are fatal to him.

This relative impotence of democratic republics is, perhaps, the greatest obstacle to the foundation of a republic of this kind in Europe. In order that such a State should subsist in one country of the Old World, it would be necessary that similar institutions should be introduced into all the other nations.

I am of opinion that a democratic government tends in the end to increase the real strength of society; but it can never combine, upon a single point and at a given time, so much power as an aristocracy or a monarchy. If a democratic country remained during a whole century subject to a republican government, it would probably at the end of that period be more populous and more prosperous than the neighbouring despotic States. But it would have incurred the risk of being conquered much oftener than they would in that lapse of years.

Self-Control of the American Democracy

The American people acquiesces slowly, or frequently does not acquiesce, in what is beneficial to its interests—The faults of the American democracy are for the most part reparable.

The difficulty which a democracy has in conquering the passions and in subduing the exigencies of the moment, with a view to the future, is conspicuous in the most trivial occurrences of the United States. The people, which is surrounded by flatterers, has great difficulty in surmounting its inclinations, and whenever it is solicited to undergo a privation or any kind of inconvenience, even to attain an end which is sanctioned by its own rational conviction, it almost always refuses to comply at first. The deference of the Americans to the laws has been very justly applauded; but it must be added that in America the legislation is made by the people and for the people. Consequently, in the United States the law favours those classes which are most interested in evading it elsewhere. It may therefore be supposed that an offensive law, which should not be acknowledged to be one of immediate utility, would either not be enacted or would not be obeyed.

In America there is no law against fraudulent bankruptcies; not because they are few, but because there are a great number of bankruptcies. The dread of being prosecuted as a bankrupt acts with more intensity upon the mind of the majority of the people than the fear of being involved in losses or ruin by the failure of other parties, and a sort of guilty tolerance is extended by the public conscience to an offence which everyone condemns in his individual capacity. In the new States of the South-West the citizens generally take justice into their own hands, and murders are of very frequent occurrence. This arises from the rude manners and the ignorance of the inhabitants of those deserts, who do not perceive the utility of investing the law with adequate force, and who prefer duels to prosecutions.

Some one observed to me one day, in Philadelphia, that almost all crimes in America are caused by the abuse of intoxicating liquors, which the lower classes can procure in great abundance, from their excessive cheapness. 'How comes it,' said I, 'that you do not put a duty upon brandy?' 'Our legislators,' rejoined my informant, 'have frequently thought of this expedient; but the task of putting it in operation is a difficult one; a revolt might be apprehended, and the members who should vote for a law of this kind would be sure of losing their seats.' 'Whence I am to infer,' replied I, 'that the drinking population constitutes the majority in your country, and that temperance is somewhat unpopular.'

When these things are pointed out to the American statesmen, they content themselves with assuring you that time will operate the necessary change, and that the experience of evil will teach the people its true interests. This is frequently true, although a democracy is more liable to error than a monarch or a body of nobles; the chances of its regaining the right path when once it has acknowledged its mistake, are greater also; because it is rarely embarrassed by internal interests, which conflict with those of the majority, and resist the authority of reason. But a democracy can only obtain truth as the result of experience, and many nations may forfeit their existence whilst they are awaiting the consequences of their errors.

The great privilege of the Americans does not simply consist in their being more enlightened than other nations, but in their being able to repair the faults they may commit. To which it must be added, that a democracy cannot derive substantial benefit from past experience, unless it be arrived at a certain pitch of knowledge and civilization. There are tribes and peoples whose education has been so vicious, and whose character presents so strange a mixture of passion, of ignorance, and of erroneous notions upon all subjects, that they are unable to discern the causes of their own wretchedness, and they fall a sacrifice to ills with which they are unacquainted.

I have crossed vast tracts of country that were formerly inhabited by powerful Indian nations which are now extinct; I have myself passed some time in the midst of mutilated tribes, which witness the daily decline of their numerical strength and of the glory of their independence; and I have heard these Indians themselves anticipate the impending doom of their race. Every European can perceive means which would rescue these unfortunate beings from inevitable destruction. They alone are insensible to the expedient; they feel the woe which year after year heaps upon their heads, but they will perish to a man without accepting the remedy. It would be necessary to employ force to induce them to submit to the protection and the constraint of civilization.

The incessant revolutions which have convulsed the South American provinces for the last quarter of a century have frequently been adverted to with astonishment, and expectations have been expressed that those nations would speedily return to their *natural state*. But can it be affirmed that the turmoil of revolution is not actually the most natural state of the South American Spaniards at the present time? In that country society is plunged into difficulties from which all its efforts are insufficient to rescue it. The inhabitants of that fair portion of the Western Hemisphere seem obstinately bent on pursuing the work of inward havoc. If they fall into a momentary repose from the effects of exhaustion, that repose prepares them for a fresh state of frenzy. When I consider their condition, which alternates between misery and crime, I should be inclined to believe that despotism itself would be a benefit to them, if it were possible that the words despotism and benefit could ever be united in my mind.

Conduct of Foreign Affairs by the American Democracy

Direction given to the foreign policy of the United States by Washington and Jefferson—Almost all the defects inherent in democratic institutions are brought to light in the conduct of foreign affairs—Their advantages are less perceptible.

We have seen that the Federal Constitution entrusts the permanent direction of the external interests of the nation to the President and the Senate,[18] which tends in some degree to detach the general foreign policy of the Union from the control of the people. It cannot therefore be asserted with truth that the external affairs of State are conducted by the democracy.

The policy of America owes its rise to Washington, and after him to Jefferson, who established those principles which it observes at the present day. Washington said in the admirable letter which he addressed to his fellow-citizens, and which may be looked upon as his political bequest to the country: 'The great rule of conduct for us in regard to foreign nations is, in extending our commercial relations, to have with them as little *political* connection as possible. So far as we have already formed engagements, let them be fulfilled with perfect good faith. Here let us stop. Europe has a set of primary interests which to us have none, or a very remote relation. Hence, she must be engaged in frequent controversies, the causes of which are essentially foreign to our concerns. Hence, therefore, it must be unwise in us to implicate ourselves, by artificial ties, in the ordinary vicissitudes of her politics, or the ordinary combinations and collisions of her friendships or enmities. Our detached and distant situation invites and enables us to pursue a different course. If we remain one people, under an efficient government, the period is not far off when we may defy material injury from external annoyance; when we may take such an attitude as will cause the neutrality we may at any time resolve upon to be scrupulously respected; when belligerent nations, under the impossibility of making acquisitions upon us, will not lightly hazard the giving us provocation; when we may choose peace or war, as our interest, guided by justice, shall counsel. Why forego the advantages of so peculiar a situation? Why quit our own to stand upon foreign ground? Why, by interweaving our destiny with that of any part of Europe, entangle our peace and prosperity in the toils of European ambition, rivalship, interest, humour, or caprice? It is our true policy to steer clear of permanent alliances with any portion of the foreign world; so far, I mean, as we are now at liberty to do it; for let me not be understood as capable of patronizing infidelity to existing engagements. I hold the maxim no less applicable to public than to private affairs, that honesty is always the best policy. I repeat it; therefore, let those engagements be observed in their genuine sense; but in my opinion it is unnecessary, and would be unwise, to extend them. Taking care always to keep ourselves, by suitable establishments, in a respectable defensive posture, we may safely trust to temporary alliances for extraordinary emergencies.' In a previous part of the same letter Washington makes the following admirable and just remark: 'The nation which indulges towards another an habitual hatred or an habitual fondness is in some degree a slave. It is a slave to its animosity or to its affection, either of which is sufficient to lead it astray from its duty and its interest.'

The political conduct of Washington was always guided by these maxims. He succeeded in maintaining his country in a state of peace whilst all the other nations of the globe were at war; and he laid it down as a fundamental doctrine, that the true interest of the Americans consisted in a perfect neutrality with regard to the internal dissensions of the European Powers.

Jefferson went still further, and he introduced a maxim into the policy of the Union, which affirms that 'the Americans ought never to solicit any privileges from foreign nations, in order not to be obliged to grant similar privileges themselves.'

These two principles, which were so plain and so just as to be adapted to the capacity of the populace, have greatly simplified the foreign policy of the United States. As the Union takes no part in the affairs of Europe, it has, properly speaking, no foreign interests to discuss, since it has at present no powerful neighbours on the American continent. The country is as much removed from the passions of the Old World by its position as by the line of policy which it has chosen, and it is neither called upon to repudiate nor to espouse the conflicting interests of Europe; whilst the dissensions of the New World are still concealed within the bosom of the future.

The Union is free from all pre-existing obligations, and it is consequently enabled to profit by the experience of the old nations of Europe, without being obliged, as they are, to make the best of the past, and to adapt it to their present circumstances; or to accept that immense inheritance which they derive from their forefathers—an inheritance of glory mingled with calamities, and of alliances conflicting with national antipathies. The foreign policy of the United States is reduced by its very nature to await the chances of the future history of the nation, and for the present it consists more in abstaining from interference than in exerting its activity.

It is therefore very difficult to ascertain, at present, what degree of sagacity the American democracy will display in the conduct of the foreign policy of the country; and upon this point its adversaries, as well as its advocates, must suspend their judgment. As for myself I have no hesitation in avowing my conviction, that it is most especially in the conduct of foreign relations that democratic governments appear to me to be decidedly inferior to governments carried on upon different principles. Experience, instruction, and habit may almost always succeed in creating a species of practical discretion in democracies, and that science of the daily occurrences of life which is called good sense. Good sense may suffice to direct the ordinary course of society; and amongst a people whose education has been provided for, the advantages of democratic liberty in the internal affairs of the country may more than compensate for the evils inherent in a democratic government. But such is not always the case in the mutual relations of foreign nations.

Foreign politics demand scarcely any of those qualities which a democracy possesses; and they require, on the contrary, the perfect use of almost all those faculties in which it is deficient. Democracy is favourable to the increase of the internal resources of the State; it tends to diffuse a moderate independence; it promotes the growth of public spirit, and fortifies the respect which is entertained for law in all classes of society; and these are advantages which only exercise an indirect influence over the relations which one people bears to another. But a democracy is unable to regulate the details of an important undertaking, to persevere in a design, and to work out its execution in the presence of serious obstacles. It cannot combine its measures with secrecy, and it will not await their consequences with patience. These are qualities which more especially belong to an individual or to an aristocracy; and they are precisely the means by which an individual people attains to a predominant position.

If, on the contrary, we observe the natural defects of aristocracy, we shall find that their influence is comparatively innoxious in the direction of the external affairs of a State. The capital fault of which aristocratic bodies may be accused is that they are more apt to contrive their own advantage than that of the mass of the people. In foreign politics it is rare for the interest of the aristocracy to be in any way distinct from that of the people.

The propensity which democracies have to obey the impulse of passion rather than the suggestions of prudence, and to abandon a mature design for the gratification of a momentary caprice, was very clearly seen in America on the breaking out of the French Revolution. It was then as evident to the simplest capacity as it is at the present time that the interest of the Americans forbade them to take any part in the contest which was about to deluge Europe with blood, but which could by no means injure the welfare of their own country. Nevertheless the sympathies of the people declared themselves with so much violence in behalf of France that nothing but the inflexible character of Washington, and the immense popularity which he enjoyed, could have prevented the Americans from declaring war against England. And even then, the exertions which the austere reason of that great man made to repress the generous but imprudent passions of his fellow-citizens, very nearly deprived him of the sole recompense which he had ever claimed—that of his country's love. The majority then reprobated the line of policy which he adopted, and which has since been unanimously approved by the nation.[19] If the Constitution and the favour of the public had not entrusted the direction of the foreign affairs of the country to Washington, it is certain that the American nation would at that time have taken the very measures which it now condemns.

Almost all the nations which have ever exercised a powerful influence upon the destinies of the world by conceiving, following up, and executing vast designs—from the Romans to the English—have been governed by aristocratic institutions. Nor will this be a subject of wonder when we recollect that nothing in the world has so absolute a fixity of purpose as an aristocracy. The mass of the people may be led astray by ignorance or passion; the mind of a king may be biased, and his perseverance in his designs may be shaken—besides which a king is not immortal but an aristocratic body is too numerous to be led astray by the blandishments of intrigue, and yet not numerous enough to yield readily to the intoxicating influence of unreflecting passion: it has the energy of a firm and enlightened individual, added to the power which it derives from perpetuity.

Chapter XIV

What the Real Advantages Are Which American Society Derives from the Government of the Democracy

Before I enter upon the subject of the present chapter I am induced to remind the reader of what I have more than once adverted to in the course of this book. The political institutions of the United States appear to me to be one of the forms of government which a democracy may adopt; but I do not regard the American Constitution as the best, or as the only one, which a democratic people may establish. In showing the advantages which the Americans derive from the government of democracy, I am therefore very far from meaning, or from believing, that similar advantages can only be obtained from the same laws.

General Tendency of the Laws under the Rule of the American Democracy, and Habits of Those Who Apply Them

Defects of a democratic government easy to be discovered—Its advantages only to be discerned by long observation—Democracy in America often inexpert, but the general tendency of the laws advantageous—In the American democracy public officers have no permanent interests distinct from those of the majority—Result of this state of things.

The defects and the weaknesses of a democratic government may very readily be discovered; they are demonstrated by the most flagrant instances, whilst its beneficial influence is less perceptibly exercised. A single glance suffices to detect its evil consequences, but its good qualities can only be discerned by long observation. The laws of the American democracy are frequently defective or incomplete; they sometimes attack vested rights, or give a sanction to others which are dangerous to the community; but even if they were good, the frequent changes which they undergo would be an evil. How comes it, then, that the American republics prosper and maintain their position?

In the consideration of laws a distinction must be carefully observed between the end at which they aim and the means by which they are directed to that end, between their absolute and their relative excellence. If it be the intention of the legislator to favour the interests of the minority at the expense of the majority, and if the measures he takes are so combined as to accomplish the object he has in view with the least possible expense of time and exertion, the law may be well drawn up, although its purpose be bad; and the more efficacious it is, the greater is the mischief which it causes.

Democratic laws generally tend to promote the welfare of the greatest possible number; for they emanate from the majority of the citizens, who are subject to error, but who cannot have an interest opposed to their own advantage. The laws of an aristocracy tend, on the contrary, to concentrate wealth and power in the hands of the minority, because an aristocracy, by its very nature, constitutes a minority. It may therefore be asserted, as a general proposition, that the purpose of a democracy in the conduct of its legislation is useful to a greater number of citizens than that of an aristocracy. This is, however, the sum total of its advantages.

Aristocracies are infinitely more expert in the science of legislation than democracies ever can be. They are possessed of a self-control which protects them from the errors of temporary excitement, and they form lasting designs which they mature with the assistance of favourable opportunities. Aristocratic government proceeds with the dexterity of art; it understands how to make the collective force of all its laws converge at the same time to a given point. Such is not the case with democracies, whose laws are almost always ineffective or inopportune. The means of democracy are therefore more imperfect than those of aristocracy, and the measures which it unwittingly adopts are frequently opposed to its own cause; but the object it has in view is more useful.

Let us now imagine a community so organized by nature, or by its constitution, that it can support the transitory action of bad laws, and that it can await, without destruction, the general tendency of the legislation: we shall then be able to conceive that a democratic government, notwithstanding its defects, will be most fitted to conduce to the prosperity of this community. This is precisely what has occurred in the United States; and I repeat, what I have before remarked, that the great advantage of the Americans consists in their being able to commit faults which they may afterward repair.

An analogous observation may be made respecting public officers. It is easy to perceive that the American democracy frequently errs in the choice of the individuals to whom it entrusts the power of the administration; but it is more difficult to say why the State prospers under their rule. In the first place it is to be remarked, that if in a democratic State the governors have less honesty and less capacity than elsewhere, the governed, on the other hand, are more enlightened and more attentive to their interests. As the people in democracies is more incessantly vigilant in its affairs and more jealous of its rights, it prevents its representatives from abandoning that general line of conduct which its own interest prescribes. In the second place, it must be remembered that if the democratic magistrate is more apt to misuse his power, he possesses it for a shorter period of time. But there is yet another reason which is still more general and conclusive. It is no doubt of importance to the welfare of nations that they should be governed by men of talents and virtue; but it is perhaps still more important that the interests of those men should not differ from the interests of the community at large; for, if such were the case, virtues of a high order might become useless, and talents might be turned to a bad account. I say that it is important that the interests of the persons in authority should not conflict with or oppose the interests of the community at large; but I do not insist upon their having the same interests as the *whole* population, because I am not aware that such a state of things ever existed in any country.

No political form has hitherto been discovered which is equally favourable to the prosperity and the development of all the classes into which society is divided. These classes continue to form, as it were, a certain number of distinct nations in the same nation; and experience has shown that it is no less dangerous to place the fate of these classes exclusively in the hands of any one of them than it is to make one people the arbiter of the destiny of another. When the rich alone govern, the interest of the poor is always endangered; and when the poor make the laws, that of the rich incurs very serious risks. The advantage of democracy does not consist, therefore, as has sometimes been asserted, in favouring the prosperity of all, but simply in contributing to the well-being of the greatest possible number.

The men who are entrusted with the direction of public affairs in the United States are frequently inferior, both in point of capacity and of morality, to those whom aristocratic institutions would raise to power. But their interest is identified and confounded with that of the majority of their fellow-citizens. They may frequently be faithless and frequently mistaken, but they will never systematically adopt a line of conduct opposed to the will of the majority; and it is impossible that they should give a dangerous or an exclusive tendency to the government.

The mal-administration of a democratic magistrate is a mere isolated fact, which only occurs during the short period for which he is elected. Corruption and incapacity do not act as common interests, which may connect men permanently with one another. A corrupt or an incapable magistrate will not concert his measures with another magistrate, simply because that individual is as corrupt and as incapable as himself; and these two men will never unite their endeavours to promote the corruption and inaptitude of their remote posterity. The ambition and the manœuvres of the one will serve, on the contrary, to unmask the other. The vices of a magistrate, in democratic states, are usually peculiar to his own person.

But under aristocratic governments public men are swayed by the interest of their order, which, if it is sometimes confounded with the interests of the majority, is very frequently distinct from them. This interest is the common and lasting bond which unites them together; it induces them to coalesce, and to combine their efforts in order to attain an end which does not always ensure the greatest happiness of the greatest number; and it serves not only to connect the persons in authority, but to unite them to a considerable portion of the community, since a numerous body of citizens belongs to the aristocracy, without being invested with official functions. The aristocratic magistrate is therefore constantly supported by a portion of the community, as well as by the Government of which he is a member.

The common purpose which connects the interest of the magistrates in aristocracies with that of a portion of their cotemporaries identifies it with that of future generations; their influence belongs to the future as much as to the present. The aristocratic magistrate is urged at the same time toward the same point by the passions of the community, by his own, and I may almost add by those of his posterity. Is it, then, wonderful that he does not resist such repeated impulses? And indeed aristocracies are often carried away by the spirit of their order without being corrupted by it; and they unconsciously fashion society to their own ends, and prepare it for their own descendants.

The English aristocracy is perhaps the most liberal which ever existed, and no body of men has ever, uninterruptedly, furnished so many honourable and enlightened individuals to the government of a country. It cannot, however, escape observation that in the legislation of England the good of the poor has been sacrificed to the advantage of the rich, and the rights of the majority to the privileges of the few. The consequence is, that England, at the present day, combines the extremes of fortune in the bosom of her society, and her perils and calamities are almost equal to her power and her renown.[1]

In the United States, where the public officers have no interests to promote connected with their, caste, the general and constant influence of the Government is beneficial, although the individuals who conduct it are frequently unskilful and sometimes contemptible. There is indeed a secret tendency in democratic institutions to render the exertions of the citizens subservient to the prosperity of the community, notwithstanding their private vices and mistakes; whilst in aristocratic institutions there is a secret propensity which, notwithstanding the talents and the virtues of those who conduct the government, leads them to contribute to the evils which oppress their fellow-creatures. In aristocratic governments public men may frequently do injuries which they do not intend, and in democratic states they produce advantages which they never thought of.

Public Spirit in the United States

Patriotism of instinct—Patriotism of reflection—Their different characteristics—Nations ought to strive to acquire the second when the first has disappeared—Efforts of the Americans to acquire it—Interest of the individual intimately connected with that of the country.

There is one sort of patriotic attachment which principally arises from that instinctive, disinterested, and undefinable feeling which connects the affections of man with his birthplace. This natural fondness is united to a taste for ancient customs, and to a reverence for ancestral traditions of the past; those who cherish it love their country as they love the mansions of their fathers. They enjoy the tranquillity which it affords them; they cling to the peaceful habits which they have contracted within its bosom; they are attached to the reminiscences which it awakens, and they are even pleased by the state of obedience in which they are placed. This patriotism is sometimes stimulated by religious enthusiasm, and then it is capable of making the most prodigious efforts. It is in itself a kind of religion; it does not reason, but it acts from the impulse of faith and of sentiment. By some nations the monarch has been regarded as a personification of the country; and the fervour of patriotism being converted into the fervour of loyalty, they took a sympathetic pride in his conquests, and gloried in his power. At one time, under the ancient monarchy, the French felt a sort of satisfaction in the sense of their dependence upon the arbitrary pleasure of their king, and they were wont to say with pride, 'We are the subjects of the most powerful king in the world.'

But, like all instinctive passions, this kind of patriotism is more apt to prompt transient exertion than to supply the motives of continuous endeavour. It may save the State in critical circumstances, but it will not unfrequently allow the nation to decline in the midst of peace. Whilst the manners of a people are simple and its faith unshaken, whilst society is steadily based upon traditional institutions whose legitimacy has never been contested, this instinctive patriotism is wont to endure.

But there is another species of attachment to a country which is more rational than the one we have been describing. It is perhaps less generous and less ardent, but it is more fruitful and more lasting; it is coeval with the spread of knowledge, it is nurtured by the laws, it grows by the exercise of civil rights, and, in the end, it is confounded with the personal interest of the citizen. A man comprehends the influence which the prosperity of his country has upon his own welfare; he is aware that the laws authorize him to contribute his assistance to that prosperity, and he labours to promote it as a portion of his interest in the first place, and as a portion of his right in the second.

But epochs sometimes occur, in the course of the existence of a nation, at which the ancient customs of a people are changed, public morality destroyed, religious belief disturbed, and the spell of tradition broken, whilst the diffusion of knowledge is yet imperfect, and the civil rights of the community are ill secured, or confined within very narrow limits. The country then assumes a dim and dubious shape in the eyes of the citizens; they no longer behold it in the soil which they inhabit, for that soil is to them a dull inanimate clod; nor in the usages of their forefathers, which they have been taught to look upon as a debasing yoke; nor in religion, for of that they doubt; nor in the laws, which do not originate in their own authority; nor in the legislator, whom they fear and despise. The country is lost to their senses, they can neither discover it under its own nor under borrowed features, and they entrench themselves within the dull precincts of a narrow egotism. They are emancipated from prejudice without having acknowledged the empire of reason; they are neither animated by the instinctive patriotism of monarchical subjects nor by the thinking patriotism of republican citizens; but they have stopped halfway between the two, in the midst of confusion and of distress.

In this predicament, to retreat is impossible; for a people cannot restore the vivacity of its earlier times, any more than a man can return to the innocence and the bloom of childhood; such things may be regretted, but they cannot be renewed. The only thing, then, which remains to be done is to proceed, and to accelerate the union of private with public interests, since the period of disinterested patriotism is gone by forever.

I am certainly very far from averring that, in order to obtain this result, the exercise of political rights should be immediately granted to all the members of the community. But I maintain that the most powerful, and perhaps the only, means of interesting men in the welfare of their country which we still possess is to make them partakers in the Government. At the present time civic zeal seems to me to be inseparable from the exercise of political rights; and I hold that the number of citizens will be found to augment or to decrease in Europe in proportion as those rights are extended.

In the United States the inhabitants were thrown but as yesterday upon the soil which they now occupy, and they brought neither customs nor traditions with them there; they meet each other for the first time with no previous acquaintance; in short, the instinctive love of their country can scarcely exist in their minds; but everyone takes as zealous an interest in the affairs of his township, his county, and of the whole State, as if they were his own, because everyone, in his sphere, takes an active part in the government of society.

The lower orders in the United States are alive to the perception of the influence exercised by the general prosperity upon their own welfare; and simple as this observation is, it is one which is but too rarely made by the people. But in America the people regards this prosperity as the result of its own exertions; the citizen looks upon the fortune of the public as his private interest, and he cooperates in its success, not so much from a sense of pride or of duty, as from what I shall venture to term cupidity.

It is unnecessary to study the institutions and the history of the Americans in order to discover the truth of this remark, for their manners render it sufficiently evident. As the American participates in all that is done in his country, he thinks himself obliged to defend whatever may be censured; for it is not only his country which is attacked upon these occasions, but it is himself. The consequence is, that his national pride resorts to a thousand artifices, and to all the petty tricks of individual vanity.

Nothing is more embarrassing in the ordinary intercourse of life than this irritable patriotism of the Americans. A stranger may be very well inclined to praise many of the institutions of their country, but he begs permission to blame some of the peculiarities which he observes—a permission which is, however, inexorably refused. America is therefore a free country, in which, lest anybody should be hurt by your remarks, you are not allowed to speak freely of private individuals, or of the State, of the citizens or of the authorities, of public or of private undertakings, or, in short, of anything at all, except it be of the climate and the soil; and even then Americans will be found ready to defend either the one or the other, as if they had been contrived by the inhabitants of the country.

In our times option must be made between the patriotism of all and the government of a few; for the force and activity which the first confers are irreconcilable with the guarantees of tranquillity which the second furnishes.

Notion of Rights in the United States

No great people without a notion of rights—How the notion of rights can be given to people—Respect of rights in the United States—Whence it arises.

After the idea of virtue, I know no higher principle than that of right; or, to speak more accurately, these two ideas are commingled in one. The idea of right is simply that of virtue introduced into the political world. It is the idea of right which enabled men to define anarchy and tyranny; and which taught them to remain independent without arrogance, as well as to obey without servility. The man who submits to violence is debased by his compliance; but when he obeys the mandate of one who possesses that right of authority which he acknowledges in a fellow-creature, he rises in some measure above the person who delivers the command. There are no great men without virtue, and there are no great nations—it may almost be added that there would be no society—without the notion of rights; for what is the condition of a mass of rational and intelligent beings who are only united together by the bond of force?

I am persuaded that the only means which we possess at the present time of inculcating the notion of rights, and of rendering it, as it were, palpable to the senses, is to invest all the members of the community with the peaceful exercise of certain rights: this is very clearly seen in children, who are men without the strength and the experience of manhood. When a child begins to move in the midst of the objects which surround him, he is instinctively led to turn everything which he can lay his hands upon to his own purposes; he has no notion of the property of others; but as he gradually learns the value of things, and begins to perceive that he may in his turn be deprived of his possessions, he becomes more circumspect, and he observes those rights in others which he wishes to have respected in himself. The principle which the child derives from the possession of his toys is taught to the man by the objects which he may call his own. In America those complaints against property in general which are so frequent in Europe are never heard, because in America there are no paupers; and as everyone has property of his own to defend, everyone recognizes the principle upon which he holds it.

The same thing occurs in the political world. In America the lowest classes have conceived a very high notion of political rights, because they exercise those rights; and they refrain from attacking those of other people, in order to ensure their own from attack. Whilst in Europe the same classes sometimes recalcitrate even against the supreme power, the American submits without a murmur to the authority of the pettiest magistrate.

This truth is exemplified by the most trivial details of national peculiarities. In France very few pleasures are exclusively reserved for the higher classes; the poor are admitted wherever the rich are received, and they consequently behave with propriety, and respect whatever contributes to the enjoyments in which they themselves participate. In England, where wealth has a monopoly of amusement as well as of power, complaints are made that whenever the poor happen to steal into the enclosures which are reserved for the pleasures of the rich, they commit acts of wanton mischief: can this be wondered at, since care has been taken that they should have nothing to lose?[2]

The government of democracy brings the notion of political rights to the level of the humblest citizens, just as the dissemination of wealth brings the notion of property within the reach of all the members of the community; and I confess that, to my mind, this is one of its greatest advantages. I do not assert that it is easy to teach men to exercise political rights; but I maintain that, when it is possible, the effects which result from it are highly important; and I add that, if there ever was a time at which such an attempt ought to be made, that time is our own. It is clear that the influence of religious belief is shaken, and that the notion of divine rights is declining; it is evident that public morality is vitiated, and the notion of moral rights is also disappearing: these are general symptoms of the substitution of argument for faith, and of calculation for the impulses of sentiment. If, in the midst of this general disruption, you do not succeed in connecting the notion of rights with that of personal interest, which is the only immutable point in the human heart, what means will you have of governing the world except by fear? When I am told that, since the laws are weak and the populace is wild, since passions are excited and the authority of virtue is paralyzed, no measures must be taken to increase the rights of the democracy, I reply, that it is for these very reasons that some measures of the kind must be taken; and I am persuaded that governments are still more interested in taking them than society at large, because governments are liable to be destroyed and society cannot perish.

I am not, however, inclined to exaggerate the example which America furnishes. In those States the people are invested with political rights at a time when they could scarcely be abused, for the citizens were few in number and simple in their manners. As they have increased, the Americans have not augmented the power of the democracy, but they have, if I may use the expression, extended its dominions. It cannot be doubted that the moment at which political rights are granted to a people that had before been without them is a very critical, though it be a necessary one. A child may kill before he is aware of the value of life; and he may deprive another person of his property before he is aware that his own may be taken away from him. The lower orders, when first they are invested with political rights, stand, in relation to those rights, in the same position as the child does to the whole of nature, and the celebrated adage may then be applied to them, *Homo puer robustus.* This truth may even be perceived in America. The States in which the citizens have enjoyed their rights longest are those in which they make the best use of them.

It cannot be repeated too often that nothing is more fertile in prodigies than the art of being free; but there is nothing more arduous than the apprenticeship of liberty. Such is not the case with despotic institutions: despotism often promises to make amends for a thousand previous ills; it supports the right, it protects the oppressed, and it maintains public order. The nation is lulled by the temporary prosperity which accrues to it, until it is roused to a sense of its own misery. Liberty, on the contrary, is generally established in the midst of agitation, it is perfected by civil discord, and its benefits cannot be appreciated until it is already old.

Respect for the Law in the United States

Respect of the Americans for the law—Parental affection which they entertain for it—Personal interest of everyone to increase the authority of the law.

It is not always feasible to consult the whole people, either directly or indirectly, in the formation of the law; but it cannot be denied that, when such a measure is possible the authority of the law is very much augmented. This popular origin, which impairs the excellence and the wisdom of legislation, contributes prodigiously to increase its power. There is an amazing strength in the expression of the determination of a whole people, and when it declares itself the imagination of those who are most inclined to contest it is overawed by its authority. The truth of this fact is very well known by parties, and they consequently strive to make out a majority whenever they can. If they have not the greater number of voters on their side, they assert that the true majority abstained from voting; and if they are foiled even there, they have recourse to the body of those persons who had no votes to give.

In the United States, except slaves, servants, and paupers in the receipt of relief from the townships, there is no class of persons who do not exercise the elective franchise, and who do not indirectly contribute to make the laws. Those who design to attack the laws must consequently either modify the opinion of the nation or trample upon its decision.

A second reason, which is still more weighty, may be further adduced; in the United States everyone is personally interested in enforcing the obedience of the whole community to the law; for as the minority may shortly rally the majority to its principles, it is interested in professing that respect for the decrees of the legislator which it may soon have occasion to claim for its own. However irksome an enactment may be, the citizen of the United States complies with it, not only because it is the work of the majority, but because it originates in his own authority, and he regards it as a contract to which he is himself a party.

In the United States, then, that numerous and turbulent multitude does not exist which always looks upon the law as its natural enemy, and accordingly surveys it with fear and with distrust. It is impossible, on the other hand, not to perceive that all classes display the utmost reliance upon the legislation of their country, and that they are attached to it by a kind of parental affection.

I am wrong, however, in saying all classes; for as in America the European scale of authority is inverted, the wealthy are there placed in a position analogous to that of the poor in the Old World, and it is the opulent classes which frequently look upon the law with suspicion. I have already observed that the advantage of democracy is not, as has been sometimes asserted, that it protects the interests of the whole community, but simply that it protects those of the majority. In the United States, where the poor rule, the rich have always some reason to dread the abuses of their power. This natural anxiety of the rich may produce a sullen dissatisfaction, but society is not disturbed by it; for the same reason which induces the rich to withhold their confidence in the legislative authority makes them obey its mandates; their wealth, which prevents them from making the law, prevents them from withstanding it. Amongst civilized nations revolts are rarely excited, except by such persons as have nothing to lose by them; and if the laws of a democracy are not always worthy of respect, at least they always obtain it; for those who usually infringe the laws have no excuse for not complying with the enactments they have themselves made, and by which they are themselves benefited, whilst the citizens whose interests might be promoted by the infraction of them are induced, by their character and their stations, to submit to the decisions of the legislature, whatever they may be. Besides which, the people in America obeys the law not only because it emanates from the popular authority, but because that authority may modify it in any points which may prove vexatory; a law is observed because it is a self-imposed evil in the first place, and an evil of transient duration in the second.

Activity Which Pervades All the Branches of the Body Politic in the United States; Influence Which It Exercises upon Society

More difficult to conceive the political activity which pervades the United States than the freedom and equality which reign there—The great activity which perpetually agitates the legislative bodies is only an episode to the general activity—Difficult for an American to confine himself to his own business—Political agitation extends to all social intercourse—Commercial activity of the Americans partly attributable to this cause—Indirect advantages which society derives from a democratic government.

On passing from a country in which free institutions are established to one where they do not exist, the traveller is struck by the change; in the former all is bustle and activity, in the latter everything is calm and motionless. In the one, amelioration and progress are the general topics of inquiry; in the other, it seems as if the community only aspired to repose in the enjoyment of the advantages which it has acquired. Nevertheless, the country which exerts itself so strenuously to promote its welfare is generally more wealthy and more prosperous than that which appears to be so contented with its lot; and when we compare them together, we can scarcely conceive how so many new wants are daily felt in the former, whilst so few seem to occur in the latter.

If this remark is applicable to those free countries in which monarchical and aristocratic institutions subsist, it is still more striking with regard to democratic republics. In these States it is not only a portion of the people which is busied with the amelioration of its social condition, but the whole community is engaged in the task; and it is not the exigencies and the convenience of a single class for which a provision is to be made, but the exigencies and the convenience of all ranks of life.

It is not impossible to conceive the surpassing liberty which the Americans enjoy; some idea may likewise be formed of the extreme equality which subsists amongst them, but the political activity which pervades the United States must be seen in order to be understood. No sooner do you set foot upon the American soil than you are stunned by a kind of tumult; a confused clamour is heard on every side; and a thousand simultaneous voices demand the immediate satisfaction of their social wants. Everything is in motion around you; here, the people of one quarter of a town are met to decide upon the building of a church; there, the election of a representative is going on; a little further the delegates of a district are posting to the town in order to consult upon some local improvements; or in another place the labourers of a village quit their ploughs to deliberate upon the project of a road or a public school. Meetings are called for the sole purpose of declaring their disapprobation of the line of conduct pursued by the Government; whilst in other assemblies the citizens salute the authorities of the day as the fathers of their country. Societies are formed which regard drunkenness as the principal cause of the evils under which the State labours, and which solemnly bind themselves to give a constant example of temperance.[3]

The great political agitation of the American legislative bodies, which is the only kind of excitement that attracts the attention of foreign countries, is a mere episode or a sort of continuation of that universal movement which originates in the lowest classes of the people and extends successively to all the ranks of society. It is impossible to spend more efforts in the pursuit of enjoyment.

The cares of political life engross a most prominent place in the occupation of a citizen in the United States, and almost the only pleasure of which an American has any idea is to take a part in the Government, and to discuss the part he has taken. This feeling pervades the most trifling habits of life; even the women frequently attend public meetings and listen to political harangues as a recreation after their household labours. Debating clubs are to a certain extent a substitute for theatrical entertainments: an American cannot converse, but he can discuss; and when he attempts to talk he falls into a dissertation. He speaks to you as if he was addressing a meeting; and if he should chance to warm in the course of the discussion, he will infallibly say, 'Gentlemen,' to the person with whom he is conversing.

In some countries the inhabitants display a certain repugnance to avail themselves of the political privileges with which the law invests them; it would seem that they set too high a value upon their time to spend it on the interests of the community; and they prefer to withdraw within the exact limits of a wholesome egotism, marked out by four sunk fences and a quickset hedge. But if an American were condemned to confine his activity to his own affairs, he would be robbed of one-half of his existence; he would feel an immense void in the life which he is accustomed to lead, and his wretchedness would be unbearable.[4] I am persuaded that, if ever a despotic government is established in America, it will find it more difficult to surmount the habits which free institutions have engendered than to conquer the attachment of the citizens to freedom.

This ceaseless agitation which democratic government has introduced into the political world influences all social intercourse. I am not sure that upon the whole this is not the greatest advantage of democracy. And I am much less inclined to applaud it for what it does than for what it causes to be done.

It is incontestable that the people frequently conducts public business very ill; but it is impossible that the lower orders should take a part in public business without extending the circle of their ideas, and without quitting the ordinary routine of their mental acquirements. The humblest individual who is called upon to cooperate in the government of society acquires a certain degree of self-respect; and as he possesses authority, he can command the services of minds much more enlightened than his own. He is canvassed by a multitude of applicants, who seek to deceive him in a thousand different ways, but who instruct him by their deceit. He takes a part in political undertakings which did not originate in his own conception, but which give him a taste for undertakings of the kind. New ameliorations are daily pointed out in the property which he holds in common with others, and this gives him the desire of improving that property which is more peculiarly his own. He is perhaps neither happier nor better than those who came before him, but he is better informed and more active. I have no doubt that the democratic institutions of the United States, joined to the physical constitution of the country, are the cause (not the direct, as is so often asserted, but the indirect cause) of the prodigious commercial activity of the inhabitants. It is not engendered by the laws, but the people learns how to promote it by the experience derived from legislation.

When the opponents of democracy assert that a single individual performs the duties which he undertakes much better than the government of the community, it appears to me that they are perfectly right. The government of an individual, supposing an equality of instruction on either side, is more consistent, more persevering, and more accurate than that of a multitude, and it is much better qualified judiciously to discriminate the characters of the men it employs. If any deny what I advance, they have certainly never seen a democratic government, or have formed their opinion upon very partial evidence. It is true that even when local circumstances and the disposition of the people allow democratic institutions to subsist, they never display a regular and methodical system of government. Democratic liberty is far from accomplishing all the projects it undertakes, with the skill of an adroit despotism. It frequently abandons them before they have borne their fruits, or risks them when the consequences may prove dangerous; but in the end it produces more than any absolute government, and if it do fewer things well, it does a greater number of things. Under its sway the transactions of the public administration are not nearly so important as what is done by private exertion. Democracy does not confer the most skilful kind of government upon the people, but it produces that which the most skilful governments are frequently unable to awaken, namely, an all-pervading and restless activity, a superabundant force, and an energy which is inseparable from it, and which may, under favourable circumstances, beget the most amazing benefits. These are the true advantages of democracy.

In the present age, when the destinies of Christendom seem to be in suspense, some hasten to assail democracy as its foe whilst it is yet in its early growth; and others are ready with their vows of adoration for this new deity which is springing forth from chaos: but both parties are very imperfectly acquainted with the object of their hatred or of their desires; they strike in the dark, and distribute their blows by mere chance.

We must first understand what the purport of society and the aim of government is held to be. If it be your intention to confer a certain elevation upon the human mind, and to teach it to regard the things of this world with generous feelings, to inspire men with a scorn of mere temporal advantage, to give birth to living convictions, and to keep alive the spirit of honourable devotedness; if you hold it to be a good thing to refine the habits, to embellish the manners, to cultivate the arts of a nation, and to promote the love of poetry, of beauty, and of renown; if you would constitute a people not unfitted to act with power upon all other nations, nor unprepared for those high enterprises which, whatever be the result of its efforts, will leave a name forever famous in time—if you believe such to be the principal object of society, you must avoid the government of democracy, which would be a very uncertain guide to the end you have in view.

But if you hold it to be expedient to divert the moral and intellectual activity of man to the production of comfort, and to the acquirement of the necessaries of life; if a clear understanding be more profitable to man than genius; if your object be not to stimulate the virtues of heroism, but to create habits of peace; if you had rather witness vices than crimes and are content to meet with fewer noble deeds, provided offences be diminished in the same proportion; if, instead of living in the midst of a brilliant state of society, you are contented to have prosperity around you;

if, in short, you are of opinion that the principal object of a Government is not to confer the greatest possible share of power and of glory upon the body of the nation, but to ensure the greatest degree of enjoyment and the least degree of misery to each of the individuals who compose it—if such be your desires, you can have no surer means of satisfying them than by equalizing the conditions of men, and establishing democratic institutions.

But if the time be passed at which such a choice was possible, and if some superhuman power impel us towards one or the other of these two governments without consulting our wishes, let us at least endeavour to make the best of that which is allotted to us; and let us so inquire into its good and its evil propensities as to be able to foster the former and repress the latter to the utmost.

Chapter XV

Unlimited Power of the Majority in the United States, and Its Consequences

Natural strength of the majority in democracies—Most of the American Constitutions have increased this strength by artificial means—How this has been done—Pledged delegates—Moral power of the majority—Opinion as to its infallibility—Respect for its rights, how augmented in the United States.

The very essence of democratic government consists in the absolute sovereignty of the majority; for there is nothing in democratic States which is capable of resisting it. Most of the American Constitutions have sought to increase this natural strength of the majority by artificial means.[1]

The legislature is, of all political institutions, the one which is most easily swayed by the wishes of the majority. The Americans determined that the members of the legislature should be elected by the people immediately, and for a very brief term, in order to subject them, not only to the general convictions, but even to the daily passions, of their constituents. The members of both houses are taken from the same class in society, and are nominated in the same manner; so that the modifications of the legislative bodies are almost as rapid and quite as irresistible as those of a single assembly. It is to a legislature thus constituted that almost all the authority of the government has been entrusted.

But whilst the law increased the strength of those authorities which of themselves were strong, it enfeebled more and more those which were naturally weak. It deprived the representatives of the executive of all stability and independence, and by subjecting them completely to the caprices of the legislature, it robbed them of the slender influence which the nature of a democratic government might have allowed them to retain. In several States the judicial power was also submitted to the elective discretion of the majority, and in all of them its existence was made to depend on the pleasure of the legislative authority, since the representatives were empowered annually to regulate the stipend of the judges.

Custom, however, has done even more than law. A proceeding which will in the end set all the guarantees of representative government at naught is becoming more and more general in the United States; it frequently happens that the electors, who choose a delegate, point out a certain line of conduct to him, and impose upon him a certain number of positive obligations which he is pledged to fulfil. With the exception of the tumult, this comes to the same thing as if the majority of the populace held its deliberations in the market-place.

Several other circumstances concur in rendering the power of the majority in America not only preponderant, but irresistible. The moral authority of the majority is partly based upon the notion that there is more intelligence and more wisdom in a great number of men collected together than in a single individual, and that the quantity of legislators is more important than their quality. The theory of equality is in fact applied to the intellect of man: and human pride is thus assailed in its last retreat by a doctrine which the minority hesitate to admit, and in which they very slowly concur. Like all other powers, and perhaps more than all other powers, the authority of the many requires the sanction of time; at first it enforces obedience by constraint, but its laws are not respected until they have long been maintained.

The right of governing society, which the majority supposes itself to derive from its superior intelligence, was introduced into the United States by the first settlers, and this idea, which would be sufficient of itself to create a free nation, has now been amalgamated with the manners of the people and the minor incidents of social intercourse.

The French, under the old monarchy, held it for a maxim (which is still a fundamental principle of the English Constitution) that the King could do no wrong; and if he did do wrong, the blame was imputed to his advisers. This notion was highly favourable to habits of obedience, and it enabled the subject to complain of the law without ceasing to love and honour the lawgiver. The Americans entertain the same opinion with respect to the majority.

The moral power of the majority is founded upon yet another principle, which is, that the interests of the many are to be preferred to those of the few. It will readily be perceived that the respect here professed for the rights of the majority must naturally increase or diminish according to the state of parties. When a nation is divided into several irreconcilable factions, the privilege of the majority is often overlooked, because it is intolerable to comply with its demands.

If there existed in America a class of citizens whom the legislating majority sought to deprive of exclusive privileges which they had possessed for ages, and to bring down from an elevated station to the level of the ranks of the multitude, it is probable that the minority would be less ready to comply with its laws. But as the United States were colonized by men holding equal rank amongst themselves, there is as yet no natural or permanent source of dissension between the interests of its different inhabitants.

There are certain communities in which the persons who constitute the minority can never hope to draw over the majority to their side, because they must then give up the very point which is at issue between them. Thus, an aristocracy can never become a majority whilst it retains its exclusive privileges, and it cannot cede its privileges without ceasing to be an aristocracy.

In the United States political questions cannot be taken up in so general and absolute a manner, and all parties are willing to recognize the rights of the majority, because they all hope to turn those rights to their own advantage at some future time. The majority therefore in that country exercises a prodigious actual authority, and a moral influence which is scarcely less preponderant; no obstacles exist which can impede or so much as retard its progress, or which can induce it to heed the complaints of those whom it crushes upon its path. This state of things is fatal in itself and dangerous for the future.

How the Unlimited Power of the Majority Increases in America the Instability of Legislation and Administration Inherent in Democracy

The Americans increase the mutability of the laws which is inherent in democracy by changing the legislature every year, and by investing it with unbounded authority—The same effect is produced upon the administration—In America social amelioration is conducted more energetically but less perseveringly than in Europe.

I have already spoken of the natural defects of democratic institutions, and they all of them increase at the exact ratio of the power of the majority. To begin with the most evident of them all; the mutability of the laws is an evil inherent in democratic government, because it is natural to democracies to raise men to power in very rapid succession. But this evil is more or less sensible in proportion to the authority and the means of action which the legislature possesses.

In America the authority exercised by the legislative bodies is supreme; nothing prevents them from accomplishing their wishes with celerity, and with irresistible power, whilst they are supplied by new representatives every year. That is to say, the circumstances which contribute most powerfully to democratic instability, and which admit of the free application of caprice to every object in the State, are here in full operation. In conformity with this principle, America is, at the present day, the country in the world where laws last the shortest time. Almost all the American constitutions have been amended within the course of thirty years: there is therefore not a single American State which has not modified the principles of its legislation in that lapse of time. As for the laws themselves, a single glance upon the archives of the different States of the Union suffices to convince one that in America the activity of the legislator never slackens. Not that the American democracy is naturally less stable than any other, but that it is allowed to follow its capricious propensities in the formation of the laws.[2]

The omnipotence of the majority, and the rapid as well as absolute manner in which its decisions are executed in the United States, has not only the effect of rendering the law unstable, but it exercises the same influence upon the execution of the law and the conduct of the public administration. As the majority is the only power which it is important to court, all its projects are taken up with the greatest ardour, but no sooner is its attention distracted than all this ardour ceases; whilst in the free States of Europe the administration is at once independent and secure, so that the projects of the legislature are put into execution, although its immediate attention may be directed to other objects.

In America certain ameliorations are undertaken with much more zeal and activity than elsewhere; in Europe the same ends are promoted by much less social effort, more continuously applied.

Some years ago several pious individuals undertook to ameliorate the condition of the prisons. The public was excited by the statements which they put forward, and the regeneration of criminals became a very popular undertaking. New prisons were built, and for the first time the idea of reforming as well as of punishing the delinquent formed a part of prison discipline. But this happy alteration, in which the public had taken so hearty an interest, and which the exertions of the citizens had irresistibly accelerated, could not be completed in a moment. Whilst the new penitentiaries were being erected (and it was the pleasure of the majority that they should be terminated with all possible celerity), the old prisons existed, which still contained a great number of offenders. These gaols became more unwholesome and more corrupt in proportion as the new establishments were beautified and improved, forming a contrast which may readily be understood. The majority was so eagerly employed in founding the new prisons that those which already existed were forgotten; and as the general attention was diverted to a novel object, the care which had hitherto been bestowed upon the others ceased. The salutary regulations of discipline were first relaxed, and afterwards broken; so that in the immediate neighbourhood of a prison which bore witness to the mild and enlightened spirit of our time, dungeons might be met with which reminded the visitor of the barbarity of the Middle Ages.

Tyranny of the Majority

How the principle of the sovereignty of the people is to be understood—Impossibility of conceiving a mixed government—The sovereign power must centre somewhere—Precautions to be taken to control its action—These precautions have not been taken in the United States—Consequences.

I hold it to be an impious and an execrable maxim that, politically speaking, a people has a right to do whatsoever it pleases, and yet I have asserted that all authority originates in the will of the majority. Am I then, in contradiction with myself?

A general law—which bears the name of Justice—has been made and sanctioned, not only by a majority of this or that people, but by a majority of mankind. The rights of every people are consequently confined within the limits of what is just. A nation may be considered in the light of a jury which is empowered to represent society at large, and to apply the great and general law of justice. Ought such a jury, which represents society, to have more power than the society in which the laws it applies originate?

When I refuse to obey an unjust law, I do not contest the right which the majority has of commanding, but I simply appeal from the sovereignty of the people to the sovereignty of mankind. It has been asserted that a people can never entirely outstep the boundaries of justice and of reason in those affairs which are more peculiarly its own, and that consequently, full power may fearlessly be given to the majority by which it is represented. But this language is that of a slave.

A majority taken collectively may be regarded as a being whose opinions, and most frequently whose interests, are opposed to those of another being, which is styled a minority. If it be admitted that a man, possessing absolute power, may misuse that power by wronging his adversaries, why should a majority not be liable to the same reproach? Men are not apt to change their characters by agglomeration; nor does their patience in the presence of obstacles increase with the consciousness of their strength.[3] And for these reasons I can never willingly invest any number of my fellow-creatures with that unlimited authority which I should refuse to any one of them.

I do not think that it is possible to combine several principles in the same government, so as at the same time to maintain freedom, and really to oppose them to one another. The form of government which is usually termed *mixed* has always appeared to me to be a mere chimera. Accurately speaking there is no such thing as a mixed government (with the meaning usually given to that word), because in all communities some one principle of action may be discovered which preponderates over the others. England in the last century, which has been more especially cited as an example of this form of Government, was in point of fact an essentially aristocratic State, although it comprised very powerful elements of democracy; for the laws and customs of the country were such that the aristocracy could not but preponderate in the end, and subject the direction of public affairs to its own will. The error arose from too much attention being paid to the actual struggle which was going on between the nobles and the people, without considering the probable issue of the contest, which was in reality the important point. When a community really has a mixed government, that is to say, when it is equally divided between two adverse principles, it must either pass through a revolution or fall into complete dissolution.

I am therefore of opinion that some one social power must always be made to predominate over the others; but I think that liberty is endangered when this power is checked by no obstacles which may retard its course, and force it to moderate its own vehemence.

Unlimited power is in itself a bad and dangerous thing; human beings are not competent to exercise it with discretion, and God alone can be omnipotent, because His wisdom and His justice are always equal to His power. But no power upon earth is so worthy of honour for itself, or of reverential obedience to the rights which it represents, that I would consent to admit its uncontrolled and all-predominant authority. When I see that the right and the means of absolute command are conferred on a people or upon a king, upon an aristocracy or a democracy, a monarchy or a republic, I recognise the germ of tyranny, and I journey onward to a land of more hopeful institutions.

In my opinion the main evil of the present democratic institutions of the United States does not arise, as is often asserted in Europe, from their weakness, but from their overpowering strength; and I am not so much alarmed at the excessive liberty which reigns in that country as at the very inadequate securities which exist against tyranny.

When an individual or a party is wronged in the United States, to whom can he apply for redress? If to public opinion, public opinion constitutes the majority; if to the legislature, it represents the majority, and implicitly obeys its injunctions; if to the executive power, it is appointed by the majority, and remains a passive tool in its hands; the public troops consist of the majority under arms; the jury is the majority invested with the right of hearing judicial cases; and in certain States even the judges are elected by the majority. However iniquitous or absurd the evil of which you complain may be, you must submit to it as well as you can.[4]

I said one day to an inhabitant of Pennsylvania, 'Be so good as to explain to me how it happens that in a State founded by Quakers, and celebrated for its toleration, freed Blacks are not allowed to exercise civil rights. They pay the taxes; is it not fair that they should have a vote?'

'You insult us,' replied my informant, 'if you imagine that our legislators could have committed so gross an act of injustice and intolerance.'

'What! then the Blacks possess the right of voting in this country?'

'Without the smallest doubt.'

'How comes it, then, that at the polling-booth this morning I did not perceive a single Negro in the whole meeting?'

'This is not the fault of the law: the Negroes have an undisputed right of voting, but they voluntarily abstain from making their appearance.'

'A very pretty piece of modesty on their parts!' rejoined I.

'Why, the truth is, that they are not disinclined to vote, but they are afraid of being maltreated; in this country the law is sometimes unable to maintain its authority without the support of the majority. But in this case the majority entertains very strong prejudices against the Blacks, and the magistrates are unable to protect them in the exercise of their legal privileges.'

'What! then the majority claims the right not only of making the laws, but of breaking the laws it has made?'

If, on the other hand, a legislative power could be so constituted as to represent the majority without necessarily being the slave of its passions; an executive, so as to retain a certain degree of uncontrolled authority; and a judiciary, so as to remain independent of the two other powers; a government would be formed which would still be democratic without incurring any risk of tyrannical abuse.

I do not say that tyrannical abuses frequently occur in America at the present day, but I maintain that no sure barrier is established against them, and that the causes which mitigate the government are to be found in the circumstances and the manners of the country more than in its laws.

Effects of the Unlimited Power of the Majority upon the Arbitrary Authority of the American Public Officers

Liberty left by the American laws to public officers within a certain sphere—Their power.

A distinction must be drawn between tyranny and arbitrary power. Tyranny may be exercised by means of the law, and in that case it is not arbitrary; arbitrary power may be exercised for the good of the community at large, in which case it is not tyrannical. Tyranny usually employs arbitrary means, but, if necessary, it can rule without them.

In the United States the unbounded power of the majority, which is favourable to the legal despotism of the legislature, is likewise favourable to the arbitrary authority of the magistrate. The majority has an entire control over the law when it is made and when it is executed; and as it possesses an equal authority over those who are in power and the community at large, it considers public officers as its passive agents, and readily confides the task of serving its designs to their vigilance. The details of their office and the privileges which they are to enjoy are rarely defined beforehand; but the majority treats them as a master does his servants when they are always at work in his sight, and he has the power of directing or reprimanding them at every instant.

In general the American functionaries are far more independent than the French civil officers within the sphere which is prescribed to them. Sometimes, even, they are allowed by the popular authority to exceed those bounds; and as they are protected by the opinion, and backed by the cooperation, of the majority, they venture upon such manifestations of their power as astonish a European. By this means habits are formed in the heart of a free country which may some day prove fatal to its liberties.

Power Exercised by the Majority in America upon Opinion

In America, when the majority has once irrevocably decided a question, all discussion ceases—Reason of this—Moral power exercised by the majority upon opinion—Democratic republics have deprived despotism of its physical instruments—Their despotism sways the minds of men.

It is in the examination of the display of public opinion in the United States that we clearly perceive how far the power of the majority surpasses all the powers with which we are acquainted in Europe. Intellectual principles exercise an influence which is so invisible, and often so inappreciable, that they baffle the toils of oppression. At the present time the most absolute monarchs in Europe are unable to prevent certain notions, which are opposed to their authority, from circulating in secret throughout their dominions, and even in their courts. Such is not the case in America; as long as the majority is still undecided, discussion is carried on; but as soon as its decision is irrevocably pronounced, a submissive silence is observed, and the friends, as well as the opponents, of the measure unite in assenting to its propriety. The reason of this is perfectly clean no monarch is so absolute as to combine all the powers of society in his own hands, and to conquer all opposition with the energy of a majority which is invested with the right of making and of executing the laws.

The authority of a king is purely physical, and it controls the actions of the subject without subduing his private will; but the majority possesses a power which is physical and moral at the same time; it acts upon the will as well as upon the actions of men, and it represses not only all contest, but all controversy.

I know no country in which there is so little true independence of mind and freedom of discussion as in America. In any constitutional state in Europe every sort of religious and political theory may be advocated and propagated abroad; for there is no country in Europe so subdued by any single authority as not to contain citizens who are ready to protect the man who raises his voice in the cause of truth from the consequences of his hardihood. If he is unfortunate enough to live under an absolute government, the people is upon his side; if he inhabits a free country, he may find a shelter behind the authority of the throne, if he require one. The aristocratic part of society supports him in some countries, and the democracy in others. But in a nation where democratic institutions exist, organized like those of the United States, there is but one sole authority, one single element of strength and of success, with nothing beyond it.

In America the majority raises very formidable barriers to the liberty of opinion: within these barriers an author may write whatever he pleases, but he will repent it if he ever step beyond them. Not that he is exposed to the terrors of an auto-da-fe, but he is tormented by the slights and persecutions of daily obloquy. His political career is closed forever, since he has offended the only authority which is able to promote his success. Every sort of compensation, even that of celebrity, is refused to him. Before he published his opinions he imagined that he held them in common with many others; but no sooner has he declared them openly than he is loudly censured by his overbearing opponents, whilst those who think without having the courage to speak, like him, abandon him in silence. He yields at length, oppressed by the daily efforts he has been making, and he subsides into silence, as if he was tormented by remorse for having spoken the truth.

Fetters and headsmen were the coarse instruments which tyranny formerly employed; but the civilization of our age has refined the arts of despotism which seemed, however, to have been sufficiently perfected before. The excesses of monarchical power had devised a variety of physical means of oppression: the democratic republics of the present day have rendered it as entirely an affair of the mind as that will which it is intended to coerce. Under the absolute sway of an individual despot the body was attacked in order to subdue the soul, and the soul escaped the blows which were directed against it and rose superior to the attempt; but such is not the course adopted by tyranny in democratic republics; there the body is left free, and the soul is enslaved. The sovereign can no longer say, 'You shall think as I do on pain of death;' but he says, 'You are free to think differently from me, and to retain your life, your property, and all that you possess; but if such be your determination, you are henceforth an alien among your people. You may retain your civil rights, but they will be useless to you, for you will never be chosen by your fellow-citizens if you solicit their suffrages, and they will affect to scorn you if you solicit their esteem. You will remain among men, but you will be deprived of the rights of mankind. Your fellow-creatures will shun you like an impure being, and those who are most persuaded of your innocence will abandon you too, lest they should be shunned in their turn. Go in peace! I have given you your life, but it is an existence incomparably worse than death.'

Monarchical institutions have thrown an odium upon despotism; let us beware lest democratic republics should restore oppression, and should render it less odious and less degrading in the eyes of the many, by making it still more onerous to the few.

Works have been published in the proudest nations of the Old World expressly intended to censure the vices and deride the follies of the times: Labruyère inhabited the palace of Louis XIV. when he composed his chapter upon the Great, and Molière criticised the courtiers in the very pieces which were acted before the Court. But the ruling power in the United States is not to be made game of; the smallest reproach irritates its sensibility, and the slightest joke which has any foundation in truth renders it indignant; from the style of its language to the more solid virtues of its character, everything must be made the subject of encomium. No writer, whatever be his eminence, can escape from this tribute of adulation to his fellow-citizens. The majority lives in the perpetual practice of self-applause, and there are certain truths which the Americans can only learn from strangers or from experience.

If great writers have not at present existed in America, the reason is very simply given in these facts; there can be no literary genius without freedom of opinion, and freedom of opinion does not exist in America. The Inquisition has never been able to prevent a vast number of anti-religious books from circulating in Spain. The empire of the majority succeeds much better in the United States, since it actually removes the wish of publishing them. Unbelievers are to be met with in America, but, to say the truth, there is no public organ of infidelity. Attempts have been made by some governments to protect the morality of nations by prohibiting licentious books. In the United States no one is punished for this sort of works, but no one is induced to write them; not because all the citizens are immaculate in their manners, but because the majority of the community is decent and orderly.

In these cases the advantages derived from the exercise of this power are unquestionable, and I am simply discussing the nature of the power itself. This irresistible authority is a constant fact, and its judicious exercise is an accidental occurrence.

Effects of the Tyranny of the Majority upon the National Character of the Americans

Effects of the tyranny of the majority more sensibly felt hitherto in the manners than in the conduct of society—They check the development of leading characters—Democratic republics organized like the United States bring the practice of courting favor within the reach of the many—Proofs of this spirit in the United States—Why there is more patriotism in the people than in those who govern in its name.

The tendencies which I have just alluded to are as yet very slightly perceptible in political society, but they already begin to exercise an unfavourable influence upon the national character of the Americans. I am inclined to attribute the singular paucity of distinguished political characters to the ever-increasing activity of the despotism of the majority in the United States. When the American Revolution broke out they arose in great numbers, for public opinion then served, not to tyrannize over, but to direct the exertions of individuals. Those celebrated men took a full part in the general agitation of mind common at that period, and they attained a high degree of personal fame, which was reflected back upon the nation, but which was by no means borrowed from it.

In absolute governments the great nobles who are nearest to the throne flatter the passions of the sovereign, and voluntarily truckle to his caprices. But the mass of the nation does not degrade itself by servitude: it often submits from weakness, from habit, or from ignorance, and sometimes from loyalty. Some nations have been known to sacrifice their own desires to those of the sovereign with pleasure and with pride, thus exhibiting a sort of independence in the very act of submission. These peoples are miserable, but they are not degraded. There is a great difference between doing what one does not approve and feigning to approve what one does; the one is the necessary case of a weak person, the other befits the temper of a lackey.

In free countries, where everyone is more or less called upon to give his opinion in the affairs of state; in democratic republics, where public life is incessantly commingled with domestic affairs, where the sovereign authority is accessible on every side, and where its attention can almost always be attracted by vociferation, more persons are to be met with who speculate upon its foibles and live at the cost of its passions than in absolute monarchies. Not because men are naturally worse in these States than elsewhere, but the temptation is stronger, and of easier access at the same time. The result is a far more extensive debasement of the characters of citizens.

Democratic republics extend the practice of currying favour with the many, and they introduce it into a greater number of classes at once: this is one of the most serious reproaches that can be addressed to them. In democratic States organized on the principles of the American republics, this is more especially the case, where the authority of the majority is so absolute and so irresistible that a man must give up his rights as a citizen, and almost abjure his quality as a human being, if he intends to stray from the track which it lays down.

In that immense crowd which throngs the avenues to power in the United States I found very few men who displayed any of that manly candor and that masculine independence of opinion which frequently distinguished the Americans in former times, and which constitutes the leading feature in distinguished characters, wheresoever they may be found. It seems, at first sight, as if all the minds of the Americans were formed upon one model, so accurately do they correspond in their manner of judging. A stranger does, indeed, sometimes meet with Americans who dissent from these rigorous formularies; with men who deplore the defects of the laws, the mutability and the ignorance of democracy; who even go so far as to observe the evil tendencies which impair the national character, and to point out such remedies as it might be possible to apply; but no one is there to hear these things besides yourself, and you, to whom these secret reflections are confided, are a stranger and a bird of passage. They are very ready to communicate truths which are useless to you, but they continue to hold a different language in public.

If ever these lines are read in America, I am well assured of two things: in the first place, that all who peruse them will raise their voices to condemn me; and in the second place, that very many of them will acquit me at the bottom of their conscience.

I have heard of patriotism in the United States, and it is a virtue which may be found among the people, but never among the leaders of the people. This may be explained by analogy; despotism debases the oppressed much more than the oppressor: in absolute monarchies the king has often great virtues, but the courtiers are invariably servile. It is true that the American courtiers do not say 'Sire,' or 'Your Majesty' a distinction without a difference. They are for ever talking of the natural intelligence of the populace they serve; they do not debate the question as to which of the virtues of their master is pre-eminently worthy of admiration, for they assure him that he possesses all the virtues under heaven without having acquired them, or without caring to acquire them; they do not give him their daughters and their wives to be raised at his pleasure to the rank of his concubines, but, by sacrificing their opinions, they prostitute themselves. Moralists and philosophers in America are not obliged to conceal their opinions under the veil of allegory; but, before they venture upon a harsh truth, they say, 'We are aware that the people which we are addressing is too superior to all the weaknesses of human nature to lose the command of its temper for an instant; and we should not hold this language if we were not speaking to men whom their virtues and their intelligence render more worthy of freedom than all the rest of the world.' It would have been impossible for the sycophants of Louis XIV. to flatter more dexterously. For my part, I am persuaded that in all governments, whatever their nature may be, servility will cower to force, and adulation will cling to power. The only means of preventing men from degrading themselves is to invest no one with that unlimited authority which is the surest method of debasing them.

The Greatest Dangers of the American Republics Proceed from the Unlimited Power of the Majority

Democratic republics liable to perish from a misuse of their power, and not by impotence—The Governments of the American republics are more centralized and more energetic than those of the monarchies of Europe—Dangers resulting from this—Opinions of Hamilton and Jefferson upon this point.

Governments usually fall a sacrifice to impotence or to tyranny. In the former case their power escapes from them; it is wrested from their grasp in the latter. Many observers, who have witnessed the anarchy of democratic States, have imagined that the government of those States was naturally weak and impotent. The truth is, that when once hostilities are begun between parties, the government loses its control over society. But I do not think that a democratic power is naturally without force or without resources: say, rather, that it is almost always by the abuse of its force and the misemployment of its resources that a democratic government fails. Anarchy is almost always produced by its tyranny or its mistakes, but not by its want of strength.

It is important not to confound stability with force, or the greatness of a thing with its duration. In democratic republics, the power which directs[5] society is not stable; for it often changes hands and assumes a new direction. But whichever way it turns, its force is almost irresistible. The Governments of the American republics appear to me to be as much centralised as those of the absolute monarchies of Europe, and more energetic than they are. I do not, therefore, imagine that they will perish from weakness.[6]

If ever the free institutions of America are destroyed, that event may be attributed to the unlimited authority of the majority, which may at some future time urge the minorities to desperation, and oblige them to have recourse to physical force. Anarchy will then be the result, but it will have been brought about by despotism.

Mr. Hamilton expresses the same opinion in the 'Federalist,' No. 51. 'It is of great importance in a republic not only to guard the society against the oppression of its rulers, but to guard one part of the society against the injustice of the other part. Justice is the end of government. It is the end of civil society. It ever has been, and ever will be, pursued until it be obtained, or until liberty be lost in the pursuit. In a society, under the forms of which the stronger faction can readily unite and oppress the weaker, anarchy may as truly be said to reign as in a state of nature, where the weaker individual is not secured against the violence of the stronger: and as in the latter state even the stronger individuals are prompted by the uncertainty of their condition to submit to a government which may protect the weak as well as themselves, so in the former state will the more powerful factions be gradually induced by a like motive to wish for a government which will protect all parties, the weaker as well as the more powerful. It can be little doubted that, if the State of Rhode Island was separated from the Confederacy and left to itself, the insecurity of right under the popular form of government within such narrow limits would be displayed by such reiterated oppressions of the factious majorities, that some power altogether independent of the people would soon be called for by the voice of the very factions whose misrule had proved the necessity of it.'

Jefferson has also thus expressed himself in a letter to Madison:[7] "The executive power in our Government is not the only, perhaps not even the principal, object of my solicitude. The tyranny of the Legislature is really the danger most to be feared, and will continue to be so for many years to come. The tyranny of the executive power will come in its turn, but at a more distant period." I am glad to cite the opinion of Jefferson upon this subject rather than that of another, because I consider him to be the most powerful advocate democracy has ever sent forth.

Chapter XVI

Causes Which Mitigate the Tyranny of the Majority in the United States

Absence of Central Administration

The national majority does not pretend to conduct all business—Is obliged to employ the town and county magistrates to execute its supreme decisions.

I have already pointed out the distinction which is to be made between a centralised government and a centralised administration. The former exists in America, but the latter is nearly unknown there. If the directing power of the American communities had both these instruments of government at its disposal, and united the habit of executing its own commands to the right of commanding; if, after having established the general principles of government, it descended to the details of public business; and if, having regulated the great interests of the country, it could penetrate into the privacy of individual interests, freedom would soon be banished from the New World.

But in the United States the majority, which so frequently displays the tastes and the propensities of a despot, is still destitute of the more perfect instruments of tyranny. In the American republics the activity of the central Government has never as yet been extended beyond a limited number of objects sufficiently prominent to call forth its attention. The secondary affairs of society have never been regulated by its authority, and nothing has hitherto betrayed its desire of interfering in them. The majority is become more and more absolute, but it has not increased the prerogatives of the central government; those great prerogatives have been confined to a certain sphere; and although the despotism of the majority may be galling upon one point, it cannot be said to extend to all. However the predominant party in the nation may be carried away by its passions, however ardent it may be in the pursuit of its projects, it cannot oblige all the citizens to comply with its desires in the same manner and at the same time throughout the country. When the central Government which represents that majority has issued a decree, it must entrust the execution of its will to agents, over whom it frequently has no control, and whom it cannot perpetually direct. The townships, municipal bodies, and counties may therefore be looked upon as concealed breakwaters, which check or part the tide of popular excitement. If an oppressive law were passed, the liberties of the people would still be protected by the means by which that law would be put in execution: the majority cannot descend to the details and (as I will venture to style them) the puerilities of administrative tyranny. Nor does the people entertain that full consciousness of its authority which would prompt it to interfere in these matters; it knows the extent of its natural powers, but it is unacquainted with the increased resources which the art of government might furnish.

This point deserves attention, for if a democratic republic similar to that of the United States were ever founded in a country where the power of a single individual had previously subsisted, and the effects of a centralised administration had sunk deep into the habits and the laws of the people, I do not hesitate to assert, that in that country a more insufferable despotism would prevail than any which now exists in the monarchical States of Europe, or indeed than any which could be found on this side of the confines of Asia.

The Profession of the Law in the United States Serves to Counterpoise the Democracy

Utility of discriminating the natural propensities of the members of the legal profession—These men called upon to act a prominent part in future society—In what manner the peculiar pursuits of lawyers give an aristocratic turn to their ideas—Accidental causes which may check this tendency—Ease with which the aristocracy coalesces with legal men—Use of lawyers to a despot—The profession of the law constitutes the only aristocratic element with which the natural elements of democracy will combine—Peculiar causes which tend to give an aristocratic turn of mind to the English and American lawyers—The aristocracy of America is on the bench and at the bar—Influence of lawyers upon American society—Their peculiar magisterial habits affect the legislature, the administration, and even the people.

In visiting the Americans and in studying their laws we perceive that the authority they have entrusted to members of the legal profession, and the influence which these individuals exercise in the Government, is the most powerful existing security against the excesses of democracy. This effect seems to me to result from a general cause which it is useful to investigate, since it may produce analogous consequences elsewhere.

The members of the legal profession have taken an important part in all the vicissitudes of political society in Europe during the last five hundred years. At one time they have been the instruments of those who were invested with political authority, and at another they have succeeded in converting political authorities into their instrument. In the Middle Ages they afforded a powerful support to the Crown, and since that period they have exerted themselves to the utmost to limit the royal prerogative. In England they have contracted a close alliance with the aristocracy; in France they have proved to be the most dangerous enemies of that class. It is my object to inquire whether, under all these circumstances, the members of the legal profession have been swayed by sudden and momentary impulses; or whether they have been impelled by principles which are inherent in their pursuits, and which will always recur in history. I am incited to this investigation by reflecting that this particular class of men will most likely play a prominent part in that order of things to which the events of our time are giving birth.

Men who have more especially devoted themselves to legal pursuits derive from those occupations certain habits of order, a taste for formalities, and a kind of instinctive regard for the regular connection of ideas, which naturally render them very hostile to the revolutionary spirit and the unreflecting passions of the multitude.

The special information which lawyers derive from their studies ensures them a separate station in society, and they constitute a sort of privileged body in the scale of intelligence. This notion of their superiority perpetually recurs to them in the practice of their profession: they are the masters of a science which is necessary, but which is not very generally known; they serve as arbiters between the citizens; and the habit of directing the blind passions of parties in litigation to their purpose inspires them with a certain contempt for the judgment of the multitude. To this it may be added that they naturally constitute *a body,* not by any previous understanding, or by an agreement which directs them to a common end; but the analogy of their studies and the uniformity of their proceedings connect their minds together, as much as a common interest could combine their endeavours.

A portion of the tastes and of the habits of the aristocracy may consequently be discovered in the characters of men in the profession of the law. They participate in the same instinctive love of order and of formalities; and they entertain the same repugnance to the actions of the multitude, and the same secret contempt of the government of the people. I do not mean to say that the natural propensities of lawyers are sufficiently strong to sway them irresistibly; for they, like most other men, are governed by their private interests and the advantages of the moment.

In a state of society in which the members of the legal profession are prevented from holding that rank in the political world which they enjoy in private life, we may rest assured that they will be the foremost agents of revolution. But it must then be inquired whether the cause which induces them to innovate and to destroy is accidental, or whether it belongs to some lasting purpose which they entertain. It is true that lawyers mainly contributed to the overthrow of the French Monarchy in 1789; but it remains to be seen whether they acted thus because they had studied the laws, or because they were prohibited from cooperating in the work of legislation.

Five hundred years ago the English nobles headed the people, and spoke in its name; at the present time the aristocracy supports the throne, and defends the royal prerogative. But aristocracy has, notwithstanding this, its peculiar instincts and propensities. We must be careful not to confound isolated members of a body with the body itself. In all free governments, of whatsoever form they may be, members of the legal profession will be found at the head of all parties. The same remark is also applicable to the aristocracy; for almost all the democratic convulsions which have agitated the world have been directed by nobles.

A privileged body can never satisfy the ambition of all its members; it has always more talents and more passions to content and to employ than it can find places; so that a considerable number of individuals are usually to be met with who are inclined to attack those very privileges which they find it impossible to turn to their own account.

I do not, then, assert that *all* the members of the legal profession are at *all* times the friends of order and the opponents of innovation, but merely that most of them usually are so. In a community in which lawyers are allowed to occupy, without opposition, that high station which naturally belongs to them, their general spirit will be eminently conservative and anti-democratic. When an aristocracy excludes the leaders of that profession from its ranks, it excites enemies which are the more formidable to its security as they are independent of the nobility by their industrious pursuits; and they feel themselves to be its equal in point of intelligence, although they enjoy less opulence and less power. But whenever an aristocracy consents to impart some of its privileges to these same individuals, the two classes coalesce very readily, and assume, as it were, the consistency of a single order of family interests.

I am, in like manner, inclined to believe that a monarch will always be able to convert legal practitioners into the most serviceable instruments of his authority. There is a far greater affinity between this class of individuals and the executive power than there is between them and the people; just as there is a greater natural affinity between the nobles and the monarch than between the nobles and the people, although the higher orders of society have occasionally resisted the prerogative of the Crown in concert with the lower classes.

Lawyers are attached to public order beyond every other consideration, and the best security of public order is authority. It must not be forgotten that, if they prize the free institutions of their country much, they nevertheless value the legality of those institutions far more: they are less afraid of tyranny than of arbitrary power; and provided that the legislature take upon itself to deprive men of their independence, they are not dissatisfied.

I am therefore convinced that the prince who, in presence of an encroaching democracy, should endeavour to impair the judicial authority in his dominions, and to diminish the political influence of lawyers, would commit a great mistake. He would let slip the substance of authority to grasp at the shadow. He would act more wisely in introducing men connected with the law into the government; and if he entrusted them with the conduct of a despotic power, bearing some marks of violence, that power would most likely assume the external features of justice and of legality in their hands.

The government of democracy is favourable to the political power of lawyers; for when the wealthy, the noble, and the prince are excluded from the government, they are sure to occupy the highest stations, in their own right, as it were, since they are the only men of information and sagacity, beyond the sphere of the people, who can be the object of the popular choice. If, then, they are led by their tastes to combine with the aristocracy and to support the Crown, they are naturally brought into contact with the people by their interests. They like the government of democracy, without participating in its propensities and without imitating its weaknesses; whence they derive a twofold authority, from it and over it. The people in democratic states does not mistrust the members of the legal profession, because it is well known that they are interested in serving the popular cause; and it listens to them without irritation, because it does not attribute to them any sinister designs. The object of lawyers is not, indeed, to overthrow the institutions of democracy, but they constantly endeavour to give it an impulse which diverts it from its real tendency, by means which are foreign to its nature. Lawyers belong to the people by birth and interest, to the aristocracy by habit and by taste, and they may be looked upon as the natural bond and connecting link of the two great classes of society.

The profession of the law is the only aristocratic element which can be amalgamated without violence with the natural elements of democracy, and which can be advantageously and permanently combined with them. I am not unacquainted with the defects which are inherent in the character of that body of men; but without this admixture of lawyer-like sobriety with the democratic principle, I question whether democratic institutions could long be maintained, and I cannot believe that a republic could subsist at the present time if the influence of lawyers in public business did not increase in proportion to the power of the people.

This aristocratic character, which I hold to be common to the legal profession, is much more distinctly marked in the United States and in England than in any other country. This proceeds not only from the legal studies of the English and American lawyers, but from the nature of the legislation, and the position which those persons occupy in the two countries. The English and the Americans have retained the law of precedents; that is to say, they continue to found their legal opinions and the decisions of their courts upon the opinions and the decisions of their forefathers. In the mind of an English or American lawyer a taste and a reverence for what is old is almost always united to a love of regular and lawful proceedings.

This predisposition has another effect upon the character of the legal profession and upon the general course of society. The English and American lawyers investigate what has been done; the French advocate inquires what should have been done; the former produce precedents, the latter reasons. A French observer is surprised to hear how often an English or an American lawyer quotes the opinions of others, and how little he alludes to his own; whilst the reverse occurs in France. There the most trifling litigation is never conducted without the introduction of an entire system of ideas peculiar to the counsel employed; and the fundamental principles of law are discussed in order to obtain a perch of land by the decision of the court. This abnegation of his own opinion, and this implicit deference to the opinion of his forefathers, which are common to the English and American lawyer, this subjection of thought which he is obliged to profess, necessarily give him more timid habits and more sluggish inclinations in England and America than in France.

The French Codes are often difficult of comprehension, but they can be read by every one; nothing, on the other hand, can be more impenetrable to the uninitiated than a legislation founded upon precedents. The indispensable want of legal assistance which is felt in England and in the United States, and the high opinion which is generally entertained of the ability of the legal profession, tend to separate it more and more from the people, and to place it in a distinct class. The French lawyer is simply a man extensively acquainted with the statutes of his country; but the English or American lawyer resembles the hierophants of Egypt, for, like them, he is the sole interpreter of an occult science.

The station which lawyers occupy in England and America exercises no less an influence upon their habits and their opinions. The English aristocracy, which has taken care to attract to its sphere whatever is at all analogous to itself, has conferred a high degree of importance and of authority upon the members of the legal profession. In English society lawyers do not occupy the first rank, but they are contented with the station assigned to them; they constitute, as it were, the younger branch of the English aristocracy, and they are attached to their elder brothers, although they do not enjoy all their privileges. The English lawyers consequently mingle the taste and the ideas of the aristocratic circles in which they move with the aristocratic interests of their profession.

And indeed the lawyer-like character which I am endeavouring to depict is most distinctly to be met with in England: there laws are esteemed not so much because they are good as because they are old; and if it be necessary to modify them in any respect, or to adapt them to the changes which time operates in society, recourse is had to the most inconceivable contrivances in order to uphold the traditionary fabric, and to maintain that nothing has been done which does not square with the intentions and complete the labours of former generations. The very individuals who conduct these changes disclaim all intention of innovation, and they had rather resort to absurd expedients than plead guilty to so great a crime. This spirit appertains more especially to the English lawyers; they seem indifferent to the real meaning of what they treat, and they direct all their attention to the letter, seeming inclined to infringe the rules of common sense and of humanity rather than to swerve one title from the law. The English legislation may be compared to the stock of an old tree, upon which lawyers have engrafted the most various shoots, with the hope that, although their fruits may differ, their foliage at least will be confounded with the venerable trunk which supports them all.

In America there are no nobles or men of letters, and the people is apt to mistrust the wealthy; lawyers consequently form the highest political class, and the most cultivated circle of society. They have therefore nothing to gain by innovation, which adds a conservative interest to their natural taste for public order. If I were asked where I place the American aristocracy, I should reply without hesitation that it is not composed of the rich, who are united together by no common tie, but that it occupies the judicial bench and the bar.

The more we reflect upon all that occurs in the United States the more shall we be persuaded that the lawyers as a body form the most powerful, if not the only, counterpoise to the democratic element. In that country we perceive how eminently the legal profession is qualified by its powers, and even by its defects, to neutralize the vices which are inherent in popular government. When the American people is intoxicated by passion, or carried away by the impetuosity of its ideas, it is checked and stopped by the almost invisible influence of its legal counsellors, who secretly oppose their aristocratic propensities to its democratic instincts, their superstitious attachment to what is antique to its love of novelty, their narrow views to its immense designs, and their habitual procrastination to its ardent impatience.

The courts of justice are the most visible organs by which the legal profession is enabled to control the democracy. The judge is a lawyer, who, independently of the taste for regularity and order which he has contracted in the study of legislation, derives an additional love of stability from his own inalienable functions. His legal attainments have already raised him to a distinguished rank amongst his fellow-citizens; his political power completes the distinction of his station, and gives him the inclinations natural to privileged classes.

Armed with the power of declaring the laws to be unconstitutional,[1] the American magistrate perpetually interferes in political affairs. He cannot force the people to make laws, but at least he can oblige it not to disobey its own enactments; or to act inconsistently with its own principles. I am aware that a secret tendency to diminish the judicial power exists in the United States, and by most of the constitutions of the several States the Government can, upon the demand of the two houses of the legislature, remove the judges from their station. By some other constitutions the members of the tribunals are elected, and they are even subjected to frequent reelections. I venture to predict that these innovations will sooner or later be attended with fatal consequences, and that it will be found out at some future period that the attack which is made upon the judicial power has affected the democratic republic itself.

It must not, however, be supposed that the legal spirit of which I have been speaking has been confined, in the United States, to the courts of justice; it extends far beyond them. As the lawyers constitute the only enlightened class which the people does not mistrust, they are naturally called upon to occupy most of the public stations. They fill the legislative assemblies, and they conduct the administration; they consequently exercise a powerful influence upon the formation of the law, and upon its execution. The lawyers are, however, obliged to yield to the current of public opinion, which is too strong for them to resist it, but it is easy to find indications of what their conduct would be if they were free to act as they chose. The Americans, who have made such copious innovations in their political legislation, have introduced very sparing alterations in their civil laws, and that with great difficulty, although those laws are frequently repugnant to their social condition. The reason of this is, that in matters of civil law the majority is obliged to defer to the authority of the legal profession, and that the American lawyers are disinclined to innovate when they are left to their own choice.

It is curious for a Frenchman, accustomed to a very different state of things, to hear the perpetual complaints which are made in the United States against the stationary propensities of legal men, and their prejudices in favour of existing institutions.

The influence of the legal habits which are common in America extends beyond the limits I have just pointed out. Scarcely any question arises in the United States which does not become, sooner or later, a subject of judicial debate; hence all parties are obliged to borrow the ideas, and even the language, usual in judicial proceedings in their daily controversies. As most public men are, or have been, legal practitioners, they introduce the customs and technicalities of their profession into the affairs of the country. The jury extends this habitude to all classes. The language of the law thus becomes, in some measure, a vulgar tongue; the spirit of the law, which is produced in the schools and courts of justice, gradually penetrates beyond their walls into the bosom of society, where it descends to the lowest classes, so that the whole people contracts the habits and the tastes of the magistrate. The lawyers of the United States form a party which is but little feared and scarcely perceived, which has no badge peculiar to itself, which adapts itself with great flexibility to the exigencies of the time, and accommodates itself to all the movements of the social body; but this party extends over the whole community, and it penetrates into all classes of society; it acts upon the country imperceptibly, but it finally fashions it to suit its purposes.

Trial by Jury in the United States Considered as a Political Institution

Trial by jury, which is one of the instruments of the sovereignty of the people, deserves to be compared with the other laws which establish that sovereignty—Composition of the jury in the United States—Effect of trial by jury upon the national character—It educates the people—It tends to establish the authority of the magistrates and to extend a knowledge of law among the people.

Since I have been led by my subject to recur to the administration of justice in the United States, I will not pass over this point without adverting to the institution of the jury. Trial by jury may be considered in two separate points of view, as a judicial and as a political institution. If it entered into my present purpose to inquire how far trial by jury (more especially in civil cases) contributes to insure the best administration of justice, I admit that its utility might be contested. As the jury was first introduced at a time when society was in an uncivilised state, and when courts of justice were merely called upon to decide on the evidence of facts, it is not an easy task to adapt it to the wants of a highly civilised community when the mutual relations of men are multiplied to a surprising extent, and have assumed the enlightened and intellectual character of the age.[2]

My present object is to consider the jury as a political institution, and any other course would divert me from my subject. Of trial by jury, considered as a judicial institution, I shall here say but very few words. When the English adopted trial by jury they were a semi-barbarous people; they are become, in course of time, one of the most enlightened nations of the earth; and their attachment to this institution seems to have increased with their increasing cultivation. They soon spread beyond their insular boundaries to every corner of the habitable globe; some have formed colonies, others independent states; the mother-country has maintained its monarchical constitution; many of its offspring have founded powerful republics; but wherever the English have been they have boasted of the privilege of trial by jury.[3] They have established it, or hastened to reestablish it, in all their settlements. A judicial institution which obtains the suffrages of a great people for so long a series of ages, which is zealously renewed at every epoch of civilisation, in all the climates of the earth and under every form of human government, cannot be contrary to the spirit of justice.[4]

I turn, however, from this part of the subject. To look upon the jury as a mere judicial institution is to confine our attention to a very narrow view of it; for however great its influence may be upon the decisions of the law courts, that influence is very subordinate to the powerful effects which it produces on the destinies of the community at large. The jury is above all a political institution, and it must be regarded in this light in order to be duly appreciated.

By the jury I mean a certain number of citizens chosen indiscriminately, and invested with a temporary right of judging. Trial by jury, as applied to the repression of crime, appears to me to introduce an eminently republican element into the Government upon the following grounds:—

The institution of the jury may be aristocratic or democratic, according to the class of society from which the jurors are selected; but it always preserves its republican character, inasmuch as it places the real direction of society in the hands of the governed, or of a portion of the governed, instead of leaving it under the authority of the Government. Force is never more than a transient element of success; and after force comes the notion of right. A government which should only be able to crush its enemies upon a field of battle would very soon be destroyed. The true sanction of political laws is to be found in penal legislation, and if that sanction be wanting the law will sooner or later lose its cogency. He who punishes infractions of the law is therefore the real master of society. Now the institution of the jury raises the people itself, or at least a class of citizens, to the bench of judicial authority. The institution of the jury consequently invests the people, or that class of citizens, with the direction of society.[5]

In England the jury is returned from the aristocratic portion of the nation;[6] the aristocracy makes the laws, applies the laws, and punishes all infractions of the laws; everything is established upon a consistent footing, and England may with truth be said to constitute an aristocratic republic. In the United States the same system is applied to the whole people. Every American citizen is qualified to be an elector, a juror, and is eligible to office.[7] The system of the jury, as it is understood in America, appears to me to be as direct and as extreme a consequence of the sovereignty of the people as universal suffrage. These institutions are two instruments of equal power, which contribute to the supremacy of the majority. All the sovereigns who have chosen to govern by their own authority, and to direct society instead of obeying its directions, have destroyed or enfeebled the institution of the jury. The monarchs of the House of Tudor sent to prison jurors who refused to convict, and Napoleon caused them to be returned by his agents.

However clear most of these truths may seem to be, they do not command universal assent, and in France, at least, the institution of trial by jury is still very imperfectly understood. If the question arises as to the proper qualification of jurors, it is confined to a discussion of the intelligence and knowledge of the citizens who may be returned, as if the jury was merely a judicial institution. This appears to me to be the least part of the subject. The jury is pre-eminently a political institution; it must be regarded as one form of the sovereignty of the people; when that sovereignty is repudiated, it must be rejected, or it must be adapted to the laws by which that sovereignty is established. The jury is that portion of the nation to which the execution of the laws is entrusted, as the Houses of Parliament constitute that part of the nation which makes the laws; and in order that society may be governed with consistency and uniformity, the list of citizens qualified to serve on juries must increase and diminish with the list of electors. This I hold to be the point of view most worthy of the attention of the legislator, and all that remains is merely accessory.

I am so entirely convinced that the jury is pre-eminently a political institution that I still consider it in this light when it is applied in civil causes. Laws are always unstable unless they are founded upon the manners of a nation; manners are the only durable and resisting power in a people. When the jury is reserved for criminal offences, the people only witnesses its occasional action in certain particular cases; the ordinary course of life goes on without its interference, and it is considered as an instrument, but not as the only instrument, of obtaining justice. This is true *à fortiori* when the jury is only applied to certain criminal causes.

When, on the contrary, the influence of the jury is extended to civil causes, its application is constantly palpable; it affects all the interests of the community; everyone cooperates in its work: it thus penetrates into all the usages of life, it fashions the human mind to its peculiar forms, and is gradually associated with the idea of justice itself.

The institution of the jury, if confined to criminal causes, is always in danger, but when once it is introduced into civil proceedings it defies the aggressions of time and of man. If it had been as easy to remove the jury from the manners as from the laws of England, it would have perished under Henry VIII. and Elizabeth, and the civil jury did in reality, at that period, save the liberties of the country. In whatever manner the jury be applied, it cannot fail to exercise a powerful influence upon the national character; but this influence is prodigiously increased when it is introduced into civil causes. The jury, and more especially the jury in civil cases, serves to communicate the spirit of the judges to the minds of all the citizens; and this spirit, with the habits which attend it, is the soundest preparation for free institutions. It imbues all classes with a respect for the thing judged, and with the notion of right. If these two elements be removed, the love of independence is reduced to a mere destructive passion. It teaches men to practise equity, every man learns to judge his neighbour as he would himself be judged; and this is especially true of the jury in civil causes, for, whilst the number of persons who have reason to apprehend a criminal prosecution is small, every one is liable to have a civil action brought against him. The jury teaches every man not to recoil before the responsibility of his own actions, and impresses him with that manly confidence without which political virtue cannot exist. It invests each citizen with a kind of magistracy, it makes them all feel the duties which they are bound to discharge towards society, and the part which they take in the Government. By obliging men to turn their attention to affairs which are not exclusively their own, it rubs off that individual egotism which is the rust of society.

The jury contributes most powerfully to form the judgement and to increase the natural intelligence of a people, and this is, in my opinion, its greatest advantage. It may be regarded as a gratuitous public school ever open, in which every juror learns to exercise his rights, enters into daily communication with the most learned and enlightened members of the upper classes, and becomes practically acquainted with the laws of his country, which are brought within the reach of his capacity by the efforts of the bar, the advice of the judge, and even by the passions of the parties. I think that the practical intelligence and political good sense of the Americans are mainly attributable to the long use which they have made of the jury in civil causes. I do not know whether the jury is useful to those who are in litigation; but I am certain it is highly beneficial to those who decide the litigation; and I look upon it as one of the most efficacious means for the education of the people which society can employ.

What I have hitherto said applies to all nations, but the remark I am now about to make is peculiar to the Americans and to democratic peoples. I have already observed that in democracies the members of the legal profession and the magistrates constitute the only aristocratic body which can check the irregularities of the people. This aristocracy is invested with no physical power, but it exercises its conservative influence upon the minds of men, and the most abundant source of its authority is the institution of the civil jury. In criminal causes, when society is armed against a single individual, the jury is apt to look upon the judge as the passive instrument of social power, and to mistrust his advice. Moreover, criminal causes are entirely founded upon the evidence of facts which common sense can readily appreciate; upon this ground the judge and the jury are equal. Such, however, is not the case in civil causes; then the judge appears as a disinterested arbiter between the conflicting passions of the parties. The jurors look up to him with confidence and listen to him with respect, for in this instance their intelligence is completely under the control of his learning. It is the judge who sums up the various arguments with which their memory has been wearied out, and who guides them through the devious course of the proceedings; he points their attention to the exact question of fact which they are called upon to solve, and he puts the answer to the question of law into their mouths. His influence upon their verdict is almost unlimited.

If I am called upon to explain why I am but little moved by the arguments derived from the ignorance of jurors in civil causes, I reply, that in these proceedings, whenever the question to be solved is not a mere question of fact, the jury has only the semblance of a judicial body. The jury sanctions the decision of the judge, they by the authority of society which they represent, and he by that of reason and of law.[8]

In England and in America the judges exercise an influence upon criminal trials which the French judges have never possessed. The reason of this difference may easily be discovered; the English and American magistrates establish their authority in civil causes, and only transfer it afterwards to tribunals of another kind, where that authority was not acquired. In some cases (and they are frequently the most important ones) the American judges have the right of deciding causes alone.[9] Upon these occasions they are accidentally placed in the position which the French judges habitually occupy, but they are invested with far more power than the latter; they are still surrounded by the reminiscence of the jury, and their judgment has almost as much authority as the voice of the community at large, represented by that institution. Their influence extends beyond the limits of the courts; in the recreations of private life as well as in the turmoil of public business, abroad and in the legislative assemblies, the American judge is constantly surrounded by men who are accustomed to regard his intelligence as superior to their own, and after having exercised his power in the decision of causes, he continues to influence the habits of thought and the characters of the individuals who took a part in his judgment.

The jury, then, which seems to restrict the rights of magistracy, does in reality consolidate its power, and in no country are the judges so powerful as there, where the people partakes their privileges. It is more especially by means of the jury in civil causes that the American magistrates imbue all classes of society with the spirit of their profession. Thus the jury; which is the most energetic means of making the people rule, is also the most efficacious means of teaching it to rule well.

Chapter XVII

Principal Causes Which Tend to Maintain the Democratic Republic in the United States

A Democratic republic subsists in the United States, and the principal object of this book has been to account for the fact of its existence. Several of the causes which contribute to maintain the institutions of America have been involuntarily passed by or only hinted at as I was borne along by my subject. Others I have been unable to discuss, and those on which I have dwelt most are, as it were, buried in the details of the former parts of this work. I think, therefore, that before I proceed to speak of the future, I cannot do better than collect within a small compass the reasons which best explain the present. In this retrospective chapter I shall be succinct, for I shall take care to remind the reader very summarily of what he already knows; and I shall only select the most prominent of those facts which I have not yet pointed out.

All the causes which contribute to the maintenance of the democratic republic in the United States are reducible to three heads:—

I. The peculiar and accidental situation in which Providence has placed the Americans.

II. The laws.

III. The manners and customs of the people.

Accidental or Providential Causes Which Contribute to the Maintenance of the Democratic Republic in the United States

The Union has no neighbours—No metropolis—The Americans have had the chances of birth in their favour—America an empty country—

How this circumstance contributes powerfully to the maintenance of the democratic republic in America—How the American wilds are peopled—Avidity of the Anglo-Americans in taking possession of the solitudes of the New World—Influence of physical prosperity upon the political opinions of the Americans.

A thousand circumstances, independent of the will of man, concur to facilitate the maintenance of a democratic republic in the United States. Some of these peculiarities are known, the others may easily be pointed out; but I shall confine myself to the most prominent amongst them.

The Americans have no neighbours, and consequently they have no great wars, or financial crises, or inroads, or conquest to dread; they require neither great taxes, nor great armies, nor great generals; and they have nothing to fear from a scourge which is more formidable to republics than all these evils combined, namely, military glory. It is impossible to deny the inconceivable influence which military glory exercises upon the spirit of a nation. General Jackson, whom the Americans have twice elected to the head of their Government, is a man of a violent temper and mediocre talents; no one circumstance in the whole course of his career ever proved that he is qualified to govern a free people, and indeed the majority of the enlightened classes of the Union has always been opposed to him. But he was raised to the Presidency, and has been maintained in that lofty station, solely by the recollection of a victory which he gained twenty years ago under the walls of New Orleans, a victory which was, however, a very ordinary achievement, and which could only be remembered in a country where battles are rare. Now the people which is thus carried away by the illusions of glory is unquestionably the most cold and calculating, the most unmilitary (if I may use the expression), and the most prosaic of all the peoples of the earth.

America has no great capital[1] city, whose influence is directly or indirectly felt over the whole extent of the country, which I hold to be one of the first causes of the maintenance of republican institutions in the United States. In cities men cannot be prevented from concerting together, and from awakening a mutual excitement which prompts sudden and passionate resolutions. Cities may be looked upon as large assemblies, of which all the inhabitants are members; their populace exercises a prodigious influence upon the magistrates, and frequently executes its own wishes without their intervention.

To subject the provinces to the metropolis is therefore not only to place the destiny of the empire in the hands of a portion of the community, which may be reprobated as unjust, but to place it in the hands of a populace acting under its own impulses, which must be avoided as dangerous. The preponderance of capital cities is therefore a serious blow upon the representative system, and it exposes modern republics to the same defect as the republics of antiquity, which all perished from not having been acquainted with that form of government.

It would be easy for me to adduce a great number of secondary causes which have contributed to establish, and which concur to maintain, the democratic republic of the United States. But I discern two principal circumstances amongst these favourable elements, which I hasten to point out. I have already observed that the origin of the American settlements may be looked upon as the first and most efficacious cause to which the present prosperity of the United States may be attributed. The Americans had the chances of birth in their favour, and their forefathers imported that equality of conditions into the country whence the democratic republic has very naturally taken its rise. Nor was this all they did; for besides this republican condition of society, the early settlers bequeathed to their descendants those customs, manners, and opinions which contribute most to the success of a republican form of government. When I reflect upon the consequences of this primary circumstance, methinks I see the destiny of America embodied in the first Puritan who landed on those shores, just as the human race was represented by the first man.

The chief circumstance which has favoured the establishment and the maintenance of a democratic republic in the United States is the nature of the territory which the American inhabit. Their ancestors gave them the love of equality and of freedom, but God himself gave them the means of remaining equal and free, by placing them upon a boundless continent, which is open to their exertions. General prosperity is favourable to the stability of all governments, but more particularly of a democratic constitution, which depends upon the dispositions of the majority, and more particularly of that portion of the community which is most exposed to feel the pressure of want. When the people rules, it must be rendered happy, or it will overturn the State, and misery is apt to stimulate it to those excesses to which ambition rouses kings. The physical causes, independent of the laws, which contribute to promote general prosperity, are more numerous in America than they have ever been in any other country in the world, at any other period of history. In the United States not only is legislation democratic, but Nature herself favours the cause of the people.

In what part of human tradition can be found anything at all similar to that which is occurring under our eyes in North America? The celebrated communities of antiquity were all founded in the midst of hostile nations, which they were obliged to subjugate before they could flourish in their place. Even the moderns have found, in some parts of South America, vast regions inhabited by a people of inferior civilization, but which occupied and cultivated the soil. To found their new States it was necessary to extirpate or to subdue a numerous population, until civilization has been made to blush for their success. But North America was only inhabited by wandering tribes, who took no thought of the natural riches of the soil, and that vast country was still, properly speaking, an empty continent, a desert land awaiting its inhabitants.

Everything is extraordinary in America, the social condition of the inhabitants, as well as the laws; but the soil upon which these institutions are founded is more extraordinary than all the rest. When man was first placed upon the earth by the Creator, the earth was inexhaustible in its youth, but man was weak and ignorant; and when he had learned to explore the treasures which it contained, hosts of his fellow creatures covered its surface, and he was obliged to earn an asylum for repose and for freedom by the sword. At that same period North America was discovered, as if it had been kept in reserve by the Deity, and had just risen from beneath the waters of the deluge.

That continent still presents, as it did in the primaeval time, rivers which rise from never-failing sources, green and moist solitudes, and fields which the ploughshare of the husbandman has never turned. In this state it is offered to man, not in the barbarous and isolated condition of the early ages, but to a being who is already in possession of the most potent secrets of the natural world, who is united to his fellow-men, and instructed by the experience of fifty centuries. At this very time thirteen millions of civilised Europeans are peaceably spreading over those fertile plains, with whose resources and whose extent they are not yet themselves accurately acquainted. Three or four thousand soldiers drive the wandering races of the aborigines before them; these are followed by the pioneers, who pierce the woods, scare off the beasts of prey, explore the courses of the inland streams, and make ready the triumphal procession of civilisation across the waste.

The favourable influence of the temporal prosperity of America upon the institutions of that country has been so often described by others, and adverted to by myself, that I shall not enlarge upon it beyond the addition of a few facts. An erroneous notion is generally entertained that the deserts of America are peopled by European emigrants, who annually disembark upon the coasts of the New World, whilst the American population increases and multiplies upon the soil which its forefathers tilled. The European settler, however, usually arrives in the United States without friends, and sometimes without resources; in order to subsist he is obliged to work for hire, and he rarely proceeds beyond that belt of industrious population which adjoins the ocean. The desert cannot be explored without capital or credit; and the body must be accustomed to the rigours of a new climate before it can be exposed to the chances of forest life. It is the Americans themselves who daily quit the spots which gave them birth to acquire extensive domains in a remote country. Thus the European leaves his cottage for the transatlantic shores; and the American, who is born on that very coast, plunges in his turn into the wilds of Central America. This double emigration is incessant; it begins in the remotest parts of Europe, it crosses the Atlantic Ocean, and it advances over the solitudes of the New World. Millions of men are marching at once towards the same horizon; their language, their religion, their manners differ, their object is the same. The gifts of fortune are promised in the West, and to the West they bend their course.[2]

No event can be compared with this continuous removal of the human race, except perhaps those irruptions which preceded the fall of the Roman Empire. Then, as well as now, generations of men were impelled forwards in the same direction to meet and struggle on the same spot; but the designs of Providence were not the same; then, every new comer was the harbinger of destruction and of death; now, every adventurer brings with him the elements of prosperity and of life. The future still conceals from us the ulterior consequences of this emigration of the Americans towards the West; but we can readily apprehend its more immediate results. As a portion of the inhabitants annually leave the States in which they were born, the population of these States increases very slowly, although they have long been established: thus in Connecticut, which only contains fifty-nine inhabitants to the square mile, the population has not increased by more than one quarter in forty years, whilst that of England has been augmented by one-third in the lapse of the same period. The European emigrant always lands, therefore, in a country which is but half full, and where hands are in request: he becomes a workman in easy circumstances; his son goes to seek his fortune in unpeopled regions, and he becomes a rich landowner. The former amasses the capital which the latter invests, and the stranger as well as the native is unacquainted with want.

The laws of the United States are extremely favourable to the division of property; but a cause which is more powerful than the laws prevents property from being divided to excess.[3] This is very perceptible in the States which are beginning to be thickly peopled; Massachusetts is the most populous part of the Union, but it contains only 80 inhabitants to the square mile, which is must less than in France, where 162 are reckoned to the same extent of country. But in Massachusetts estates are very rarely divided; the eldest son takes the land, and the others go to seek their fortune in the desert. The law has abolished the rights of primogeniture, but circumstances have concurred to reestablish it under a form of which none can complain, and by which no just rights are impaired.

A single fact will suffice to show the prodigious number of individuals who leave New England, in this manner, to settle themselves in the wilds. We were assured in 1830 that thirty-six of the members of Congress were born in the little State of Connecticut. The population of Connecticut, which constitutes only one forty-third part of that of the United States, thus furnished one-eighth of the whole body of representatives. The States of Connecticut, however, only sends five delegates to Congress; and the thirty-one others sit for the new Western States. If these thirty-one individuals had remained in Connecticut, it is probable that instead of becoming rich landowners they would have remained humble labourers, that they would have lived in obscurity without being able to rise into public life, and that, far from becoming useful members of the legislature, they might have been unruly citizens.

These reflections do not escape the observation of the Americans any more than of ourselves. 'It cannot be doubted,' says Chancellor Kent in his Treatise on American Law, 'that the division of landed estates must produce great evils when it is carried to such excess as that each parcel of land is insufficient to support a family; but these disadvantages have never been felt in the United States, and many generations must elapse before they can be felt. The extent of our inhabited territory, the abundance of adjacent land, and the continual stream of emigration flowing from the shores of the Atlantic towards the interior of the country, suffice as yet, and will long suffice, to prevent the parcelling out of estates.'

It is difficult to describe the rapacity with which the American rushes forward to secure the immense booty which fortune proffers to him. In the pursuit he fearlessly braves the arrow of the Indian and the distempers of the forest; he is unimpressed by the silence of the woods; the approach of beasts of prey does not disturb him; for he is goaded onwards by a passion more intense than the love of life. Before him lies a boundless continent, and he urges onwards as if time pressed, and he was afraid of finding no room for his exertions. I have spoken of the emigration from the older States, but how shall I describe that which takes place from the more recent ones? Fifty years have scarcely elapsed since that of Ohio was founded; the greater part of its inhabitants were not born within its confines; its capital has only been built thirty years, and its territory is still covered by an immense extent of uncultivated fields; nevertheless the population of Ohio is already proceeding westward, and most of the settlers who descend to the fertile savannahs of Illinois are citizens of Ohio. These men left their first country to improve their condition;

they quit their resting-place to ameliorate it still more; fortune awaits them everywhere, but happiness they cannot attain. The desire of prosperity is become an ardent and restless passion in their minds which grows by what it gains. They early broke the ties which bound them to their natal earth, and they have contracted no fresh ones on their way. Emigration was at first necessary to them as a means of subsistence; and it soon becomes a sort of game of chance, which they pursue for the emotions it excites as much as for the gain it procures.

Sometimes the progress of man is so rapid that the desert reappears behind him. The woods stoop to give him a passage, and spring up again when he has passed. It is not uncommon in crossing the new States of the West to meet with deserted dwellings in the midst of the wilds; the traveller frequently discovers the vestiges of a log-house in the most solitary retreats, which bear witness to the power, and no less to the inconstancy of man. In these abandoned fields, and over these ruins of a day, the primaeval forest soon scatters a fresh vegetation, the beasts resume the haunts which were once their own, and Nature covers the traces of man's path with branches and with flowers, which obliterate his evanescent track.

I remember that, in crossing one of the woodland districts which still cover the State of New York, I reached the shores of a lake embosomed in forests coeval with the world. A small island, covered with woods whose thick foliage concealed its banks, rose from the centre of the waters. Upon the shores of the lake no object attested the presence of man, except a column of smoke which might be seen on the horizon rising from the tops of the trees to the clouds, and seeming to hang from heaven rather than to be mounting to the sky. An Indian shallop was hauled up on the sand, which tempted me to visit the islet that had first attracted my attention, and in a few minutes I set foot upon its banks. The whole island formed one of those delicious solitudes of the New World which almost lead civilized man to regret the haunts of the savage. A luxuriant vegetation bore witness to the incomparable fruitfulness of the soil. The deep silence which is common to the wilds of North America was only broken by the hoarse cooing of the wood-pigeon, and the tapping of the woodpecker upon the bark of trees. I was far from supposing that this spot had ever been inhabited, so completely did Nature seem to be left to her own caprices; but when I reached the centre of the isle I thought that I discovered some traces of man. I then proceeded to examine the surrounding objects with care, and I soon perceived that a European had undoubtedly been led to seek a refuge in this retreat. Yet what changes had taken place in the scene of his labours! The logs which he had hastily hewn to build himself a shed had sprouted afresh; the very props were intertwined with living verdure, and his cabin was transformed into a bower. In the midst of these shrubs a few stones were to be seen, blackened with fire and sprinkled with thin ashes; here the hearth had no doubt been, and the chimney in falling had covered it with rubbish. I stood for some time in silent admiration of the exuberance of Nature and the littleness of man: and when I was obliged to leave that enchanting solitude, I exclaimed with melancholy, "Are ruins, then, already here?"

In Europe we are wont to look upon a restless disposition, an unbounded desire of riches, and an excessive love of independence, as propensities very formidable to society. Yet these are the very elements which ensure a long and peaceful duration to the republics of America. Without these unquiet passions the population would collect in certain spots, and would soon be subject to wants like those of the Old World, which it is difficult to satisfy; for such is the present good fortune of the New World, that the vices of its inhabitants are scarcely less favourable to society than their virtues. These circumstances exercise a great influence on the estimation in which human actions are held in the two hemispheres. The Americans frequently term what we should call cupidity a laudable industry; and they blame as faint-heartedness what we consider to be the virtue of moderate desires.

In France, simple tastes, orderly manners, domestic affections, and the attachments which men feel to the place of their birth, are looked upon as great guarantees of the tranquillity and happiness of the State. But in America nothing seems to be more prejudicial to society than these virtues. The French Canadians, who have faithfully preserved the traditions of their pristine manners, are already embarrassed for room upon their small territory; and this little community, which has so recently begun to exist, will shortly be a prey to the calamities incident to old nations. In Canada, the most enlightened, patriotic, and humane inhabitants make extraordinary efforts to render the people dissatisfied with those simple enjoyments which still content it. There, the seductions of wealth are vaunted with as much zeal as the charms of an honest but limited income in the Old World, and more exertions are made to excite the passions of the citizens there than to calm them elsewhere. If we listen to their eulogies, we shall hear that nothing is more praiseworthy than to exchange the pure and homely pleasures which even the poor man tastes in his own country for the dull delights of prosperity under a foreign sky; to leave the patrimonial hearth and the turf beneath which his forefathers sleep; in short, to abandon the living and the dead in quest of fortune.

At the present time America presents a field for human effort far more extensive than any sum of labour which can be applied to work it. In America too much knowledge cannot be diffused; for all knowledge, whilst it may serve him who possesses it, turns also to the advantage of those who are without it. New wants are not to be feared, since they can be satisfied without difficulty; the growth of human passions need not be dreaded, since all passions may find an easy and a legitimate object; nor can men be put in possession of too much freedom, since they are scarcely ever tempted to misuse their liberties.

The American republics of the present day are like companies of adventurers formed to explore in common the waste lands of the New World, and busied in a flourishing trade. The passions which agitate the Americans most deeply are not their political but their commercial passions; or, to speak more correctly, they introduce the habits they contract in business into their political life. They love order, without which affairs do not prosper; and they set an especial value upon a regular conduct, which is the foundation of a solid business; they prefer the good sense which amasses large fortunes to that enterprising spirit which frequently dissipates them; general ideas alarm their minds, which are accustomed to positive calculations, and they hold practice in more honour than theory.

It is in America that one learns to understand the influence which physical prosperity exercises over political actions, and even over opinions which ought to acknowledge no sway but that of reason; and it is more especially amongst strangers that this truth is perceptible. Most of the European emigrants to the New World carry with them that wild love of independence and of change which our calamities are so apt to engender. I sometimes met with Europeans in the United States who had been obliged to leave their own country on account of their political opinions. They all astonished me by the language they held, but one of them surprised me more than all the rest. As I was crossing one of the most remote districts of Pennsylvania I was benighted, and obliged to beg for hospitality at the gate of a wealthy planter, who was a Frenchman by birth. He bade me sit down beside his fire, and we began to talk with that freedom which befits persons who meet in the backwoods, two thousand leagues from their native country. I was aware that my host had been a great leveller and an ardent demagogue forty years ago, and that his name was not unknown to fame. I was therefore not a little surprised to hear him discuss the rights of property as an economist or a landowner might have done: he spoke of the necessary gradations which fortune establishes among men, of obedience to established laws, of the influence of good morals in commonwealths, and of the support which religious opinions give to order and to freedom; he even went so far as to quote an evangelical authority in corroboration of one of his political tenets.

I listened, and marvelled at the feebleness of human reason. A proposition is true or false, but no art can prove it to be one or the other, in the midst of the uncertainties of science and the conflicting lessons of experience, until a new incident disperses the clouds of doubt; I was poor, I become rich, and I am not to expect that prosperity will act upon my conduct, and leave my judgment free; my opinions change with my fortune, and the happy circumstances which I turn to my advantage furnish me with that decisive argument which was before wanting.

The influence of prosperity acts still more freely upon the American than upon strangers. The American has always seen the connection of public order and public prosperity, intimately united as they are, go on before his eyes; he does not conceive that one can subsist without the other; he has therefore nothing to forget; nor has he, like so many Europeans, to unlearn the lessons of his early education.

Influence of the Laws upon the Maintenance of the Democratic Republic in the United States

Three principal causes of the maintenance of the democratic republic—Federal Constitutions—Municipal institutions—Judicial power.

The principal aim of this book has been to make known the laws of the United States; if this purpose has been accomplished, the reader is already enabled to judge for himself which are the laws that really tend to maintain the democratic republic, and which endanger its existence. If I have not succeeded in explaining this in the whole course of my work, I cannot hope to do so within the limits of a single chapter. It is not my intention to retrace the path I have already pursued, and a very few lines will suffice to recapitulate what I have previously explained.

Three circumstances seem to me to contribute most powerfully to the maintenance of the democratic republic in the United States.

The first is that Federal form of Government which the Americans have adopted, and which enables the Union to combine the power of a great empire with the security of a small State.

The second consists in those municipal institutions which limit the despotism of the majority, and at the same time impart a taste for freedom and a knowledge of the art of being free to the people.

The third is to be met with in the constitution of the judicial power. I have shown in what manner the courts of justice serve to repress the excesses of democracy, and how they check and direct the impulses of the majority without stopping its activity.

Influence of Manners upon the Maintenance of the Democratic Republic in the United States

I have previously remarked that the manners of the people may be considered as one of the general causes to which the maintenance of a democratic republic in the United States is attributable. I here used the word *manners* with the meaning which the ancients attached to the word *mores;* for I apply it not only to manners in their proper sense of what constitutes the character of social intercourse, but I extend it to the various notions and opinions current among men, and to the mass of those ideas which constitute their character of mind. I comprise, therefore, under this term the whole moral and intellectual condition of a people. My intention is not to draw a picture of American manners, but simply to point out such features of them as are favourable to the maintenance of political institutions.

Religion Considered as a Political Constitution, Which Powerfully Contributes to the Maintenance of the Democratic Republic amongst the Americans

North America peopled by men who professed a democratic and republican Christianity—Arrival of the Catholics—For what reason the Catholics form the most democratic and the most republican class at the present time.

Every religion is to be found in juxtaposition to a political opinion which is connected with it by affinity. If the human mind be left to follow its own bent, it will regulate the temporal and spiritual institutions of society upon one uniform principle; and man will endeavour, if I may use the expression, to harmonise the state in which he lives upon earth with the state which he believes to await him in heaven. The greatest part of British America was peopled by men who, after having shaken off the authority of the Pope, acknowledged no other religious supremacy; they brought with them into the New World a form of Christianity which I cannot better describe than by styling it a democratic and republican religion. This sect contributed powerfully to the establishment of a democracy and a republic, and from the earliest settlement of the emigrants politics and religion contracted an alliance which has never been dissolved.

About fifty years ago Ireland began to pour a Catholic population into the United States; on the other hand, the Catholics of America made proselytes, and at the present moment more than a million of Christians professing the truths of the Church of Rome are to be met with in the Union.[4] The Catholics are faithful to the observances of their religion; they are fervent and zealous in the support and belief of their doctrines. Nevertheless they constitute the most republican and the most democratic class of citizens which exists in the United States; and although this fact may surprise the observer at first, the causes by which it is occasioned may easily be discovered upon reflection.

I think that the Catholic religion has erroneously been looked upon as the natural enemy of democracy. Amongst the various sects of Christians, Catholicism seems to me, on the contrary, to be one of those which are most favourable to the equality of conditions. In the Catholic Church, the religious community is composed of only two elements, the priest and the people. The priest alone rises above the rank of his flock, and all below him are equal.

On doctrinal points the Catholic faith places all human capacities upon the same level; it subjects the wise and ignorant, the man of genius and the vulgar crowd, to the details of the same creed; it imposes the same observances upon the rich and needy, it inflicts the same austerities upon the strong and the weak, it listens to no compromise with mortal man, but, reducing all the human race to the same standard, it confounds all the distinctions of society at the foot of the same altar, even as they are confounded in the sight of God. If Catholicism predisposes the faithful to obedience, it certainly does not prepare them for inequality; but the contrary may be said of Protestantism, which generally tends to make men independent, more than to render them equal.

Catholicism is like an absolute monarchy; if the sovereign be removed, all the other classes of society are more equal than they are in republics. It has not unfrequently occurred that the Catholic priest has left the service of the altar to mix with the governing powers of society, and to take his place amongst the civil gradations of men. This religious influence has sometimes been used to secure the interests of that political state of things to which he belonged. At other times Catholics have taken the side of aristocracy from a spirit of religion.

But no sooner is the priesthood entirely separated from the government, as is the case in the United States, than it is found that no class of men are more naturally disposed than the Catholics to transfuse the doctrine of the equality of conditions into the political world. If, then, the Catholic citizens of the United States are not forcibly led by the nature of their tenets to adopt democratic and republican principles, at least they are not necessarily opposed to them; and their social position, as well as their limited number, obliges them to adopt these opinions. Most of the Catholics are poor, and they have no chance of taking a part in the Government unless it be open to all the citizens. They constitute a minority, and all rights must be respected in order to insure to them the free exercise of their own privileges. These two causes induce them, unconsciously, to adopt political doctrines which they would perhaps support with less zeal if they were rich and preponderant.

The Catholic clergy of the United States has never attempted to oppose this political tendency, but it seeks rather to justify its results. The priests in America have divided the intellectual world into two parts: in the one they place the doctrines of revealed religion, which command their assent; in the other they leave those truths which they believe to have been freely left open to the researches of political inquiry. Thus the Catholics of the United States are at the same time the most faithful believers and the most zealous citizens.

It may be asserted that in the United States no religious doctrine displays the slightest hostility to democratic and republican institutions. The clergy of all the different sects hold the same language, their opinions are consonant to the laws, and the human intellect flows onwards in one sole current.

I happened to be staying in one of the largest towns in the Union, when I was invited to attend a public meeting which had been called for the purpose of assisting the Poles, and of sending them supplies of arms and money. I found two or three thousand persons collected in a vast hall which had been prepared to receive them. In a short time a priest in his ecclesiastical robes advanced to the front of the hustings: the spectators rose, and stood uncovered, whilst he spoke in the following terms:—

'Almighty God! the God of Armies! Thou who didst strengthen the hearts and guide the arms of our fathers when they were fighting for the sacred rights of national independence; Thou who didst make them triumph over a hateful oppression, and hast granted to our people the benefits of liberty and peace; Turn, O Lord, a favourable eye upon the other hemisphere; pitifully look down upon that heroic nation which is even now struggling as we did in the former time, and for the same rights which we defended with our blood. Thou, who didst create Man in the likeness of the same image, let not tyranny mar Thy work, and establish inequality upon the earth. Almighty God! do Thou watch over the destiny of the Poles, and render them worthy to be free. May Thy wisdom direct their councils, and may Thy strength sustain their arms! Shed forth Thy terror over their enemies, scatter the powers which take counsel against them; and vouchsafe that the injustice which the world has witnessed for fifty years, be not consummated in our time. O Lord, who holdest alike the hearts of nations and of men in Thy powerful hand; raise up allies to the sacred cause of right; arouse the French nation from the apathy in which its rulers retain it, that it go forth again to fight for the liberties of the world.'

'Lord, turn not Thou Thy face from us, and grant that we may always be the most religious as well as the freest people of the earth. Almighty God, hear our supplications this day. Save the Poles, we beseech Thee, in the name of Thy well-beloved Son, our Lord Jesus Christ, who died upon the cross for the salvation of men. Amen.'

The whole meeting responded 'Amen!' with devotion.

Indirect Influence of Religious Opinions upon Political Society in the United States

Christian morality common to all sects—Influence of religion upon the manners of the Americans—Respect for the marriage tie—In what manner religion confines the imagination of the Americans within certain limits, and checks the passion of innovation—Opinion of the Americans on the political utility of religion—Their exertions to extend and secure its predominance.

I have just shown what the direct influence of religion upon politics is in the United States, but its indirect influence appears to me to be still more considerable, and it never instructs the Americans more fully in the art of being free than when it says nothing of freedom.

The sects which exist in the United States are innumerable. They all differ in respect to the worship which is due from man to his Creator, but they all agree in respect to the duties which are due from man to man. Each sect adores the Deity in its own peculiar manner, but all the sects preach the same moral law in the name of God. If it be of the highest importance to man, as an individual, that his religion should be true, the case of society is not the same. Society has no future life to hope for or to fear; and provided the citizens profess a religion, the peculiar tenets of that religion are of very little importance to its interests. Moreover, almost all the sects of the United States are comprised within the great unity of Christianity, and Christian morality is everywhere the same.

It may be believed without unfairness that a certain number of Americans pursue a peculiar form of worship, from habit more than from conviction. In the United States the sovereign authority is religious, and consequently hypocrisy must be common; but there is no country in the whole world in which the Christian religion retains a greater influence over the souls of men than in America; and there can be no greater proof of its utility, and of its conformity to human nature, than that its influence is most powerfully felt over the most enlightened and free nation of the earth.

I have remarked that the members of the American clergy in general, without even excepting those who do not admit religious liberty, are all in favour of civil freedom; but they do not support any particular political system. They keep aloof from parties and from public affairs. In the United States religion exercises but little influence upon the laws and upon the details of public opinion, but it directs the manners of the community, and by regulating domestic life it regulates the State.

I do not question that the great austerity of manners which is observable in the United States, arises, in the first instance, from religious faith. Religion is often unable to restrain man from the numberless temptations of fortune; nor can it check that passion for gain which every incident of his life contributes to arouse, but its influence over the mind of woman is supreme, and women are the protectors of morals. There is certainly no country in the world where the tie of marriage is so much respected as in America, or where conjugal happiness is more highly or worthily appreciated. In Europe almost all the disturbances of society arise from the irregularities of domestic life. To despise the natural bonds and legitimate pleasures of home, is to contract a taste for excesses, a restlessness of heart, and the evil of fluctuating desires. Agitated by the tumultuous passions which frequently disturb his dwelling, the European is galled by the obedience which the legislative powers of the State exact. But when the American retires from the turmoil of public life to the bosom of his family, he finds in it the image of order and of peace. There his pleasures are simple and natural, his joys are innocent and calm; and as he finds that an orderly life is the surest path to happiness, he accustoms himself without difficulty to moderate his opinions as well as his tastes. Whilst the European endeavours to forget his domestic troubles by agitating society, the American derives from his own home that love of order which he afterwards carries with him into public affairs.

In the United States the influence of religion is not confined to the manners, but it extends to the intelligence of the people. Amongst the Anglo-Americans, there are some who profess the doctrines of Christianity from a sincere belief in them, and others who do the same because they are afraid to be suspected of unbelief. Christianity, therefore, reigns without any obstacle, by universal consent; the consequence is, as I have before observed, that every principle of the moral world is fixed and determinate, although the political world is abandoned to the debates and the experiments of men. Thus the human mind is never left to wander across a boundless field; and, whatever may be its pretensions, it is checked from time to time by barriers which it cannot surmount. Before it can perpetrate innovation, certain primal and immutable principles are laid down, and the boldest conceptions of human device are subjected to certain forms which retard and stop their completion.

The imagination of the Americans, even in its greatest flights, is circumspect and undecided; its impulses are checked, and its works unfinished. These habits of restraint recur in political society, and are singularly favourable both to the tranquillity of the people and to the durability of the institutions it has established. Nature and circumstances concurred to make the inhabitants of the United States bold men, as is sufficiently attested by the enterprising spirit with which they seek for fortune. If the mind of the Americans were free from all trammels, they would very shortly become the most daring innovators and the most implacable disputants in the world. But the revolutionists of America are obliged to profess an ostensible respect for Christian morality and equity, which does not easily permit them to violate the laws that oppose their designs; nor would they find it easy to surmount the scruples of their partisans, even if they were able to get over their own. Hitherto no one in the United States has dared to advance the maxim, that everything is permissible with a view to the interests of society; an impious adage which seems to have been invented in an age of freedom to shelter all the tyrants of future ages. Thus whilst the law permits the Americans to do what they please, religion prevents them from conceiving, and forbids them to commit, what is rash or unjust.

Religion in America takes no direct part in the government of society, but it must nevertheless be regarded as the foremost of the political institutions of that country; for if it does not impart a taste for freedom, it facilitates the use of free institutions. Indeed, it is in this same point of view that the inhabitants of the United States themselves look upon religious belief. I do not know whether all the Americans have a sincere faith in their religion, for who can search the human heart? but I am certain that they hold it to be indispensable to the maintenance of republican institutions. This opinion is not peculiar to a class of citizens or to a party, but it belongs to the whole nation, and to every rank of society.

In the United States, if a political character attacks a sect, this may not prevent even the partisans of that very sect from supporting him; but if he attacks all the sects together, everyone abandons him, and he remains alone.

Whilst I was in America, a witness, who happened to be called at the assizes of the county of Chester (State of New York), declared that he did not believe in the existence of God, or in the immortality of the soul. The judge refused to admit his evidence, on the ground that the witness had destroyed beforehand all the confidence of the Court in what he was about to say.[5] The newspapers related the fact without any further comment.

The Americans combine the notions of Christianity and of liberty so intimately in their minds, that it is impossible to make them conceive the one without the other; and with them this conviction does not spring from that barren traditionary faith which seems to vegetate in the soul rather than to live.

I have known of societies formed by the Americans to send out ministers of the Gospel into the new Western States to found schools and churches there, lest religion should be suffered to die away in those remote settlements, and the rising States be less fitted to enjoy free institutions than the people from which they emanated. I met with wealthy New Englanders who abandoned the country in which they were born in order to lay the foundations of Christianity and of freedom on the banks of the Missouri, or in the prairies of Illinois. Thus religious zeal is perpetually stimulated in the United States by the duties of patriotism. These men do not act from an exclusive consideration of the promises of a future life; eternity is only one motive of their devotion to the cause; and if you converse with these missionaries of Christian civilisation, you will be surprised to find how much value they set upon the goods of this world, and that you meet with a politician where you expected to find a priest. They will tell you that 'all the American republics are collectively involved with each other; if the republics of the West were to fall into anarchy, or to be mastered by a despot, the republican institutions which now flourish upon the shores of the Atlantic Ocean would be in great peril. It is, therefore, our interest that the new States should be religious, in order to maintain our liberties.'

Such are the opinions of the Americans, and if any hold that the religious spirit which I admire is the very thing most amiss in America, and that the only element wanting to the freedom and happiness of the human race is to believe in some blind cosmogony, or to assert with Cabanis the secretion of thought by the brain, I can only reply that those who hold this language have never been in America, and that they have never seen a religious or a free nation. When they return from their expedition, we shall hear what they have to say.

There are persons in France who look upon republican institutions as a temporary means of power, of wealth, and distinction; men who are the *condottieri* of liberty, and who fight for their own advantage, whatever be the colours they wear: it is not to these that I address myself. But there are others who look forward to the republican form of government as a tranquil and lasting state, towards which modern society is daily impelled by the ideas and manners of the time, and who sincerely desire to prepare men to be free. When these men attack religious opinions, they obey the dictates of their passions to the prejudice of their interests. Despotism may govern without faith, but liberty cannot. Religion is much more necessary in the republic which they set forth in glowing colours than in the monarchy which they attack; and it is more needed in democratic republics than in any others. How is it possible that society should escape destruction if the moral tie be not strengthened in proportion as the political tie is relaxed? and what can be done with a people which is its own master, if it be not submissive to the Divinity?

Principal Causes Which Render Religion Powerful in America

Care taken by the Americans to separate the Church from the State—The laws, public opinion, and even the exertions of the clergy concur to promote this end—Influence of religion upon the mind in the United States attributable to this cause—Reason of this—What is the natural state of men with regard to religion at the present time—What are the peculiar and incidental causes which prevent men, in certain countries, from arriving at this state.

The philosophers of the eighteenth century explained the gradual decay of religious faith in a very simple manner. Religious zeal, said they, must necessarily fail, the more generally liberty is established and knowledge diffused. Unfortunately, facts are by no means in accordance with their theory. There are certain populations in Europe whose unbelief is only equalled by their ignorance and their debasement, whilst in America one of the freest and most enlightened nations in the world fulfils all the outward duties of religious fervour.

Upon my arrival in the United States, the religious aspect of the country was the first thing that struck my attention; and the longer I stayed there the more did I perceive the great political consequences resulting from this state of things, to which I was unaccustomed. In France I had almost always seen the spirit of religion and the spirit of freedom pursuing courses diametrically opposed to each other; but in America I found that they were intimately united, and that they reigned in common over the same country. My desire to discover the causes of this phenomenon increased from day to day. In order to satisfy it I questioned the members of all the different sects; and I more especially sought the society of the clergy, who are the depositaries of the different persuasions, and who are more especially interested in their duration. As a member of the Roman Catholic Church I was more particularly brought into contact with several of its priests, with whom I became intimately acquainted. To each of these men I expressed my astonishment and I explained my doubts; I found that they differed upon matters of detail alone; and that they mainly attributed the peaceful dominion of religion in their country to the separation of Church and State. I do not hesitate to affirm that during my stay in America I did not meet with a single individual, of the clergy or of the laity, who was not of the same opinion upon this point.

This led me to examine more attentively than I had hitherto done, the station which the American clergy occupy in political society. I learned with surprise that they filled no public appointments;[6] not one of them is to be met with in the administration, and they are not even represented in the legislative assemblies. In several States[7] the law excludes them from political life, public opinion in all. And when I came to inquire into the prevailing spirit of the clergy I found that most of its members seemed to retire of their own accord from the exercise of power, and that they made it the pride of their profession to abstain from politics.

I heard them inveigh against ambition and deceit, under whatever political opinions these vices might chance to lurk; but I learned from their discourses that men are not guilty in the eye of God for any opinions concerning political government which they may profess with sincerity, any more than they are for their mistakes in building a house or in driving a furrow. I perceived that these ministers of the gospel eschewed all parties with the anxiety attendant upon personal interest. These facts convinced me that what I had been told was true; and it then became my object to investigate their causes, and to inquire how it happened that the real authority of religion was increased by a state of things which diminished its apparent force: these causes did not long escape my researches.

The short space of threescore years can never content the imagination of man; nor can the imperfect joys of this world satisfy his heart. Man alone, of all created beings, displays a natural contempt of existence, and yet a boundless desire to exist; he scorns life, but he dreads annihilation. These different feelings incessantly urge his soul to the contemplation of a future state, and religion directs his musings thither. Religion, then, is simply another form of hope; and it is no less natural to the human heart than hope itself. Men cannot abandon their religious faith without a kind of aberration of intellect, and a sort of violent distortion of their true natures; but they are invincibly brought back to more pious sentiments; for unbelief is an accident, and faith is the only permanent state of mankind. If we only consider religious institutions in a purely human point of view, they may be said to derive an inexhaustible element of strength from man himself, since they belong to one of the constituent principles of human nature.

I am aware that at certain times religion may strengthen this influence, which originates in itself, by the artificial power of the laws, and by the support of those temporal institutions which direct society. Religions, intimately united to the governments of the earth, have been known to exercise a sovereign authority derived from the twofold source of terror and of faith; but when a religion contracts an alliance of this nature, I do not hesitate to affirm that it commits the same error as a man who should sacrifice his future to his present welfare; and in obtaining a power to which it has no claim, it risks that authority which is rightfully its own. When a religion founds its empire upon the desire of immortality which lives in every human heart, it may aspire to universal dominion; but when it connects itself with a government, it must necessarily adopt maxims which are only applicable to certain nations. Thus, in forming an alliance with a political power, religion augments its authority over a few, and forfeits the hope of reigning over all.

As long as a religion rests upon those sentiments which are the consolation of all affliction, it may attract the affections of mankind. But if it be mixed up with the bitter passions of the world, it may be constrained to defend allies whom its interests, and not the principle of love, have given to it; or to repel as antagonists men who are still attached to its own spirit, however opposed they may be to the powers to which it is allied. The Church cannot share the temporal power of the State without being the object of a portion of that animosity which the latter excites.

The political powers which seem to be most firmly established have frequently no better guarantee for their duration than the opinions of a generation, the interests of the time, or the life of an individual. A law may modify the social condition which seems to be most fixed and determinate; and with the social condition everything else must change. The powers of society are more or less fugitive, like the years which we spend upon the earth; they succeed each other with rapidity, like the fleeting cares of life; and no government has ever yet been founded upon an invariable disposition of the human heart, or upon an imperishable interest.

As long as a religion is sustained by those feelings, propensities, and passions which are found to occur under the same forms, at all the different periods of history, it may defy the efforts of time; or at least it can only be destroyed by another religion. But when religion clings to the interests of the world, it becomes almost as fragile a thing as the powers of earth. It is the only one of them all which can hope for immortality; but if it be connected with their ephemeral authority, it shares their fortunes, and may fall with those transient passions which supported them for a day. The alliance which religion contracts with political powers must needs be onerous to itself; since it does not require their assistance to live, and by giving them its assistance to live, and by giving them its assistance it may be exposed to decay.

The danger which I have just pointed out always exists, but it is not always equally visible. In some ages governments seem to be imperishable; in others, the existence of society appears to be more precarious than the life of man. Some constitutions plunge the citizens into a lethargic somnolence, and others rouse them to feverish excitement. When governments appear to be so strong, and laws so stable, men do not perceive the dangers which may accrue from a union of Church and State. When governments display so much weakness, and laws so much inconstancy, the danger is self-evident, but it is no longer possible to avoid it; to be effectual, measures must be taken to discover its approach.

In proportion as a nation assumes a democratic condition of society, and as communities display democratic propensities, it becomes more and more dangerous to connect religion with political institutions; for the time is coining when authority will be bandied from hand to hand, when political theories will succeed each other, and when men, laws, and constitutions will disappear, or be modified from day to day, and this, not for a season only, but unceasingly. Agitation and mutability are inherent in the nature of democratic republics, just as stagnation and inertness are the law of absolute monarchies.

If the Americans, who change the head of the Government once in four years, who elect new legislators every two years, and renew the provincial officers every twelvemonth; if the Americans, who have abandoned the political world to the attempts of innovators, had not placed religion beyond their reach, where could it abide in the ebb and flow of human opinions? where would that respect which belongs to it be paid, amidst the struggles of faction? and what would become of its immortality, in the midst of perpetual decay? The American clergy were the first to perceive this truth, and to act in conformity with it. They saw that they must renounce their religious influence, if they were to strive for political power; and they chose to give up the support of the State, rather than to share its vicissitudes.

In America, religion is perhaps less powerful than it has been at certain periods in the history of certain peoples; but its influence is more lasting. It restricts itself to its own resources, but of those none can deprive it: its circle is limited to certain principles, but those principles are entirely its own, and under its undisputed control.

On every side in Europe we hear voices complaining of the absence of religious faith, and inquiring the means of restoring to religion some remnant of its pristine authority. It seems to me that we must first attentively consider what ought to be *the natural state* of men with regard to religion at the present time; and when we know what we have to hope and to fear, we may discern the end to which our efforts ought to be directed.

The two great dangers which threaten the existence of religions are schism and indifference. In ages of fervent devotion, men sometimes abandon their religion, but they only shake it off in order to adopt another. Their faith changes the objects to which it is directed, but it suffers no decline. The old religion then excites enthusiastic attachment or bitter enmity in either party; some leave it with anger, others cling to it with increased devotedness, and although persuasions differ, irreligion is unknown. Such, however, is not the case when a religious belief is secretly undermined by doctrines which may be termed negative, since they deny the truth of one religion without affirming that of any other. Progidious revolutions then take place in the human mind, without the apparent cooperation of the passions of man, and almost without his knowledge. Men lose the objects of their fondest hopes, as if through forgetfulness. They are carried away by an imperceptible current which they have not the courage to stem, but which they follow with regret, since it bears them from a faith they love, to a scepticism that plunges them into despair.

In ages which answer to this description, men desert their religious opinions from lukewarmness rather than from dislike; they do not reject them, but the sentiments by which they were once fostered disappear. But if the unbeliever does not admit religion to be true, he still considers it useful. Regarding religious institutions in a human point of view, he acknowledges their influence upon manners and legislation. He admits that they may serve to make men live in peace with one another, and to prepare them gently for the hour of death. He regrets the faith which he has lost; and as he is deprived of a treasure which he has learned to estimate at its full value, he scruples to take it from those who still possess it.

On the other hand, those who continue to believe are not afraid openly to avow their faith. They look upon those who do not share their persuasion as more worthy of pity than of opposition; and they are aware that to acquire the esteem of the unbelieving, they are not obliged to follow their example. They are hostile to no one in the world; and as they do not consider the society in which they live as an arena in which religion is bound to face its thousand deadly foes, they love their contemporaries, whilst they condemn their weaknesses and lament their errors.

As those who do not believe, conceal their incredulity; and as those who believe, display their faith, public opinion pronounces itself in favour of religion: love, support, and honour are bestowed upon it, and it is only by searching the human soul that we can detect the wounds which it has received. The mass of mankind, who are never without the feeling of religion, do not perceive anything at variance with the established faith. The instinctive desire of a future life brings the crowd about the altar, and opens the hearts of men to the precepts and consolations of religion.

But this picture is not applicable to us: for there are men amongst us who have ceased to believe in Christianity, without adopting any other religion; others who are in the perplexities of doubt, and who already affect not to believe; and others, again, who are afraid to avow that Christian faith which they still cherish in secret.

Amidst these lukewarm partisans and ardent antagonists a small number of believers exist, who are ready to brave all obstacles and to scorn all dangers in defence of their faith. They have done violence to human weakness, in order to rise superior to public opinion. Excited by the effort they have made, they scarcely knew where to stop; and as they know that the first use which the French made of independence was to attack religion, they look upon their contemporaries with dread, and they recoil in alarm from the liberty which their fellow-citizens are seeking to obtain. As unbelief appears to them to be a novelty, they comprise all that is new in one indiscriminate animosity. They are at war with their age and country, and they look upon every opinion which is put forth there as the necessary enemy of the Faith.

Such is not the natural state of men with regard to religion at the present day; and some extraordinary or incidental cause must be at work in France to prevent the human mind from following its original propensities and to drive it beyond the limits at which it ought naturally to stop. I am intimately convinced that this extraordinary and incidental cause is the close connexion of politics and religion. The unbelievers of Europe attack the Christians as their political opponents, rather than as their religious adversaries; they hate the Christian religion as the opinion of a party, much more than as an error of belief; and they reject the clergy less because they are the representatives of the Divinity than because they are the allies of authority.

In Europe, Christianity has been intimately united to the powers of the earth. Those powers are now in decay, and it is, as it were, buried under their ruins. The living body of religion has been bound down to the dead corpse of superannuated polity: cut but the bonds which restrain it, and that which is alive will rise once more. I know not what could restore the Christian Church of Europe to the energy of its earlier days; that power belongs to God alone; but it may be the effect of human policy to leave the Faith in the full exercise of the strength which it still retains.

How the Instruction, the Habits, and the Practical Experience of the Americans Promote the Success of Their Democratic Institutions

What is to be understood by the instruction of the American people—The human mind more superficially instructed in the United States than in Europe—No one completely uninstructed—

Reason of this—Rapidity with which opinions are diffused even in the uncultivated States of the West—Practical experience more serviceable to the Americans than book-learning.

I have but little to add to what I have already said concerning the influence which the instruction and the habits of the Americans exercise upon the maintenance of their political institutions.

America has hitherto produced very few writers of distinction; it possesses no great historians, and not a single eminent poet. The inhabitants of that country look upon what are properly styled literary pursuits with a kind of disapprobation; and there are towns of very second-rate importance in Europe in which more literary works are annually published than in the twenty-four States of the Union put together. The spirit of the Americans is averse to general ideas; and it does not seek theoretical discoveries. Neither politics nor manufactures direct them to these occupations; and although new laws are perpetually enacted in the United States, no great writers have hitherto inquired into the general principles of their legislation. The Americans have lawyers and commentators, but no jurists;[8] and they furnish examples rather than lessons to the world. The same observation applies to the mechanical arts. In America, the inventions of Europe are adopted with sagacity; they are perfected, and adapted with admirable skill to the wants of the country. Manufactures exist, but the science of manufacture is not cultivated; and they have good workmen, but very few inventors. Fulton was obliged to proffer his services to foreign nations for a long time before he was able to devote them to his own country.

The observer who is desirous of forming an opinion on the state of instruction amongst the Anglo-Americans must consider the same object from two different points of view. If he only singles out the learned, he will be astonished to find how rare they are; but if he counts the ignorant, the American people will appear to be the most enlightened community in the world. The whole population, as I observed in another place, is situated between these two extremes. In New England, every citizen receives the elementary notions of human knowledge; he is moreover taught the doctrines and the evidences of his religion, the history of his country, and the leading features of its Constitution. In the States of Connecticut and Massachusetts, it is extremely rare to find a man imperfectly acquainted with all these things, and a person wholly ignorant of them is a sort of phenomenon.

When I compare the Greek and Roman republics with these American States; the manuscript libraries of the former, and their rude population, with the innumerable journals and the enlightened people of the latter; when I remember all the attempts which are made to judge the modern republics by the assistance of those of antiquity, and to infer what will happen in our time from what took place two thousand years ago, I am tempted to burn my books, in order to apply none but novel ideas to so novel a condition of society.

What I have said of New England must not, however, be applied indistinctly to the whole Union; as we advance towards the West of the South, the instruction of the people diminishes. In the States which are adjacent to the Gulf of Mexico, a certain number of individuals may be found, as in our own countries, who are devoid of the rudiments of instruction. But there is not a single district in the United States sunk in complete ignorance; and for a very simple reason: the peoples of Europe started from the darkness of a barbarous condition, to advance toward the light of civilization; their progress has been unequal; some of them have improved apace, whilst others have loitered in their course, and some have stopped, and are still sleeping upon the way.[9]

Such has not been the case in the United States. The Anglo-Americans settled in a state of civilization, upon that territory which their descendants occupy; they had not to begin to learn, and it was sufficient for them not to forget. Now the children of these same Americans are the persons who, year by year, transport their dwellings into the wilds; and with their dwellings their acquired information and their esteem for knowledge. Education has taught them the utility of instruction, and has enabled them to transmit that instruction to their posterity. In the United States society has no infancy, but it is born in man's estate.

The Americans never use the word 'peasant,' because they have no idea of the peculiar class which that term denotes; the ignorance of more remote ages, the simplicity of rural life, and the rusticity of the villager have not been preserved amongst them; and they are alike unacquainted with the virtues, the vices, the coarse habits, and the simple graces of an early stage of civilisation. At the extreme borders of the confederate States, upon the confines of society and of the wilderness, a population of bold adventurers have taken up their abode, who pierce the solitudes of the American woods, and seek a country there, in order to escape that poverty which awaited them in their native provinces. As soon as the pioneer arrives upon the spot which is to serve him for a retreat, he fells a few trees and builds a log-house. Nothing can offer a more miserable aspect than these isolated dwellings. The traveller who approaches one of them towards nightfall, sees the flicker of the hearth-flame through the chinks in the walls; and at night, if the wind rises, he hears the roof of boughs shake to and fro in the midst of the great forest trees. Who would not suppose that this poor hut is the asylum of rudeness and ignorance? Yet no sort of comparison can be drawn between the pioneer and the dwelling which shelters him. Everything about him is primitive and unformed, but he is himself the result of the labour and the experience of eighteen centuries. He wears the dress, and he speaks the language of cities; he is acquainted with the past, curious of the future, and ready for argument upon the present; he is, in short, a highly civilised being, who consents, for a time, to inhabit the backwoods, and who penetrates into the wilds of the New World with the Bible, an axe, and a file of newspapers.

It is difficult to imagine the incredible rapidity with which public opinion circulates in the midst of these deserts.[10] I do not think that so much intellectual intercourse takes place in the most enlightened and populous districts of France.[11] It cannot be doubted that, in the United States, the instruction of the people powerfully contributes to the support of a democratic republic; and such must always be the case, I believe, where instruction which awakens the understanding is not separated from moral education which amends the heart. But I by no means exaggerate this benefit, and I am still further from thinking, as so many people do think in Europe, that men can be instantaneously made citizens by teaching them to read and write. True information is mainly derived from experience; and if the Americans had not been gradually accustomed to govern themselves, their book-learning would not assist them much at the present day.

I have lived a great deal with the people in the United States, and I cannot express how much I admire their experience and their good sense. An American should never be allowed to speak of Europe; for he will then probably display a vast deal of presumption and very foolish pride. He will take up with those crude and vague notions which are so useful to the ignorant all over the world. But if you question him respecting his own country, the cloud which dimmed his intelligence will immediately disperse; his language will become as clear and as precise as his thoughts. He will inform you what his rights are, and by what means he exercises them; he will be able to point out the customs which obtain in the political world. You will find that he is well acquainted with the rules of the administration, and that he is familiar with the mechanism of the laws. The citizen of the United States does not acquire his practical science and his positive notions from books; the instruction he has acquired may have prepared him for receiving those ideas, but it did not furnish them. The American learns to know the laws by participating in the act of legislation; and he takes a lesson in the forms of government from governing. The great work of society is ever going on beneath his eyes, and, as it were, under his hands.

In the United States politics are the end and aim of education; in Europe its principal object is to fit men for private life. The interference of the citizens in public affairs is too rare an occurrence for it to be anticipated beforehand. Upon casting a glance over society in the two hemispheres, these differences are indicated even by its external aspect.

In Europe we frequently introduce the ideas and the habits of private life into public affairs; and as we pass at once from the domestic circle to the government of the State, we may frequently be heard to discuss the great interests of society in the same manner in which we converse with our friends. The Americans, on the other hand, transfuse the habits of public life into their manners in private; and in their country the jury is introduced into the games of schoolboys, and parliamentary forms are observed in the order of a feast.

The Laws Contribute More to the Maintenance of the Democratic Republic in the United States Than the Physical Circumstances of the Country, and the Manners More Than the Laws

All the nations of America have a democratic state of society—Yet democratic institutions only subsist amongst the Anglo-Americans—The Spaniards of South America, equally favoured by physical causes as the Anglo-Americans, unable to maintain a democratic republic—Mexico, which has adopted the Constitution of the United States, in the same predicament—The Anglo-Americans of the West less able to maintain it than those of the East—Reason of these different results.

I have remarked that the maintenance of democratic institutions in the United States is attributable to the circumstances, the laws, and the manners of that country.[12] Most Europeans are only acquainted with the first of these three causes, and they are apt to give it a preponderating importance which it does not really possess.

It is true that the Anglo-Saxons settled in the New World in a state of social equality; the low-born and the noble were not to be found amongst them; and professional prejudices were always as entirely unknown as the prejudices of birth. Thus, as the condition of society was democratic, the empire of democracy was established without difficulty. But this circumstance is by no means peculiar to the United States; almost all the transatlantic colonies were founded by men equal amongst themselves, or who became so by inhabiting them. In no one part of the New World have Europeans been able to create an aristocracy. Nevertheless, democratic institutions prosper nowhere but in the United States.

The American Union has no enemies to contend with; it stands in the wilds like an island in the ocean. But the Spaniards of South America were no less isolated by nature; yet their position has not relieved them from the charge of standing armies. They make war upon each other when they have no foreign enemies to oppose; and the Anglo-American democracy is the only one which has hitherto been able to maintain itself in peace.[13]

The territory of the Union presents a boundless field to human activity, and inexhaustible materials for industry and labour. The passion of wealth takes the place of ambition, and the warmth of faction is mitigated by a sense of prosperity. But in what portion of the globe shall we meet with more fertile plains, with mightier rivers, or with more unexplored and inexhaustible riches than in South America?

Nevertheless, South America has been unable to maintain democratic institutions. If the welfare of nations depended on their being placed in a remote position, with an unbounded space of habitable territory before them, the Spaniards of South America would have no reason to complain of their fate. And although they might enjoy less prosperity than the inhabitants of the United States, their lot might still be such as to excite the envy of some nations in Europe. There are, however, no nations upon the face of the earth more miserable than those of South America.

Thus, not only are physical causes inadequate to produce results analogous to those which occur in North America, but they are unable to raise the population of South America above the level of European States, where they act in a contrary direction. Physical causes do not, therefore, affect the destiny of nations so much as has been supposed.

I have met with men in New England who were on the point of leaving a country, where they might have remained in easy circumstances, to go to seek their fortune in the wilds. Not far from that district I found a French population in Canada, which was closely crowded on a narrow territory, although the same wilds were at hand; and whilst the emigrant from the United States purchased an extensive estate with the earnings of a short term of labour, the Canadian paid as much for land as he would have done in France. Nature offers the solitudes of the New World to Europeans; but they are not always acquainted with the means of turning her gifts to account. Other peoples of America have the same physical conditions of prosperity as the Anglo-Americans, but without their laws and their manners; and these peoples are wretched. The laws and manners of the Anglo-Americans are therefore that efficient cause of their greatness which is the object of my inquiry.

I am far from supposing that the American laws are preeminently good in themselves; I do not hold them to be applicable to all democratic peoples; and several of them seem to me to be dangerous, even in the United States. Nevertheless, it cannot be denied that the American legislation, taken collectively, is extremely well adapted to the genius of the people and the nature of the country which it is intended to govern. The American laws are therefore good, and to them must be attributed a large portion of the success which attends the government of democracy in America: but I do not believe them to be the principal cause of that success; and if they seem to me to have more influence upon the social happiness of the Americans than the nature of the country, on the other hand there is reason to believe that their effect is still inferior to that produced by the manners of the people.

The Federal laws undoubtedly constitute the most important part of the legislation of the United States. Mexico, which is not less fortunately situated than the Anglo-American Union, has adopted the same laws, but is unable to accustom itself to the government of democracy. Some other cause is therefore at work, independently of those physical circumstances and peculiar laws which enable the democracy to rule in the United States.

Another still more striking proof may be adduced. Almost all the inhabitants of the territory of the Union are the descendants of a common stock; they speak the same language, they worship God in the same manner, they are affected by the same physical causes, and they obey the same laws. Whence, then, do their characteristic differences arise? Why, in the Eastern States of the Union, does the republican Government display vigour and regularity, and proceed with mature deliberation? Whence does it derive the wisdom and the durability which mark its acts, whilst in the Western States, on the contrary, society seems to be ruled by the powers of chance? There, public business is conducted with an irregularity and a passionate and feverish excitement, which does not announce a long or sure duration.

I am no longer comparing the Anglo-American States to foreign nations; but I am contrasting them with each other, and endeavouring to discover why they are so unlike. The arguments which are derived from the nature of the country and the difference of legislation are here all set aside. Recourse must be had to some other cause; and what other cause can there be except the manners of the people?

It is in the Eastern States that the Anglo-Americans have been longest accustomed to the government of democracy, and that they have adopted the habits and conceived the notions most favourable to its maintenance. Democracy has gradually penetrated into their customs, their opinions, and the forms of social intercourse; it is to be found in all the details of daily life equally as in the laws. In the Eastern States the instruction and practical education of the people have been most perfected, and religion has been most thoroughly amalgamated with liberty. Now these habits, opinions, customs, and convictions are precisely the constituent elements of that which I have denominated manners.

In the Western States, on the contrary, a portion of the same advantages is still wanting. Many of the Americans of the West were born in the woods, and they mix the ideas and the customs of savage life with the civilisation of their parents. Their passions are more intense; their religious morality less authoritative; and their convictions less secure. The inhabitants exercise no sort of control over their fellow-citizens, for they are scarcely acquainted with each other. The nations of the West display, to a certain extent, the inexperience and the rude habits of a people in its infancy; for although they are composed of old elements, their assemblage is of recent date.

The manners of the Americans of the United States are, then, the real cause which renders that people the only one of the American nations that is able to support a democratic Government; and it is the influence of manners which produces the different degrees of order and of prosperity that may be distinguished in the several Anglo-American democracies. Thus the effect which the geographical position of a country may have upon the duration of democratic institutions is exaggerated in Europe. Too much importance is attributed to legislation, too little to manners. These three great causes serve, no doubt, to regulate and direct the American democracy; but if they were to be classed in their proper order, I should say that the physical circumstances are less efficient than the laws, and the laws very subordinate to the manners of the people. I am convinced that the most advantageous situation and the best possible laws cannot maintain a constitution in spite of the manners of a country; whilst the latter may turn the most unfavourable positions and the worst laws to some advantage. The importance of manners is a common truth to which study and experience incessantly direct our attention. It may be regarded as a central point in the range of human observation, and the common termination of all inquiry. So seriously do I insist upon this head, that if I have hitherto failed in making the reader feel the important influence which I attribute to the practical experience, the habits, the opinions, in short, to the manners of the Americans, upon the maintenance of their institutions, I have failed in the principal object of my work.

Whether Laws and Manners Are Sufficient to Maintain Democratic Institutions in Other Countries Besides America

The Anglo-Americans, if transported into Europe, would be obliged to modify their laws—Distinction to be made between democratic institutions and American institutions—Democratic laws may be conceived better than, or at least different from, those which the American democracy has adopted—The example of America only proves that it is possible to regulate democracy by the assistance of manners and legislation.

I have asserted that the success of democratic institutions in the United States is more intimately connected with the laws themselves, and the manners of the people, than with the nature of the country. But does it follow that the same causes would of themselves produce the same results, if they were put into operation elsewhere; and if the country is no adequate substitute for laws and manners, can laws and manners in their turn prove a substitute for the country? It will readily be understood that the necessary elements of a reply to this question are wanting: other peoples are to be found in the New World besides the Anglo-Americans, and as these people are affected by the same physical circumstances as the latter, they may fairly be compared together. But there are no nations out of America which have adopted the same laws and manners, being destitute of the physical advantages peculiar to the Anglo-Americans. No standard of comparison therefore exists, and we can only hazard an opinion upon this subject.

It appears to me, in the first place, that a careful distinction must be made between the institutions of the United States and democratic institutions in general. When I reflect upon the state of Europe, its mighty nations, its populous cities, its formidable armies, and the complex nature of its politics, I cannot suppose that even the Anglo-Americans, if they were transported to our hemisphere, with their ideas, their religion, and their manners, could exist without considerably altering their laws. But a democratic nation may be imagined, organised differently from the American people. It is not impossible to conceive a government really established upon the will of the majority; but in which the majority, repressing its natural propensity to equality, should consent, with a view to the order and the stability of the State, to invest a family or an individual with all the prerogatives of the executive. A democratic society might exist, in which the forces of the nation would be more centralised than they are in the United States; the people would exercise a less direct and less irresistible influence upon public affairs, and yet every citizen invested with certain rights would participate, within his sphere, in the conduct of the government. The observations I made amongst the Anglo-Americans induce me to believe that democratic institutions of this kind, prudently introduced into society, so as gradually to mix with the habits and to be interfused with the opinions of the people, might subsist in other countries besides America. If the laws of the United

States were the only imaginable democratic laws, or the most perfect which it is possible to conceive, I should admit that the success of those institutions affords no proof of the success of democratic institutions in general, in a country less favoured by natural circumstances. But as the laws of America appear to me to be defective in several respects, and as I can readily imagine others of the same general nature, the peculiar advantages of that country do not prove that democratic institutions cannot succeed in a nation less favoured by circumstances, if ruled by better laws.

If human nature were different in America from what it is elsewhere; or if the social condition of the Americans engendered habits and opinions amongst them different from those which originate in the same social condition in the Old World, the American democracies would afford no means of predicting what may occur in other democracies. If the Americans displayed the same propensities as all other democratic nations, and if their legislators had relied upon the nature of the country and the favour of circumstances to restrain those propensities within due limits, the prosperity of the United States would be exclusively attributable to physical causes, and it would afford no encouragement to a people inclined to imitate their example, without sharing their natural advantages. But neither of these suppositions is borne out by facts.

In America the same passions are to be met with as in Europe; some originating in human nature, others in the democratic condition of society. Thus in the United States I found that restlessness of heart which is natural to men, when all ranks are nearly equal and the chances of elevation are the same to all. I found the democratic feeling of envy expressed under a thousand different forms. I remarked that the people frequently displayed, in the conduct of affairs, a consummate mixture of ignorance and presumption; and I inferred that in America, men are liable to the same failings and the same absurdities as amongst ourselves. But upon examining the state of society more attentively, I speedily discovered that the Americans had made great and successful efforts to counteract these imperfections of human nature, and to correct the natural defects of democracy. Their divers municipal laws appeared to me to be a means of restraining the ambition of the citizens within a narrow sphere, and of turning those same passions which might have worked havoc in the State, to the good of the township or the parish. The American legislators have succeeded to a certain extent in opposing the notion of rights to the feelings of envy; the permanence of the religious world to the continual shifting of politics; the experience of the people to its theoretical ignorance; and its practical knowledge of business to the impatience of its desires.

The Americans, then, have not relied upon the nature of their country to counterpoise those dangers which originate in their Constitution and in their political laws. To evils which are common to all democratic peoples they have applied remedies which none but themselves had ever thought of before; and although they were the first to make the experiment, they have succeeded in it.

The manners and laws of the Americans are not the only ones which may suit a democratic people; but the Americans have shown that it would be wrong to despair of regulating democracy by the aid of manners and of laws. If other nations should borrow this general and pregnant idea from the Americans, without however intending to imitate them in the peculiar application which they have made of it; if they should attempt to fit themselves for that social condition, which it seems to be the will of Providence to impose upon the generations of this age, and so to escape from the despotism or the anarchy which threatens them; what reason is there to suppose that their efforts would not be crowned with success? The organisation and the establishment of democracy in Christendom is the great political problem of the time. The Americans, unquestionably, have not resolved this problem, but they furnish useful data to those who undertake the task.

Importance of What Precedes with Respect to the State of Europe

It may readily be discovered with what intention I undertook the foregoing inquiries. The question here discussed is interesting not only to the United States, but to the whole world; it concerns, not a nation, but all mankind. If those nations whose social condition is democratic could only remain free as long as they are inhabitants of the wilds, we could not but despair of the future destiny of the human race; for democracy is rapidly acquiring a more extended sway, and the wilds are gradually peopled with men. If it were true that laws and manners are insufficient to maintain democratic institutions, what refuge would remain open to the nations, except the despotism of a single individual? I am aware that there are many worthy persons at the present time who are not alarmed at this latter alternative, and who are so tired of liberty as to be glad of repose, far from those storms by which it is attended. But these individuals are ill acquainted with the haven towards which they are bound. They are so deluded by their recollections, as to judge the tendency of absolute power by what it was formerly, and not by what it might become at the present time.

If absolute power were reestablished amongst the democratic nations of Europe, I am persuaded that it would assume a new form, and appear under features unknown to our forefathers. There was a time in Europe when the laws and the consent of the people had invested princes with almost unlimited authority; but they scarcely ever availed themselves of it. I do not speak of the prerogatives of the nobility, of the authority of supreme courts of justice, of corporations and their chartered rights, or of provincial privileges, which served to break the blows of the sovereign authority, and to maintain a spirit of resistance in the nation. Independently of these political institutions—which, however opposed they might be to personal liberty, served to keep alive the love of freedom in the mind of the public, and which may be esteemed to have been useful in this respect—the manners and opinions of the nation confined the royal authority within barriers which were not less powerful, although they were less conspicuous. Religion, the affections of the people, the benevolence of the prince, the sense of honour, family pride, provincial prejudices, custom, and public opinion limited the power of kings, and restrained their authority within an invisible circle. The constitution of nations was despotic at that time, but their manners were free. Princes had the right, but they had neither the means nor the desire, of doing whatever they pleased.

But what now remains of those barriers which formerly arrested the aggressions of tyranny? Since religion has lost its empire over the souls of men, the most prominent boundary which divided good from evil is overthrown; the very elements of the moral world are indeterminate; the princes and the peoples of the earth are guided by chance, and none can define the natural limits of despotism and the bounds of licence. Long revolutions have for ever destroyed the respect which surrounded the rulers of the State; and since they have been relieved from the burden of public esteem, princes may henceforward surrender themselves without fear to the seductions of arbitrary power.

When kings find that the hearts of their subjects are turned towards them, they are clement, because they are conscious of their strength, and they are chary of the affection of their people, because the affection of their people is the bulwark of the throne. A mutual interchange of good-will then takes place between the prince and the people, which resembles the gracious intercourse of domestic society. The subjects may murmur at the sovereign's decree, but they are grieved to displease him; and the sovereign chastises his subjects with the light hand of parental affection.

But when once the spell of royalty is broken in the tumult of revolution; when successive monarchs have crossed the throne, so as alternately to display to the people the weakness of their right and the harshness of their power, the sovereign is no longer regarded by any as the Father of the State, and he is feared by all as its master. If he be weak, he is despised; if he be strong, he is detested. He is himself full of animosity and alarm; he finds that he is as a stranger in his own country, and he treats his subjects like conquered enemies.

When the provinces and the towns formed so many different nations in the midst of their common country, each of them had a will of its own, which was opposed to the general spirit of subjection; but now that all the parts of the same empire, after having lost their immunities, their customs, their prejudices, their traditions, and their names, are subjected and accustomed to the same laws, it is not more difficult to oppress them collectively than it was formerly to oppress them singly.

Whilst the nobles enjoyed their power, and indeed long after that power was lost, the honour of aristocracy conferred an extraordinary degree offeree upon their personal opposition. They afford instances of men who, notwithstanding their weakness, still entertained a high opinion of their personal value, and dared to cope single-handed with the efforts of the public authority. But at the present day, when all ranks are more and more confounded, when the individual disappears in the throng, and is easily lost in the midst of a common obscurity, when the honour of monarchy has almost lost its empire without being succeeded by public virtue, and when nothing can enable man to rise above himself, who shall say at what point the exigencies of power and the servility of weakness will stop?

As long as family feeling was kept alive, the antagonist of oppression was never alone; he looked about him, and found his clients, his hereditary friends, and his kinsfolk. If this support was wanting, he was sustained by his ancestors and animated by his posterity. But when patrimonial estates are divided, and when a few years suffice to confound the distinctions of a race, where can family feeling be found? What force can there be in the customs of a country which has changed and is still perpetually changing, its aspect; in which every act of tyranny has a precedent, and every crime an example; in which there is nothing so old that its antiquity can save it from destruction, and nothing so unparalleled that its novelty can prevent it from being done? What resistance can be offered by manners of so pliant a make that they have already often yielded? What strength can even public opinion have retained, when no twenty persons are connected by a common tie; when not a man, nor a family, nor chartered corporation, nor class, nor free institution, has the power of representing or exerting that opinion; and when every citizen—being equally weak, equally poor, and equally dependent—has only his personal impotence to oppose to the organized force of the Government?

The annals of France furnish nothing analogous to the condition in which that country might then be thrown. But it may more aptly be assimilated to the times of old, and to those hideous eras of Roman oppression, when the manners of the people were corrupted, their traditions obliterated, their habits destroyed, their opinions shaken, and freedom, expelled from the laws, could find no refuge in the land; when nothing protected the citizens, and the citizens no longer protected themselves; when human nature was the sport of man, and princes wearied out the clemency of Heaven before they exhausted the patience of their subjects. Those who hope to revive the monarchy of Henry IV. or of Louis XIV., appear to me to be afflicted with mental blindness; and when I consider the present condition of several European nations—a condition to which all the others tend—I am led to believe that they will soon be left with no other alternative than democratic liberty, or the tyranny of the Cassars.[14]

And indeed it is deserving of consideration, whether men are to be entirely emancipated or entirely enslaved; whether their rights are to be made equal, or wholly taken away from them. If the rulers of society were reduced either gradually to raise the crowd to their own level, or to sink the citizens below that of humanity, would not the doubts of many be resolved, the consciences of many be healed, and the community prepared to make great sacrifices with little difficulty? In that case, the gradual growth of democratic manners and institutions should be regarded, not as the best, but as the only means of preserving freedom; and without liking the government of democracy, it might be adopted as the most applicable and the fairest remedy for the present ills of society.

It is difficult to associate a people in the work of government; but it is still more difficult to supply it with experience, and to inspire it with the feelings which it requires in order to govern well. I grant that the caprices of democracy are perpetual; its instruments are rude; its laws imperfect. But if it were true that soon no just medium would exist between the empire of democracy and the dominion of a single arm, should we not rather incline towards the former than submit voluntarily to the latter? And if complete equality be our fate, is it not better to be levelled by free institutions than by despotic power?

Those who, after having read this book, should imagine that my intention in writing it has been to propose the laws and manners of the Anglo-Americans for the imitation of all democratic peoples, would commit a very great mistake; they must have paid more attention to the form than to the substance of my ideas. My aim has been to show, by the example of America, that laws, and especially manners, may exist which will allow a democratic people to remain free. But I am very far from thinking that we ought to follow the example of the American democracy, and copy the means which it has employed to attain its ends; for I am well aware of the influence which the nature of a country and its political precedents exercise upon a constitution; and I should regard it as a great misfortune for mankind if liberty were to exist all over the world under the same forms.

But I am of opinion that if we do not succeed in gradually introducing democratic institutions into France, and if we despair of imparting to the citizens those ideas and sentiments which first prepare them for freedom, and afterwards allow them to enjoy it, there will be no independence at all, either for the middling classes or the nobility, for the poor or for the rich, but an equal tyranny over all; and I foresee that if the peaceable empire of the majority be not founded amongst us in time, we shall sooner or later arrive at the unlimited authority of a single despot.

Chapter XVIII

The Present and Probable Future Condition of the Three Races Which Inhabit the Territory of the United States

The principal part of the task which I had imposed upon myself is now performed. I have shown, as far as I was able, the laws and the manners of the American democracy. Here I might stop; but the reader would perhaps feel that I had not satisfied his expectations.

The absolute supremacy of democracy is not all that we meet with in America; the inhabitants of the New World may be considered from more than one point of view. In the course of this work my subject has often led me to speak of the Indians and the Negroes; but I have never been able to stop in order to show what place these two races occupy in the midst of the democratic people whom I was engaged in describing. I have mentioned in what spirit, and according to what laws, the Anglo-American Union was formed; but I could only glance at the dangers which menace that confederation, whilst it was equally impossible for me to give a detailed account of its chances of duration, independently of its laws and manners. When speaking of the United republican States, I hazarded no conjectures upon the permanence of republican forms in the New World, and when making frequent allusion to the commercial activity which reigns in the Union, I was unable to inquire into the future condition of the Americans as a commercial people.

These topics are collaterally connected with my subject without forming a part of it; they are American without being democratic; and to portray democracy has been my principal aim. It was therefore necessary to postpone these questions, which I now take up as the proper termination of my work.

The territory now occupied or claimed by the American Union spreads from the shores of the Atlantic to those of the Pacific Ocean. On the East and West its limits are those of the continent itself. On the South it advances nearly to the Tropic, and it extends upwards to the icy regions of the North. The human beings who are scattered over this space do not form, as in Europe, so many branches of the same stock. Three races, naturally distinct, and, I might almost say, hostile to each other, are discoverable amongst them at the first glance. Almost insurmountable barriers had been raised between them by education and by law, as well as by their origin and outward characteristics; but fortune has brought them together on the same soil, where, although they are mixed, they do not amalgamate, and each race fulfils its destiny apart.

Amongst these widely differing families of men, the first which attracts attention, the superior in intelligence, in power and in enjoyment, is the White or European, the MAN pre-eminent; and in subordinate grades, the Negro and the Indian. These two unhappy races have nothing in common; neither birth, nor features, nor language, nor habits. Their only resemblance lies in their misfortunes. Both of them occupy an inferior rank in the country they inhabit; both suffer from tyranny; and if their wrongs are not the same, they originate, at any rate, with the same authors.

If we reasoned from what passes in the world, we should almost say that the European is to the other races of mankind, what man is to the lower animals;—he makes them subservient to his use; and when he cannot subdue, he destroys them. Oppression has, at one stroke, deprived the descendants of the Africans of almost all the privileges of humanity. The Negro of the United States has lost all remembrance of his country; the language which his forefathers spoke is never heard around him; he abjured their religion and forgot their customs when he ceased to belong to Africa, without acquiring any claim to European privileges. But he remains half way between the two communities; sold by the one, repulsed by the other; finding not a spot in the universe to call by the name of country, except the faint image of a home which the shelter of his master's roof affords.

The Negro has no family; woman is merely the temporary companion of his pleasures, and his children are upon an equality with himself from the moment of their birth. Am I to call it a proof of God's mercy or a visitation of his wrath, that man in certain states appears to be insensible to his extreme wretchedness, and almost affects, with a depraved taste, the cause of his misfortunes? The Negro, who is plunged in this abyss of evils, scarcely feels his own calamitous situation. Violence made him a slave, and the habit of servitude gives him the thoughts and desires of a slave; he admires his tyrants more than he hates them, and finds his joy and his pride in the servile imitation of those who oppress him: his understanding is degraded to the level of his soul.

The Negro enters upon slavery as soon as he is born: nay, he may have been purchased in the womb, and have begun his slavery before he began his existence. Equally devoid of wants and of enjoyment, and useless to himself, he learns, with his first notions of existence, that he is the property of another, who has an interest in preserving his life, and that the care of it does not devolve upon himself; even the power of thought appears to him a useless gift of Providence, and he quietly enjoys the privileges of his debasement. If he becomes free, independence is often felt by him to be a heavier burden than slavery; for having learned, in the course of his life, to submit to everything except reason, he is too much unacquainted with her dictates to obey them. A thousand new desires beset him, and he is destitute of the knowledge and energy necessary to resist them: these are masters which it is necessary to contend with, and he has learnt only to submit and obey. In short, he sinks to such a depth of wretchedness, that while servitude brutalises, liberty destroys him.

Oppression has been no less fatal to the Indian than to the Negro race, but its effects are different. Before the arrival of white men in the New World, the inhabitants of North America lived quietly in their woods, enduring the vicissitudes and practising the virtues and vices common to savage nations. The Europeans, having dispersed the Indian tribes and driven them into the deserts, condemned them to a wandering life full of inexpressible sufferings.

Savage nations are only controlled by opinion and by custom. When the North American Indians had lost the sentiment of attachment to their country; when their families were dispersed, their traditions obscured, and the chain of their recollections broken; when all their habits were changed, and their wants increased beyond measure, European tyranny rendered them more disorderly and less civilized than they were before. The moral and physical condition of these tribes continually grew worse, and they became more barbarous as they became more wretched. Nevertheless, the Europeans have not been able to metamorphose the character of the Indians; and though they have had power to destroy them, they have never been able to make them submit to the rules of civilised society.

The lot of the Negro is placed on the extreme limit of servitude, while that of the Indian lies on the uttermost verge of liberty; and slavery does not produce more fatal effects upon the first, than independence upon the second. The Negro has lost all property in his own person, and he cannot dispose of his existence without committing a sort of fraud: but the savage is his own master as soon as he is able to act; parental authority is scarcely known to him; he has never bent his will to that of any of his kind, nor learned the difference between voluntary obedience and a shameful subjection; and the very name of law is unknown to him. To be free, with him, signifies to escape from all the shackles of society. As he delights in this barbarous independence, and would rather perish than sacrifice the least part of it, civilisation has little power over him.

The Negro makes a thousand fruitless efforts to insinuate himself amongst men who repulse him; he conforms to the tastes of his oppressors, adopts their opinions, and hopes by imitating them to form a part of their community. Having been told from infancy that his race is naturally inferior to that of the Whites, he assents to the proposition and is ashamed of his own nature. In each of his features he discovers a trace of slavery, and, if it were in his power, he would willingly rid himself of everything that makes him what he is.

The Indian, on the contrary, has his imagination inflated with the pretended nobility of his origin, and lives and dies in the midst of these dreams of pride. Far from desiring to conform his habits to ours, he loves his savage life as the distinguishing mark of his race, and he repels every advance to civilisation, less perhaps from the hatred which he entertains for it, than from a dread of resembling the Europeans.[1] While he has nothing to oppose to our perfection in the arts but the resources of the desert, to our tactics nothing but undisciplined courage; whilst our well-digested plans are met by the spontaneous instincts of savage life, who can wonder if he fails in this unequal contest?

The Negro, who earnestly desires to mingle his race with that of the European, cannot effect if; while the Indian, who might succeed to a certain extent, disdains to make the attempt. The servility of the one dooms him to slavery, the pride of the other to death.

I remember that while I was travelling through the forests which still cover the State of Alabama, I arrived one day at the log-house of a pioneer. I did not wish to penetrate into the dwelling of the American, but retired to rest myself for a while on the margin of a spring, which was not far off, in the woods. While I was in this place (which was in the neighbourhood of the Creek territory), an Indian woman appeared, followed by a Negress, and holding by the hand a little white girl of five or six years old, whom I took to be the daughter of the pioneer. A sort of barbarous luxury set off the costume of the Indian; rings of metal were hanging from her nostrils and ears; her hair, which was adorned with glass beads, fell loosely upon her shoulders; and I saw that she was not married, for she still wore that necklace of shells which the bride always deposits on the nuptial couch. The Negress was clad in squalid European garments. They all three came and seated themselves upon the banks of the fountain; and the young Indian, taking the child in her arms, lavished upon her such fond caresses as mothers give; while the Negress endeavoured by various little artifices to attract the attention of the young Creole.

The child displayed in her slightest gestures a consciousness of superiority which formed a strange contrast with her infantine weakness; as if she received the attentions of her companions with a sort of condescension. The Negress was seated on the ground before her mistress, watching her smallest desires, and apparently divided between strong affection for the child and servile fear; whilst the savage displayed, in the midst of her tenderness, an air of freedom and of pride which was almost ferocious. I had approached the group, and I contemplated them in silence; but my curiosity was probably displeasing to the Indian woman, for she suddenly rose, pushed the child roughly from her, and giving me an angry look plunged into the thicket. I had often chanced to see individuals met together in the same place, who belonged to the three races of men which people North America. I had perceived from many different results the preponderance of the Whites. But in the picture which I have just been describing there was something peculiarly touching; a bond of affection here united the oppressors with the oppressed, and the effort of Nature to bring them together rendered still more striking the immense distance placed between them by prejudice and by law.

The Present and Probable Future Condition of the Indian Tribes Which Inhabit the Territory Possessed by the Union

Gradual disappearance of the native tribes—Manner in which it takes place—Miseries accompanying the forced migrations of the Indians—The savages of North America had only two ways of escaping destruction; war or civilisation—They are no longer able to make war—Reasons why they refused to become civilised when it was in their power, and why they cannot become so now that they desire it—Instance of the Creeks and Cherokees Policy of the particular States towards these Indians—Policy of the Federal Government.

None of the Indian tribes which formerly inhabited the territory of New England—the Naragansetts, the Mohicans, the Pecots—have any existence but in the recollection of man. The Lenapes, who received William Penn, a hundred and fifty years ago, upon the banks of the Delaware, have disappeared; and I myself met with the last of the Iroquois, who were begging alms. The nations I have mentioned formerly covered the country to the sea-coast; but a traveller at the present day must penetrate more than a hundred leagues into the interior of the continent to find an Indian. Not only have these wild tribes receded, but they are destroyed;[2] and as they give way or perish, an immense and increasing people fills their place. There is no instance upon record of so prodigious a growth, or so rapid a destruction: the manner in which the latter change takes place is not difficult to describe.

When the Indians were the sole inhabitants of the wilds from whence they have since been expelled, their wants were few. Their arms were of their own manufacture, their only drink was the water of the brook, and their clothes consisted of the skins of animals, whose flesh furnished them with food.

The Europeans introduced amongst the savages of North America fire-arms, ardent spirits, and iron: they taught them to exchange for manufactured stuffs, the rough garments which had previously satisfied their untutored simplicity. Having acquired new tastes, without the arts by which they could be gratified, the Indians were obliged to have recourse to the workmanship of the Whites; but in return for their productions the savage had nothing to offer except the rich furs which still abounded in his woods. Hence the chase became necessary, not merely to provide for his subsistence, but in order to procure the only objects of barter which he could furnish to Europe.[3] Whilst the wants of the natives were thus increasing, their resources continued to diminish.

From the moment when a European settlement is formed in the neighbourhood of the territory occupied by the Indians, the beasts of chase take the alarm.[4] Thousands of savages, wandering in the forests and destitute of any fixed dwelling, did not disturb them; but as soon as the continuous sounds of European labour are heard in their neighbourhood, they begin to flee away, and retire to the West, where their instinct teaches them that they will find deserts of immeasurable extent. 'The buffalo is constantly receding,' say Messrs. Clarke and Cass in their Report of the year 1829; 'a few years since they approached the base of the Alleghany; and a few years hence they may even be rare upon the immense plains which extend to the base of the Rocky Mountains.' I have been assured that this effect of the approach of the Whites is often felt at two hundred leagues' distance from their frontier. Their influence is thus exerted over tribes whose name is unknown to them; and who suffer the evils of usurpation long before they are acquainted with the authors of their distress.[5]

Bold adventurers soon penetrate into the country the Indians have deserted, and when they have advanced about fifteen or twenty leagues from the extreme frontiers of the Whites, they begin to build habitations for civilised beings in the midst of the wilderness. This is done without difficulty, as the territory of a hunting-nation is ill-defined; it is the common property of the tribe, and belongs to no one in particular, so that individual interests are not concerned in the protection of any part of it.

A few European families, settled in different situations at a considerable distance from each other, soon drive away the wild animals which remain between their places of abode. The Indians, who had previously lived in a sort of abundance, then find it difficult to subsist, and still more difficult to procure the articles of barter which they stand in need of.

To drive away their game is to deprive them of the means of existence, as effectually as if the fields of our agriculturists were stricken with barrenness; and they are reduced, like famished wolves, to prowl through the forsaken woods in quest of prey. Their instinctive love of their country attaches them to the soil which gave them birth,[6] even after it has ceased to yield anything but misery and death. At length they are compelled to acquiesce, and to depart: they follow the traces of the elk, the buffalo, and the beaver, and are guided by these wild animals in the choice of their future country. Properly speaking, therefore, it is not the Europeans who drive away the native inhabitants of America; it is famine which compels them to recede; a happy distinction which had escaped the casuists of former times, and for which we are indebted to modern discovery!

It is impossible to conceive the extent of the sufferings which attend these forced emigrations. They are undertaken by a people already exhausted and reduced; and the countries to which the new comers betake themselves are inhabited by other tribes which receive them with jealous hostility. Hunger is in the rear; war awaits them, and misery besets them on all sides. In the hope of escaping from such a host of enemies, they separate, and each individual endeavours to procure the means of supporting his existence in solitude and secrecy, living in the immensity of the desert like an outcast in civilised society. The social tie, which distress had long since weakened, is then dissolved; they have lost their country, and their people soon desert them: their very families are obliterated; the names they bore in common are forgotten, their language perishes, and all traces of their origin disappear. Their nation has ceased to exist, except in the recollection of the antiquaries of America and a few of the learned of Europe.

I should be sorry to have my reader suppose that I am colouring the picture too highly; I saw with my own eyes several of the cases of misery which I have been describing; and I was the witness of sufferings which I have not the power to portray.

At the end of the year 1831, whilst I was on the left bank of the Mississippi at a place named by Europeans, Memphis, there arrived a numerous band of Choctaws (or Chactas, as they are called by the French in Louisiana). These savages had left their country, and were endeavouring to gain the right bank of the Mississippi, where they hoped to find an asylum which had been promised them by the American Government. It was then the middle of winter, and the cold was unusually severe; the snow had frozen hard upon the ground, and the river was drifting huge masses of ice. The Indians had their families with them; and they brought in their train the wounded and sick, with children newly born, and old men upon the verge of death. They possessed neither tents nor wagons, but only their arms and some provisions. I saw them embark to pass the mighty river, and never will that solemn spectacle fade from my remembrance. No cry, no sob was heard amongst the assembled crowd; all were silent. Their calamities were of ancient date, and they knew them to be irremediable. The Indians had all stepped into the bark which was to carry them across, but their dogs remained upon the bank. As soon as these animals perceived that their masters were finally leaving the shore, they set up a dismal howl, and, plunging all together into the icy waters of the Mississippi, they swam after the boat.

The ejectment of the Indians very often takes place at the present day, in a regular, and, as it were, a legal manner. When the European population begins to approach the limit of the desert inhabited by a savage tribe, the Government of the United States usually dispatches envoys to them, who assemble the Indians in a large plain, and having first eaten and drunk with them, accost them in the following manner: 'What have you to do in the land of your fathers? Before long, you must dig up their bones in order to live. In what respect is the country you inhabit better than another? Are there no woods, marshes, or prairies, except where you dwell? And can you live nowhere but under your own sun? Beyond those mountains which you see at the horizon, beyond the lake which bounds your territory on the West, there lie vast countries where beasts of chase are found in great abundance; sell your lands to us, and go to live happily in those solitudes.' After holding this language, they spread before the eyes of the Indians fire-arms, woollen garments, kegs of brandy, glass necklaces, bracelets of tinsel, ear-rings, and looking glasses.[7] If, when they have beheld all these riches, they still hesitate, it is insinuated that they have not the means of refusing their required consent, and that the Government itself will not long have the power of protecting them in their rights. What are they to do? Half convinced, and half compelled, they go to inhabit new deserts, where the importunate Whites will not let them remain ten years in tranquillity. In this manner do the Americans obtain, at a very low price, whole provinces, which the richest sovereigns of Europe could not purchase.[8]

These are great evils; and it must be added that they appear to me to be irremediable. I believe that the Indian nations of North America are doomed to perish; and that whenever the Europeans shall be established on the shores of the Pacific Ocean, that race of men will be no more.[9] The Indians had only the two alternatives of war or civilisation; in other words, they must either have destroyed the Europeans or become their equals.

At the first settlement of the colonies they might have found it possible, by uniting their forces, to deliver themselves from the small bodies of strangers who landed on their continent.[10] They several times attempted to do it, and were on the point of succeeding; but the disproportion of their resources, at the present day, when compared with those of the Whites, is too great to allow such an enterprise to be thought of. Nevertheless, there do arise from time to time among the Indians men of penetration, who foresee the final destiny which awaits the native population, and who exert themselves to unite all the tribes in common hostility to the Europeans; but their efforts are unavailing. Those tribes which are in the neighbourhood of the Whites, are too much weakened to offer an effectual resistance; whilst the others, giving way to that childish carelessness of the morrow which characterises savage life, wait for the near approach of danger before they prepare to meet it; some are unable, the others are unwilling, to exert themselves.

It is easy to foresee that the Indians will never conform to civilisation; or that it will be too late, whenever they may be inclined to make the experiment.

Civilisation is the result of a long social process which takes place in the same spot, and is handed down from one generation to another, each one profiting by the experience of the last. Of all nations, those submit to civilisation with the most difficulty which habitually live by the chase. Pastoral tribes, indeed, often change their place of abode; but they follow a regular order in their migrations, and often return again to their old stations, whilst the dwelling of the hunter varies with that of the animals he pursues.

Several attempts have been made to diffuse knowledge amongst the Indians, without controlling their wandering propensities; by the Jesuits in Canada, and by the Puritans in New England;[11] but none of these endeavours were crowned by any lasting success. Civilisation began in the cabin, but it soon retired to expire in the woods. The great error of these legislators of the Indians was their not understanding that, in order to succeed in civilising a people, it is first necessary to fix it; which cannot be done without inducing it to cultivate the soil; the Indians ought in the first place to have been accustomed to agriculture. But not only are they destitute of this indispensable preliminary to civilisation, they would even have great difficulty in acquiring it. Men who have once abandoned themselves to the restless and adventurous life of the hunter, feel an insurmountable disgust for the constant and regular labour which tillage requires. We see this proved in the bosom of our own society; but it is far more visible among peoples whose partiality for the chase is a part of their national character.

Independently of this general difficulty, there is another, which applies peculiarly to the Indians; they consider labour not merely as an evil, but as a disgrace; so that their pride prevents them from becoming civilised, as much as their indolence.[12]

There is no Indian so wretched as not to retain under his hut of bark a lofty idea of his personal worth; he considers the cares of industry and labour as degrading occupations; he compares the husbandman to the ox which traces the furrow; and even in our most ingenious handicraft, he can see nothing but the labour of slaves. Not that he is devoid of admiration for the power and intellectual greatness of the Whites; but although the result of our efforts surprises him, he contemns the means by which we obtain it; and while he acknowledges our ascendancy, he still believes in his superiority. War and hunting are the only pursuits which appear to him worthy to be the occupations of a man.[13] The Indian, in the dreary solitudes of his woods, cherishes the same ideas, the same opinions as the noble of the Middle ages in his castle, and he only requires to become a conqueror to complete the resemblance; thus, however strange it may seem, it is in the forests of the New World, and not amongst the Europeans who people its coasts, that the ancient prejudices of Europe are still in existence.

More than once, in the course of this work, I have endeavoured to explain the prodigious influence which the social condition appears to exercise upon the laws and the manners of men; and I beg to add a few words on the same subject.

When I perceive the resemblance which exists between the political institutions of our ancestors, the Germans, and of the wandering tribes of North America; between the customs described by Tacitus, and those of which I have sometimes been a witness, I cannot help thinking that the same cause has brought about the same results in both hemispheres; and that in the midst of the apparent diversity of human affairs, a certain number of primary facts may be discovered, from which all the others are derived. In what we usually call the German institutions, then, I am inclined only to perceive barbarian habits; and the opinions of savages in what we style feudal principles.

However strongly the vices and prejudices of the North American Indians may be opposed to their becoming agricultural and civilised, necessity sometimes obliges them to it. Several of the Southern nations, and amongst others the Cherokees and the Creeks,[14] were surrounded by Europeans, who had landed on the shores of the Atlantic; and who, either descending the Ohio or proceeding up the Mississippi, arrived simultaneously upon their borders. These tribes have not been driven from place to place, like their Northern brethren; but they have been gradually enclosed within narrow limits, like the game within the thicket, before the huntsmen plunge into the interior. The Indians who were thus placed between civilisation and death, found themselves obliged to live by ignominious labour like the Whites. They took to agriculture, and without entirely forsaking their old habits or manners, sacrificed only as much as was necessary to their existence.

The Cherokees went further; they created a written language; established a permanent form of government; and as everything proceeds rapidly in the New World, before they had all of them clothes, they set up a newspaper.[15]

The growth of European habits has been remarkably accelerated among these Indians by the mixed race which has sprung up.[16] Deriving intelligence from their father's side, without entirely losing the savage customs of the mother, the half-blood forms the natural link between civilisation and barbarism. Wherever this race has multiplied the savage state has become modified, and a great change has taken place in the manners of the people.[17]

The success of the Cherokees proves that the Indians are capable of civilisation, but it does not prove that they will succeed in it. This difficulty which the Indians find in submitting to civilisation proceeds from the influence of a general cause, which it is almost impossible for them to escape. An attentive survey of history demonstrates that, in general, barbarous nations have raised themselves to civilisation by degrees, and by their own efforts. Whenever they derive knowledge from a foreign people, they stood towards it in the relation of conquerors, and not of a conquered nation. When the conquered nation is enlightened, and the conquerors are half savage, as in the case of the invasion of Rome by the Northern nations or that of China by the Mongols, the power which victory bestows upon the barbarian is sufficient to keep up his importance among civilised men, and permit him to rank as their equal, until he becomes their rival: the one has might on his side, the other has intelligence; the former admires the knowledge and the arts of the conquered, the latter envies the power of the conquerors. The barbarians at length admit civilised man into their palaces, and he in turn opens his schools to the barbarians. But when the side on which the physical force lies, also possesses an intellectual preponderance, the conquered party seldom become civilised; it retreats, or is destroyed. It may therefore be said, in a general way, that savages go forth in arms to seek knowledge, but that they do not receive it when it comes to them.

If the Indian tribes which now inhabit the heart of the continent could summon up energy enough to attempt to civilise themselves, they might possibly succeed.

Superior already to the barbarous nations which surround them, they would gradually gain strength and experience, and when the Europeans should appear upon their borders, they would be in a state, if not to maintain their independence, at least to assert their right to the soil, and to incorporate themselves with the conquerors. But it is the misfortune of Indians to be brought into contact with a civilised people, which is also (it must be owned) the most avaricious nation on the globe, whilst they are still semi-barbarian: to find despots in their instructors, and to receive knowledge from the hand of oppression. Living in the freedom of the woods, the North American Indian was destitute, but he had no feeling of inferiority towards anyone; as soon, however, as he desires to penetrate into the social scale of the Whites, he takes the lowest rank in society, for he enters, ignorant and poor, within the pale of science and wealth. After having led a life of agitation, beset with evils and dangers, but at the same time filled with proud emotions,[18] he is obliged to submit to a wearisome, obscure, and degraded state; and to gain the bread which nourishes him by hard and ignoble labour; such are in his eyes the only results of which civilisation can boast: and even this much he is not sure to obtain.

When the Indians undertake to imitate their European neighbours, and to till the earth like the settlers, they are immediately exposed to a very formidable competition. The white man is skilled in the craft of agriculture; the Indian is a rough beginner in an art with which he is unacquainted. The former reaps abundant crops without difficulty, the latter meets with a thousand obstacles in raising the fruits of the earth.

The European is placed amongst a population whose wants he knows and partakes. The savage is isolated in the midst of a hostile people, with whose manners, language, and laws he is imperfectly acquainted, but without whose assistance he cannot live. He can only procure the materials of comfort by bartering his commodities against the goods of the European, for the assistance of his countrymen is wholly insufficient to supply his wants. When the Indian wishes to sell the produce of his labour, he cannot always meet with a purchaser, whilst the European readily finds a market; and the former can only produce at a considerable cost that which the latter vends at a very low rate. Thus the Indian has no sooner escaped those evils to which barbarous nations are exposed, than he is subjected to the still greater miseries of civilised communities; and he finds it scarcely less difficult to live in the midst of our abundance, than in the depth of his own wilderness.

He has not yet lost the habits of his erratic life; the traditions of his fathers and his passion for the chase are still alive within him. The wild enjoyments which formerly animated him in the woods, painfully excite his troubled imagination; and his former privations appear to be less keen, his former perils less appalling. He contrasts the independence which he possessed amongst his equals with the servile position which he occupies in civilised society. On the other hand, the solitudes which were so long his free home are still at hand; a few hours' march will bring him back to them once more. The Whites offer him a sum, which seems to him to be considerable, for the ground which he has begun to clear. This money of the Europeans may possibly furnish him with the means of a happy and peaceful subsistence in remoter regions; and he quits the plough, resumes his native arms, and returns to the wilderness forever.[19] The condition of the Creeks and Cherokees, to which I have already alluded, sufficiently corroborates the truth of this deplorable picture.

The Indians, in the little which they have done, have unquestionably displayed as much natural genius as the peoples of Europe in their most important designs; but nations as well as men require time to learn, whatever may be their intelligence and their zeal. Whilst the savages were engaged in the work of civilisation, the Europeans continued to surround them on every side, and to confine them within narrower limits; the two races gradually met, and they are now in immediate juxtaposition to each other. The Indian is already superior to his barbarous parent, but he is still very far below his white neighbour. With their resources and acquired knowledge, the Europeans soon appropriated to themselves most of the advantages which the natives might have derived from the possession of the soil; they have settled in the country, they have purchased land at a very low rate or have occupied it by force, and the Indians have been ruined by a competition which they had not the means of resisting. They were isolated in their own country, and their race only constituted a colony of troublesome aliens in the midst of a numerous and domineering people.[20]

Washington said in one of his messages to Congress, 'We are more enlightened and more powerful than the Indian nations, we are therefore bound in honour to treat them with kindness and even with generosity.' But this virtuous and high-minded policy has not been followed. The rapacity of the settlers is usually backed by the tyranny of the Government. Although the Cherokees and the Creeks are established upon the territory which they inhabited before the settlement of the Europeans, and although the Americans have frequently treated with them as with foreign nations, the surrounding States have not consented to acknowledge them as independent peoples, and attempts have been made to subject these children of the woods to Anglo-American magistrates, laws, and customs.[21] Destitution had driven these unfortunate Indians to civilization, and oppression now drives them back to their former condition: many of them abandon the soil which they had begun to clear, and return to their savage course of life.

If we consider the tyrannical measures which have been adopted by the legislatures of the Southern States, the conduct of their Governors, and the decrees of their courts of justice, we shall be convinced that the entire expulsion of the Indians is the final result to which the efforts of their policy are directed. The Americans of that part of the Union look with jealousy upon the Aborigines,[22] they are aware that these tribes have not yet lost the traditions of savage life, and before civilisation has permanently fixed them to the soil, it is intended to force them to recede by reducing them to despair. The Creeks and Cherokees, oppressed by the several States, have appealed to the central Government, which is by no means insensible to their misfortunes, and is sincerely desirous of saving the remnant of the natives, and of maintaining them in the free possession of that territory, which the Union is pledged to respect.[23] But the several States oppose so formidable a resistance to the execution of this design, that the government is obliged to consent to the extirpation of a few barbarous tribes in order not to endanger the safety of the American Union.

But the federal government, which is not able to protect the Indians, would fain mitigate the hardships of their lot; and, with this intention, proposals have been made to transport them into more remote regions at the public cost.

Between the 33rd and 37th degrees of north latitude, a vast tract of country lies, which has taken the name of Arkansas, from the principal river that waters its extent. It is bounded on the one side by the confines of Mexico, on the other by the Mississippi. Numberless streams cross it in every direction; the climate is mild, and the soil productive, but it is only inhabited by a few wandering hordes of savages. The Government of the Union wishes to transport the broken remnants of the indigenous population of the South to the portion of this country which is nearest to Mexico, and at a great distance from the American settlements.

We were assured, towards the end of the year 1831, that 10,000 Indians had already gone down to the shores of the Arkansas; and fresh detachments were constantly following them; but Congress has been unable to excite a unanimous determination in those whom it is disposed to protect. Some, indeed, are willing to quit the seat of oppression, but the most enlightened members of the community refuse to abandon their recent dwellings and their springing crops; they are of opinion that the work of civilisation, once interrupted, will never be resumed; they fear that those domestic habits which have been so recently contracted, may be irrevocably lost in the midst of a country which is still barbarous, and where nothing is prepared for the subsistence of an agricultural people; they know that their entrance into those wilds will be opposed by inimical hordes, and that they have lost the energy of barbarians, without acquiring the resources of civilisation to resist their attacks. Moreover, the Indians readily discover that the settlement which is proposed to them is merely a temporary expedient. Who can assure them that they will at length be allowed to dwell in peace in their new retreat? The United States pledge themselves to the observance of the obligation; but the territory which they at present occupy was formerly secured to them by the most solemn oaths of Anglo-American faith.[24] The American Government does not indeed rob them of their lands, but it allows perpetual incursions to be made on them. In a few years the same white population which now flocks around them, will track them to the solitudes of the Arkansas; they will then be exposed to the same evils without the same remedies, and as the limits of the earth will at last fail them, their only refuge is the grave.

The Union treats the Indians with less cupidity and rigour than the policy of the several States, but the two Governments are alike destitute of good faith. The States extend what they are pleased to term the benefits of their laws to the Indians, with a belief that the tribes will recede rather than submit; and the central Government, which promises a permanent refuge to these unhappy beings is well aware of its inability to secure it to them.[25]

Thus the tyranny of the States obliges the savages to retire, the Union, by its promises and resources, facilitates their retreat; and these measures tend to precisely the same end.[26] 'By the will of our Father in Heaven, the Governor of the whole world,' said the Cherokees in their petition to Congress,[27] 'the red man of America has become small, and the white man great and renowned. When the ancestors of the people of these United States first came to the shores of America they found the red man strong: though he was ignorant and savage, yet he received them kindly, and gave them dry land to rest their weary feet. They met in peace, and shook hands in token of friendship. Whatever the white man wanted and asked of the Indian, the latter willingly gave. At that time the Indian was the lord, and the white man the suppliant. But now the scene has changed. The strength of the red man has become weakness. As his neighbours increased in numbers his power became less and less, and now, of the many and powerful tribes who once covered these United States, only a few are to be seen—a few whom a sweeping pestilence has left. The northern tribes, who were once so numerous and powerful, are now nearly extinct. Thus it has happened to the red man of America. Shall we, who are remnants, share the same fate?

'The land on which we stand we have received as an inheritance from our fathers, who possessed it from time immemorial, as a gift from our common Father in Heaven. They bequeathed it to us as their children, and we have sacredly kept it, as containing the remains of our beloved men. This right of inheritance we have never ceded nor ever forfeited. Permit us to ask what better right can the people have to a country than the right of inheritance and immemorial peaceable possession? We know it is said of late by the State of Georgia and by the Executive of the United States, that we have forfeited this right; but we think this is said gratuitously. At what time have we made the forfeit? What great crime have we committed, whereby we must forever be divested of our country and rights? Was it when we were hostile to the United States, and took part with the King of Great Britain, during the struggle for independence? If so, why was not this forfeiture declared in the first treaty of peace between the United States and our beloved men? Why was not such an article as the following inserted in the treaty:—"The United States give peace to the Cherokees, but, for the part they took in the late war, declare them to be but tenants at will, to be removed when the convenience of the States, within whose chartered limits they live, shall require it?" That was the proper time to assume such a possession. But it was not thought of, nor would our forefathers have agreed to any treaty whose tendency was to deprive them of their rights and their country.'

Such is the language of the Indians: their assertions are true, their forebodings inevitable. From whichever side we consider the destinies of the aborigines of North America, their calamities appear to be irremediable: if they continue barbarous, they are forced to retire; if they attempt to civilise their manners, the contact of a more civilised community subjects them to oppression and destitution. They perish if they continue to wander from waste to waste, and if they attempt to settle they still must perish; the assistance of Europeans is necessary to instruct them, but the approach of Europeans corrupts and repels them into savage life; they refuse to change their habits as long as their solitudes are their own, and it is too late to change them when they are constrained to submit.

The Spaniards pursued the Indians with blood-hounds, like wild beasts; they sacked the New World with no more temper or compassion than a city taken by storm; but destruction must cease, and frenzy be stayed; the remnant of the Indian population which had escaped the massacre mixed with its conquerors, and adopted in the end their religion and their manners.[28] The conduct of the Americans of the United States towards the aborigines is characterised, on the other hand, by a singular attachment to the formalities of law. Provided that the Indians retain their barbarous condition, the Americans take no part in their affairs; they treat them as independent nations, and do not possess themselves of their hunting grounds without a treaty of purchase; and if an Indian nation happens to be so encroached upon as to be unable to subsist upon its territory, they afford it brotherly assistance in transporting it to a grave sufficiently remote from the land of its fathers.

The Spaniards were unable to exterminate the Indian race by those unparalleled atrocities which brand them with indelible shame, nor did they even succeed in wholly depriving it of its rights; but the Americans of the United States have accomplished this twofold purpose with singular felicity; tranquilly, legally, philanthropically, without shedding blood, and without violating a single great principle of morality in the eyes of the world.[29] It is impossible to destroy men with more respect for the laws of humanity.

[I leave this chapter wholly unchanged, for it has always appeared to me to be one of the most eloquent and touching parts of this book. But it has ceased to be prophetic; the destruction of the Indian race in the United States is already consummated. In 1870 there remained but 25,731 Indians in the whole territory of the Union, and of these by far the largest part exist in California, Michigan, Wisconsin, Dakota, and New Mexico and Nevada. In New England, Pennsylvania, and New York the race is extinct; and the predictions of M. de Tocqueville are fulfilled.]

Situation of the Black Population in the United States, and Dangers with Which Its Presence Threatens the Whites

Why it is more difficult to abolish slavery, and to efface all vestiges of it amongst the moderns than it was amongst the ancients—In the United States the prejudices of the Whites against the Blacks seem to increase in proportion as slavery is abolished—Situation of the Negroes in the Northern and Southern States—Why the Americans abolish slavery—Servitude, which debases the slave, impoverishes the master. Contrast between the left and the right bank of the Ohio—To what attributable—The Black race, as well as slavery, recedes towards the South—Explanation of this fact—Difficulties attendant upon the abolition of slavery in the South—Dangers to come—General anxiety—Foundation of a Black colony in Africa—Why the Americans of the South increase the hardships of slavery, whilst they are distressed at its continuance.

The Indians will perish in the same isolated condition in which they have lived; but the destiny of the negroes is in some measure interwoven with that of the Europeans. These two races are attached to each other without intermingling, and they are alike unable entirely to separate or to combine. The most formidable of all the ills which threaten the future existence of the Union arises from the presence of a black population upon its territory; and in contemplating the cause of the present embarrassments or of the future dangers of the United States, the observer is invariably led to consider this as a primary fact.

The permanent evils to which mankind is subjected are usually produced by the vehement or the increasing efforts of men; but there is one calamity which penetrated furtively into the world, and which was at first scarcely distinguishable amidst the ordinary abuses of power; it originated with an individual whose name history has not preserved; it was wafted like some accursed germ upon a portion of the soil, but it afterwards nurtured itself, grew without effort, and spreads naturally with the society to which it belongs. I need scarcely add that this calamity is slavery. Christianity suppressed slavery, but the Christians of the sixteenth century reestablished it—as an exception, indeed, to their social system, and restricted to one of the races of mankind; but the wound thus inflicted upon humanity, though less extensive, was at the same time rendered far more difficult of cure.

It is important to make an accurate distinction between slavery itself and its consequences. The immediate evils which are produced by slavery were very nearly the same in antiquity as they are amongst the moderns; but the consequences of these evils were different. The slave, amongst the ancients, belonged to the same race as his master, and he was often the superior of the two in education[30] and instruction. Freedom was the only distinction between them; and when freedom was conferred they were easily confounded together. The ancients, then, had a very simple means of avoiding slavery and its evil consequences, which was that of affranchisement; and they succeeded as soon as they adopted this measure generally. Not but, in ancient States, the vestiges of servitude subsisted for some time after servitude itself was abolished. There is a natural prejudice which prompts men to despise whomsoever has been their inferior long after he is become their equal; and the real inequality which is produced by fortune or by law is always succeeded by an imaginary inequality which is implanted in the manners of the people. Nevertheless, this secondary consequence of slavery was limited to a certain term amongst the ancients, for the freedman bore so entire a resemblance to those born free, that it soon became impossible to distinguish him from amongst them.

The greatest difficulty in antiquity was that of altering the law; amongst the moderns it is that of altering the manners; and, as far as we are concerned, the real obstacles begin where those of the ancients left off. This arises from the circumstance that, amongst the moderns, the abstract and transient fact of slavery is fatally united to the physical and permanent fact of colour. The tradition of slavery dishonours the race, and the peculiarity of the race perpetuates the tradition of slavery. No African has ever voluntarily emigrated to the shores of the New World; whence it must be inferred, that all the blacks who are now to be found in that hemisphere are either slaves or freedmen. Thus the negro transmits the eternal mark of his ignominy to all his descendants; and although the law may abolish slavery, God alone can obliterate the traces of its existence.

The modern slave differs from his master not only in his condition, but in his origin. You may set the negro free, but you cannot make him otherwise than an alien to the European. Nor is this all; we scarcely acknowledge the common features of mankind in this child of debasement whom slavery has brought amongst us. His physiognomy is to our eyes hideous, his understanding weak, his tastes low; and we are almost inclined to look upon him as a being intermediate between man and the brutes.[31] The moderns, then, after they have abolished slavery, have three prejudices to contend against, which are less easy to attack and far less easy to conquer than the mere fact of servitude: the prejudice of the master, the prejudice of the race, and the prejudice of color.

It is difficult for us, who have had the good fortune to be born amongst men like ourselves by nature, and equal to ourselves by law, to conceive the irreconcilable differences which separate the negro from the European in America. But we may derive some faint notion of them from analogy. France was formerly a country in which numerous distinctions of rank existed, that had been created by the legislation. Nothing can be more fictitious than a purely legal inferiority; nothing more contrary to the instinct of mankind than these permanent divisions which had been established between beings evidently similar. Nevertheless these divisions subsisted for ages; they still subsist in many places; and on all sides they have left imaginary vestiges, which time alone can efface. If it be so difficult to root out an inequality which solely originates in the law, how are those distinctions to be destroyed which seem to be based upon the immutable laws of Nature herself? When I remember the extreme difficulty with which aristocratic bodies, of whatever nature they may be, are commingled with the mass of the people; and the exceeding care which they take to preserve the ideal boundaries of their caste inviolate, I despair of seeing an aristocracy disappear which is founded upon visible and indelible signs. Those who hope that the Europeans will ever mix with the negroes, appear to me to delude themselves; and I am not led to any such conclusion by my own reason, or by the evidence of facts.

Hitherto, wherever the whites have been the most powerful, they have maintained the blacks in a subordinate or a servile position; wherever the negroes have been strongest they have destroyed the whites; such has been the only retribution which has ever taken place between the two races.

I see that in a certain portion of the territory of the United States at the present day, the legal barrier which separated the two races is tending to fall away, but not that which exists in the manners of the country; slavery recedes, but the prejudice to which it has given birth remains stationary. Whosoever has inhabited the United States must have perceived that in those parts of the Union in which the negroes are no longer slaves, they have in no wise drawn nearer to the whites. On the contrary, the prejudice of the race appears to be stronger in the States which have abolished slavery, than in those where it still exists; and nowhere is it so intolerant as in those States where servitude has never been known.

It is true, that in the North of the Union, marriages may be legally contracted between negroes and whites; but public opinion would stigmatise a man who should connect himself with a negress as infamous, and it would be difficult to meet with a single instance of such a union. The electoral franchise has been conferred upon the negroes in almost all the States in which slavery has been abolished; but if they come forward to vote, their lives are in danger. If oppressed, they may bring an action at law, but they will find none but whites amongst their judges; and although they may legally serve as jurors, prejudice repulses them from that office. The same schools do not receive the child of the black and of the European. In the theatres, gold cannot procure a seat for the servile race beside their former masters; in the hospitals they lie apart; and although they are allowed to invoke the same Divinity as the whites, it must be at a different altar, and in their own churches, with their own clergy. The gates of Heaven are not closed against these unhappy beings; but their inferiority is continued to the very confines of the other world; when the negro is defunct, his bones are cast aside, and the distinction of condition prevails even in the equality of death. The negro is free, but he can share neither the rights, nor the pleasures, nor the labour, nor the afflictions, nor the tomb of him whose equal he has been declared to be; and he cannot meet him upon fair terms in life or in death.

In the South, where slavery still exists, the negroes are less carefully kept apart; they sometimes share the labour and the recreations of the whites; the whites consent to intermix with them to a certain extent, and although the legislation treats them more harshly, the habits of the people are more tolerant and compassionate. In the South the master is not afraid to raise his slave to his own standing, because he knows that he can in a moment reduce him to the dust at pleasure. In the North the white no longer distinctly perceives the barrier which separates him from the degraded race, and he shuns the negro with the more pertinacity, since he fears lest they should some day be confounded together.

Amongst the Americans of the South, Nature sometimes reasserts her rights, and restores a transient equality between the blacks and the whites; but in the North pride restrains the most imperious of human passions. The American of the Northern States would perhaps allow the negress to share his licentious pleasures, if the laws of his country did not declare that she may aspire to be the legitimate partner of his bed; but he recoils with horror from her who might become his wife.

Thus it is, in the United States, that the prejudice which repels the negroes seems to increase in proportion as they are emancipated, and inequality is sanctioned by the manners whilst it is effaced from the laws of the country. But if the relative position of the two races which inhabit the United States is such as I have described, it may be asked why the Americans have abolished slavery in the North of the Union, why they maintain it in the South, and why they aggravate its hardships there? The answer is easily given. It is not for the good of the negroes, but for that of the whites, that measures are taken to abolish slavery in the United States.

The first negroes were imported into Virginia about the year 1621.[32] In America, therefore, as well as in the rest of the globe, slavery originated in the South. Thence it spread from one settlement to another; but the number of slaves diminished towards the Northern States, and the negro population was always very limited in New England.[33]

A century had scarcely elapsed since the foundation of the colonies, when the attention of the planters was struck by the extraordinary fact, that the provinces which were comparatively destitute of slaves, increased in population, in wealth, and in prosperity more rapidly than those which contained the greatest number of negroes. In the former, however, the inhabitants were obliged to cultivate the soil themselves, or by hired labourers; in the latter they were furnished with hands for which they paid no wages; yet although labour and expense were on the one side, and ease with economy on the other, the former were in possession of the most advantageous system. This consequence seemed to be the more difficult to explain, since the settlers, who all belonged to the same European race, had the same habits, the same civilisation, the same laws, and their shades of difference were extremely slight.

Time, however, continued to advance, and the Anglo-Americans, spreading beyond the coasts of the Atlantic Ocean, penetrated farther and farther into the solitudes of the West; they met with a new soil and an unwonted climate; the obstacles which opposed them were of the most various character; their races intermingled, the inhabitants of the South went up towards the North, those of the North descended to the South; but in the midst of all these causes, the same result occurred at every step, and in general, the colonies in which there were no slaves became more populous and more rich than those in which slavery flourished. The more progress was made, the more was it shown that slavery, which is so cruel to the slave, is prejudicial to the master.

But this truth was most satisfactorily demonstrated when civilisation reached the banks of the Ohio. The stream which the Indians had distinguished by the name of Ohio, or Beautiful River, waters one of the most magnificent valleys that has ever been made the abode of man. Undulating lands extend upon both shores of the Ohio, whose soil affords inexhaustible treasures to the labourer; on either bank the air is wholesome and the climate mild, and each of them forms the extreme frontier of a vast State: That which follows the numerous windings of the Ohio upon the left is called Kentucky, that upon the right bears the name of the river. These two States only differ in a single respect; Kentucky has admitted slavery, but the State of Ohio has prohibited the existence of slaves within its borders.[34]

Thus the traveller who floats down the current of the Ohio to the spot where that river falls into the Mississippi, may be said to sail between liberty and servitude; and a transient inspection of the surrounding objects will convince him as to which of the two is most favourable to mankind. Upon the left bank of the stream the population is rare; from time to time one descries a troop of slaves loitering in the half-desert fields; the primaeval forest recurs at every turn; society seems to be asleep, man to be idle, and nature alone offers a scene of activity and of life. From the right bank, on the contrary, a confused hum is heard which proclaims the presence of industry; the fields are covered with abundant harvests, the elegance of the dwellings announces the taste and activity of the labourer, and man appears to be in the enjoyment of that wealth and contentment which is the reward of labour.[35]

The State of Kentucky was founded in 1775, the State of Ohio only twelve years later; but twelve years are more in America than half a century in Europe, and, at the present day, the population of Ohio exceeds that of Kentucky by two hundred and fifty thousand souls.[36] These opposite consequences of slavery and freedom may readily be understood, and they suffice to explain many of the differences which we remark between the civilisation of antiquity and that of our own time.

Upon the left bank of the Ohio labour is confounded with the idea of slavery, upon the right bank it is identified with that of prosperity and improvement; on the one side it is degraded, on the other it is honoured; on the former territory no white laborers can be found, for they would be afraid of assimilating themselves to the negroes; on the latter no one is idle, for the white population extends its activity and its intelligence to every kind of employment. Thus the men whose task it is to cultivate the rich soil of Kentucky are ignorant and lukewarm; whilst those who are active and enlightened either do nothing or pass over into the State of Ohio, where they may work without dishonour.

It is true that in Kentucky the planters are not obliged to pay wages to the slaves whom they employ; but they derive small profits from their labour, whilst the wages paid to free workmen would be returned with interest in the value of their services. The free workman is paid, but he does his work quicker than the slave, and rapidity of execution is one of the great elements of economy. The white sells his services, but they are only purchased at the times at which they may be useful; the black can claim no remuneration for his toil, but the expense of his maintenance is perpetual; he must be supported in his old age as well as in the prime of manhood, in his profitless infancy as well as in the productive years of youth. Payment must equally be made in order to obtain the services of either class of men: the free workman receives his wages in money, the slave in education, in food, in care, and in clothing. The money which a master spends in the maintenance of his slaves goes gradually and in detail, so that it is scarcely perceived; the salary of the free workman is paid in a round sum, which appears only to enrich the individual who receives it, but in the end the slave has cost more than the free servant, and his labour is less productive.[37]

The influence of slavery extends still further; it affects the character of the master, and imparts a peculiar tendency to his ideas and his tastes. Upon both banks of the Ohio, the character of the inhabitants is enterprising and energetic; but this vigour is very differently exercised in the two States. The white inhabitant of Ohio, who is obliged to subsist by his own exertions, regards temporal prosperity as the principal aim of his existence; and as the country which he occupies presents inexhaustible resources to his industry and ever-varying lures to his activity, his acquisitive ardour surpasses the ordinary limits of human cupidity: he is tormented by the desire of wealth, and he boldly enters upon every path which fortune opens to him; he becomes a sailor, a pioneer, an artisan, or a labourer with the same indifference, and he supports, with equal constancy, the fatigues and the dangers incidental to these various professions; the resources of his intelligence are astonishing, and his avidity in the pursuit of gain amounts to a species of heroism.

But the Kentuckian scorns not only labour, but all the undertakings which labour promotes; as he lives in an idle independence, his tastes are those of an idle man; money loses a portion of its value in his eyes; he covets wealth much less than pleasure and excitement; and the energy which his neighbour devotes to gain, turns with him to a passionate love of field sports and military exercises; he delights in violent bodily exertion, he is familiar with the use of arms, and is accustomed from a very early age to expose his life in single combat. Thus slavery not only prevents the whites from becoming opulent, but even from desiring to become so.

As the same causes have been continually producing opposite effects for the last two centuries in the British colonies of North America, they have established a very striking difference between the commercial capacity of the inhabitants of the South and those of the North. At the present day it is only the Northern States which are in possession of shipping, manufactures, railroads, and canals. This difference is perceptible not only in comparing the North with the South, but in comparing the several Southern States. Almost all the individuals who carry on commercial operations, or who endeavour to turn slave-labour to account in the most Southern districts of the Union, have emigrated from the North. The natives of the Northern States are constantly spreading over that portion of the American territory where they have less to fear from competition; they discover resources there which escaped the notice of the inhabitants; and, as they comply with a system which they do not approve, they succeed in turning it to better advantage than those who first founded and who still maintain it.

Were I inclined to continue this parallel, I could easily prove that almost all the differences which may be remarked between the characters of the Americans in the Southern and in the Northern States have originated in slavery; but this would divert me from my subject, and my present intention is not to point out all the consequences of servitude, but those effects which it has produced upon the prosperity of the countries which have admitted it.

The influence of slavery upon the production of wealth must have been very imperfectly known in antiquity, as slavery then obtained throughout the civilised world; and the nations which were unacquainted with it were barbarous. And indeed Christianity only abolished slavery by advocating the claims of the slave; at the present time it may be attacked in the name of the master, and, upon this point, interest is reconciled with morality.

As these truths became apparent in the United States, slavery receded before the progress of experience. Servitude had begun in the South, and had thence spread towards the North; but it now retires again. Freedom, which started from the North, now descends uninterruptedly towards the South. Amongst the great States, Pennsylvania now constitutes the extreme limit of slavery to the North: but even within those limits the slave system is shaken: Maryland, which is immediately below Pennsylvania, is preparing for its abolition; and Virginia, which comes next to Maryland, is already discussing its utility and its dangers.[38]

No great change takes place in human institutions without involving amongst its causes the law of inheritance. When the law of primogeniture obtained in the South, each family was represented by a wealthy individual, who was neither compelled nor induced to labour; and he was surrounded, as by parasitic plants, by the other members of his family who were then excluded by law from sharing the common inheritance, and who led the same kind of life as himself. The very same thing then occurred in all the families of the South as still happens in the wealthy families of some countries in Europe, namely, that the younger sons remain in the same state of idleness as their elder brother, without being as rich as he is. This identical result seems to be produced in Europe and in America by wholly analogous causes. In the South of the United States the whole race of whites formed an aristocratic body, which was headed by a certain number of privileged individuals, whose wealth was permanent, and whose leisure was hereditary. These leaders of the American nobility kept alive the traditional prejudices of the white race in the body of which they were the representatives, and maintained the honour of inactive life. This aristocracy contained many who were poor, but none who would work; its members preferred want to labour, consequently no competition was set on foot against negro labourers and slaves, and, whatever opinion might be entertained as to the utility of their efforts, it was indispensable to employ them, since there was no one else to work.

No sooner was the law of primogeniture abolished than fortunes began to diminish, and all the families of the country were simultaneously reduced to a state in which labour became necessary to procure the means of subsistence: several of them have since entirely disappeared, and all of them learned to look forward to the time at which it would be necessary for everyone to provide for his own wants. Wealthy individuals are still to be met with, but they no longer constitute a compact and hereditary body, nor have they been able to adopt a line of conduct in which they could persevere, and which they could infuse into all ranks of society. The prejudice which stigmatised labour was in the first place abandoned by common consent; the number of needy men was increased, and the needy were allowed to gain a laborious subsistence without blushing for their exertions. Thus one of the most immediate consequences of the partible quality of estates has been to create a class of free labourers. As soon as a competition was set on foot between the free labourer and the slave, the inferiority of the latter became manifest, and slavery was attacked in its fundamental principle, which is the interest of the master.

As slavery recedes, the black population follows its retrograde course, and returns with it towards those tropical regions from which it originally came. However singular this fact may at first appear to be, it may readily be explained. Although the Americans abolish the principle of slavery, they do not set their slaves free. To illustrate this remark, I will quote the example of the State of New York. In 1788, the State of New York prohibited the sale of slaves within its limits, which was an indirect method of prohibiting the importation of blacks. Thenceforward the number of negroes could only increase according to the ratio of the natural increase of population. But eight years later a more decisive measure was taken, and it was enacted that all children born of slave parents after July 4, 1799, should be free. No increase could then take place, and although slaves still existed, slavery might be said to be abolished.

From the time at which a Northern State prohibited the importation of slaves, no slaves were brought from the South to be sold in its markets. On the other hand, as the sale of slaves was forbidden in that State, an owner was no longer able to get rid of his slave (who thus became a burdensome possession) otherwise than by transporting him to the South. But when a Northern State declared that the son of the slave should be born free, the slave lost a large portion of his market value, since his posterity was no longer included in the bargain, and the owner had then a strong interest in transporting him to the South. Thus the same law prevents the slaves of the South from coming to the Northern States, and drives those of the North to the South.

The want of free hands is felt in a State in proportion as the number of slaves decreases. But in proportion as labour is performed by free hands, slave-labour becomes less productive; and the slave is then a useless or onerous possession, whom it is important to export to those Southern States where the same competition is not to be feared. Thus the abolition of slavery does not set the slave free, but it merely transfers him from one master to another, and from the North to the South.

The emancipated negroes, and those born after the abolition of slavery, do not, indeed, migrate from the North to the South; but their situation with regard to the Europeans is not unlike that of the aborigines of America; they remain half civilised, and deprived of their rights in the midst of a population which is far superior to them in wealth and in knowledge; where they are exposed to the tyranny of the laws[39] and the intolerance of the people. On some accounts they are still more to be pitied than the Indians, since they are haunted by the reminiscence of slavery, and they cannot claim possession of a single portion of the soil: many of them perish miserably,[40] and the rest congregate in the great towns, where they perform the meanest offices, and lead a wretched and precarious existence.

But even if the number of negroes continued to increase as rapidly as when they were still in a state of slavery, as the number of whites augments with twofold rapidity since the abolition of slavery, the blacks would soon be, as it were, lost in the midst of a strange population.

A district which is cultivated by slaves is in general more scantily peopled than a district cultivated by free labour: moreover, America is still a new country, and a State is therefore not half peopled at the time when it abolishes slavery. No sooner is an end put to slavery than the want of free labour is felt, and a crowd of enterprising adventurers immediately arrive from all parts of the country, who hasten to profit by the fresh resources which are then opened to industry. The soil is soon divided amongst them, and a family of white settlers takes possession of each tract of country. Besides which, European emigration is exclusively directed to the free States; for what would be the fate of a poor emigrant who crosses the Atlantic in search of ease and happiness if he were to land in a country where labour is stigmatized as degrading?

Thus the white population grows by its natural increase, and at the same time by the immense influx of emigrants; whilst the black population receives no emigrants, and is upon its decline. The proportion which existed between the two races is soon inverted. The negroes constitute a scanty remnant, a poor tribe of vagrants, which is lost in the midst of an immense people in full possession of the land; and the presence of the blacks is only marked by the injustice and the hardships of which they are the unhappy victims.

In several of the Western States the negro race never made its appearance, and in all the Northern States it is rapidly declining. Thus the great question of its future condition is confined within a narrow circle, where it becomes less formidable, though not more easy of solution.

The more we descend towards the South, the more difficult does it become to abolish slavery with advantage: and this arises from several physical causes which it is important to point out.

The first of these causes is the climate; it is well known that in proportion as Europeans approach the Tropics they suffer more from labour. Many of the Americans even assert that within a certain latitude the exertions which a negro can make without danger are fatal to them;[41] but I do not think that this opinion, which is so favourable to the indolence of the inhabitants of southern regions, is confirmed by experience. The southern parts of the Union are not hotter than the South of Italy and of Spain;[42] and it may be asked why the European cannot work as well there as in the two latter countries. If slavery has been abolished in Italy and in Spain without causing the destruction of the masters, why should not the same thing take place in the Union? I cannot believe that nature has prohibited the Europeans in Georgia and the Floridas, under pain of death, from raising the means of subsistence from the soil, but their labour would unquestionably be more irksome and less productive[43] to them than to the inhabitants of New England. As the free workman thus loses a portion of his superiority over the slave in the Southern States, there are fewer inducements to abolish slavery.

All the plants of Europe grow in the northern parts of the Union; the South has special productions of its own. It has been observed that slave-labour is a very expensive method of cultivating corn. The farmer of corn land in a country where slavery is unknown habitually retains a small number of labourers in his service, and at seed-time and harvest he hires several additional hands, who only live at his cost for a short period. But the agriculturist in a slave State is obliged to keep a large number of slaves the whole year round, in order to sow his fields and to gather in his crops, although their services are only required for a few weeks; but slaves are unable to wait till they are hired, and to subsist by their own labour in the mean time like free labourers; in order to have their services they must be bought. Slavery, independently of its general disadvantages, is therefore still more inapplicable to countries in which corn is cultivated than to those which produce crops of a different kind. The cultivation of tobacco, of cotton, and especially of the sugar-cane, demands, on the other hand, unremitting attention: and women and children are employed in it, whose services are of but little use in the cultivation of wheat. Thus slavery is naturally more fitted to the countries from which these productions are derived. Tobacco, cotton, and the sugar-cane are exclusively grown in the South, and they form one of the principal sources of the wealth of those States. If slavery were abolished, the inhabitants of the South would be constrained to adopt one of two alternatives: they must either change their system of cultivation, and then they would come into competition with the more active and more experienced inhabitants of the North; or, if they continued to cultivate the same produce without slave-labour, they would have to support the competition of the other States of the South, which might still retain their slaves. Thus, peculiar reasons for maintaining slavery exist in the South which do not operate in the North.

But there is yet another motive which is more cogent than all the others: the South might indeed, rigorously speaking, abolish slavery; but how should it rid its territory of the black population? Slaves and slavery are driven from the North by the same law, but this twofold result cannot be hoped for in the South.

The arguments which I have adduced to show that slavery is more natural and more advantageous in the South than in the North, sufficiently prove that the number of slaves must be far greater in the former districts. It was to the southern settlements that the first Africans were brought, and it is there that the greatest number of them have always been imported. As we advance towards the South, the prejudice which sanctions idleness increases in power. In the States nearest to the tropics there is not a single white labourer; the negroes are consequently much more numerous in the South than in the North. And, as I have already observed, this disproportion increases daily, since the negroes are transferred to one part of the Union as soon as slavery is abolished in the other. Thus the black population augments in the South, not only by its natural fecundity, but by the compulsory emigration of the negroes from the North; and the African race has causes of increase in the South very analogous to those which so powerfully accelerate the growth of the European race in the North.

In the State of Maine there is one negro in three hundred inhabitants; in Massachusetts, one in one hundred; in New York, two in one hundred; in Pennsylvania, three in the same number; in Maryland, thirty-four; in Virginia, forty-two; and lastly, in South Carolina[44] fifty-five per cent. Such was the proportion of the black population to the whites in the year 1830. But this proportion is perpetually changing, as it constantly decreases in the North and augments in the South.

It is evident that the most Southern States of the Union cannot abolish slavery without incurring very great dangers, which the North had no reason to apprehend when it emancipated its black population. We have already shown the system by which the Northern States secure the transition from slavery to freedom, by keeping the present generation in chains, and setting their descendants free; by this means the negroes are gradually introduced into society; and whilst the men who might abuse their freedom are kept in a state of servitude, those who are emancipated may learn the art of being free before they become their own masters. But it would be difficult to apply this method in the South. To declare that all the negroes born after a certain period shall be free, is to introduce the principle and the notion of liberty into the heart of slavery; the blacks whom the law thus maintains in a state of slavery from which their children are delivered, are astonished at so unequal a fate, and their astonishment is only the prelude to their impatience and irritation. Thenceforward slavery loses, in their eyes, that kind of moral power which it derived from time and habit; it is reduced to a mere palpable abuse of force. The Northern States had nothing to fear from the contrast, because in them the blacks were few in number, and the white population was very considerable. But if this faint dawn of freedom were to show two millions of men their true position, the oppressors would have reason to tremble. After having affranchised the children of their slaves the Europeans of the Southern States would very shortly be obliged to extend the same benefit to the whole black population.

In the North, as I have already remarked, a twofold migration ensues upon the abolition of slavery, or even precedes that event when circumstances have rendered it probable; the slaves quit the country to be transported southwards; and the whites of the Northern States, as well as the emigrants from Europe, hasten to fill up their place. But these two causes cannot operate in the same manner in the Southern States. On the one hand, the mass of slaves is too great for any expectation of their ever being removed from the country to be entertained; and on the other hand, the Europeans and Anglo-Americans of the North are afraid to come to inhabit a country in which labour has not yet been reinstated in its rightful honours. Besides, they very justly look upon the States in which the proportion of the negroes equals or exceeds that of the whites, as exposed to very great dangers; and they refrain from turning their activity in that direction.

Thus the inhabitants of the South would not be able, like their Northern countrymen, to initiate the slaves gradually into a state of freedom by abolishing slavery; they have no means of perceptibly diminishing the black population, and they would remain unsupported to repress its excesses. So that in the course of a few years, a great people of free negroes would exist in the heart of a white nation of equal size.

The same abuses of power which still maintain slavery, would then become the source of the most alarming perils which the white population of the South might have to apprehend. At the present time the descendants of the Europeans are the sole owners of the land; the absolute masters of all labour; and the only persons who are possessed of wealth, knowledge, and arms. The black is destitute of all these advantages, but he subsists without them because he is a slave. If he were free, and obliged to provide for his own subsistence, would it be possible for him to remain without these things and to support life? Or would not the very instruments of the present superiority of the white, whilst slavery exists, expose him to a thousand dangers if it were abolished?

As long as the negro remains a slave, he may be kept in a condition not very far removed from that of the brutes; but, with his liberty, he cannot but acquire a degree of instruction which will enable him to appreciate his misfortunes, and to discern a remedy for them. Moreover, there exists a singular principle of relative justice which is very firmly implanted in the human heart. Men are much more forcibly struck by those inequalities which exist within the circle of the same class, than with those which may be remarked between different classes. It is more easy for them to admit slavery, than to allow several millions of citizens to exist under a load of eternal infamy and hereditary wretchedness. In the North the population of freed negroes feels these hardships and resents these indignities; but its numbers and its powers are small, whilst in the South it would be numerous and strong.

As soon as it is admitted that the whites and the emancipated blacks are placed upon the same territory in the situation of two alien communities, it will readily be understood that there are but two alternatives for the future; the negroes and the whites must either wholly part or wholly mingle. I have already expressed the conviction which I entertain as to the latter event.[45] I do not imagine that the white and black races will ever live in any country upon an equal footing. But I believe the difficulty to be still greater in the United States than elsewhere. An isolated individual may surmount the prejudices of religion, of his country, or of his race, and if this individual is a king he may effect surprising changes in society; but a whole people cannot rise, as it were, above itself. A despot who should subject the Americans and their former slaves to the same yoke, might perhaps succeed in commingling their races; but as long as the American democracy remains at the head of affairs, no one will undertake so difficult a task; and it may be foreseen that the freer the white population of the United States becomes, the more isolated will it remain.[46]

I have previously observed that the mixed race is the true bond of union between the Europeans and the Indians; just so the mulattoes are the true means of transition between the white and the negro; so that wherever mulattoes abound, the intermixture of the two races is not impossible. In some parts of America, the European and the negro races are so crossed by one another, that it is rare to meet with a man who is entirely black, or entirely white: when they are arrived at this point, the two races may really be said to be combined; or rather to have been absorbed in a third race, which is connected with both without being identical with either.

Of all the Europeans the English are those who have mixed least with the negroes. More mulattoes are to be seen in the South of the Union than in the North, but still they are infinitely more scarce than in any other European colony: mulattoes are by no means numerous in the United States; they have no force peculiar to themselves, and when quarrels originating in differences of colour take place, they generally side with the whites; just as the lackeys of the great, in Europe, assume the contemptuous airs of nobility to the lower orders.

The pride of origin, which is natural to the English, is singularly augmented by the personal pride which democratic liberty fosters amongst the Americans: the white citizen of the United States is proud of his race, and proud of himself. But if the whites and the negroes do not intermingle in the North of the Union, how should they mix in the South? Can it be supposed for an instant, that an American of the Southern States, placed, as he must forever be, between the white man with all his physical and moral superiority and the negro, will ever think of preferring the latter? The Americans of the Southern States have two powerful passions which will always keep them aloof; the first is the fear of being assimilated to the negroes, their former slaves; and the second the dread of sinking below the whites, their neighbours.

If I were called upon to predict what will probably occur at some future time, I should say, that the abolition of slavery in the South will, in the common course of things, increase the repugnance of the white population for the men of colour. I found this opinion upon the analogous observation which I already had occasion to make in the North. I there remarked that the white inhabitants of the North avoid the negroes with increasing care, in proportion as the legal barriers of separation are removed by the legislature; and why should not the same result take place in the South? In the North, the whites are deterred from intermingling with the blacks by the fear of an imaginary danger; in the South, where the danger would be real, I cannot imagine that the fear would be less general.

If, on the one hand, it be admitted (and the fact is unquestionable) that the coloured population perpetually accumulates in the extreme South, and that it increases more rapidly than that of the whites; and if, on the other hand, it be allowed that it is impossible to foresee a time at which the whites and the blacks will be so intermingled as to derive the same benefits from society; must it not be inferred that the blacks and the whites will, sooner or later, come to open strife in the Southern States of the Union? But if it be asked what the issue of the struggle is likely to be, it will readily be understood that we are here left to form a very vague surmise of the truth. The human mind may succeed in tracing a wide circle, as it were, which includes the course of future events; but within that circle a thousand various chances and circumstances may direct it in as many different ways; and in every picture of the future there is a dim spot, which the eye of the understanding cannot penetrate. It appears, however, to be extremely probable that in the West Indian Islands the white race is destined to be subdued, and the black population to share the same fate upon the continent.

In the West India Islands the white planters are surrounded by an immense black population; on the continent, the blacks are placed between the ocean and an innumerable people, which already extends over them in a dense mass, from the icy confines of Canada to the frontiers of Virginia, and from the banks of the Missouri to the shores of the Atlantic. If the white citizens of North America remain united, it cannot be supposed that the negroes will escape the destruction with which they are menaced; they must be subdued by want or by the sword. But the black population which is accumulated along the coast of the Gulf of Mexico, has a chance of success if the American Union is dissolved when the struggle between the two races begins. If the Federal tie were broken, the citizens of the South would be wrong to rely upon any lasting succour from their Northern countrymen. The latter are well aware that the danger can never reach them; and unless they are constrained to march to the assistance of the South by a positive obligation, it may be foreseen that the sympathy of colour will be insufficient to stimulate their exertions.

Yet, at whatever period the strife may break out, the whites of the South, even if they are abandoned to their own resources, will enter the lists with an immense superiority of knowledge and of the means of warfare; but the blacks will have numerical strength and the energy of despair upon their side, and these are powerful resources to men who have taken up arms. The fate of the white population of the Southern States will, perhaps, be similar to that of the Moors in Spain. After having occupied the land for centuries, it will perhaps be forced to retire to the country whence its ancestors came, and to abandon to the negroes the possession of a territory, which Providence seems to have more peculiarly destined for them, since they can subsist and labour in it more easily that the whites.

The danger of a conflict between the white and the black inhabitants of the Southern States of the Union—a danger which, however remote it may be, is inevitable—perpetually haunts the imagination of the Americans. The inhabitants of the North make it a common topic of conversation, although they have no direct injury to fear from the struggle; but they vainly endeavour to devise some means of obviating the misfortunes which they foresee. In the Southern States the subject is not discussed: the planter does not allude to the future in conversing with strangers; the citizen does not communicate his apprehensions to his friends; he seeks to conceal them from himself; but there is something more alarming in the tacit forebodings of the South, than in the clamorous fears of the Northern States.

This all-pervading disquietude has given birth to an undertaking which is but little known, but which may have the effect of changing the fate of a portion of the human race. From apprehension of the dangers which I have just been describing, a certain number of American citizens have formed a society for the purpose of exporting to the coast of Guinea, at their own expense, such free negroes as may be willing to escape from the oppression to which they are subject.[47] In 1820, the society to which I allude formed a settlement in Africa, upon the seventh degree of north latitude, which bears the name of Liberia. The most recent intelligence informs us that two thousand five hundred negroes are collected there; they have introduced the democratic institutions of America into the country of their forefathers; and Liberia has a representative system of government, negro jurymen, negro magistrates, and negro priests; churches have been built, newspapers established, and, by a singular change in the vicissitudes of the world, white men are prohibited from sojourning within the setdement.[48]

This is indeed a strange caprice of fortune. Two hundred years have now elapsed since the inhabitants of Europe undertook to tear the negro from his family and his home, in order to transport him to the shores of North America; at the present day, the European settlers are engaged in sending back the descendants of those very negroes to the Continent from which they were originally taken; and the barbarous Africans have been brought into contact with civilisation in the midst of bondage, and have become acquainted with free political institutions in slavery. Up to the present time Africa has been closed against the arts and sciences of the whites; but the inventions of Europe will perhaps penetrate into those regions, now that they are introduced by Africans themselves. The settlement of Liberia is founded upon a lofty and a most fruitful idea; but whatever may be its results with regard to the Continent of Africa, it can afford no remedy to the New World.

In twelve years the Colonisation Society has transported two thousand five hundred negroes to Africa; in the same space of time about seven hundred thousand blacks were born in the United States. If the colony of Liberia were so situated as to be able to receive thousands of new inhabitants every year, and if the negroes were in a state to be sent thither with advantage; if the Union were to supply the society with annual subsidies,[49] and to transport the negroes to Africa in the vessels of the State, it would still be unable to counterpoise the natural increase of population amongst the blacks; and as it could not remove as many men in a year as are born upon its territory within the same space of time, it would fail in suspending the growth of the evil which is daily increasing in the States.[50] The negro race will never leave those shores of the American continent, to which it was brought by the passions and the vices of Europeans; and it will not disappear from the New World as long as it continues to exist. The inhabitants of the United States may retard the calamities which they apprehend, but they cannot now destroy their efficient cause.

I am obliged to confess that I do not regard the abolition of slavery as a means of warding off the struggle of the two races in the United States. The negroes may long remain slaves without complaining; but if they are once raised to the level of free men, they will soon revolt at being deprived of all their civil rights; and as they cannot become the equals of the whites, they will speedily declare themselves as enemies. In the North everything contributed to facilitate the emancipation of the slaves; and slavery was abolished, without placing the free negroes in a position which could become formidable, since their number was too small for them ever to claim the exercise of their rights. But such is not the case in the South. The question of slavery was a question of commerce and manufacture for the slave-owners in the North; for those of the South, it is a question of life and death. God forbid that I should seek to justify the principle of negro slavery, as has been done by some American writers! But

I only observe that all the countries which formerly adopted that execrable principle are not equally able to abandon it at the present time.

When I contemplate the condition of the South, I can only discover two alternatives which may be adopted by the white inhabitants of those States; viz., either to emancipate the negroes, and to intermingle with them; or, remaining isolated from them, to keep them in a state of slavery as long as possible. All intermediate measures seem to me likely to terminate, and that shortly, in the most horrible of civil wars, and perhaps in the extirpation of one or other of the two races. Such is the view which the Americans of the South take of the question, and they act consistently with it. As they are determined not to mingle with the negroes, they refuse to emancipate them.

Not that the inhabitants of the South regard slavery as necessary to the wealth of the planter, for on this point many of them agree with their Northern countrymen in freely admitting that slavery is prejudicial to their interest; but they are convinced that, however prejudicial it may be, they hold their lives upon no other tenure. The instruction which is now diffused in the South has convinced the inhabitants that slavery is injurious to the slave-owner, but it has also shown them, more clearly than before, that no means exist of getting rid of its bad consequences. Hence arises a singular contrast; the more the utility of slavery is contested, the more firmly is it established in the laws; and whilst the principle of servitude is gradually abolished in the North, that self-same principle gives rise to more and more rigorous consequences in the South.

The legislation of the Southern States with regard to slaves, presents at the present day such unparalleled atrocities as suffice to show how radically the laws of humanity have been perverted, and to betray the desperate position of the community in which that legislation has been promulgated. The Americans of this portion of the Union have not, indeed, augmented the hardships of slavery; they have, on the contrary, bettered the physical condition of the slaves. The only means by which the ancients maintained slavery were fetters and death; the Americans of the South of the Union have discovered more intellectual securities for the duration of their power. They have employed their despotism and their violence against the human mind. In antiquity, precautions were taken to prevent the slave from breaking his chains; at the present day measures are adopted to deprive him even of the desire of freedom. The ancients kept the bodies of their slaves in bondage, but they placed no restraint upon the mind and no check upon education; and they acted consistently with their established principle, since a natural termination of slavery then existed, and one day or other the slave might be set free, and become the equal of his master. But the Americans of the

South, who do not admit that the negroes can ever be commingled with themselves, have forbidden them to be taught to read or to write, under severe penalties; and as they will not raise them to their own level, they sink them as nearly as possible to that of the brutes.

The hope of liberty had always been allowed to the slave to cheer the hardships of his condition. But the Americans of the South are well aware that emancipation cannot but be dangerous, when the freed man can never be assimilated to his former master. To give a man his freedom, and to leave him in wretchedness and ignominy, is nothing less than to prepare a future chief for a revolt of the slaves. Moreover, it has long been remarked that the presence of a free negro vaguely agitates the minds of his less fortunate brethren, and conveys to them a dim notion of their rights. The Americans of the South have consequently taken measures to prevent slave-owners from emancipating their slaves in most cases; not indeed by a positive prohibition, but by subjecting that step to various forms which it is difficult to comply with. I happened to meet with an old man, in the South of the Union, who had lived in illicit intercourse with one of his negresses, and had had several children by her, who were born the slaves of their father. He had indeed frequently thought of bequeathing to them at least their liberty; but years had elapsed without his being able to surmount the legal obstacles to their emancipation, and in the meanwhile his old age was come, and he was about to die. He pictured to himself his sons dragged from market to market, and passing from the authority of a parent to the rod of the stranger, until these horrid anticipations worked his expiring imagination into frenzy. When I saw him he was a prey to all the anguish of despair, and he made me feel how awful is the retribution of Nature upon those who have broken her laws.

These evils are unquestionably great; but they are the necessary and foreseen consequence of the very principle of modern slavery. When the Europeans chose their slaves from a race differing from their own, which many of them considered as inferior to the other races of mankind, and which they all repelled with horror from any notion of intimate connection, they must have believed that slavery would last forever; since there is no intermediate state which can be durable between the excessive inequality produced by servitude and the complete equality which originates in independence. The Europeans did imperfectly feel this truth, but without acknowledging it even to themselves. Whenever they have had to do with negroes, their conduct has either been dictated by their interest and their pride, or by their compassion. They first violated every right of humanity by their treatment of the negro and they afterwards informed him that those rights were precious and inviolable. They affected to open their ranks to the slaves, but the negroes who attempted to penetrate into the community were driven back with scorn; and they have incautiously and involuntarily been led to admit of freedom instead of slavery, without having the courage to be wholly iniquitous, or wholly just.

If it be impossible to anticipate a period at which the Americans of the South will mingle their blood with that of the negroes, can they allow their slaves to become free without compromising their own security? And if they are obliged to keep that race in bondage in order to save their own families, may they not be excused for availing themselves of the means best adapted to that end? The events which are taking place in the Southern States of the Union appear to me to be at once the most horrible and the most natural results of slavery. When I see the order of nature overthrown, and when I hear the cry of humanity in its vain struggle against the laws, my indignation does not light upon the men of our own time who are the instruments of these outrages; but I reserve my execration for those who, after a thousand years of freedom, brought back slavery into the world once more.

Whatever may be the efforts of the Americans of the South to maintain slavery, they will not always succeed. Slavery, which is now confined to a single tract of the civilised earth, which is attacked by Christianity as unjust, and by political economy as prejudicial; and which is now contrasted with democratic liberties and the information of our age, cannot survive. By the choice of the master, or by the will of the slave, it will cease; and in either case great calamities may be expected to ensue. If liberty be refused to the negroes of the South, they will in the end seize it for themselves by force; if it be given, they will abuse it ere long.[51]

What Are the Chances in Favour of the Duration of the American Union, and What Dangers Threaten It[52]

Reason for which the preponderating force lies in the States rather than in the Union—The Union will only last as long as all the States choose to belong to it—Causes which tend to keep them united—Utility of the Union to resist foreign enemies, and to prevent the existence of foreigners in America—No natural barriers between the several States—No conflicting interests to divide them—Reciprocal interests of the Northern, Southern, and Western States—Intellectual ties of union. Uniformity of opinions—Dangers of the Union resulting from the different characters and the passions of its citizens—Character of the citizens in the South and in the North—The rapid growth of the Union one of its greatest dangers—Progress of the population to the North-west—

Power gravitates in the same direction—Passions originating from sudden turns of fortune—Whether the existing Government of the Union tends to gain strength, or to lose it—Various signs of its decrease—Internal improvements—Waste lands—Indians—The Bank—The Tariff—General Jackson.

The maintenance of the existing institutions of the several States depends in some measure upon the maintenance of the Union itself. It is therefore important in the first instance to inquire into the probable fate of the Union. One point may indeed be assumed at once: if the present confederation were dissolved, it appears to me to be incontestable that the States of which it is now composed would not return to their original isolated condition, but that several Unions would then be formed in the place of one. It is not my intention to inquire into the principles upon which these new Unions would probably be established, but merely to show what the causes are which may effect the dismemberment of the existing confederation.

With this object I shall be obliged to retrace some of the steps which I have already taken, and to revert to topics which I have before discussed. I am aware that the reader may accuse me of repetition, but the importance of the matter which still remains to be treated is my excuse; I had rather say too much, than say too little to be thoroughly understood, and I prefer injuring the author to slighting the subject.

The legislators who formed the Constitution of 1789 endeavoured to confer a distinct and preponderating authority upon the federal power. But they were confined by the conditions of the task which they had undertaken to perform. They were not appointed to constitute the government of a single people, but to regulate the association of several States; and, whatever their inclinations might be, they could not but divide the exercise of sovereignty in the end.

In order to understand the consequences of this division, it is necessary to make a short distinction between the affairs of the Government. There are some objects which are national by their very nature, that is to say, which affect the nation as a body, and can only be intrusted to the man or the assembly of men who most completely represent the entire nation. Amongst these may be reckoned war and diplomacy. There are other objects which are provincial by their very nature, that is to say, which only affect certain localities, and which can only be properly treated in that locality. Such, for instance, is the budget of a municipality. Lastly, there are certain objects of a mixed nature, which are national inasmuch as they affect all the citizens who compose the nation, and which are provincial inasmuch as it is not necessary that the nation itself should provide for them all. Such are the rights which regulate the civil and political condition of the citizens. No society can exist without civil and political rights. These rights therefore interest all the citizens alike; but it is not always necessary to the existence and the prosperity of the nation that these rights should be uniform, nor, consequently, that they should be regulated by the central authority.

There are, then, two distinct categories of objects which are submitted to the direction of the sovereign power; and these categories occur in all well-constituted communities, whatever the basis of the political constitution may otherwise be. Between these two extremes the objects which I have termed mixed may be considered to lie. As these objects are neither exclusively national nor entirely provincial, they may be obtained by a national or by a provincial government, according to the agreement of the contracting parties, without in any way impairing the contract of association.

The sovereign power is usually formed by the union of separate individuals, who compose a people; and individual powers or collective forces, each representing a very small portion of the sovereign authority, are the sole elements which are subjected to the general Government of their choice. In this case the general Government is more naturally called upon to regulate, not only those affairs which are of essential national importance, but those which are of a more local interest; and the local governments are reduced to that small share of sovereign authority which is indispensable to their prosperity.

But sometimes the sovereign authority is composed of preorganised political bodies, by virtue of circumstances anterior to their union; and in this case the provincial Governments assume the control, not only of those affairs which more peculiarly belong to their province, but of all, or of a part of the mixed affairs to which allusion has been made. For the confederate nations which were independent sovereign States before their union, and which still represent a very considerable share of the sovereign power, have only consented to cede to the general Government the exercise of those rights which are indispensable to the Union.

When the national Government, independently of the prerogatives inherent in its nature, is invested with the right of regulating the affairs which relate partly to the general and partly to the local interests, it possesses a preponderating influence. Not only are its own rights extensive, but all the rights which it does not possess exist by its sufferance, and it may be apprehended that the provincial Governments may be deprived of their natural and necessary prerogatives by its influence.

When, on the other hand, the provincial Governments are invested with the power of regulating those same affairs of mixed interest, an opposite tendency prevails in society. The preponderating force resides in the province, not in the nation;

and it may be apprehended that the national Government may in the end be stripped of the privileges which are necessary to its existence.

Independent nations have therefore a natural tendency to centralisation, and confederations to dismemberment.

It now only remains for us to apply these general principles to the American Union. The several States were necessarily possessed of the right of regulating all exclusively provincial affairs. Moreover these same States retained the rights of determining the civil and political competency of the citizens, or regulating the reciprocal relations of the members of the community, and of dispensing justice; rights which are of a general nature, but which do not necessarily appertain to the national Government. We have shown that the Government of the Union is invested with the power of acting in the name of the whole nation in those cases in which the nation has to appear as a single and undivided power; as, for instance, in foreign relations, and in offering a common resistance to a common enemy; in short, in conducting those affairs which I have styled exclusively national.

In this division of the rights of sovereignty, the share of the Union seems at first sight to be more considerable than that of the States; but a more attentive investigation shows it to be less so. The undertakings of the Government of the Union are more vast, but their influence is more rarely felt. Those of the provincial governments are comparatively small, but they are incessant, and they serve to keep alive the authority which they represent. The Government of the Union watches the general interests of the country; but the general interests of a people have a very questionable influence upon individual happiness, whilst provincial interests produce a most immediate effect upon the welfare of the inhabitants. The Union secures the independence and the greatness of the nation, which do not immediately affect private citizens; but the several States maintain the liberty, regulate the rights, protect the fortune, and secure the life and the whole future prosperity of every citizen.

The Federal Government is very far removed from its subjects, whilst the provincial Governments are within the reach of them all, and are ready to attend to the smallest appeal. The central Government has upon its side the passions of a few superior men who aspire to conduct it; but upon the side of the provincial governments are the interests of all those second-rate individuals who can only hope to obtain power within their own State, and who nevertheless exercise the largest share of authority over the people because they are placed nearest to its level. The Americans have therefore much more to hope and to fear from the States than from the Union; and, in conformity with the natural tendency of the human mind, they are more likely to attach themselves to the former than to the latter. In this respect their habits and feelings harmonise with their interests.

When a compact nation divides its sovereignty, and adopts a confederate form of government, the traditions, the customs, and the manners of the people are for a long time at variance with their legislation; and the former tend to give a degree of influence to the central Government which the latter forbids. When a number of confederate States unite to form a single nation, the same causes operate in an opposite direction. I have no doubt that if France were to become a confederate republic like that of the United States, the Government would at first display more energy than that of the Union; and if the Union were to alter its constitution to a monarchy like that of France, I think that the American Government would be a long time in acquiring the force which now rules the latter nation. When the national existence of the Anglo-Americans began, their provincial existence was already of long standing; necessary relations were established between the townships and the individual citizens of the same States; and they were accustomed to consider some objects as common to them all, and to conduct other affairs as exclusively relating to their own special interests.

The Union is a vast body which presents no definite object to patriotic feeling. The forms and limits of the State are distinct and circumscribed; since it represents a certain number of objects which are familiar to the citizens and beloved by all. It is identified with the very soil, with the right of property and the domestic affections, with the recollections of the past, the labours of the present, and the hopes of the future. Patriotism, then, which is frequently a mere extension of individual egotism, is still directed to the State, and is not excited by the Union. Thus the tendency of the interests, the habits, and the feelings of the people is to centre political activity in the States, in preference to the Union.

It is easy to estimate the different forces of the two Governments, by remarking the manner in which they fulfil their respective functions. Whenever the government of a State has occasion to address an individual or an assembly of individuals, its language is clear and imperative; and such is also the tone of the Federal Government in its intercourse with individuals, but no sooner has it anything to do with a State than it begins to parley, to explain its motives and to justify its conduct, to argue, to advise, and, in short, anything but to command. If doubts are raised as to the limits of the constitutional powers of each Government, the provincial Government prefers its claim with boldness, and takes prompt and energetic steps to support it. In the meanwhile the Government of the Union reasons; it appeals to the interests, to the good sense, to the glory of the nation; it temporises, it negotiates, and does not consent to act until it is reduced to the last extremity. At first sight it might readily be imagined that it is the provincial Government which is armed with the authority of the nation, and that Congress represents a single State.

The Federal Government is, therefore, notwithstanding the precautions of those who founded it, naturally so weak that it more peculiarly requires the free consent of the governed to enable it to subsist. It is easy to perceive that its object is to enable the States to realise with facility their determination of remaining united; and, as long as this preliminary condition exists, its authority is great, temperate, and effective. The Constitution fits the Government to control individuals, and easily to surmount such obstacles as they may be inclined to offer; but it was by no means established with a view to the possible separation of one or more of the States from the Union.

If the Sovereignty of the Union were to engage in a struggle with that of the States at the present day, its defeat may be confidently predicted; and it is not probable that such a struggle would be seriously undertaken. As often as a steady resistance is offered to the Federal Government it will be found to yield. Experience has hitherto shown that whenever a State has demanded anything with perseverance and resolution, it has invariably succeeded; and that if a separate Government has distinctly refused to act, it was left to do as it thought fit.[53]

But even if the Government of the Union had any strength inherent in itself, the physical situation of the country would render the exercise of that strength very difficult.[54] The United States cover an immense territory; they are separated from each other by great distances; and the population is disseminated over the surface of a country which is still half a wilderness. If the Union were to undertake to enforce the allegiance of the Confederate States by military means, it would be in a position very analogous to that of England at the time of the War of Independence.

However strong a Government may be, it cannot easily escape from the consequences of a principle which it has once admitted as the foundation of its constitution. The Union was formed by the voluntary agreement of the States; and, in uniting together, they have not forfeited their nationality, nor have they been reduced to the condition of one and the same people. If one of the States chose to withdraw its name from the contract, it would be difficult to disprove its right of doing so; and the Federal Government would have no means of maintaining its claims directly, either by force or by right. In order to enable the Federal Government easily to conquer the resistance which may be offered to it by any one of its subjects, it would be necessary that one or more of them should be specially interested in the existence of the Union, as has frequently been the case in the history of confederations.

If it be supposed that amongst the States which are united by the Federal tie there are some which exclusively enjoy the principal advantages of union, or whose prosperity depends on the duration of that union, it is unquestionable that they will always be ready to support the central Government in enforcing the obedience of the others. But the Government would then be exerting a force not derived from itself, but from a principle contrary to its nature. States form confederations in order to derive equal advantages from their union; and in the case just alluded to, the Federal Government would derive its power from the unequal distribution of those benefits amongst the States.

If one of the confederate States have acquired a preponderance sufficiently great to enable it to take exclusive possession of the central authority, it will consider the other States as subject provinces, and it will cause its own supremacy to be respected under the borrowed name of the sovereignty of the Union. Great things may then be done in the name of the Federal Government, but in reality that Government will have ceased to exist.[55] In both these cases, the power which acts in the name of the confederation becomes stronger the more it abandons the natural state and the acknowledged principles of confederations.

In America the existing Union is advantageous to all the States, but it is not indispensable to any one of them. Several of them might break the Federal tie without compromising the welfare of the others, although their own prosperity would be lessened. As the existence and the happiness of none of the States are wholly dependent on the present Constitution, they would none of them be disposed to make great personal sacrifices to maintain it. On the other hand, there is no State which seems hitherto to have its ambition much interested in the maintenance of the existing Union. They certainly do not all exercise the same influence in the Federal Councils, but no one of them can hope to domineer over the rest, or to treat them as its inferiors or as its subjects.

It appears to me unquestionable that if any portion of the Union seriously desired to separate itself from the other States, they would not be able, nor indeed would they attempt, to prevent it; and that the present Union will only last as long as the States which compose it choose to continue members of the confederation. If this point be admitted, the question becomes less difficult; and our object is, not to inquire whether the States of the existing Union are capable of separating, but whether they will choose to remain united.

Amongst the various reasons which tend to render the existing Union useful to the Americans, two principal causes are peculiarly evident to the observer. Although the Americans are, as it were, alone upon their continent, their commerce makes them the neighbours of all the nations with which they trade. Notwithstanding their apparent isolation, the Americans require a certain degree of strength, which they cannot retain otherwise than by remaining united to each other. If the States were to split, they would not only diminish the strength which they are now able to display towards foreign nations, but they would soon create foreign powers upon their own territory. A system of inland custom-houses would then be established; the valleys would be divided by imaginary boundary lines; the courses of the rivers would be confined by territorial distinctions; and a multitude of hindrances would prevent the Americans from exploring the whole of that vast continent which Providence has allotted to them for a dominion. At present they have no invasion to fear, and consequently no standing armies to maintain, no taxes to levy. If the Union were dissolved, all these burdensome measures might ere long be required. The Americans are then very powerfully interested in the maintenance of their Union. On the other hand, it is almost impossible to discover any sort of material interest which might at present tempt a portion of the Union to separate from the other States.

When we cast our eyes upon the map of the United States, we perceive the chain of the Alleghany Mountains, running from the north-east to the south-west, and crossing nearly one thousand miles of country; and we are led to imagine that the design of Providence was to raise between the valley of the Mississippi and the coast of the Atlantic Ocean one of those natural barriers which break the mutual intercourse of men, and form the necessary limits of different States. But the average height of the Alleghanies does not exceed 2,500 feet; their greatest elevation is not above 4,000 feet; their rounded summits, and the spacious valleys which they conceal within their passes, are of easy access from several sides. Besides which, the principal rivers which fall into the Atlantic Ocean—the Hudson, the Susquehanna, and the Potomac—take their rise beyond the Alleghanies, in an open district, which borders upon the valley of the Mississippi. These streams quit this tract of country, make their way through the barrier which would seem to turn them westward, and as they wind through the mountains they open an easy and natural passage to man. No natural barrier exists in the regions which are now inhabited by the Anglo-Americans; the Alleghanies are so far from serving as a boundary to separate nations, that they do not even serve as a frontier to the States. New York, Pennsylvania, and Virginia comprise them within their borders, and extend as much to the west as to the east of the line. The territory now occupied by the twenty-four States of the Union, and the three great districts which have not yet acquired the rank of States, although they already contain inhabitants, covers a surface of 1,002,600 square miles,[56] which is about equal to five times the extent of France. Within these limits the qualities of the soil, the temperature, and the produce of the country, are extremely various. The vast extent of territory occupied by the Anglo-American republics has given rise to doubts as to the maintenance of their Union. Here a distinction must be made; contrary interests sometimes arise in the different provinces of a vast empire, which often terminate in open dissensions; and the extent of the country is then most prejudicial to the power of the State. But if the inhabitants of these vast regions are not divided by contrary interests, the extent of the territory may be favourable to their prosperity; for the unity of the Government promotes the interchange of the different productions of the soil, and increases their value by facilitating their consumption.

It is indeed easy to discover different interests in the different parts of the Union, but I am unacquainted with any which are hostile to each other. The Southern States are almost exclusively agricultural. The Northern States are more peculiarly commercial and manufacturing. The States of the West are at the same time agricultural and manufacturing. In the South the crops consist of tobacco, of rice, of cotton, and of sugar; in the North and the West, of wheat and maize. These are different sources of wealth; but union is the means by which these sources are opened to all, and rendered equally advantageous to the several districts.

The North, which ships the produce of the Anglo-Americans to all parts of the world, and brings back the produce of the globe to the Union, is evidently interested in maintaining the confederation in its present condition, in order that the number of American producers and consumers may remain as large as possible. The North is the most natural agent of communication between the South and the West of the Union on the one hand, and the rest of the world upon the other; the North is therefore interested in the union and prosperity of the South and the West, in order that they may continue to furnish raw materials for its manufactures, and cargoes for its shipping.

The South and the West, on their side, are still more directly interested in the preservation of the Union, and the prosperity of the North. The produce of the South is, for the most part, exported beyond seas; the South and the West consequently stand in need of the commercial resources of the North. They are likewise interested in the maintenance of a powerful fleet by the Union, to protect them efficaciously. The South and the West have no vessels, but they cannot refuse a willing subsidy to defray the expenses of the navy; for if the fleets of Europe were to blockade the ports of the South and the delta of the Mississippi, what would become of the rice of the Carolinas, the tobacco of Virginia, and the sugar and cotton which grow in the valley of the Mississippi? Every portion of the federal budget does therefore contribute to the maintenance of material interests which are common to all the confederate States.

Independently of this commercial utility, the South and the West of the Union derive great political advantages from their connection with the North. The South contains an enormous slave population; a population which is already alarming, and still more formidable for the future. The States of the West lie in the remotest parts of a single valley; and all the rivers which intersect their territory rise in the Rocky Mountains or in the Alleghanies, and fall into the Mississippi, which bears them onwards to the Gulf of Mexico. The Western States are consequently entirely cut off, by their position, from the traditions of Europe and the civilisation of the Old World. The inhabitants of the South, then, are induced to support the Union in order to avail themselves of its protection against the blacks; and the inhabitants of the West in order not to be excluded from a free communication with the rest of the globe, and shut up in the wilds of central America. The North cannot but desire the maintenance of the Union, in order to remain, as it now is, the connecting link between that vast body and the other parts of the world.

The temporal interests of all the several parts of the Union are, then, intimately connected; and the same assertion holds true respecting those opinions and sentiments which may be termed the immaterial interests of men.

The inhabitants of the United States talk a great deal of their attachment to their country; but I confess that I do not rely upon that calculating patriotism which is founded upon interest, and which a change in the interests at stake may obliterate. Nor do I attach much importance to the language of the Americans, when they manifest, in their daily conversations, the intention of maintaining the Federal system adopted by their forefathers. A government retains its sway over a great number of citizens, far less by the voluntary and rational consent of the multitude, than by that instinctive, and to a certain extent involuntary agreement, which results from similarity of feelings and resemblances of opinion. I will never admit that men constitute a social body, simply because they obey the same head and the same laws. Society can only exist when a great number of men consider a great number of things in the same point of view; when they hold the same opinions upon many subjects, and when the same occurrences suggest the same thoughts and impressions to their minds.

The observer who examines the present condition of the United States upon this principle, will readily discover, that although the citizens are divided into twenty-four distinct sovereignties, they nevertheless constitute a single people; and he may perhaps be led to think that the state of the Anglo-American Union is more truly a state of society than that of certain nations of Europe which live under the same legislation and the same prince.

Although the Anglo-Americans have several religious sects, they all regard religion in the same manner. They are not always agreed upon the measures which are most conducive to good government, and they vary upon some of the forms of government which it is expedient to adopt; but they are unanimous upon the general principles which ought to rule human society. From Maine to the Floridas, and from the Missouri to the Atlantic Ocean, the people is held to be the legitimate source of all power. The same notions are entertained respecting liberty and equality, the liberty of the press, the right of association, the jury, and the responsibility of the agents of Government.

If we turn from their political and religious opinions to the moral and philosophical principles which regulate the daily actions of life and govern their conduct, we shall still find the same uniformity. The Anglo-Americans[57] acknowledge the absolute moral authority of the reason of the community, as they acknowledge the political authority of the mass of citizens; and they hold that public opinion is the surest arbiter of what is lawful or forbidden, true or false. The majority of them believe that a man will be led to do what is just and good by following his own interest rightly understood. They hold that every man is born in possession of the right of self-government, and that no one has the right of constraining his fellow-creatures to be happy. They have all a lively faith in the perfectibility of man; they are of opinion that the effects of the diffusion of knowledge must necessarily be advantageous, and the consequences of ignorance fatal; they all consider society as a body in a state of improvement, humanity as a changing scene, in which nothing is, or ought to be, permanent; and they admit that what appears to them to be good today may be superseded by something better-tomorrow. I do not give all these opinions as true, but I quote them as characteristic of the Americans.

The Anglo-Americans are not only united together by these common opinions, but they are separated from all other nations by a common feeling of pride. For the last fifty years no pains have been spared to convince the inhabitants of the United States that they constitute the only religious, enlightened, and free people. They perceive that, for the present, their own democratic institutions succeed, whilst those of other countries fail; hence they conceive an overweening opinion of their superiority, and they are not very remote from believing themselves to belong to a distinct race of mankind.

The dangers which threaten the American Union do not originate in the diversity of interests or of opinions, but in the various characters and passions of the Americans. The men who inhabit the vast territory of the United States are almost all the issue of a common stock; but the effects of the climate, and more especially of slavery, have gradually introduced very striking differences between the British settler of the Southern States and the British settler of the North. In Europe it is generally believed that slavery has rendered the interests of one part of the Union contrary to those of another part; but I by no means remarked this to be the case: slavery has not created interests in the South contrary to those of the North, but it has modified the character and changed the habits of the natives of the South.

I have already explained the influence which slavery has exercised upon the commercial ability of the Americans in the South; and this same influence equally extends to their manners. The slave is a servant who never remonstrates, and who submits to everything without complaint. He may sometimes assassinate, but he never withstands, his master. In the South there are no families so poor as not to have slaves. The citizen of the Southern States of the Union is invested with a sort of domestic dictatorship, from his earliest years; the first notion he acquires in life is that he is born to command, and the first habit which he contracts is that of being obeyed without resistance. His education tends, then, to give him the character of a supercilious and a hasty man; irascible, violent, and ardent in his desires, impatient of obstacles, but easily discouraged if he cannot succeed upon his first attempt.

The American of the Northern States is surrounded by no slaves in his childhood; he is even unattended by free servants, and is usually obliged to provide for his own wants. No sooner does he enter the world than the idea of necessity assails him on every side: he soon learns to know exactly the natural limit of his authority; he never expects to subdue those who withstand him, by force; and he knows that the surest means of obtaining the support of his fellow-creatures, is to win their favour. He therefore becomes patient, reflecting, tolerant, slow to act, and persevering in his designs.

In the Southern States the more immediate wants of life are always supplied; the inhabitants of those parts are not busied in the material cares of life, which are always provided for by others; and their imagination is diverted to more captivating and less definite objects. The American of the South is fond of grandeur, luxury, and renown, of gaiety, of pleasure, and above all of idleness; nothing obliges him to exert himself in order to subsist; and as he has no necessary occupations, he gives way to indolence, and does not even attempt what would be useful.

But the equality of fortunes, and the absence of slavery in the North, plunge the inhabitants in those same cares of daily life which are disdained by the white population of the South. They are taught from infancy to combat want, and to place comfort above all the pleasures of the intellect or the heart. The imagination is extinguished by the trivial details of life, and the ideas become less numerous and less general, but far more practical and more precise. As prosperity is the sole aim of exertion, it is excellently well attained; nature and mankind are turned to the best pecuniary advantage, and society is dexterously made to contribute to the welfare of each of its members, whilst individual egotism is the source of general happiness.

The citizen of the North has not only experience, but knowledge: nevertheless he sets but little value upon the pleasures of knowledge; he esteems it as the means of attaining a certain end, and he is only anxious to seize its more lucrative applications. The citizen of the South is more given to act upon impulse; he is more clever, more frank, more generous, more intellectual, and more brilliant. The former, with a greater degree of activity, of common sense, of information, and of general aptitude, has the characteristic good and evil qualities of the middle classes. The latter has the tastes, the prejudices, the weaknesses, and the magnanimity of all aristocracies. If two men are united in society, who have the same interests, and to a certain extent the same opinions, but different characters, different acquirements, and a different style of civilisation, it is probable that these men will not agree. The same remark is applicable to a society of nations. Slavery, then, does not attack the American Union directly in its interests, but indirectly in its manners.

The States which gave their assent to the Federal Contract in 1790 were thirteen in number; the Union now consists of thirty-four members. The population, which amounted to nearly four millions in 1790, had more than tripled in the space of forty years; and in 1830 it amounted to nearly thirteen millions.[58] Changes of such magnitude cannot take place without some danger.

A society of nations, as well as a society of individuals, derives its principal chances of duration from the wisdom of its members, their individual weakness, and their limited number. The Americans who quit the coasts of the Atlantic Ocean to plunge into the western wilderness, are adventurers impatient of restraint, greedy of wealth, and frequently men expelled from the States in which they were born. When they arrive in the deserts they are unknown to each other, and they have neither traditions, family feeling, nor the force of example to check their excesses. The empire of the laws is feeble amongst them; that of morality is still more powerless. The settlers who are constantly peopling the valley of the Mississippi are, then, in every respect very inferior to the Americans who inhabit the older parts of the Union. Nevertheless, they already exercise a great influence in its councils; and they arrive at the government of the commonwealth before they have learnt to govern themselves.[59]

The greater the individual weakness of each of the contracting parties, the greater are the chances of the duration of the contract; for their safety is then dependent upon their union. When, in 1790, the most populous of the American republics did not contain 500,000 inhabitants,[60] each of them felt its own insignificance as an independent people, and this feeling rendered compliance with the Federal authority more easy. But when one of the confederate States reckons, like the State of New York, two millions of inhabitants, and covers an extent of territory equal in surface to a quarter of France,[61] it feels its own strength; and although it may continue to support the Union as advantageous to its prosperity, it no longer regards that body as necessary to its existence, and as it continues to belong to the Federal compact, it soon aims at preponderance in the Federal assemblies. The probable unanimity of the States is diminished as their number increases. At present the interests of the different parts of the Union are not at variance; but who is able to foresee the multifarious changes of the future, in a country in which towns are founded from day to day, and States almost from year to year?

Since the first settlement of the British colonies, the number of inhabitants has about doubled every twenty-two years. I perceive no causes which are likely to check this progressive increase of the Anglo-American population for the next hundred years; and before that space of time has elapsed, I believe that the territories and dependencies of the United States will be covered by more than a hundred millions of inhabitants, and divided into forty States.[62] I admit that these hundred millions of men have no hostile interests. I suppose, on the contrary, that they are all equally interested in the maintenance of the Union; but I am still of opinion that where there are a hundred millions of men, and forty distinct nations, unequally strong, the continuance of the Federal Government can only be a fortunate accident.

Whatever faith I may have in the perfectibility of man, until human nature is altered, and men wholly transformed, I shall refuse to believe in the duration of a government which is called upon to hold together forty different peoples, disseminated over a territory equal to one-half of Europe in extent; to avoid all rivalry, ambition, and struggles between them, and to direct their independent activity to the accomplishment of the same designs.

But the greatest peril to which the Union is exposed by its increase arises from the continual changes which take place in the position of its internal strength. The distance from Lake Superior to the Gulf of Mexico extends from the 47th to the 30th degree of latitude, a distance of more than twelve hundred miles as the bird flies. The frontier of the United States winds along the whole of this immense line, sometimes falling within its limits, but more frequently extending far beyond it, into the waste. It has been calculated that the whites advance every year a mean distance of seventeen miles along the whole of his vast boundary.[63] Obstacles, such as an unproductive district, a lake or an Indian nation unexpectedly encountered, are sometimes met with. The advancing column then halts for a while; its two extremities fall back upon themselves, and as soon as they are reunited they proceed onwards. This gradual and continuous progress of the European race towards the Rocky Mountains has the solemnity of a providential event; it is like a deluge of men rising unabatedly, and daily driven onwards by the hand of God.

Within this first line of conquering settlers towns are built, and vast States founded. In 1790 there were only a few thousand pioneers sprinkled along the valleys of the Mississippi; and at the present day these valleys contain as many inhabitants as were to be found in the whole Union in 1790. Their population amounts to nearly four millions.[64] The city of Washington was founded in 1800, in the very centre of the Union; but such are the changes which have taken place, that it now stands at one of the extremities; and the delegates of the most remote Western States are already obliged to perform a journey as long as that from Vienna to Paris.[65]

All the States are borne onwards at the same time in the path of fortune, but of course they do not all increase and prosper in the same proportion. To the North of the Union the detached branches of the Alleghany chain, which extend as far as the Atlantic Ocean, form spacious roads and ports, which are constantly accessible to vessels of the greatest burden. But from the Potomac to the mouth of the Mississippi the coast is sandy and flat. In this part of the Union the mouths of almost all the rivers are obstructed; and the few harbours which exist amongst these lagoons afford much shallower water to vessels, and much fewer commercial advantages than those of the North.

This first natural cause of inferiority is united to another cause proceeding from the laws. We have already seen that slavery, which is abolished in the North, still exists in the South; and I have pointed out its fatal consequences upon the prosperity of the planter himself.

The North is therefore superior to the South both in commerce[66] and manufacture; the natural consequence of which is the more rapid increase of population and of wealth within its borders. The States situate upon the shores of the Atlantic Ocean are already half-peopled. Most of the land is held by an owner; and these districts cannot therefore receive so many emigrants as the Western States, where a boundless field is still open to their exertions. The valley of the Mississippi is far more fertile than the coast of the Atlantic Ocean. This reason, added to all the others, contributes to drive the Europeans westward—a fact which may be rigorously demonstrated by figures. It is found that the sum total of the population of all the United States has about tripled in the course of forty years. But in the recent States adjacent to the Mississippi, the population has increased thirty-one fold, within the same space of time.[67]

The relative position of the central Federal power is continually displaced. Forty years ago the majority of the citizens of the Union was established upon the coast of the Atlantic, in the environs of the spot upon which Washington now stands; but the great body of the people is now advancing inland and to the North, so that in twenty years the majority will unquestionably be on the western side of the Alleghanies. If the Union goes on to subsist, the basin of the Mississippi is evidently marked out, by its fertility and its extent, as the future centre of the Federal Government. In thirty or forty years, that tract of country will have assumed the rank which naturally belongs to it. It is easy to calculate that its population, compared to that of the coast of the Atlantic, will be, in round numbers, as 40 to 11. In a few years the States which founded the Union will lose the direction of its policy, and the population of the valley of the Mississippi will preponderate in the Federal assemblies.

This constant gravitation of the Federal power and influence towards the North-west is shown every ten years, when a general census of the population is made, and the number of delegates which each State sends to Congress is settled afresh.[68] In 1790 Virginia had nineteen representatives in Congress. This number continued to increase until the year 1813, when it reached to twenty-three; from that time it began to decrease, and in 1833 Virginia elected only twenty-one representatives.[69] During the same period the State of New York progressed in the contrary direction: in 1790 it had ten representatives in Congress; in 1813, twenty-seven; in 1823, thirty-four; and in 1833, forty. The State of Ohio had only one representative in 1803, and in 1833 it had already nineteen.

It is difficult to imagine a durable union of a people which is rich and strong with one which is poor and weak, even if it were proved that the strength and wealth of the one are not the causes of the weakness and poverty of the other. But union is still more difficult to maintain at a time at which one party is losing strength, and the other is gaining it. This rapid and disproportionate increase of certain States threatens the independence of the others. New York might perhaps succeed, with its two millions of inhabitants and its forty representatives, in dictating to the other States in Congress. But even if the more powerful States make no attempt to bear down the lesser ones, the danger still exists; for there is almost as much in the possibility of the act as in the act itself. The weak generally mistrust the justice and the reason of the strong. The States which increase less rapidly than the others look upon those which are more favoured by fortune with envy and suspicion. Hence arise the deep-seated uneasiness and ill-defined agitation which are observable in the South, and which form so striking a contrast to the confidence and prosperity which are common to other parts of the Union. I am inclined to think that the hostile measures taken by the Southern provinces upon a recent occasion are attributable to no other cause. The inhabitants of the Southern States are, of all the Americans, those who are most interested in the maintenance of the Union; they would assuredly suffer most from being left to themselves; and yet they are the only citizens who threaten to break the tie of confederation. But it is easy to perceive that the South, which has given four Presidents, Washington, Jefferson, Madison, and Monroe, to the Union, which perceives that it is losing its Federal influence, and that the number of its representatives in Congress is diminishing from year to year, whilst those of the Northern and Western States are increasing; the South, which is peopled with ardent and irascible beings, is becoming more and more irritated and alarmed. The citizens reflect upon their present position and remember their past influence, with the melancholy uneasiness of men who suspect oppression: if they discover a law of the Union which is not unequivocally favourable to their interests, they protest against it as an abuse of force; and if their ardent remonstrances are not listened to, they threaten to quit an association which loads them with burdens whilst it deprives them of their due profits. 'The Tariff,' said the inhabitants of Carolina in 1832, 'enriches the North, and ruins the South; for if this were not the case, to what can we attribute the continually increasing power and wealth of the North, with its inclement skies and arid soil; whilst the South, which may be styled the garden of America, is rapidly declining?'[70]

If the changes which I have described were gradual, so that each generation at least might have time to disappear with the order of things under which it had lived, the danger would be less; but the progress of society in America is precipitate, and almost revolutionary. The same citizen may have lived to see his State take the lead in the Union, and afterwards become powerless in the Federal assemblies; and an Anglo-American republic has been known to grow as rapidly as a man passing from birth and infancy to maturity in the course of thirty years. It must not be imagined, however, that the States which lose their preponderance, also lose their population or their riches: no stop is put to their prosperity, and they even go on to increase more rapidly than any kingdom in Europe.[71] But they believe themselves to be impoverished because their wealth does not augment as rapidly as that of their neighbours; and they think that their power is lost, because they suddenly come into collision with a power greater than their own:[72] thus they are more hurt in their feelings and their passions than in their interests. But this is amply sufficient to endanger the maintenance of the Union. If kings and peoples had only had their true interests in view ever since the beginning of the world, the name of war would scarcely be known among mankind.

Thus the prosperity of the United States is the source of the most serious dangers that threaten them, since it tends to create in some of the confederate States that over-excitement which accompanies a rapid increase of fortune; and to awaken in others those feelings of envy, mistrust, and regret which usually attend upon the loss of it. The Americans contemplate this extraordinary and hasty progress with exultation; but they would be wiser to consider it with sorrow and alarm. The Americans of the United States must inevitably become one of the greatest nations in the world; their offset will cover almost the whole of North America; the continent which they inhabit is their dominion, and it cannot escape them. What urges them to take possession of it so soon? Riches, power, and renown cannot fail to be theirs at some future time, but they rush upon their fortune as if but a moment remained for them to make it their own.

I think that I have demonstrated that the existence of the present confederation depends entirely on the continued assent of all the confederates; and, starting from this principle, I have inquired into the causes which may induce the several States to separate from the others. The Union may, however, perish in two different ways: one of the confederate States may choose to retire from the compact, and so forcibly to sever the Federal tie; and it is to this supposition that most of the remarks that I have made apply: or the authority of the Federal Government may be progressively entrenched on by the simultaneous tendency of the united republics to resume their independence. The central power, successively stripped of all its prerogatives, and reduced to impotence by tacit consent, would become incompetent to fulfil its purpose; and the second Union would perish, like the first, by a sort of senile inaptitude. The gradual weakening of the Federal tie, which may finally lead to the dissolution of the Union, is a distinct circumstance, that may produce a variety of minor consequences before it operates so violent a change. The confederation might still subsist, although its Government were reduced to such a degree of inanition as to paralyse the nation, to cause internal anarchy, and to check the general prosperity of the country.

After having investigated the causes which may induce the Anglo-Americans to disunite, it is important to inquire whether, if the Union continues to subsist, their Government will extend or contract its sphere of action, and whether it will become more energetic or more weak.

The Americans are evidently disposed to look upon their future condition with alarm. They perceive that in most of the nations of the world the exercise of the rights of sovereignty tends to fall under the control of a few individuals, and they are dismayed by the idea that such will also be the case in their own country. Even the statesmen feel, or affect to feel, these fears; for, in America, centralisation is by no means popular, and there is no surer means of courting the majority than by inveighing against the encroachments of the central power. The Americans do not perceive that the countries in which this alarming tendency to centralisation exists are inhabited by a single people; whilst the fact of the Union being composed of different confederate communities is sufficient to baffle all the inferences which might be drawn from analogous circumstances. I confess that I am inclined to consider the fears of a great number of Americans as purely imaginary; and far from participating in their dread of the consolidation of power in the hands of the Union, I think that the Federal Government is visibly losing strength.

To prove this assertion I shall not have recourse to any remote occurrences, but to circumstances which I have myself witnessed, and which belong to our own time.

An attentive examination of what is going on in the United States will easily convince us that two opposite tendencies exist in that country, like two distinct currents flowing in contrary directions in the same channel. The Union has now existed for forty-five years, and in the course of that time a vast number of provincial prejudices, which were at first hostile to its power, have died away. The patriotic feeling which attached each of the Americans to his own native State is become less exclusive; and the different parts of the Union have become more intimately connected the better they have become acquainted with each other. The Post,[73] that great instrument of intellectual intercourse, now reaches into the backwoods; and steamboats have established daily means of communication between the different points of the coast. An inland navigation of unexampled rapidity conveys commodities up and down the rivers of the country.[74] And to these facilities of nature and art may be added those restless cravings, that busy-mindedness, and love of pelf, which are constantly urging the American into active life, and bringing him into contact with his fellow-citizens. He crosses the country in every direction; he visits all the various populations of the land; and there is not a province in France in which the natives are so well known to each other as the thirteen millions of men who cover the territory of the United States.

But whilst the Americans intermingle, they grow in resemblance of each other; the differences resulting from their climate, their origin, and their institutions, diminish; and they all draw nearer and nearer to the common type. Every year, thousands of men leave the North to settle in different parts of the Union: they bring with them their faith, their opinions, and their manners; and as they are more enlightened than the men amongst whom they are about to dwell, they soon rise to the head of affairs, and they adapt society to their own advantage. This continual emigration of the North to the South is peculiarly favourable to the fusion of all the different provincial characters into one national character. The civilisation of the North appears to be the common standard, to which the whole nation will one day be assimilated.

The commercial ties which unite the confederate States are strengthened by the increasing manufactures of the Americans; and the union which began to exist in their opinions, gradually forms a part of their habits: the course of time has swept away the bugbear thoughts which haunted the imaginations of the citizens in 1789. The Federal power is not become oppressive; it has not destroyed the independence of the States; it has not subjected the confederates to monarchial institutions; and the Union has not rendered the lesser States dependent upon the larger ones; but the Confederation has continued to increase in population, in wealth, and in power. I am therefore convinced that the natural obstacles to the continuance of the American Union are not so powerful at the present time as they were in 1789; and that the enemies of the Union are not so numerous.

Nevertheless, a careful examination of the history of the United States for the last forty-five years will readily convince us that the Federal power is declining; nor is it difficult to explain the causes of this phenomenon.[75] When the Constitution of 1789 was promulgated, the nation was a prey to anarchy; the Union, which succeeded this confusion, excited much dread and much animosity; but it was warmly supported because it satisfied an imperious want. Thus, although it was more attacked than it is now, the Federal power soon reached the maximum of its authority, as is usually the case with a government which triumphs after having braced its strength by the struggle. At that time the interpretation of the Constitution seemed to extend, rather than to repress, the Federal sovereignty; and the Union offered, in several respects, the appearance of a single and undivided people, directed in its foreign and internal policy by a single Government. But to attain this point the people had risen, to a certain extent, above itself.

The Constitution had not destroyed the distinct sovereignty of the States; and all communities, of whatever nature they may be, are impelled by a secret propensity to assert their independence. This propensity is still more decided in a country like America, in which every village forms a sort of republic accustomed to conduct its own affairs. It therefore cost the States an effort to submit to the Federal supremacy; and all efforts, however successful they may be, necessarily subside with the causes in which they originated.

As the Federal Government consolidated its authority, America resumed its rank amongst the nations, peace returned to its frontiers, and public credit was restored; confusion was succeeded by a fixed state of things, which was favourable to the full and free exercise of industrious enterprise. It was this very prosperity which made the Americans forget the cause to which it was attributable; and when once the danger was passed, the energy and the patriotism which had enabled them to brave it disappeared from amongst them. No sooner were they delivered from the cares which oppressed them, than they easily returned to their ordinary habits, and gave themselves up without resistance to their natural inclinations. When a powerful Government no longer appeared to be necessary, they once more began to think it irksome. The Union encouraged a general prosperity, and the States were not inclined to abandon the Union; but they desired to render the action of the power which represented that body as light as possible. The general principle of Union was adopted, but in every minor detail there was an actual tendency to independence. The principle of confederation was every day more easily admitted, and more rarely applied; so that the Federal Government brought about its own decline, whilst it was creating order and peace.

As soon as this tendency of public opinion began to be manifested externally, the leaders of parties, who live by the passions of the people, began to work it to their own advantage. The position of the Federal Government then became exceedingly critical. Its enemies were in possession of the popular favour; and they obtained the right of conducting its policy by pledging themselves to lessen its influence. From that time forwards the Government of the Union has invariably been obliged to recede, as often as it has attempted to enter the lists with the governments of the States. And whenever an interpretation of the terms of the Federal Constitution has been called for, that interpretation has most frequently been opposed to the Union, and favourable to the States.

The Constitution invested the Federal Government with the right of providing for the interests of the nation; and it had been held that no other authority was so fit to superintend the 'internal improvements' which affected the prosperity of the whole Union; such, for instance, as the cutting of canals. But the States were alarmed at a power, distinct from their own, which could thus dispose of a portion of their territory; and they were afraid that the central Government would, by this means, acquire a formidable extent of patronage within their own confines, and exercise a degree of influence which they intended to reserve exclusively to their own agents. The democratic party, which has constantly been opposed to the increase of the Federal authority, then accused the Congress of usurpation, and the Chief Magistrate of ambition. The central Government was intimidated by the opposition; and it soon acknowledged its error, promising exactly to confine its influence for the future within the circle which was prescribed to it.

The Constitution confers upon the Union the right of treating with foreign nations. The Indian tribes, which border upon the frontiers of the United States, had usually been regarded in this light. As long as these savages consented to retire before the civilised settlers, the Federal right was not contested: but as soon as an Indian tribe attempted to fix its dwelling upon a given spot, the adjacent States claimed possession of the lands and the rights of sovereignty over the natives. The central Government soon recognised both these claims; and after it had concluded treaties with the Indians as independent nations, it gave them up as subjects to the legislative tyranny of the States.[76]

Some of the States which had been founded upon the coast of the Atlantic, extended indefinitely to the West, into wild regions where no European had ever penetrated. The States whose confines were irrevocably fixed, looked with a jealous eye upon the unbounded regions which the future would enable their neighbors to explore. The latter then agreed, with a view to conciliate the others, and to facilitate the Act of Union, to lay down their own boundaries, and to abandon all the territory which lay beyond those limits to the confederation at large.[77] Thenceforward the Federal Government became the owner of all the uncultivated lands which he beyond the borders of the thirteen States first confederated. It was invested with the right of parcelling and selling them, and the sums derived from this source were exclusively reserved to the public treasure of the Union, in order to furnish supplies for purchasing tracts of country from the Indians, for opening roads to the remote settlements, and for accelerating the increase of civilisation as much as possible. New States have, however, been formed in the course of time, in the midst of those wilds which were formerly ceded by the inhabitants of the shores of the Atlantic. Congress has gone on to sell, for the profit of the nation at large, the uncultivated lands which those new States contained. But the latter at length asserted that, as they were now fully constituted, they ought to enjoy the exclusive right of converting the produce of these sales to their own use. As their remonstrances became more and more threatening, Congress thought fit to deprive the Union of a portion of the privileges which it had hitherto enjoyed; and at the end of 1832 it passed a law by which the greatest part of the revenue derived from the sale of lands was made over to the new western Republics, although the lands themselves were not ceded to them.[78]

The slightest observation in the United States enables one to appreciate the advantages which the country derives from the Bank. These advantages are of several kinds, but one of them is peculiarly striking to the stranger. The banknotes of the United States are taken upon the borders of the desert for the same value as at Philadelphia, where the Bank conducts its operations.[79]

The Bank of the United States is nevertheless the object of great animosity. Its directors have proclaimed their hostility to the President: and they are accused, not without some show of probability, of having abused their influence to thwart his election. The President therefore attacks the establishment which they represent with all the warmth of personal enmity; and he is encouraged in the pursuit of his revenge by the conviction that he is supported by the secret propensities of the majority. The Bank may be regarded as the great monetary tie of the Union, just as Congress is the great legislative tie; and the same passions which tend to render the States independent of the central power, contribute to the overthrow of the Bank.

The Bank of the United States always holds a great number of the notes issued by the provincial banks, which it can at any time oblige them to convert into cash. It has itself nothing to fear from a similar demand, as the extent of its resources enables it to meet all claims. But the existence of the provincial banks is thus threatened, and their operations are restricted, since they are only able to issue a quantity of notes duly proportioned to their capital. They submit with impatience to this salutary control. The newspapers which they have bought over, and the President, whose interest renders him their instrument, attack the Bank with the greatest vehemence. They rouse the local passions and the blind democratic instinct of the country to aid their cause; and they assert that the Bank-directors form a permanent aristocratic body, whose influence must ultimately be felt in the Government, and must affect those principles of equality upon which society rests in America.

The contest between the Bank and its opponents is only an incident in the great struggle which is going on in America between the provinces and the central power; between the spirit of democratic independence and the spirit of gradation and subordination. I do not mean that the enemies of the Bank are identically the same individuals who, on other points, attack the Federal Government; but I assert that the attacks directed against the Bank of the United States originate in the same propensities which militate against the Federal Government; and that the very numerous opponents of the former afford a deplorable symptom of the decreasing support of the latter.

The Union has never displayed so much weakness as in the celebrated question of the Tariff.[80] The wars of the French Revolution and of 1812 had created manufacturing establishments in the North of the Union, by cutting off all free communication between America and Europe. When peace was concluded, and the channel of intercourse reopened by which the produce of Europe was transmitted to the New World, the Americans thought fit to establish a system of import duties, for the twofold purpose of protecting their incipient manufactures and of paying off the amount of the debt contracted during the war. The Southern States, which have no manufactures to encourage, and which are exclusively agricultural, soon complained of this measure. Such were the simple facts, and I do not pretend to examine in this place whether their complaints were well-founded or unjust.

As early as the year 1820, South Carolina declared, in a petition to Congress, that the Tariff was 'unconstitutional, oppressive, and unjust.' And the States of Georgia, Virginia, North Carolina, Alabama, and Mississippi subsequently remonstrated against it with more or less vigour. But Congress, far from lending an ear to these complaints, raised the scale of Tariff duties in the years 1824 and 1828, and recognised anew the principle on which it was founded. A doctrine was then proclaimed, or rather revived, in the South, which took the name of Nullification.

I have shown in the proper place that the object of the Federal Constitution was not to form a league, but to create a national Government. The Americans of the United States form a sole and undivided people, in all the cases which are specified by that Constitution; and upon these points the will of the nation is expressed, as it is in all constitutional nations, by the voice of the majority. When the majority has pronounced its decision, it is the duty of the minority to submit. Such is the sound legal doctrine, and the only one which agrees with the text of the Constitution, and the known intention of those who framed it.

The partisans of Nullification in the South maintain, on the contrary, that the intention of the Americans in uniting was not to reduce themselves to the condition of one and the same people; that they meant to constitute a league of independent States; and that each State, consequently retains its entire sovereignty, if not *de facto,* at least *de jure;* and has the right of putting its own construction upon the laws of Congress, and of suspending their execution within the limits of its own territory, if they are held to be unconstitutional and unjust.

The entire doctrine of Nullification is comprised in a sentence uttered by Vice-President Calhoun, the head of that party in the South, before the Senate of the United States, in the year 1833. 'The Constitution is a compact to which the States were parties in their sovereign capacity; now, whenever a compact is entered into by parties which acknowledge no tribunal above their authority to decide in the last resort, each of them has a right to judge for itself in relation to the nature, extent, and obligations of the instrument.' It is evident that a similar doctrine destroys the very basis of the Federal Constitution, and brings back all the evils of the old Confederation, from which the Americans were supposed to have had a safe deliverance.

When South Carolina perceived that Congress turned a deaf ear to its remonstrances, it threatened to apply the doctrine of nullification to the Federal Tariff bill. Congress persisted in its former system; and at length the storm broke out. In the course of 1832 the citizens of South Carolina,[81] named a National Convention, to consult upon the extraordinary measures which they were called upon to take; and on the 24th of November of the same year this Convention promulgated a law, under the form of a decree, which annulled the Federal law of the Tariff, forbade the levy of the imposts which that law commands, and refused to recognise the appeal which might be made to the Federal courts of law.[82] This decree was only to be put in execution in the ensuing month of February, and it was intimated, that if Congress modified the Tariff before that period, South Carolina might be induced to proceed no further with her menaces; and a vague desire was afterwards expressed of submitting the question to an extraordinary assembly of all the confederate States.

In the meantime South Carolina armed her militia, and prepared for war. But Congress, which had slighted its suppliant subjects, listened to their complaints as soon as they were found to have taken up arms.[83] A law was passed, by which the tariff duties were to be progressively reduced for ten years, until they were brought so low as not to exceed the amount of supplies necessary to the Government.[84] Thus Congress completely abandoned the principle of the Tariff; and substituted a mere fiscal impost to a system of protective duties.[85] The Government of the Union, in order to conceal its defeat, had recourse to an expedient which is very much in vogue with feeble governments. It yielded the point *de facto,* but it remained inflexible upon the principles in question; and whilst Congress was altering the Tariff law, it passed another bill, by which the President was invested with extraordinary powers, enabling him to overcome by force a resistance which was then no longer to be apprehended.

But South Carolina did not consent to leave the Union in the enjoyment of these scanty trophies of success: the same national Convention which had annulled the Tariff bill, met again, and accepted the proffered concession; but at the same time it declared it unabated perseverance in the doctrine of nullification: and to prove what it said, it annulled the law investing the President with extraordinary powers, although it was very certain that the clauses of that law would never be carried into effect.

Almost all the controversies of which I have been speaking have taken place under the Presidency of General Jackson; and it cannot be denied that in the question of the Tariff he has supported the claims of the Union with vigour and with skill. I am, however, of opinion that the conduct of the individual who now represents the Federal Government may be reckoned as one of the dangers which threaten its continuance.

Some persons in Europe have formed an opinion of the possible influence of General Jackson upon the affairs of his country, which appears highly extravagant to those who have seen more of the subject. We have been told that General Jackson has won sundry battles, that he is an energetic man, prone by nature and by habit to the use of force, covetous of power, and a despot by taste. All this may perhaps be true; but the inferences which have been drawn from these truths are exceedingly erroneous. It has been imagined that General Jackson is bent on establishing a dictatorship in America, on introducing a military spirit, and on giving a degree of influence to the central authority which cannot but be dangerous to provincial liberties. But in America the time for similar undertakings, and the age for men of this kind, is not yet come: if General Jackson had entertained a hope of exercising his authority in this manner, he would infallibly have forfeited his political station, and compromised his life; accordingly he has not been so imprudent as to make any such attempt.

Far from wishing to extend the Federal power, the President belongs to the party which is desirous of limiting that power to the bare and precise letter of the Constitution, and which never puts a construction upon that act favourable to the Government of the Union; far from standing forth as the champion of centralisation, General Jackson is the agent of all the jealousies of the States; and he was placed in the lofty station he occupies by the passions of the people which are most opposed to the central Government. It is by perpetually flattering these passions that he maintains his station and his popularity. General Jackson is the slave of the majority: he yields to its wishes, its propensities, and its demands; say rather, that he anticipates and forestalls them.

Whenever the Governments of the States come into collision with that of the Union, the President is generally the first to question his own rights: he almost always outstrips the legislature; and when the extent of the Federal power is controverted, he takes part, as it were, against himself; he conceals his official interests, and extinguishes his own natural inclinations. Not indeed that he is naturally weak or hostile to the Union; for when the majority decided against the claims of the partisans of nullification, he put himself at its head, asserted the doctrines which the nation held distinctly and energetically, and was the first to recommend forcible measures;

but General Jackson appears to me, if I may use the American expressions, to be a Federalist by taste, and a Republican by calculation.

General Jackson stoops to gain the favour of the majority, but when he feels that his popularity is secure, he overthrows all obstacles in the pursuit of the objects which the community approves, or of those which it does not look upon with a jealous eye. He is supported by a power with which his predecessors were unacquainted; and he tramples on his personal enemies whenever they cross his path with a facility which no former President ever enjoyed; he takes upon himself the responsibility of measures which no one before him would have ventured to attempt: he even treats the national representatives with disdain approaching to insult; he puts his veto upon the laws of Congress, and frequently neglects to reply to that powerful body. He is a favourite who sometimes treats his master roughly. The power of General Jackson perpetually increases; but that of the President declines; in his hands the Federal Government is strong, but it will pass enfeebled into the hands of his successor.

I am strangely mistaken if the Federal Government of the United States be not constantly losing strength, retiring gradually from public affairs, and narrowing its circle of action more and more. It is naturally feeble, but it now abandons even its pretensions to strength. On the other hand, I thought that I remarked a more lively sense of independence, and a more decided attachment to provincial government, in the States. The Union is to subsist, but to subsist as a shadow; it is to be strong in certain cases, and weak in all others; in time of warfare, it is to be able to concentrate all the forces of the nation and all the resources of the country in its hands; and in time of peace its existence is to be scarcely perceptible: as if this alternate debility and vigour were natural or possible.

I do not foresee anything for the present which may be able to check this general impulse of public opinion; the causes in which it originated do not cease to operate with the same effect. The change will therefore go on, and it may be predicted that, unless some extraordinary event occurs, the Government of the Union will grow weaker and weaker every day.

I think, however, that the period is still remote at which the Federal power will be entirely extinguished by its inability to protect itself and to maintain peace in the country. The Union is sanctioned by the manners and desires of the people; its results are palpable, its benefits visible. When it is perceived that the weakness of the Federal Government compromises the existence of the Union, I do not doubt that a reaction will take place with a view to increase its strength.

The Government of the United States is, of all the Federal Governments which have hitherto been established, the one which is most naturally destined to act. As long as it is only indirectly assailed by the interpretation of its laws, and as long as its substance is not seriously altered, a change of opinion, an internal crisis, or a war, may restore all the vigour which it requires. The point which I have been most anxious to put in a clear light is simply this: Many people, especially in France, imagine that a change in opinion is going on in the United States, which is favourable to a centralisation of power in the hands of the President and the Congress. I hold that a contrary tendency may distinctly be observed. So far is the Federal Government from acquiring strength, and from threatening the sovereignty of the States, as it grows older, that I maintain it to be growing weaker and weaker, and that the sovereignty of the Union alone is in danger. Such are the facts which the present time discloses. The future conceals the final result of this tendency, and the events which may check, retard, or accelerate the changes I have described; but I do not affect to be able to remove the veil which hides them from our sight.

Of the Republican Institutions of the United States, and What Their Chances of Duration Are

The Union is accidental—The Republican institutions have more prospect of permanence—A republic for the present the natural state of the Anglo-Americans—Reason of this—In order to destroy it, all the laws must be changed at the same time, and a great alteration take place in manners—Difficulties experienced by the Americans in creating an aristocracy.

The dismemberment of the Union, by the introduction of war into the heart of those States which are now confederate, with standing armies, a dictatorship, and a heavy taxation, might, eventually, compromise the fate of the republican institutions. But we ought not to confound the future prospects of the republic with those of the Union. The Union is an accident, which will only last as long as circumstances are favourable to its existence; but a republican form of Government seems to me to be the natural state of the Americans; which nothing but the continued action of hostile causes, always acting in the same direction, could change into a monarchy. The Union exists principally in the law which formed it; one revolution, one change in public opinion, might destroy it forever; but the republic has a much deeper foundation to rest upon.

What is understood by a republican government in the United States is the slow and quiet action of society upon itself. It is a regular state of things really founded upon the enlightened will of the people. It is a conciliatory government under which resolutions are allowed time to ripen; and in which they are deliberately discussed, and executed with mature judgment. The republicans in the United States set a high value upon morality, respect religious belief, and acknowledge the existence of rights. They profess to think that a people ought to be moral, religious, and temperate, in proportion as it is free. What is called the republic in the United States, is the tranquil rule of the majority, which, after having had time to examine itself, and to give proof of its existence, is the common source of all the powers of the State. But the power of the majority is not of itself unlimited. In the moral world humanity, justice, and reason enjoy an undisputed supremacy; in the political world vested rights are treated with no less deference. The majority recognises these two barriers; and if it now and then overstep them, it is because, like individuals, it has passions, and, like them, it is prone to do what is wrong, whilst it discerns what is right.

But the demagogues of Europe have made strange discoveries. A republic is not, according to them, the rule of the majority, as has hitherto been thought, but the rule of those who are strenuous partisans of the majority. It is not the people who preponderates in this kind of government, but those who are best versed in the good qualities of the people. A happy distinction, which allows men to act in the name of nations without consulting them, and to claim their gratitude whilst their rights are spurned. A republican government, moreover, is the only one which claims the right of doing whatever it chooses, and despising what men have hitherto respected, from the highest moral obligations to the vulgar rules of common sense. It had been supposed, until our time, that despotism was odious, under whatever form it appeared. But it is a discovery of modern days that there are such things as legitimate tyranny and holy injustice, provided they are exercised in the name of the people.

The ideas which the Americans have adopted respecting the republican form of government, render it easy for them to live under it, and insure its duration. If, in their country, this form be often practically bad, at least it is theoretically good; and, in the end, the people always acts in conformity to it.

It was impossible at the foundation of the States, and it would still be difficult, to establish a central administration in America. The inhabitants are dispersed over too great a space, and separated by too many natural obstacles, for one man to undertake to direct the details of their existence. America is therefore pre-eminently the country of provincial and municipal government. To this cause, which was plainly felt by all the Europeans of the New World, the Anglo-Americans added several others peculiar to themselves.

At the time of the settlement of the North American colonies, municipal liberty had already penetrated into the laws as well as the manners of the English; and the emigrants adopted it, not only as a necessary thing, but as a benefit which they knew how to appreciate. We have already seen the manner in which the Colonies were founded: every province, and almost every district, was peopled separately by men who were strangers to each other, or who associated with very different purposes. The English settlers in the United States, therefore, early perceived that they were divided into a great number of small and distinct communities which belonged to no common centre; and that it was needful for each of these little communities to take care of its own affairs, since there did not appear to be any central authority which was naturally bound and easily enabled to provide for them. Thus, the nature of the country, the manner in which the British colonies were founded, the habits of the first emigrants, in short everything, united to promote, in an extraordinary degree, municipal and provincial liberties.

In the United States, therefore, the mass of the institutions of the country is essentially republican; and in order permanently to destroy the laws which form the basis of the republic, it would be necessary to abolish all the laws at once. At the present day it would be even more difficult for a party to succeed in founding a monarchy in the United States than for a set of men to proclaim that France should henceforward be a republic. Royalty would not find a system of legislation prepared for it beforehand; and a monarchy would then exist, really surrounded by republican institutions. The monarchical principle would likewise have great difficulty in penetrating into the manners of the Americans.

In the United States, the sovereignty of the people is not an isolated doctrine bearing no relation to the prevailing manners and ideas of the people: it may, on the contrary, be regarded as the last link of a chain of opinions which binds the whole Anglo-American world. That Providence has given to every human being the degree of reason necessary to direct himself in the affairs which interest him exclusively; such is the grand maxim upon which civil and political society rests in the United States. The father of a family applies it to his children; the master to his servants; the township to its officers; the province to its townships; the State to its provinces; the Union to the States; and when extended to the nation, it becomes the doctrine of the sovereignty of the people.

Thus, in the United States, the fundamental principle of the republic is the same which governs the greater part of human actions; republican notions insinuate themselves into all the ideas, opinions, and habits of the Americans, whilst they are formerly recognised by the legislation: and before this legislation can be altered the whole community must undergo very serious changes. In the United States, even the religion of most of the citizens is republican, since it submits the truths of the other world to private judgment: as in politics the care of its temporal interests is abandoned to the good sense of the people. Thus every man is allowed freely to take that road which he thinks will lead him to heaven; just as the law permits every citizen to have the right of choosing his government.

It is evident that nothing but a long series of events, all having the same tendency, can substitute for this combination of laws, opinions, and manners, a mass of opposite opinions, manners, and laws.

If republican principles are to perish in America, they can only yield after a laborious social process, often interrupted, and as often resumed; they will have many apparent revivals, and will not become totally extinct until an entirely new people shall have succeeded to that which now exists. Now, it must be admitted that there is no symptom or presage of the approach of such a revolution. There is nothing more striking to a person newly arrived in the United States, than the kind of tumultuous agitation in which he finds political society. The laws are incessantly changing, and at first sight it seems impossible that a people so variable in its desires should avoid adopting, within a short space of time, a completely new form of government. Such apprehensions are, however, premature; the instability which affects political institutions is of two kinds, which ought not to be confounded: the first, which modifies secondary laws, is not incompatible with a very settled state of society; the other shakes the very foundations of the Constitution, and attacks the fundamental principles of legislation; this species of instability is always followed by troubles and revolutions, and the nation which suffers under it is in a state of violent transition.

Experience shows that these two kinds of legislative instability have no necessary connection; for they have been found united or separate, according to times and circumstances. The first is common in the United States, but not the second: the Americans often change their laws, but the foundation of the Constitution is respected.

In our days the republican principle rules in America, as the monarchical principle did in France under Louis XIV. The French of that period were not only friends of the monarchy, but they thought it impossible to put anything in its place; they received it as we receive the rays of the sun and the return of the seasons. Amongst them the royal power had neither advocates nor opponents. In like manner does the republican Government exist in America, without contention or opposition; without proofs and arguments, by a tacit agreement, a sort of *consensus universalis*. It is, however, my opinion that by changing their administrative forms as often as they do, the inhabitants of the United States compromise the future stability of their Government.

It may be apprehended that men, perpetually thwarted in their designs by the mutability of the legislation, will learn to look upon republican institutions as an inconvenient form of society; the evil resulting from the instability of the secondary enactments might then raise a doubt as to the nature of the fundamental principles of the Constitution, and indirectly bring about a revolution; but this epoch is still very remote.

It may, however, be foreseen even now, that when the Americans lose their republican institutions they will speedily arrive at a despotic Government, without a long interval of limited monarchy. Montesquieu remarked, that nothing is more absolute than the authority of a prince who immediately succeeds a republic, since the powers which had fearlessly been intrusted to an elected magistrate are then transferred to a hereditary sovereign. This is true in general, but it is more peculiarly applicable to a democratic republic. In the United States, the magistrates are not elected by a particular class of citizens, but by the majority of the nation; they are the immediate representatives of the passions of the multitude; and as they are wholly dependent upon its pleasure, they excite neither hatred nor fear: hence, as I have already shown, very little care has been taken to limit their influence, and they are left in possession of a vast deal of arbitrary power. This state of things has engendered habits which would outlive itself; the American magistrate would retain his power, but he would cease to be responsible for the exercise of it; and it is impossible to say what bounds could then be set to tyranny.

Some of our European politicians expect to see an aristocracy arise in America, and they already predict the exact period at which it will be able to assume the reins of government. I have previously observed, and I repeat my assertion, that the present tendency of American society appears to me to become more and more democratic. Nevertheless, I do not assert that the Americans will not, at some future time, restrict the circle of political rights in their country, or confiscate those rights to the advantage of a single individual; but I cannot imagine that they will ever bestow the exclusive exercise of them upon a privileged class of citizens, or, in other words, that they will ever found an aristocracy.

An aristocratic body is composed of a certain number of citizens who, without being very far removed from the mass of the people, are, nevertheless, permanently stationed above it: a body which it is easy to touch and difficult to strike; with which the people are in daily contact, but with which they can never combine. Nothing can be imagined more contrary to nature and to the secret propensities of the human heart than a subjection of this kind; and men who are left to follow their own bent will always prefer the arbitrary power of a king to the regular administration of an aristocracy. Aristocratic institutions cannot subsist without laying down the inequality of men as a fundamental principle, as a part and parcel of the legislation, affecting the condition of the human family as much as it affects that of society; but these are things so repugnant to natural equity that they can only be extorted from men by constraint.

I do not think a single people can be quoted, since human society began to exist, which has, by its own free will and by its own exertions, created an aristocracy within its own bosom. All the aristocracies of the Middle Ages were founded by military conquest; the conqueror was the noble, the vanquished became the serf. Inequality was then imposed by force; and after it had been introduced into the manners of the country it maintained its own authority, and was sanctioned by the legislation. Communities have existed which were aristocratic from their earliest origin, owing to circumstances anterior to that event, and which became more democratic in each succeeding age. Such was the destiny of the Romans, and of the barbarians after them. But a people, having taken its rise in civilisation and democracy, which should gradually establish an inequality of conditions, until it arrived at inviolable privileges and exclusive castes, would be a novelty in the world; and nothing intimates that America is likely to furnish so singular an example.

Reflection on the Causes of the Commercial Prosperity of the United States

The Americans destined by Nature to be a great maritime people—Extent of their coasts—Depth of their ports—Size of their rivers—The commercial superiority of the Anglo-Americans less attributable, however, to physical circumstances than to moral and intellectual causes—Reason of this opinion—Future destiny of the Anglo-Americans as a commercial nation—The dissolution of the Union would not check the maritime vigour of the States—Reason of this—Anglo-Americans will naturally supply the wants of the inhabitants of South America—They will become, like the English, the factors of a great portion of the world.

The coast of the United States, from the Bay of Fundy to the Sabine River in the Gulf of Mexico, is more than two thousand miles in extent. These shores form an unbroken line, and they are all subject to the same Government. No nation in the world possesses vaster, deeper, or more secure ports for shipping than the Americans.

The inhabitants of the United States constitute a great civilised people, which fortune has placed in the midst of an uncultivated country at a distance of three thousand miles from the central point of civilisation. America consequently stands in daily need of European trade. The Americans will, no doubt, ultimately succeed in producing or manufacturing at home most of the articles which they require; but the two continents can never be independent of each other, so numerous are the natural ties which exist between their wants, their ideas, their habits, and their manners.

The Union produces peculiar commodities which are now become necessary to us, but which cannot be cultivated, or can only be raised at an enormous expense, upon the soil of Europe. The Americans only consume a small portion of this produce, and they are willing to sell us the rest. Europe is therefore the market of America, as America is the market of Europe; and maritime commerce is no less necessary to enable the inhabitants of the United States to transport their raw materials to the ports of Europe, than it is to enable us to supply them with our manufactured produce. The United States were therefore necessarily reduced to the alternative of increasing the business of other maritime nations to a great extent, if they had themselves declined to enter into commerce, as the Spaniards of Mexico have hitherto done; or, in the second place, of becoming one of the first trading powers of the globe.

The Anglo-Americans have always displayed a very decided taste for the sea. The Declaration of Independence broke the commercial restrictions which united them to England, and gave a fresh and powerful stimulus to their maritime genius. Ever since that time, the shipping of the Union has increased in almost the same rapid proportion as the number of its inhabitants. The Americans themselves now transport to their own shores nine-tenths of the European produce which they consume.[86] And they also bring three-quarters of the exports of the New World to the European consumer.[87] The ships of the United States fill the docks of Havre and of Liverpool; whilst the number of English and French vessels which are to be seen at New York is comparatively small.[88]

Thus, not only does the American merchant face the competition of his own countrymen, but he even supports that of foreign nations in their own ports with success. This is readily explained by the fact that the vessels of the United States can cross the seas at a cheaper rate than any other vessels in the world. As long as the mercantile shipping of the United States preserves this superiority, it will not only retain what it has acquired, but it will constantly increase in prosperity.

It is difficult to say for what reason the Americans can trade at a lower rate than other nations; and one is at first led to attribute this circumstance to the physical or natural advantages which are within their reach; but this supposition is erroneous.

The American vessels cost almost as much to build as our own;[89] they are not better built, and they generally last for a shorter time. The pay of the American sailor is more considerable than the pay on board European ships; which is proved by the great number of Europeans who are to be met with in the merchant vessels of the United States. But I am of opinion that the true cause of their superiority must not be sought for in physical advantages, but that it is wholly attributable to their moral and intellectual qualities.

The following comparison will illustrate my meaning. During the campaigns of the Revolution the French introduced a new system of tactics into the art of war, which perplexed the oldest generals, and very nearly destroyed the most ancient monarchies in Europe. They undertook (what had never before been attempted) to make shift without a number of things which had always been held to be indispensable in warfare; they required novel exertions on the part of their troops which no civilised nations had ever thought of; they achieved great actions in an incredibly short space of time; and they risked human life without hesitation to obtain the object in view. The French had less money and fewer men than their enemies; their resources were infinitely inferior; nevertheless they were constantly victorious, until their adversaries chose to imitate their example. Americans have introduced a similar system into their commercial speculations; and they do for cheapness what the French did for conquest. The European sailor navigates with prudence; he only sets sail when the weather is favourable; if an unforseen accident befalls him, he puts into port; at night he furls a portion of his canvas; and when the whitening billows intimate the vicinity of land, he checks his way, and takes an observation of the sun. But the American neglects these precautions and braves these dangers. He weighs anchor in the midst of tempestuous gales; by night and by day he spreads his sheets to the wind; he repairs as he goes along such damage as his vessel may have sustained from the storm; and when he at last approaches the term of his voyage, he darts onward to the shore as if he already descried a port. The Americans are often shipwrecked, but no trader crosses the seas so rapidly. And as they perform the same distance in a shorter time, they can perform it at a cheaper rate.

The European touches several times at different ports in the course of a long voyage; he loses a good deal of precious time in making the harbour, or in waiting for a favourable wind to leave it; and he pays daily dues to be allowed to remain there. The American starts from Boston to go to purchase tea in China; he arrives at Canton, stays there a few days, and then returns. In less than two years he has sailed as far as the entire circumference of the globe, and he has seen land but once. It is true that during a voyage of eight or ten months he has drunk brackish water and lived upon salt meat; that he has been in a continual contest with the sea, with disease, and with a tedious existence; but upon his return he can sell a pound of his tea for a half-penny less than the English merchant, and his purpose is accomplished.

I cannot better explain my meaning than by saying that the Americans affect a sort of heroism in their manner of trading. But the European merchant will always find it very difficult to imitate his American competitor, who, in adopting the system which I have just described, follows not only a calculation of his gain, but an impulse of his nature.

The inhabitants of the United States are subject to all the wants and all the desires which result from an advanced stage of civilisation; but as they are not surrounded by a community admirably adapted, like that of Europe, to satisfy their wants, they are often obliged to procure for themselves the various articles which education and habit have rendered necessaries. In America it sometimes happens that the same individual tills his field, builds his dwelling, contrives his tools, makes his shoes, and weaves the coarse stuff of which his dress is composed. This circumstance is prejudicial to the excellence of the work; but it powerfully contributes to awaken the intelligence of the workman. Nothing tends to materialise man, and to deprive his work of the faintest trace of mind, more than extreme division of labour. In a country like America, where men devoted to special occupations are rare, a long apprenticeship cannot be required from anyone who embraces a profession. The Americans, therefore, change their means of gaining a livelihood very readily; and they suit their occupations to the exigencies of the moment, in the manner most profitable to themselves. Men are to be met with who have successively been barristers, farmers, merchants, ministers of the Gospel, and physicians. If the American be less perfect in each craft than the European, at least there is scarcely any trade with which he is utterly unacquainted. His capacity is more general, and the circle of his intelligence is enlarged.

The inhabitants of the United States are never fettered by the axioms of their profession; they escape from all the prejudices of their present station; they are not more attached to one line of operation than to another; they are not more prone to employ an old method than a new one; they have no rooted habits, and they easily shake off the influence which the habits of other nations might exercise upon their minds from a conviction that their country is unlike any other, and that its situation is without a precedent in the world. America is a land of wonders, in which everything is in constant motion, and every movement seems an improvement. The idea of novelty is there indissolubly connected with the idea of amelioration. No natural boundary seems to be set to the efforts of man; and what is not yet done is only what he has not yet attempted to do.

This perpetual change which goes on in the United States, these frequent vicissitudes of fortune, accompanied by such unforeseen fluctuations in private and in public wealth, serve to keep the minds of the citizens in a perpetual state of feverish agitation, which admirably invigorates their exertions, and keeps them in a state of excitement above the ordinary level of mankind. The whole life of an American is passed like a game of chance, a revolutionary crisis, or a battle. As the same causes are continually in operation throughout the country, they ultimately impart an irresistible impulse to the national character. The American, taken as a chance specimen of his countrymen, must then be a man of singular warmth in his desires, enterprising, fond of adventure, and, above all, of innovation. The same bent is manifest in all that he does; he introduces it into his political laws, his religious doctrines, his theories of social economy, and his domestic occupations; he bears it with him in the depths of the backwoods, as well as in the business of the city. It is this same passion, applied to maritime commerce, which makes him the cheapest and the quickest trader in the world.

As long as the sailors of the United States retain these inspiriting advantages, and the practical superiority which they derive from them, they will not only continue to supply the wants of the producers and consumers of their own country, but they will tend more and more to become, like the English, the factors of all other peoples.[90] This prediction has already begun to be realised; we perceive that the American traders are introducing themselves as intermediate agents in the commerce of several European nations;[91] and America will offer a still wider field to their enterprise.

The great colonies which were founded in South America by the Spaniards and the Portuguese have since become empires. Civil war and oppression now lay waste those extensive regions. Population does not increase, and the thinly-scattered inhabitants are too much absorbed in the cares of self-defense even to attempt any amelioration of their condition. Such, however, will not always be the case. Europe has succeeded by her own efforts in piercing the gloom of the Middle Ages; South America has the same Christian laws and Christian manners as we have; she contains all the germs of civilisation which have grown amidst the nations of Europe or their offsets, added to the advantages to be derived from our example: why then should she always remain uncivilised? It is clear that the question is simply one of time; at some future period, which may be more or less remote, the inhabitants of South America will constitute flourishing and enlightened nations.

But when the Spaniards and Portuguese of South America begin to feel the wants common to all civilised nations, they will still be unable to satisfy those wants for themselves; as the youngest children of civilisation, they must perforce admit the superiority of their elder brethren. They will be agriculturists long before they succeed in manufactures or commerce, and they will require the mediation of strangers to exchange their produce beyond seas for those articles for which a demand will begin to be felt.

It is unquestionable that the Americans of the North will one day supply the wants of the Americans of the South. Nature has placed them in contiguity, and has furnished the former with every means of knowing and appreciating those demands, of establishing a permanent connection with those States, and of gradually filling their markets. The merchant of the United States could only forfeit these natural advantages if he were very inferior to the merchant of Europe; to whom he is, on the contrary, superior in several respects. The Americans of the United States already exercise a very considerable moral influence upon all the peoples of the New World. They are the source of intelligence, and all the nations which inhabit the same continent are already accustomed to consider them as the most enlightened, the most powerful, and the most wealthy members of the great American family. All eyes are therefore turned towards the Union; and the States of which that body is composed are the models which the other communities try to imitate to the best of their power; it is from the United States that they borrow their political principles and their laws.

The Americans of the United States stand in precisely the same position with regard to the peoples of South America as their fathers, the English, occupy with regard to the Italians, the Spaniards, the Portuguese, and all those nations of Europe which receive their articles of daily consumption from England, because they are less advanced in civilisation and trade. England is at this time the natural emporium of almost all the nations which are within its reach; the American Union will perform the same part in the other hemisphere; and every community which is founded, or which prospers in the New World, is founded and prospers to the advantage of the Anglo-Americans.

If the Union were to be dissolved, the commerce of the States which now compose it would undoubtedly be checked for a time; but this consequence would be less perceptible than is generally supposed. It is evident that, whatever may happen, the commercial States will remain united. They are all contiguous to each other; they have identically the same opinions, interests, and manners; and they are alone competent to form a very great maritime power. Even if the South of the Union were to become independent of the North, it would still require the services of those States. I have already observed that the South is not a commercial country, and nothing intimates that it is likely to become so. The Americans of the South of the

United States will therefore be obliged, for a long time to come, to have recourse to strangers to export their produce, and to supply them with the commodities which are requisite to satisfy their wants. But the Northern States are undoubtedly able to act as their intermediate agents cheaper than any other merchants. They will therefore retain that employment, for cheapness is the sovereign law of commerce. National claims and national prejudices cannot resist the influence of cheapness. Nothing can be more virulent than the hatred which exists between the Americans of the United States and the English. But notwithstanding these inimical feelings, the Americans derive the greater part of their manufactured commodities from England, because England supplies them at a cheaper rate than any other nation. Thus the increasing prosperity of America turns, notwithstanding the grudges of the Americans, to the advantage of British manufactures.

Reason shows and experience proves that no commercial prosperity can be durable if it cannot be united, in case of need, to naval force. This truth is as well understood in the United States as it can be anywhere else: the Americans are already able to make their flag respected; in a few years they will be able to make it feared. I am convinced that the dismemberment of the Union would not have the effect of diminishing the naval power of the Americans, but that it would powerfully contribute to increase it. At the present time the commercial States are connected with others which have not the same interests, and which frequently yield an unwilling consent to the increase of a maritime power by which they are only indirectly benefited. If, on the contrary, the commercial States of the Union formed one independent nation, commerce would become the foremost of their national interests; they would consequently be willing to make very great sacrifices to protect their shipping, and nothing would prevent them from pursuing their designs upon this point.

Nations, as well as men, almost always betray the most prominent features of their future destiny in their earliest years. When I contemplate the ardour with which the Anglo-Americans prosecute commercial enterprise, the advantages which befriend them, and the success of their undertakings, I cannot refrain from believing that they will one day become the first maritime power of the globe. They are born to rule the seas, as the Romans were to conquer the world.

Conclusion

I have now nearly reached the close of my inquiry; hitherto, in speaking of the future destiny of the United States, I have endeavoured to divide my subject into distinct portions, in order to study each of them with more attention. My present object is to embrace the whole from one single point; the remarks I shall make will be less detailed, but they will be more sure. I shall perceive each object less distinctly, but I shall descry the principal facts with more certainty. A traveller who has just left the walls of an immense city, climbs the neighbouring hill; as he goes further off he loses sight of the men whom he has so recently quitted; their dwellings are confused in a dense mass; he can no longer distinguish the public squares, and he can scarcely trace out the great thoroughfares; but his eye has less difficulty in following the boundaries of the city, and for the first time he sees the shape of the vast whole. Such is the future destiny of the British race in North America to my eye; the details of the stupendous picture are overhung with shade, but I conceive a clear idea of the entire subject.

The territory now occupied or possessed by the United States of America forms about one-twentieth part of the habitable earth. But extensive as these confines are, it must not be supposed that the Anglo-American race will always remain within them; indeed, it has already far overstepped them.

There was once a time at which we also might have created a great French nation in the American wilds, to counterbalance the influence of the English upon the destinies of the New World. France formerly possessed a territory in North America, scarcely less extensive than the whole of Europe. The three greatest rivers of that continent then flowed within her dominions. The Indian tribes which dwelt between the mouth of the St. Lawrence and the delta of the Mississippi were unaccustomed to any other tongue but ours; and all the European settlements scattered over that immense region recalled the traditions of our country. Louisbourg, Montmorency, Duquesne, Saint-Louis, Vincennes, New Orleans (for such were the names they bore) are words dear to France and familiar to our ears.

But a concourse of circumstances, which it would be tedious to enumerate,[92] have deprived us of this magnificent inheritance. Wherever the French settlers were numerically weak and partially established, they have disappeared: those who remain are collected on a small extent of country, and are now subject to other laws. The 400,000 French inhabitants of Lower Canada constitute, at the present time, the remnant of an old nation lost in the midst of a new people. A foreign population is increasing around them unceasingly and on all sides, which already penetrates amongst the ancient masters of the country, predominates in their cities and corrupts their language. This population is identical with that of the United States; it is therefore with truth that I asserted that the British race is not confined within the frontiers of the Union, since it already extends to the North-east.

To the North-west nothing is to be met with but a few insignificant Russian settlements; but to the South-west, Mexico presents a barrier to the Anglo-Americans.

Thus, the Spaniards and the Anglo-Americans are, properly speaking, the only two races which divide the possession of the New World. The limits of separation between them have been settled by a treaty; but although the conditions of that treaty are exceedingly favourable to the Anglo-Americans, I do not doubt that they will shortly infringe this arrangement. Vast provinces, extending beyond the frontiers of the Union towards Mexico, are still destitute of inhabitants. The natives of the United States will forestall the rightful occupants of these solitary regions. They will take possession of the soil, and establish social institutions, so that when the legal owner arrives at length, he will find the wilderness under cultivation, and strangers quietly settled in the midst of his inheritance.[93]

The lands of the New World belong to the first occupant, and they are the natural reward of the swiftest pioneer. Even the countries which are already peopled will have some difficulty in securing themselves from this invasion. I have already alluded to what is taking place in the province of Texas. The inhabitants of the United States are perpetually migrating to Texas, where they purchase land; and although they conform to the laws of the country, they are gradually founding the empire of their own language and their own manners. The province of Texas is still part of the Mexican dominions, but it will soon contain no Mexicans; the same thing has occurred whenever the Anglo-Americans have come into contact with populations of a different origin.

It cannot be denied that the British race has acquired an amazing preponderance over all the other European races in the New World; and that it is very superior to them in civilisation, in industry, and in power. As long as it is only surrounded by desert or thinly peopled countries, as long as it encounters no dense populations upon its route, through which it cannot work its way, it will assuredly continue to spread. The lines marked out by treaties will not stop it; but it will everywhere transgress these imaginary barriers.

The geographical position of the British race in the New World is peculiarly favourable to its rapid increase. Above its northern frontiers the icy regions of the Pole extend; and a few degrees below its southern confines lies the burning climate of the Equator. The Anglo-Americans are, therefore, placed in the most temperate and habitable zone of the continent.

It is generally supposed that the prodigious increase of population in the United States is posterior to their Declaration of Independence. But this is an error: the population increased as rapidly under the colonial system as it does at the present day; that is to say, it doubled in about twenty-two years. But this proportion which is now applied to millions, was then applied to thousands of inhabitants; and the same fact which was scarcely noticeable a century ago, is now evident to every observer.

The British subjects in Canada, who are dependent on a king, augment and spread almost as rapidly as the British settlers of the United States, who live under a republican Government. During the War of Independence, which lasted eight years, the population continued to increase without intermission in the same ratio. Although powerful Indian nations allied with the English existed at that time upon the western frontiers, the emigration westward was never checked. Whilst the enemy laid waste the shores of the Atlantic, Kentucky, the western parts of Pennsylvania, and the States of Vermont and of Maine were filling with inhabitants. Nor did the unsettled state of the Constitution, which succeeded the war, prevent the increase of the population, or stop its progress across the wilds. Thus, the difference of laws, the various conditions of peace and war, of order and of anarchy, have exercised no perceptible influence upon the gradual development of the Anglo-Americans. This may be readily understood; for the fact is, that no causes are sufficiently general to exercise a simultaneous influence over the whole of so extensive a territory. One portion of the country always offers a sure retreat from the calamities which afflict another part; and however great may be the evil, the remedy which is at hand is greater still.

It must not, then, be imagined that the impulse of the British race in the new world can be arrested. The dismemberment of the Union, and the hostilities which might ensue, the abolition of republican institutions, and the tyrannical government which might succeed it, may retard this impulse, but they cannot prevent it from ultimately fulfilling the destinies to which that race is reserved. No power upon earth can close upon the emigrants that fertile wilderness which offers resources to all industry, and a refuge from all want. Future events, of whatever nature they may be, will not deprive the Americans of their climate or of their inland seas, of their great rivers or of their exuberant soil. Nor will bad laws, revolutions, and anarchy be able to obliterate that love of prosperity and that spirit of enterprise which seem to be the distinctive characteristics of their race, or to extinguish that knowledge which guides them on their way.

Thus, in the midst of the uncertain future, one event at least is sure. At a period which may be said to be near (for we are speaking of the life of a nation), the Anglo-Americans will alone cover the immense space contained between the Polar regions and the Tropics, extending from the coasts of the Atlantic to the shores of the Pacific Ocean. The territory which will probably be occupied by the Anglo-Americans at some future time, may be computed to equal three-quarters of Europe in extent.[94]

The climate of the Union is upon the whole preferable to that of Europe, and its natural advantages are not less great; it is therefore evident that its population will at some future time be proportionate to our own. Europe, divided as it is between so many different nations, and torn as it has been by incessant wars and the barbarous manners of the Middle Ages, has notwithstanding attained a population of 410 inhabitants to the square league.[95] What cause can prevent the United States from having as numerous a population in time?

Many ages must elapse before the divers offsets of the British race in America cease to present the same homogeneous characteristics: and the time cannot be foreseen at which a permanent inequality of conditions will be established in the New World. Whatever differences may arise, from peace or from war, from freedom or oppression, from prosperity or want, between the destinies of the different descendants of the great Anglo-American family, they will at least preserve an analogous social condition, and they will hold in common the customs and the opinions to which that social condition has given birth.

In the Middle Ages, the tie of religion was sufficiently powerful to imbue all the different populations of Europe with the same civilisation. The British of the New World have a thousand other reciprocal ties; and they live at a time when the tendency to equality is general amongst mankind. The Middle Ages were a period when everything was broken up; when each people, each province, each city, and each family, had a strong tendency to maintain its distinct individuality. At the present time an opposite tendency seems to prevail, and the nations seem to be advancing to unity. Our means of intellectual intercourse unite the most remote parts of the earth; and it is impossible for men to remain strangers to each other, or to be ignorant of the events which are taking place in any corner of the globe. The consequence is that there is less difference, at the present day, between the Europeans and their descendants in the New World, than there was between certain towns in the thirteenth century which were only separated by a river. If this tendency to assimilation brings foreign nations closer to each other, it must *à fortiori* prevent the descendants of the same people from becoming aliens to each other.

The time will therefore come when one hundred and fifty millions of men will be living in North America,[96] equal in condition, the progeny of one race, owing their origin to the same cause, and preserving the same civilisation, the same language, the same religion, the same habits, the same manners, and imbued with the same opinions, propagated under the same forms. The rest is uncertain, but this is certain; and it is a fact new to the world a fact fraught with such portentous consequences as to baffle the efforts even of the imagination.

There are, at the present time, two great nations in the world which seem to tend towards the same end, although they started from different points: I allude to the Russians and the Americans. Both of them have grown up unnoticed; and whilst the attention of mankind was directed elsewhere, they have suddenly assumed a most prominent place amongst the nations; and the world learned their existence and their greatness at almost the same time.

All other nations seem to have nearly reached their natural limits, and only to be charged with the maintenance of their power; but these are still in the act of growth;[97] all the others are stopped, or continue to advance with extreme difficulty; these are proceeding with ease and with celerity along a path to which the human eye can assign no term. The American struggles against the natural obstacles which oppose him; the adversaries of the Russian are men; the former combats the wilderness and savage life; the latter, civilisation with all its weapons and its arts: the conquests of the one are therefore gained by the ploughshare; those of the other by the sword. The Anglo-American relies upon personal interest to accomplish his ends, and gives free scope to the unguided exertions and common-sense of the citizens; the Russian centres all the authority of society in a single arm: the principal instrument of the former is freedom; of the latter servitude. Their starting-point is different, and their courses are not the same; yet each of them seems to be marked out by the will of Heaven to sway the destinies of half the globe.

Opinions of the Present Work

'Let me earnestly advise your perusal of the work, if you have not yet read it, of a very able and intelligent Frenchman, who has made the institutions of the United States the peculiar object of his observation and study. M. De Tocqueville is the name of the Frenchman to whom I allude, and this is the account which he gives of the results produced by republican institutions in the United States.'

Speech of the late Sir Robert Peel, at Glasgow, January 12, 1837.

'The English public may now know and read the first philosophical book ever written on Democracy, as it manifests itself in modern society; a book, the essential doctrines of which it is not likely that any future speculations will subvert, to whatever degree they may modify them; while its spirit, and the general mode in which it treats its subject, constitute it the beginning of a new era in the scientific study of politics.'

Edinburgh Review, October 1840.

'M. De Tocqueville, among the first, has set the example of analysing Democracy: of distinguishing one of its features, one of its tendencies, from another; of showing which of these tendencies is good, and which bad, in itself. He does this with so noble a field as a great nation to demonstrate upon, which field he has commenced by minutely examining; selecting by a discernment of which we have had no previous example, the material facts, and surveying these by the light of principles, drawn from no ordinary knowledge of human nature. The author's mind seems to us to resemble Montesquieu most among the great French writers. The book is such as Montesquieu might have written, if to his genius he had superadded good sense, and the lights which mankind have since gained from the experiences of a period in which they may be said to have lived centuries in fifty years.'

John Stuart Mill, in the *London Review*, vol. ii. p. 95.

'It is our opinion that M. De Tocqueville has approached the working of the American institutions in a better temper, and treated it in a far more philosophical manner, than any preceding writer.... We feel highly grateful to M. De Tocqueville for having acted towards us on this occasion the part of a travelling tutor. He has not only showed us the country, but explained to us the reasons why it exists in its present state; and for the first time, so far as we are aware of, not only the true situation of that extraordinary people, but the true causes of their social and political situation, are clearly developed. . . . Persons who seek in these pages for materials to advance any merely party, or other selfish purpose, will certainly be disappointed, for they are entirely free from envy, hatred, and malice, and from all uncharitableness. Neither is there any satire contained in them, expressed or understood; all is grave, and plain, and above-board, and withal so temperate, that even where we do not agree with his deductions our confidence in his good faith and singleness of purpose remains unbroken. This is a great charm. We cannot indeed recall to our memory any work at all similar to this, in which there is no narrative, nor any other enlivening circumstance to give it animation, and yet in which the interest is sustained from beginning to end without once flagging. . . . The work is divided into two parts. In the First Part he shows the direction which the democracy of America, almost entirely unrestrained, and let loose to follow its natural propensities, has given to the laws and to the general administration of public affairs. He, moreover, endeavours to trace its evils and advantages, and to learn what precautions have been used by statesmen in America to regulate this enormous machine, so as to render its movements subservient to the government of society. In the Second Part M. De Tocqueville examines the influence which the equality of conditions and the actual administration of affairs under a democratical government have exercised on the habits, the opinions, the sentiments, and manners of the Americans: in short, to determine how far their moral character, their intellectual attainments, their pursuit of business or of pleasure, their intercourse with one another and with foreigners, and all their other private relations, have been modified by the complete establishment of the democratic system under circumstances entirely dissimilar to any which the world had heretofore witnessed. . . . In America the people—that is to say, the mass of the people, the numerical majority—regulate all things, and, in fact, govern the country: they appoint the legislative as well as the executive power; they nominate the judges and the juries; they elect their representatives directly, and for the most part annually; and thus, in every possible way, although the government be nominally what is called representative, it is evident that the opinions of the majority, however fluctuating or inconsistent—their passions, however violent—and their prejudices, however absurd—or their interests, however selfish—do, in all cases, exercise a perpetual influence on society. Formerly there existed two great and distinct parties—the Federalists and the Republicans; but these distinctions are now quite obliterated. . . . M. De Tocqueville considers it quite a mistake, though a common one in Europe, to suppose that the democratical institutions of America are likely to perish from weakness; their chief danger, in his opinion, arising from their own enormous power. This universal and uncontrollable influence, he thinks, may at some future time urge the oppressed minorities to desperation, and oblige them to seek relief by the hazardous experiment of a physical collision. "Anarchy will then be the result, but it will have been brought about by despotism." In support of this opinion he quotes two great American names, that of General Hamilton, and that of Jefferson.... M. De Tocqueville is a firm believer in the Roman Catholic creed: and yet (we think) he has not failed to obtain some glimpses of the danger to which America is at present exposed by the spread of a Roman Catholic population within her territory.... It is due to Mr. Reeve, the translator of M. De Tocqueville's very nice and delicate language, to bear our testimony to the fidelity with which he has executed a task of considerable difficulty.... In conclusion, we once more congratulate the public on their having at last obtained a popular account of America, written in the very purest spirit of philosophy, and with such rare temperance, that persons of all parties, and of all shades of parties, may read it not only with profit, but without their patience being ruffled.'

Quarterly Review, September 1836.

'Read in the spirit in which it was conceived, the Democracy in America is no cold or abstract treatise, but the vivid representation of a patriotic mind, and it abounds in suggestive interest when collated with the subsequent course of events in France. There is indeed another train of thought which may well induce any one who meditates on the dangers and chances of human affairs, to take down these volumes from the shelf once more. In the Chapter on the Dangers which menace the American Union, will be found the most interesting prognostics of actual occurrences—the fear that the rapid and disproportionate increase of certain States will injure the independence of others—the deep-seated uneasiness and ill-defined agitation of the South, which feels its comparative strength gradually diminishing, and that of the North and West becoming preponderant—the constant envy and suspicion manifesting itself in the interpretation given to every act of the Legislature that is not unequivocally favourable to Southern interests—the belief of the Southern States that they are impoverished because their wealth does not augment as rapidly as that of their neighbours, and that their power is lapsing from them—all these motives are portrayed and analysed, regretted and reproved.'

Quarterly Review, October 1861.

SECOND PART

Preface to the Second Part

The Americans live in a democratic state of society, which has naturally suggested to them certain laws and a certain political character. This same state of society has, moreover, engendered amongst them a multitude of feelings and opinions which were unknown amongst the elder aristocratic communities of Europe: it has destroyed or modified all the relations which before existed, and established others of a novel kind. The aspect of civil society has been no less affected by these changes than that of the political world. The former subject has been treated of in the work on the Democracy of America, which I published five years ago; to examine the latter is the object of the present book; but these two parts complete each other, and form one and the same work.

I must at once warn the reader against an error which would be extremely prejudicial to me. When he finds that I attribute so many different consequences to the principle of equality, he may thence infer that I consider that principle to be the sole cause of all that takes place in the present age: but this would be to impute to me a very narrow view. A multitude of opinions, feelings, and propensities are now in existence, which owe their origin to circumstances unconnected with or even contrary to the principle of equality. Thus if I were to select the United States as an example, I could easily prove that the nature of the country, the origin of its inhabitants, the religion of its founders, their acquired knowledge, and their former habits, have exercised, and still exercise, independently of Democracy, a vast influence upon the thoughts and feelings of that people. Different causes, but no less distinct from the circumstances of the equality of conditions, might be traced in Europe, and would explain a great portion of the occurrences taking place amongst us.

I acknowledge the existence of all these different causes, and their power, but my subject does not lead me to treat of them. I have not undertaken to unfold the reason of all our inclinations and all our notions: my only object is to show in what respects the principle of equality has modified both the former and the latter.

Some readers may perhaps be astonished that, firmly persuaded, as I am that the democratic revolution which we are witnessing is an irresistible fact against which it would be neither desirable nor wise to struggle, I should often have had occasion in this book to address language of such severity to those democratic communities which this revolution has brought into being. My answer is simply, that it is because I am not an adversary of Democracy, that I have sought to speak of Democracy in all sincerity.

Men will not accept truth at the hands of their enemies, and truth is seldom offered to them by their friends: for this reason I have spoken it. I was persuaded that many would take upon themselves to announce the new blessings which the principle of equality promises to mankind, but that few would dare to point out from afar the dangers with which it threatens them. To those perils therefore I have turned my chief attention, and believing that I had discovered them clearly, I have not had the cowardice to leave them untold.

I trust that my readers will find in this Second Part that impartiality which seems to have been remarked in the former work. Placed as I am in the midst of the conflicting opinions between which we are divided, I have endeavoured to suppress within me for a time the favourable sympathies or the adverse emotions with which each of them inspires me. If those who read this book can find a single sentence intended to flatter any of the great parties which have agitated my country, or any of those petty factions which now harass and weaken it, let such readers raise their voices to accuse me.

The subject I have sought to embrace is immense, for it includes the greater part of the feelings and opinions to which the new state of society has given birth. Such a subject is doubtless above my strength, and in treating it I have not succeeded in satisfying myself. But, if I have not been able to reach the goal to which I had in view, my readers will at least do me the justice to acknowledge that I have conceived and followed up my undertaking in a spirit not unworthy of success.

A. De T.
March 1840

SECOND PART

FIRST BOOK

Influence of Democracy on the Progress of Opinion in the United States

Chapter I

Philosophical Method among the Americans

I think that in no country in the civilised world is less attention paid to philosophy than in the United States. The Americans have no philosophical school of their own; and they care but little for all the schools into which Europe is divided, the very names of which are scarcely known to them. Nevertheless it is easy to perceive that almost all the inhabitants of the United States conduct their understanding in the same manner, and govern it by the same rules; that is to say, that without ever having taken the trouble to define the rules of a philosophical method, they are in possession of one, common to the whole people. To evade the bondage of system and habit, of family-maxims, class-opinions, and, in some degree, of national prejudices; to accept tradition only as a means of information, and existing facts only as a lesson used in doing otherwise and doing better; to seek the reason of things for one's self, and in one's self alone; to tend to results without being bound to means, and to aim at the substance through the form;—such are the principal characteristics of what I shall call the philosophical method of the Americans. But if I go further, and if I seek amongst these characteristics that which predominates over and includes almost all the rest, I discover that in most of the operations of the mind, each American appeals to the individual exercise of his own understanding alone. America is therefore one of the countries in the world where philosophy is least studied, and where the precepts of Descartes are best applied. Nor is this surprising. The Americans do not read the works of Descartes, because their social condition deters them from speculative studies; but they follow his maxims because this very social condition naturally disposes their understanding to adopt them. In the midst of the continual movement which agitates a democratic community, the tie which unites one generation to another is relaxed or broken; every man readily loses the trace of the ideas of his forefathers or takes no care about them. Nor can men living in this state of society derive their belief from the opinions of the class to which they belong, for, so to speak, there are no longer any classes, or those which still exist are composed of such mobile elements, that their body can never exercise a real control over its members. As to the influence which the intelligence of one man has on that of another, it must necessarily be very limited in a country where the citizens, placed on the footing of a general similitude, are all closely seen by each other; and where, as no signs of incontestable greatness or superiority are perceived in any one of them, they are constantly brought back to their own reason as the most obvious and proximate source of truth. It is not only confidence in this or that man which is then destroyed, but the taste for trusting the *ipse dixit* of any man whatsoever. Everyone shuts himself up in his own breast, and affects from that point to judge the world.

The practice which obtains amongst the Americans of fixing the standard of their judgment in themselves alone, leads them to other habits of mind. As they perceive that they succeed in resolving without assistance all the little difficulties which their practical life presents, they readily conclude that everything in the world may be explained, and that nothing in it transcends the limits of the understanding. Thus they fall to denying what they cannot comprehend; which leaves them but little faith for whatever is extraordinary, and an almost insurmountable distaste for whatever is supernatural. As it is on their own testimony that they are accustomed to rely, they like to discern the object which engages their attention with extreme clearness; they therefore strip off as much as possible all that covers it, they rid themselves of whatever separates them from it, they remove whatever conceals it from sight, in order to view it more closely and in the broad light of day. This disposition of the mind soon leads them to contemn forms, which they regard as useless and inconvenient veils placed between them and the truth.

The Americans then have not required to extract their philosophical method from books; they have found it in themselves. The same thing may be remarked in what has taken place in Europe. This same method has only been established and made popular in Europe in proportion as the condition of society has become more equal, and men have grown more like each other. Let us consider for a moment the connection of the periods in which this change may be traced. In the sixteenth century the Reformers subjected some of the dogmas of the ancient faith to the scrutiny of private judgment; but they still withheld from it the judgment of all the rest. In the seventeenth century, Bacon in the natural sciences, and Descartes in the study of philosophy in the strict sense of the term, abolished recognised formulas, destroyed the empire of tradition, and overthrew the authority of the schools. The philosophers of the eighteenth century, generalising at length the same principle, undertook to submit to the private judgment of each man all the objects of his belief.

Who does not perceive that Luther, Descartes, and Voltaire employed the same method, and that they differed only in the greater or less use which they professed should be made of it? Why did the Reformers confine themselves so closely within the circle of religious ideas? Why did Descartes, choosing only to apply his method to certain matters, though he had made it fit to be applied to all, declare that men might judge for themselves in matters philosophical but not in matters political? How happened it that in the eighteenth century those general applications were all at once drawn from this same method, which Descartes and his predecessors had either not perceived or had rejected? To what, lastly, is the fact to be attributed, that at this period the method we are speaking of suddenly emerged from the schools, to penetrate into society and become the common standard of intelligence; and that, after it had become popular among the French, it has been ostensibly adopted or secretly followed by all the nations of Europe?

The philosophical method here designated may have been engendered in the sixteenth century—it may have been more accurately defined and more extensively applied in the seventeenth; but neither in the one nor in the other could it be commonly adopted. Political laws, the condition of society, and the habits of mind which are derived from these causes, were as yet opposed to it. It was discovered at a time when men were beginning to equalise and assimilate their conditions. It could only be generally followed in ages when those conditions had at length become nearly equal, and men nearly alike.

The philosophical method of the eighteenth century is then not only French, but it is democratic; and this explains why it was so readily admitted throughout Europe, where it has contributed so powerfully to change the face of society. It is not because the French have changed their former opinions, and altered their former manners, that they have convulsed the world; but because they were the first to generalise and bring to light a philosophical method, by the assistance of which it became easy to attack all that was old, and to open a path to all that was new.

If it be asked why, at the present day, this same method is more rigorously followed and more frequently applied by the French than by the Americans, although the principle of equality be no less complete, and of more ancient date, amongst the latter people, the fact may be attributed to two circumstances, which it is essential to have clearly understood in the first instance.

It must never be forgotten that religion gave birth to Anglo-American society. In the United States religion is therefore commingled with all the habits of the nation and all the feelings of patriotism; whence it derives a peculiar force. To this powerful reason another of no less intensity may be added: in American religion has, as it were, laid down its own limits. Religious institutions have remained wholly distinct from political institutions, so that former laws have been easily changed whilst former belief has remained unshaken. Christianity has therefore retained a strong hold on the public mind in America; and, I would more particularly remark, that its sway is not only that of a philosophical doctrine which has been adopted upon inquiry, but of a religion which is believed without discussion. In the United States Christian sects are infinitely diversified and perpetually modified; but Christianity itself is a fact so irresistibly established, that no one undertakes either to attack or to defend it. The Americans, having admitted the principal doctrines of the Christian religion without inquiry, are obliged to accept in like manner a great number of moral truths originating in it and connected with it. Hence the activity of individual analysis is restrained within narrow limits, and many of the most important of human opinions are removed from the range of its influence.

The second circumstance to which I have alluded is the following: the social condition and the constitution of the Americans are democratic, but they have not had a democratic revolution. They arrived upon the soil they occupy in nearly the condition in which we see them at the present day; and this is of very considerable importance.

There are no revolutions which do not shake existing belief, enervate authority, and throw doubts over commonly received ideas. The effect of all revolutions is therefore, more or less, to surrender men to their own guidance, and to open to the mind of every man a void and almost unlimited range of speculation. When equality of conditions succeeds a protracted conflict between the different classes of which the elder society was composed, envy, hatred, and uncharitableness, pride, and exaggerated self-confidence are apt to seize upon the human heart, and plant their sway there for a time. This, independently of equality itself, tends powerfully to divide men—to lead them to mistrust the judgment of others, and to seek the light of truth nowhere but in their own understandings. Every one then attempts to be his own sufficient guide, and makes it his boast to form his own opinions on all subjects. Men are no longer bound together by ideas, but by interests; and it would seem as if human opinions were reduced to a sort of intellectual dust, scattered on every side, unable to collect, unable to cohere.

Thus, that independence of mind which equality supposes to exist, is never so great, nor ever appears so excessive, as at the time when equality is beginning to establish itself, and in the course of that painful labour by which it is established.

That sort of intellectual freedom which equality may give ought, therefore, to be very carefully distinguished from the anarchy which revolution brings. Each of these two things must be severally considered, in order not to conceive exaggerated hopes or fears of the future.

I believe that the men who will live under the new forms of society will make frequent use of their private judgment; but I am far from thinking that they will often abuse it. This is attributable to a cause of more general application to all democratic countries, and which, in the long run, must needs restrain in them the independence of individual speculation within fixed, and sometimes narrow, limits. I shall proceed to point out this cause in the next chapter.

Chapter II

Of the Principal Source of Belief among Democratic Nations

At different periods dogmatical belief is more or less abundant. It arises in different ways, and it may change its object or its form; but under no circumstances will dogmatical belief cease to exist, or, in other words, men will never cease to entertain some implicit opinions without trying them by actual discussion. If everyone undertook to form his own opinions and to seek for truth by isolated paths struck out by himself alone, it is not to be supposed that any considerable number of men would ever unite in any common belief. But obviously without such common belief no society can prosper—say rather no society can subsist; for without ideas held in common, there is no common action, and without common action, there may still be men, but there is no social body. In order that society should exist, and, *à fortiori,* that a society should prosper, it is required that all the minds of the citizens should be rallied and held together by certain predominant ideas; and this cannot be the case, unless each of them sometimes draws his opinions from the common source, and consents to accept certain matters of belief at the hands of the community.

If I now consider man in his isolated capacity, I find that dogmatical belief is not less indispensable to him in order to live alone, than it is to enable him to cooperate with his fellow-creatures. If man were forced to demonstrate to himself all the truths of which he makes daily use, his task would never end. He would exhaust his strength in preparatory exercises, without advancing beyond them. As, from the shortness of his life, he has not the time, nor, from the limits of his intelligence, the capacity, to accomplish this, he is reduced to take upon trust a number of facts and opinions which he has not had either the time or the power to verify himself, but which men of greater ability have sought out, or which the world adopts. On this groundwork he raises for himself the structure of his own thoughts; nor is he led to proceed in this manner by choice so much as he is constrained by the inflexible law of his condition. There is no philosopher of such great parts in the world, but that he believes a million of things on the faith of other people, and supposes a great many more truths than he demonstrates. This is not only necessary but desirable. A man who should undertake to inquire into everything for himself, could devote to each thing but little time and attention. His task would keep his mind in perpetual unrest, which would prevent him from penetrating to the depth of any truth, or of grappling his mind indissolubly to any conviction. His intellect would be at once independent and powerless. He must therefore make his choice from amongst the various objects of human belief, and he must adopt many opinions without discussion, in order to search the better into that smaller number which he sets apart for investigation. It is true that whoever receives an opinion on the word of another, does so far enslave his mind; but it is a salutary servitude which allows him to make a good use of freedom.

A principle of authority must then always occur, under all circumstances, in some part or other of the moral and intellectual world. Its place is variable, but a place it necessarily has. The independence of individual minds may be greater, or it may be less: unbounded it cannot be. Thus the question is, not to know whether any intellectual authority exists in the ages of democracy, but simply where it resides and by what standard it is to be measured.

I have shown in the preceding chapter how the equality of conditions leads men to entertain a sort of instinctive incredulity of the supernatural, and a very lofty and often exaggerated opinion of the human understanding. The men who live at a period of social equality are not therefore easily led to place that intellectual authority to which they bow either beyond or above humanity. They commonly seek for the sources of truth in themselves, or in those who are like themselves. This would be enough to prove that at such periods no new religion could be established, and that all schemes for such a purpose would be not only impious but absurd and irrational. It may be foreseen that a democratic people will not easily give credence to divine missions; that they will turn modern prophets to a ready jest; and they that will seek to discover the chief arbiter of their belief within, and not beyond, the limits of their kind.

When the ranks of society are unequal, and men unlike each other in condition, there are some individuals invested with all the power of superior intelligence, learning, and enlightenment, whilst the multitude is sunk in ignorance and prejudice. Men living at these aristocratic periods are therefore naturally induced to shape their opinions by the superior standard of a person or a class of persons, whilst they are averse to recognise the infallibility of the mass of the people.

The contrary takes place in ages of equality. The nearer the citizens are drawn to the common level of an equal and similar condition, the less prone does each man become to place implicit faith in a certain man or a certain class of men. But his readiness to believe the multitude increases, and opinion is more than ever mistress of the world. Not only is common opinion the only guide which private judgment retains amongst a democratic people, but amongst such a people it possesses a power infinitely beyond what it has elsewhere. At periods of equality men have no faith in one another, by reason of their common resemblance; but this very resemblance gives them almost unbounded confidence in the judgment of the public; for it would not seem probable, as they are all endowed with equal means of judging, but that the greater truth should go with the greater number.

When the inhabitant of a democratic country compares himself individually with all those about him, he feels with pride that he is the equal of any one of them; but when he comes to survey the totality of his fellows, and to place himself in contrast to so huge a body, he is instantly overwhelmed by the sense of his own insignificance and weakness. The same equality which renders him independent of each of his fellow-citizens taken severally, exposes him alone and unprotected to the influence of the greater number. The public has therefore among a democratic people a singular power, of which aristocratic nations could never so much as conceive an idea; for it does not persuade to certain opinions, but it enforces them, and infuses them into the faculties by a sort of enormous pressure of the minds of all upon the reason of each.

In the United States the majority undertakes to supply a multitude of ready-made opinions for the use of individuals, who are thus relieved from the necessity of forming opinions of their own. Everybody there adopts great numbers of theories, on philosophy, morals, and politics, without inquiry, upon public trust; and if we look to it very narrowly, it will be perceived that religion herself holds her sway there, much less as a doctrine of revelation than as a commonly received opinion. The fact that the political laws of the Americans are such that the majority rules the community with sovereign sway, materially increases the power which that majority naturally exercises over the mind. For nothing is more customary in man than to recognise superior wisdom in the person of his oppressor. This political omnipotence of the majority in the United States doubtless augments the influence which public opinion would obtain without it over the mind of each member of the community; but the foundations of that influence do not rest upon it. They must be sought for in the principle of equality itself, not in the more or less popular institutions which men living under that condition may give themselves. The intellectual dominion of the greater number would probably be less absolute amongst a democratic people governed by a king than in the sphere of a pure democracy, but it will always be extremely absolute; and by whatever political laws men are governed in the ages of equality, it may be foreseen that faith in public opinion will become a species of religion there, and the majority its ministering prophet.

Thus intellectual authority will be different, but it will not be diminished; and far from thinking that it will disappear, I augur that it may readily acquire too much preponderance, and confine the action of private judgment within narrower limits than are suited either to the greatness or the happiness of the human race. In the principle of equality I very clearly discern two tendencies; the one leading the mind of every man to untried thoughts, the other inclined to prohibit him from thinking at all. And I perceive how, under the dominion of certain laws, democracy would extinguish that liberty of the mind to which a democratic social condition is favourable; so that, after having broken all the bondage once imposed on it by ranks or by men, the human mind would be closely fettered to the general will of the greatest number.

If the absolute power of the majority were to be substituted by democratic nations, for all the different powers which checked or retarded overmuch the energy of individual minds, the evil would only have changed its symptoms. Men would not have found the means of independent life; they would simply have invented (no easy task) a new dress for servitude. There is—and I cannot repeat it too often—there is in this matter for profound reflection for those who look on freedom as a holy thing, and who hate not only the despot, but despotism. For myself, when I feel the hand of power he heavy on my brow, I care but little to know who oppresses me; and I am not the more disposed to pass beneath the yoke, because it is held out to me by the arms of a million of men.

Chapter III

Why the Americans Display More Readiness and More Taste for General Ideas Than Their Forefathers the English

The Deity does not regard the human race collectively. He surveys at one glance and severally all the beings of whom mankind is composed, and he discerns in each man the resemblances which assimilate him to all his fellows, and the differences which distinguish him from them. God, therefore, stands in no need of general ideas; that is to say, he is never sensible of the necessity of collecting a considerable number of analogous objects under the same form for greater convenience in thinking. Such is, however, not the case with man. If the human mind were to attempt to examine and pass a judgment on all the individual cases before it, the immensity of detail would soon lead it astray and bewilder its discernment: in this strait, man has recourse to an imperfect but necessary expedient, which at once assists and demonstrates his weakness. Having superficially considered a certain number of objects, and remarked their resemblance, he assigns to them a common name, sets them apart, and proceeds onwards.

General ideas are no proof of the strength, but rather of the insufficiency of the human intellect; for there are in nature no beings exactly alike, no things precisely identical, nor any rules indiscriminately and alike applicable to several objects at once. The chief merit of general ideas is, that they enable the human mind to pass a rapid judgment on a great many objects at once; but, on the other hand, the notions they convey are never otherwise than incomplete, and they always cause the mind to lose as much in accuracy as it gains in comprehensiveness. As social bodies advance in civilisation, they acquire the knowledge of new facts, and they daily lay hold almost unconsciously of some particular truths. The more truths of this kind a man apprehends, the more general ideas is he naturally led to conceive. A multitude of particular facts cannot be seen separately, without at last discovering the common tie which connects them. Several individuals lead to the perception of the species; several species to that of the genus. Hence the habit and the taste for general ideas will always be greatest amongst a people of ancient cultivation and extensive knowledge.

But there are other reasons which impel men to generalise their ideas, or which restrain them from it.

The Americans are much more addicted to the use of general ideas than the English, and entertain a much greater relish for them: this appears very singular at first sight, when it is remembered that the two nations have the same origin, that they lived for centuries under the same laws, and that they still incessantly interchange their opinions and their manners. This contrast becomes much more striking still, if we fix our eyes on our own part of the world, and compare together the two most enlightened nations which inhabit it. It would seem as if the mind of the English could only tear itself reluctantly and painfully away from the observation of particular facts, to rise from them to their causes; and that it only generalizes in spite of itself. Amongst the French, on the contrary, the taste for general ideas would seem to have grown to so ardent a passion, that it must be satisfied on every occasion. I am informed, every morning when I wake, that some general and eternal law has just been discovered, which I never heard mentioned before. There is not a mediocre scribbler who does not try his hand at discovering truths applicable to a great kingdom, and who is very ill pleased with himself if he does not succeed in compressing the human race into the compass of an article. So great a dissimilarity between two very enlightened nations surprises me. If I again turn my attention to England, and observe the events which have occurred there in the last half-century, I think I may affirm that a taste for general ideas increases in that country in proportion as its ancient constitution is weakened.

The state of civilisation is therefore insufficient by itself to explain what suggests to the human mind the love of general ideas, or diverts it from them. When the conditions of men are very unequal, and inequality itself is the permanent state of society, individual men gradually become so dissimilar that each class assumes the aspect of a distinct race: only one of these classes is ever in view at the same instant; and losing sight of that general tie which binds them all within the vast bosom of mankind, the observation invariably rests not on man, but on certain men. Those who live in this aristocratic state of society never, therefore, conceive very general ideas respecting themselves, and that is enough to imbue them with an habitual distrust of such ideas, and an instinctive aversion of them.

He, on the contrary, who inhabits a democratic country, sees around him, on every hand, men differing but little from each other; he cannot turn his mind to any one portion of mankind, without expanding and dilating his thought till it embrace the whole. All the truths which are applicable to himself, appear to him equally and similarly applicable to each of his fellow-citizens and fellowmen. Having contracted the habit of generalising his ideas in the study which engages him most, and interests him more than others, he transfers the same habit to all his pursuits; and thus it is that the craving to discover general laws in everything, to include a great number of objects under the same formula, and to explain a mass of facts by a single cause, becomes an ardent, and sometimes an undiscerning, passion in the human mind.

Nothing shows the truth of this proposition more clearly than the opinions of the ancients respecting their slaves. The most profound and capacious minds of Rome and Greece were never able to reach the idea, at once so general and so simple, of the common likeness of men, and of the common birthright of each to freedom: they strove to prove that slavery was in the order of nature, and that it would always exist. Nay, more, everything shows that those of the ancients who had passed from the servile to the free condition, many of whom have left us excellent writings, did themselves regard servitude in no other light.

All the great writers of antiquity belonged to the aristocracy of masters, or at least they saw that aristocracy established and uncontested before their eyes. Their mind, after it had expanded itself in several directions, was barred from further progress in this one; and the advent of Jesus Christ upon earth was required to teach that all the members of the human race are by nature equal and alike.

In the ages of equality all men are independent of each other, isolated and weak. The movements of the multitude are not permanently guided by the will of any individuals; at such times humanity seems always to advance of itself. In order, therefore, to explain what is passing in the world, man is driven to seek for some great causes, which, acting in the same manner on all our fellow-creatures, thus impel them all involuntarily to pursue the same track. This again naturally leads the human mind to conceive general ideas, and superinduces a taste for them.

I have already shown in what way the equality of conditions leads every man to investigate truths for himself. It may readily be perceived that a method of this kind must insensibly beget a tendency to general ideas in the human mind. When I repudiate the traditions of rank, profession, and birth; when I escape from the authority of example, to seek out, by the single effort of my reason, the path to be followed, I am inclined to derive the motives of my opinions from human nature itself; which leads me necessarily, and almost unconsciously, to adopt a great number of very general notions.

All that I have here said explains the reasons for which the English display much less readiness and taste for the generalisation of ideas than their American progeny, and still less again than their French neighbours; and likewise the reason for which the English of the present day display more of these qualities than their forefathers did. The English have long been a very enlightened and a very aristocratic nation; their enlightened condition urged them constantly to generalise, and their aristocratic habits confined them to particularise. Hence arose that philosophy, at once bold and timid, broad and narrow, which has hitherto prevailed in England, and which still obstructs and stagnates in so many minds in that country.

Independently of the causes I have pointed out in what goes before, others may be discerned less apparent, but no less efficacious, which engender amongst almost every democratic people a taste, and frequently a passion, for general ideas. An accurate distinction must be taken between ideas of this kind. Some are the result of slow, minute, and conscientious labour of the mind, and these extend the sphere of human knowledge; others spring up at once from the first rapid exercise of the wits, and beget none but very superficial and very uncertain notions. Men who live in ages of equality have a great deal of curiosity and very little leisure; their life is so practical, so confused, so excited, so active, that but little time remains to them for thought. Such men are prone to general ideas because they spare them the trouble of studying particulars; they contain, if I may so speak, a great deal in a little compass, and give, in a little time, a great return. If then, upon a brief and inattentive investigation, a common relation is thought to be detected between certain objects, inquiry is not pushed any further; and without examining in detail how far these different objects differ or agree, they are hastily arranged under one formulary, in order to pass to another subject.

One of the distinguishing characteristics of a democratic period is the taste all men have at such ties for easy success and present enjoyment. This occurs in the pursuits of the intellect as well as in all others. Most of those who live at a time of equality are full of an ambition at once aspiring and relaxed: they would fain succeed brilliantly and at once, but they would be dispensed from great efforts to obtain success. These conflicting tendencies lead straight to the research of general ideas, by aid of which they flatter themselves that they can figure very importantly at a small expense, and draw the attention of the public with very little trouble. And I know not whether they be wrong in thinking thus. For their readers are as much averse to investigating anything to the bottom as they can be themselves; and what is generally sought in the productions of the mind is easy pleasure and information without labour.

If aristocratic nations do not make sufficient use of general ideas, and frequently treat them with inconsiderate disdain, it is true, on the other hand, that a democratic people is ever ready to carry ideas of this kind to excess, and to espouse them with injudicious warmth.

Chapter IV

Why the Americans Have Never Been So Eager as the French for General Ideas in Political Matters

I observed in the last chapter, that the Americans show a less decided taste for general ideas than the French; this is more especially true in political matters. Although the Americans infuse into their legislation infinitely more general ideas than the English, and although they pay much more attention than the latter people to the adjustment of the practice of affairs to theory, no political bodies in the United States have ever shown so warm an attachment to general ideas as the Constituent Assembly and the Convention in France. At no time has the American people laid hold on ideas of this kind with the passionate energy of the French people in the eighteenth century, or displayed the same blind confidence in the value and absolute truth of any theory. This difference between the Americans and the French originates in several causes, but principally in the following one. The Americans form a democratic people, which has always itself directed public affairs. The French are a democratic people, who, for a long time, could only speculate on the best manner of conducting them. The social condition of France led that people to conceive very general ideas on the subject of government, whilst its political constitution prevented it from correcting those ideas by experiment, and from gradually detecting their insufficiency; whereas in America the two things constantly balance and correct each other.

It may seem, at first sight, that this is very much opposed to what I have said before, that democratic nations derive their love of theory from the excitement of their active life. A more attentive examination will show that there is nothing contradictory in the proposition. Men living in democratic countries eagerly lay hold of general ideas because they have but little leisure, and because these ideas spare them the trouble of studying particulars. This is true; but it is only to be understood to apply to those matters which are not the necessary and habitual subjects of their thoughts. Mercantile men will take up very eagerly, and without any very close scrutiny, all the general ideas on philosophy, politics, science, or the arts, which may be presented to them; but for such as relate to commerce, they will not receive them without inquiry, or adopt them without reserve. The same thing applies to statesmen with regard to general ideas in politics. If, then, there be a subject upon which a democratic people is peculiarly liable to abandon itself, blindly and extravagantly, to general ideas, the best corrective that can be used will be to make that subject a part of the daily practical occupation of that people. The people will then be compelled to enter upon its details, and the details will teach them the weak points of the theory. This remedy may frequently be a painful one, but its effect is certain.

Thus it happens, that the democratic institutions which compel every citizen to take a practical part in the government, moderate that excessive taste for general theories in politics which the principle of equality suggests.

Chapter V

Of the Manner in Which Religion in the United States Avails Itself of Democratic Tendencies

I have laid it down in a preceding chapter that men cannot do without dogmatical belief; and even that it is very much to be desired that such belief should exist amongst them. I now add, that of all the kinds of dogmatical belief the most desirable appears to me to be dogmatical belief in matters of religion; and this is a very clear inference, even from no higher consideration than the interests of this world. There is hardly any human action, however particular a character be assigned to it, which does not originate in some very general idea men have conceived of the Deity, of His relation to mankind, of the nature of their own souls, and of their duties to their fellow-creatures. Nor can anything prevent these ideas from being the common spring from which everything else emanates. Men are therefore immeasurably interested in acquiring fixed ideas of God, of the soul, and of their common duties to their Creator and to their fellowmen; for doubt on these first principles would abandon all their actions to the impulse of chance, and would condemn them to live, to a certain extent, powerless and undisciplined.

This is then the subject on which it is most important for each of us to entertain fixed ideas; and unhappily it is also the subject on which it is most difficult for each of us, left to himself, to settle his opinions by the sole force of his reason. None but minds singularly free from the ordinary anxieties of life—minds at once penetrating, subtle, and trained by thinking—can even with the assistance of much time and care, sound the depth of these most necessary truths. And, indeed, we see that these philosophers are themselves almost always enshrouded in uncertainties; that at every step the natural light which illuminates their path grows dimmer and less secure; and that, in spite of all their efforts, they have as yet only discovered a small number of conflicting notions, on which the mind of man has been tossed about for thousands of years, without either laying a firmer grasp on truth, or finding novelty even in its errors. Studies of this nature are far above the average capacity of men; and even if the majority of mankind were capable of such pursuits, it is evident that leisure to cultivate them would still be wanting. Fixed ideas of God and human nature are indispensable to the daily practice of men's lives; but the practice of their lives prevents them from acquiring such ideas.

The difficulty appears to me to be without a parallel. Amongst the sciences there are some which are useful to the mass of mankind, and which are within its reach; others can only be approached by the few, and are not cultivated by the many, who require nothing beyond their more remote applications: but the daily practice of the science I speak of is indispensable to all, although the study of it is inaccessible to the far greater number.

General ideas respecting God and human nature are therefore the ideas above all others which it is most suitable to withdraw from the habitual action of private judgment, and in which there is most to gain and least to lose by recognising a principle of authority. The first object and one of the principal advantages of religions, is to furnish to each of these fundamental questions a solution which is at once clear, precise, intelligible to the mass of mankind, and lasting. There are religions which are very false and very absurd; but it may be affirmed, that any religion which remains within the circle I have just traced, without aspiring to go beyond it (as many religions have attempted to do, for the purpose of enclosing on every side the free progress of the human mind), imposes a salutary restraint on the intellect; and it must be admitted that, if it do not save men in another world, such religion is at least very conducive to their happiness and their greatness in this. This is more especially true of men living in free countries. When the religion of a people is destroyed, doubt gets hold of the highest portions of the intellect, and half paralyses all the rest of its powers. Every man accustoms himself to entertain none but confused and changing notions on the subjects most interesting to his fellow-creatures and himself. His opinions are ill-defended and easily abandoned: and, despairing of ever resolving by himself the hardest problems of the destiny of man, he ignobly submits to think no more about them. Such a condition cannot but enervate the soul, relax the springs of the will, and prepare a people for servitude. Nor does it only happen, in such a case, that they allow their freedom to be wrested from them; they frequency themselves surrender it. When there is no longer any principle of authority in religion any more than in politics, men are speedily frightened at the aspect of this unbounded independence. The constant agitation of all surrounding things alarms and exhausts them. As everything is at sea in the sphere of the intellect, they determine at least that the mechanism of society should be firm and fixed; and as they cannot resume their ancient belief, they assume a master.

For my own part, I doubt whether man can ever support at the same time complete religious independence and entire public freedom. And I am inclined to think, that if faith be wanting in him, he must serve; and if he be free, he must believe.

Perhaps, however, this great utility of religions is still more obvious amongst nations where equality of conditions prevails than amongst others. It must be acknowledged that equality, which brings great benefits into the world, nevertheless suggests to men (as will be shown hereafter) some very dangerous propensities. It tends to isolate them from each other, to concentrate every man's attention upon himself; and it lays open the soul to an inordinate love of material gratification. The greatest advantage of religion is to inspire diametrically contrary principles. There is no religion which does not place the object of man's desires above and beyond the treasures of earth, and which does not naturally raise his soul to regions far above those of the senses. Nor is there any which does not impose on man some sort of duties to his kind, and thus draws him at times from the contemplation of himself. This occurs in religions the most false and dangerous. Religious nations are therefore naturally strong on the very point on which democratic nations are weak; which shows of what importance it is for men to preserve their religion as their conditions become more equal.

I have neither the right nor the intention of examining the supernatural means which God employs to infuse religious belief into the heart of man. I am at this moment considering religions in a purely human point of view: my object is to inquire by what means they may most easily retain their sway in the democratic ages upon which we are entering. It has been shown that, at times of general cultivation and equality, the human mind does not consent to adopt dogmatical opinions without reluctance, and feels their necessity acutely in spiritual matters only. This proves, in the first place, that at such times religions ought, more cautiously than at any other, to confine themselves within their own precincts; for in seeking to extend their power beyond religious matters, they incur a risk of not being believed at all. The circle within which they seek to bound the human intellect ought therefore to be carefully traced, and beyond its verge the mind should be left in entire freedom to its own guidance. Mahommed professed to derive from Heaven, and he has inserted in the Koran, not only a body of religious doctrines, but political maxims, civil and criminal laws, and theories of science. The Gospel, on the contrary, only speaks of the general relations of men to God and to each other—beyond which it inculcates and imposes no point of faith. This alone, besides a thousand other reasons, would suffice to prove that the former of these religions will never long predominate in a cultivated and democratic age, whilst the latter is destined to retain its sway at these as at all other periods.

But in continuation of this branch of the subject, I find that in order for religions to maintain their authority, humanly speaking, in democratic ages, they must not only confine themselves strictly within the circle of spiritual matters: their power also depends very much on the nature of the belief they inculcate, on the external forms they assume, and on the obligations they impose. The preceding observation, that equality leads men to very general and very extensive notions, is principally to be understood as applied to the question of religion. Men living in a similar and equal condition in the world readily conceive the idea of the one God, governing every man by the same laws, and granting to every man future happiness on the same conditions. The idea of the unity of mankind constantly leads them back to the idea of the unity of the Creator; whilst, on the contrary, in a state of society where men are broken up into very unequal ranks, they are apt to devise as many deities as there are nations, castes, classes, or families, and to trace a thousand private roads to Heaven.

It cannot be denied that Christianity itself has felt, to a certain extent, the influence which social and political conditions exercise on religious opinions. At the epoch at which the Christian religion appeared upon earth, Providence, by whom the world was doubtless prepared for its coming, had gathered a large portion of the human race, like an immense flock, under the sceptre of the Cæsars. The men of whom this multitude was composed were distinguished by numerous differences; but they had thus much in common, that they all obeyed the same laws, and that every subject was so weak and insignificant in relation to the imperial potentate, that all appeared equal when their condition was contrasted with his. This novel and peculiar state of mankind necessarily predisposed men to listen to the general truths which Christianity teaches, and may serve to explain the facility and rapidity with which they then penetrated into the human mind. The counterpart of this state of things was exhibited after the destruction of the Empire. The Roman World being then as it were shattered into a thousand fragments, each nation resumed its pristine individuality. An infinite scale of ranks very soon grew up in the bosom of these nations; the different races were more sharply defined, and each nation was divided by castes into several peoples. In the midst of this common effort, which seemed to be urging human society to the greatest conceivable amount of voluntary subdivision, Christianity did not lose sight of the leading general ideas which it had brought into the world. But it appeared, nevertheless, to lend itself, as much as was possible, to those new tendencies to which the fractional distribution of mankind had given birth. Men continued to worship an only God, the Creator and Preserver of all things; but every people, every city, and, so to speak, every man, thought to obtain some distinct privilege, and win the favour of an especial patron at the foot of the Throne of Grace. Unable to subdivide the Deity, they multiplied and improperly enhanced the importance of the divine agents. The homage due to Saints and Angels became almost idolatrous worship amongst the majority of the Christian world; and apprehensions might be entertained for a moment lest the religion of Christ should retrograde towards the superstitions which it had subdued. It seems evident, that the more the barriers are removed which separate nation from nation amongst mankind, and citizen from citizen amongst a people, the stronger is the bent of the human mind, as if by its own impulse, towards the idea of an only and all-powerful Being, dispensing equal laws in the same manner to every man. In democratic ages, then, it is more particularly important not to allow the homage paid to secondary agents to be confounded with the worship due to the Creator alone.

Another truth is no less clear—that religions ought to assume fewer external observances in democratic periods than at any others. In speaking of philosophical method among the Americans, I have shown that nothing is more repugnant to the human mind in an age of equality than the idea of subjection to forms. Men living at such times are impatient of figures; to their eyes symbols appear to be the puerile artifice which is used to conceal or to set off truths, which should more naturally be bared to the light of open day: they are unmoved by ceremonial observances, and they are predisposed to attach a secondary importance to the details of public worship. Those whose care it is to regulate the external forms of religion in a democratic age should pay a close attention to these natural propensities of the human mind, in order not unnecessarily to run counter to them. I firmly believe in the necessity of forms, which fix the human mind in the contemplation of abstract truths, and stimulate its ardour in the pursuit of them, whilst they invigorate its powers of retaining them steadfastly. Nor do I suppose that it is possible to maintain a religion without external observances; but, on the other hand, I am persuaded that, in the ages upon which we are entering, it would be peculiarly dangerous to multiply them beyond measure; and that they ought rather to be limited to as much as is absolutely necessary to perpetuate the doctrine itself, which is the substance of religions of which the ritual is only the form.[1] A religion which should become more minute, more peremptory, and more surcharged with small observances at a time in which men are becoming more equal, would soon find itself reduced to a band of fanatical zealots in the midst of an infidel people.

I anticipate the objection, that as all religions have general and eternal truths for their object, they cannot thus shape themselves to the shifting spirit of every age without forfeiting their claim to certainty in the eyes of mankind. To this I reply again, that the principal opinions which constitute belief, and which theologians call articles of faith, must be very carefully distinguished from the accessories connected with them. Religions are obliged to hold fast to the former, whatever be the peculiar spirit of the age; but they should take good care not to bind themselves in the same manner to the latter at a time when everything is in transition, and when the mind, accustomed to the moving pageant of human affairs, reluctantly endures the attempt to fix it to any given point. The fixity of external and secondary things can only afford a chance of duration when civil society is itself fixed; under any other circumstances I hold it to be perilous.

We shall have occasion to see that, of all the passions which originate in, or are fostered by, equality, there is one which it renders peculiarly intense, and which it infuses at the same time into the heart of every man: I mean the love of well-being. The taste for well-being is the prominent and indelible feature of democratic ages. It may be believed that a religion which should undertake to destroy so deep-seated a passion, would meet its own destruction thence in the end; and if it attempted to wean men entirely from the contemplation of the good things of this world, in order to devote their faculties exclusively to the thought of another, it may be foreseen that the soul would at length escape from its grasp, to plunge into the exclusive enjoyment of present and material pleasures. The chief concern of religions is to purify, to regulate, and to restrain the excessive and exclusive taste for well-being which men feel at periods of equality; but they would err in attempting to control it completely or to eradicate it. They will not succeed in curing men of the love of riches: but they may still persuade men to enrich themselves by none but honest means.

This brings me to a final consideration, which comprises, as it were, all the others. The more the conditions of men are equalised and assimilated to each other, the more important is it for religions, whilst they carefully abstain from the daily turmoil of secular affairs, not needlessly to run counter to the ideas which generally prevail, and the permanent interests which exist in the mass of the people. For as public opinion grows to be more and more evidently the first and most irresistible of existing powers, the religious principle has no external support strong enough to enable it long to resist its attacks. This is not less true of a democratic people, ruled by a despot, than in a republic. In ages of equality, kings may often command obedience, but the majority always commands belief: to the majority, therefore, deference is to be paid in whatsoever is not contrary to the faith.

I showed in my former volumes how the American clergy stand aloof from secular affairs. This is the most obvious, but it is not the only, example of their self-restraint. In America religion is a distinct sphere, in which the priest is sovereign, but out of which he takes care never to go. Within its limits he is the master of the mind; beyond them, he leaves men to themselves, and surrenders them to the independence and instability which belong to their nature and their age. I have seen no country in which Christianity is clothed with fewer forms, figures, and observances than in the United States; or where it presents more distinct, more simple, or more general notions to the mind. Although the Christians of America are divided into a multitude of sects, they all look upon their religion in the same light. This applies to Roman Catholicism as well as to the other forms of belief. There are no Romish priests who show less taste for the minute individual observances, for extraordinary or peculiar means of salvation, or who cling more to the spirit, and less to the letter of the law, than the Roman Catholic priests of the United States. Nowhere is that doctrine of the Church, which prohibits the worship reserved to God alone from being offered to the saints, more clearly inculcated or more generally followed. Yet the Roman Catholics of America are very submissive and very sincere.

Another remark is applicable to the clergy of every communion. The American ministers of the gospel do not attempt to draw or to fix all the thoughts of man upon the life to come; they are willing to surrender a portion of his heart to the cares of the present; seeming to consider the goods of this world as important, although as secondary, objects. If they take no part themselves in productive labour, they are at least interested in its progression, and ready to applaud its results; and whilst they never cease to point to the other world as the great object of the hopes and fears of the believer, they do not forbid him honestly to court prosperity in this. Far from attempting to show that these things are distinct and contrary to one another, they study rather to find out on what point they are most nearly and closely connected.

All the American clergy know and respect the intellectual supremacy exercised by the majority; they never sustain any but necessary conflicts with it. They take no share in the altercations of parties, but they readily adopt the general opinions of their country and their age; and they allow themselves to be borne away without opposition in the current of feeling and opinion by which everything around them is carried along. They endeavour to amend their contemporaries, but they do not quit fellowship with them. Public opinion is therefore never hostile to them; it rather supports and protects them; and their belief owes its authority at the same time to the strength which is its own, and to that which they borrow from the opinions of the majority.

Thus it is that, by respecting all democratic tendencies not absolutely contrary to herself, and by making use of several of them for her own purposes, Religion sustains an advantageous struggle with that spirit of individual independence which is her most dangerous antagonist.

Chapter VI

Of the Progress of Roman Catholicism in the United States

America is the most democratic country in the world, and it is at the same time (according to reports worthy of belief) the country in which the Roman Catholic religion makes most progress. At first sight this is surprising. Two things must here be accurately distinguished: equality inclines men to wish to form their own opinions; but, on the other hand, it imbues them with the taste and the idea of unity, simplicity, and impartiality in the power which governs society. Men living in democratic ages are therefore very prone to shake off all religious authority; but if they consent to subject themselves to any authority of this kind, they choose at least that it should be single and uniform. Religious powers not radiating from a common centre are naturally repugnant to their minds; and they almost as readily conceive that there should be no religion, as that there should be several. At the present time, more than in any preceding one, Roman Catholics are seen to lapse into infidelity, and Protestants to be converted to Roman Catholicism. If the Roman Catholic faith be considered within the pale of the Church, it would seem to be losing ground; without that pale, to be gaining it. Nor is this circumstance difficult of explanation. The men of our days are naturally disposed to believe; but, as soon as they have any religion, they immediately find in themselves a latent propensity which urges them unconsciously towards Catholicism. Many of the doctrines and the practices of the Romish Church astonish them; but they feel a secret admiration for its discipline, and its great unity attracts them. If Catholicism could at length withdraw itself from the political animosities to which it has given rise, I have hardly any doubt but that the same spirit of the age, which appears to be so opposed to it, would become so favourable as to admit of its great and sudden advancement. One of the most ordinary weaknesses of the human intellect is to seek to reconcile contrary principles, and to purchase peace at the expense of logic. Thus there have ever been, and will ever be, men who, after having submitted some portion of their religious belief to the principle of authority, will seek to exempt several other parts of their faith from its influence, and to keep their minds floating at random between liberty and obedience. But I am inclined to believe that the number of these thinkers will be less in democratic than in other ages; and that our posterity will tend more and more to a single division into two parts—some relinquishing Christianity entirely, and others returning to the bosom of the Church of Rome.

Chapter VII

Of the Cause of a Leaning to Pantheism amongst Democratic Nations

I shall take occasion hereafter to show under what form the preponderating taste of a democratic people for very general ideas manifests itself in politics; but I would point out, at the present stage of my work, its principal effect on philosophy. It cannot be denied that pantheism has made great progress in our age. The writings of a part of Europe bear visible marks of it: the Germans introduce it into philosophy, and the French into literature. Most of the works of imagination published in France contain some opinions or some tinge caught from pantheistical doctrines, or they disclose some tendency to such doctrines in their authors. This appears to me not only to proceed from an accidental, but from a permanent cause.

When the conditions of society are becoming more equal, and each individual man becomes more like all the rest, more weak and more insignificant, a habit grows up of ceasing to notice the citizens to consider only the people, and of overlooking individuals to think only of their kind. At such times the human mind seeks to embrace a multitude of different objects at once; and it constantly strives to succeed in connecting a variety of consequences with a single cause. The idea of unity so possesses itself of man, and is sought for by him so universally, that if he thinks he has found it, he readily yields himself up to repose in that belief. Nor does he content himself with the discovery that nothing is in the world but a creation and a Creator; still embarrassed by this primary division of things, he seeks to expand and to simplify his conception by including God and the Universe in one great Whole. If there be a philosophical system which teaches that all things material and immaterial, visible and invisible, which the world contains, are only to be considered as the several parts of an immense Being, which alone remains unchanged amidst the continual change and ceaseless transformation of all that constitutes it, we may readily infer that such a system, although it destroy the individuality of man—nay, rather because it destroys that individuality—will have secret charms for men living in democracies. All their habits of thought prepare them to conceive it, and predispose them to adopt it. It naturally attracts and fixes their imagination; it fosters the pride, whilst it soothes the indolence, of their minds. Amongst the different systems by whose aid Philosophy endeavours to explain the Universe, I believe pantheism to be one of those most fitted to seduce the human mind in democratic ages. Against it all who abide in their attachment to the true greatness of man should struggle and combine.

Chapter VIII

The Principle of Equality Suggests to the Americans the Idea of the Indefinite Perfectibility of Man

Equality suggests to the human mind several ideas which would not have originated from any other source, and it modifies almost all those previously entertained. I take as an example the idea of human perfectibility, because it is one of the principal notions that the intellect can conceive, and because it constitutes of itself a great philosophical theory, which is every instant to be traced by its consequences in the practice of human affairs. Although man has many points of resemblance with the brute creation, one characteristic is peculiar to himself—he improves: they are incapable of improvement. Mankind could not fail to discover this difference from its earliest period. The idea of perfectibility is therefore as old as the world; equality did not give birth to it, although it has imparted to it a novel character.

When the citizens of a community are classed according to their rank, their profession, or their birth, and when all men are constrained to follow the career which happens to open before them, every one thinks that the utmost limits of human power are to be discerned in proximity to himself, and none seeks any longer to resist the inevitable law of his destiny. Not indeed that an aristocratic people absolutely contests man's faculty of self-improvement, but they do not hold it to be indefinite; amelioration they conceive, but not change: they imagine that the future condition of society may be better, but not essentially different; and whilst they admit that mankind has made vast strides in improvement, and may still have some to make, they assign to it beforehand certain impassable limits. Thus they do not presume that they have arrived at the supreme good or at absolute truth (what people or what man was ever wild enough to imagine it?) but they cherish a persuasion that they have pretty nearly reached that degree of greatness and knowledge which our imperfect nature admits of; and as nothing moves about them they are willing to fancy that everything is in its fit place. Then it is that the legislator affects to lay down eternal laws; that kings and nations will raise none but imperishable monuments; and that the present generation undertakes to spare generations to come the care of regulating their destinies.

In proportion as castes disappear and the classes of society approximate—as manners, customs, and laws vary, from the tumultuous intercourse of men—as new facts arise—as new truths are brought to light—as ancient opinions are dissipated, and others take their place—the image of an ideal perfection, for ever on the wing, presents itself to the human mind. Continual changes are then every instant occurring under the observation of every man: the position of some is rendered worse; and he learns but too well, that no people and no individual, how enlightened soever they may be, can lay claim to infallibility;—the condition of others is improved; whence he infers that man is endowed with an indefinite faculty of improvement. His reverses teach him. that none may hope to have discovered absolute good—his success stimulates him to the never-ending pursuit of it. Thus, for ever seeking—for ever falling, to rise again—often disappointed, but not discouraged—he tends unceasingly towards that unmeasured greatness so indistinctly visible at the end of the long track which humanity has yet to tread. It can hardly be believed how many facts naturally flow from the philosophical theory of the indefinite perfectibility of man, or how strong an influence it exercises even on men who, living entirely for the purposes of action and not of thought, seem to conform their actions to it, without knowing anything about it. I accost an American sailor, and I inquire why the ships of his country are built so as to last but for a short time; he answers without hesitation that the art of navigation is every day making such rapid progress, that the finest vessel would become almost useless if it lasted beyond a certain number of years. In these words, which fell accidentally and on a particular subject from a man of rude attainments, I recognise the general and systematic idea upon which a great people directs all its concerns.

Aristocratic nations are naturally too apt to narrow the scope of human perfectibility; democratic nations to expand it beyond compass.

Chapter IX

The Example of the Americans Does Not Prove That a Democratic People Can Have No Aptitude and No Taste for Science, Literature, or Art

It must be acknowledged that amongst few of the civilised nations of our time have the higher sciences made less progress than in the United States; and in fewhave great artists, fine poets, or celebrated writers been more rare. Many Europeans, struck by this fact, have looked upon it as a natural and inevitable result of equality; and they have supposed that if a democratic state of society and democratic institutions were ever to prevail over the whole earth, the human mind would gradually find its beacon-lights grow dim, and men would relapse into a period of darkness. To reason thus is, I think, to confound several ideas which it is important to divide and to examine separately: it is to mingle, unintentionally, what is democratic with what is only American.

The religion professed by the first emigrants, and bequeathed by them to their descendants, simple in its form of worship, austere and almost harsh in its principles, and hostile to external symbols and to ceremonial pomp, is naturally unfavourable to the fine arts, and only yields a reluctant sufferance to the pleasures of literature. The Americans are a very old and a very enlightened people, who have fallen upon a new and unbounded country, where they may extend themselves at pleasure, and which they may fertilise without difficulty. This state of things is without a parallel in the history of the world. In America, then, every one finds facilities, unknown elsewhere, for making or increasing his fortune. The spirit of gain is always on the stretch, and the human mind, constantly diverted from the pleasures of imagination and the labours of the intellect, is there swayed by no impulse but the pursuit of wealth. Not only are manufacturing and commercial classes to be found in the United States, as they are in all other countries; but what never occurred elsewhere, the whole community is simultaneously engaged in productive industry and commerce. I am convinced that, if the Americans had been alone in the world, with the freedom and the knowledge acquired by their forefathers, and the passions which are their own, they would not have been slow to discover that progress cannot long be made in the application of the sciences without cultivating the theory of them; that all the arts are perfected by one another: and, however absorbed they might have been by the pursuit of the principal object of their desires, they would speedily have admitted, that it is necessary to turn aside from it occasionally, in order the better to attain it in the end.

The taste for the pleasures of the mind is moreover so natural to the heart of civilised man, that amongst the polite nations, which are least disposed to give themselves up to these pursuits, a certain number of citizens are always to be found who take part in them. This intellectual craving, when once felt, would very soon have been satisfied. But at the very time when the Americans were naturally inclined to require nothing of science but its special applications to the useful arts and the means of rendering life comfortable, learned and literary Europe was engaged in exploring the common sources of truth, and in improving at the same time all that can minister to the pleasures or satisfy the wants of man. At the head of the enlightened nations of the Old World the inhabitants of the United States more particularly distinguished one, to which they were closely united by a common origin and by kindred habits. Amongst this people they found distinguished men of science, artists of skill, writers of eminence, and they were enabled to enjoy the treasures of the intellect without requiring to labour in amassing them. I cannot consent to separate America from Europe, in spite of the ocean which intervenes. I consider the people of the United States as that portion of the English people which is commissioned to explore the wilds of the New World; whilst the rest of the nation, enjoying more leisure and less harassed by the drudgery of life, may devote its energies to thought, and enlarge in all directions the empire of the mind.

The position of the Americans is therefore quite exceptional, and it may be believed that no democratic people will ever be placed in a similar one. Their strictly Puritanical origin—their exclusively commercial habits—even the country they inhabit, which seems to divert their minds from the pursuit of science, literature, and the arts—the proximity of Europe, which allows them to neglect these pursuits without relapsing into barbarism—a thousand special causes, of which I have only been able to point out the most important—have singularly concurred to fix the mind of the American upon purely practical objects. His passions, his wants, his education, and everything about him seem to unite in drawing the native of the United States earthward: his religion alone bids him turn, from time to time, a transient and distracted glance to heaven. Let us cease then to view all democratic nations under the mask of the American people, and let us attempt to survey them at length with their own proper features.

It is possible to conceive a people not subdivided into any castes or scale of ranks; in which the law, recognising no privileges, should divide inherited property into equal shares; but which, at the same time, should be without knowledge and without freedom. Nor is this an empty hypothesis: a despot may find that it is his interest to render his subjects equal and to leave them ignorant, in order more easily to keep them slaves. Not only would a democratic people of this kind show neither aptitude nor taste for science, literature, or art, but it would probably never arrive at the possession of them. The law of descent would of itself provide for the destruction of fortunes at each succeeding generation; and new fortunes would be acquired by none. The poor man, without either knowledge or freedom, would not so much as conceive the idea of raising himself to wealth; and the rich man would allow himself to be degraded to poverty, without a notion of self-defence. Between these two members of the community complete and invincible equality would soon be established.

No one would then have time or taste to devote himself to the pursuits or pleasures of the intellect; but all men would remain paralysed by a state of common ignorance and equal servitude. When I conceive a democratic society of this kind, I fancy myself in one of those low, close, and gloomy abodes, where the light which breaks in from without soon faints and fades away. A sudden heaviness overpowers me, and I grope through the surrounding darkness, to find the aperture which will restore me to daylight and the air.

But all this is not applicable to men already enlightened who retain their freedom, after having abolished from amongst them those peculiar and hereditary rights which perpetuated the tenure of property in the hands of certain individuals or certain bodies. When men living in a democratic state of society are enlightened, they readily discover that they are confined and fixed within no limits which constrain them to take up with their present fortune. They all therefore conceive the idea of increasing it; if they are free, they all attempt it, but all do not succeed in the same manner. The legislature, it is true, no longer grants privileges, but they are bestowed by nature. As natural inequality is very great, fortunes become unequal as soon as every man exerts all his faculties to get rich. The law of descent prevents the establishment of wealthy families; but it does not prevent the existence of wealthy individuals.

It constantly brings back the members of the community to a common level, from which they as constantly escape: and the inequality of fortunes augments in proportion as knowledge is diffused and liberty increased.

A sect which arose in our time, and was celebrated for its talents and its extravagance, proposed to concentrate all property into the hands of a central power, whose function it should afterwards be to parcel it out to individuals, according to their capacity. This would have been a method of escaping from that complete and eternal equality which seems to threaten democratic society. But it would be a simpler and less dangerous remedy to grant no privilege to any, giving to all equal cultivation and equal independence, and leaving everyone to determine his own position. Natural inequality will very soon make way for itself, and wealth will spontaneously pass into the hands of the most capable.

Free and democratic communities, then, will always contain a considerable number of people enjoying opulence or competency. The wealthy will not be so closely linked to each other as the members of the former aristocratic class of society: their propensities will be different, and they will scarcely ever enjoy leisure as secure or as complete: but they will be far more numerous than those who belonged to that class of society could ever be. These persons will not be strictly confined to the cares of practical life, and they will still be able, though in different degrees, to indulge in the pursuits and pleasures of the intellect. In those pleasures they will indulge; for if it be true that the human mind leans on one side to the narrow, the practical, and the useful, it naturally rises on the other to the infinite, the spiritual, and the beautiful. Physical wants confine it to the earth; but, as soon as the tie is loosened, it will unbend itself again.

Not only will the number of those who can take an interest in the productions of the mind be enlarged, but the taste for intellectual enjoyment will descend, step by step, even to those who, in aristocratic societies, seem to have neither time nor ability to indulge in them. When hereditary wealth, the privileges of rank, and the prerogatives of birth have ceased to be, and when every man derives his strength from himself alone, it becomes evident that the chief cause of disparity between the fortunes of men is the mind. Whatever tends to invigorate, to extend, or to adorn the mind, instantly rises to great value. The utility of knowledge becomes singularly conspicuous even to the eyes of the multitude: those who have no taste for its charms set store upon its results, and make some efforts to acquire it.

In free and enlightened democratic ages, there is nothing to separate men from each other or to retain them in their peculiar sphere; they rise or sink with extreme rapidity. All classes live in perpetual intercourse from their great proximity to each other. They communicate and intermingle every day—they imitate and envy one other: this suggests to the people many ideas, notions, and desires which it would never have entertained if the distinctions of rank had been fixed and society at rest. In such nations the servant never considers himself as an entire stranger to the pleasures and toils of his master, nor the poor man to those of the rich; the rural population assimilates itself to that of the towns, and the provinces to the capital. No one easily allows himself to be reduced to the mere material cares of life; and the humblest artisan casts at times an eager and a furtive glance into the higher regions of the intellect. People do not read with the same notions or in the same manner as they do in an aristocratic community; but the circle of readers is unceasingly expanded, till it includes all the citizens.

As soon as the multitude begins to take an interest in the labours of the mind, it finds out that to excel in some of them is a powerful method of acquiring fame, power, or wealth. The restless ambition which equality begets instantly takes this direction as it does all others. The number of those who cultivate science, letters, and the arts, becomes immense. The intellectual world starts into prodigious activity: everyone endeavours to open for himself a path there, and to draw the eyes of the public after him. Something analogous occurs to what happens in society in the United States, politically considered. What is done is often imperfect, but the attempts are innumerable; and, although the results of individual effort are commonly very small, the total amount is always very large.

It is therefore not true to assert that men living in democratic ages are naturally indifferent to science, literature, and the arts: only it must be acknowledged that they cultivate them after their own fashion, and bring to the task their own peculiar qualifications and deficiencies.

Chapter X

Why the Americans Are More Addicted to Practical Than to Theoretical Science

If a democratic state of society and democratic institutions do not stop the career of the human mind, they incontestably guide it in one direction in preference to another. Their effects, thus circumscribed, are still exceedingly great; and I trust I may be pardoned if I pause for a moment to survey them. We had occasion, in speaking of the philosophical method of the American people, to make several remarks which must here be turned to account.

Equality begets in man the desire of judging of everything for himself: it gives him, in all things, a taste for the tangible and the real, a contempt for tradition and for forms. These general tendencies are principally discernible in the peculiar subject of this chapter. Those who cultivate the sciences amongst a democratic people are always afraid of losing their way in visionary speculation. They mistrust systems; they adhere closely to facts and the study of facts with their own senses. As they do not easily defer to the mere name of any fellow-man, they are never inclined to rest upon any man's authority; but, on the contrary, they are unremitting in their efforts to point out the weaker points of their neighbours' opinions. Scientific precedents have very little weight with them; they are never long detained by the subtility of the schools, nor ready to accept big words for sterling coin; they penetrate, as far as they can, into the principal parts of the subject which engages them, and they expound them in the vernacular tongue. Scientific pursuits then follow a freer and a safer course, but a less lofty one.

The mind may, as it appears to me, divide science into three parts. The first comprises the most theoretical principles, and those more abstract notions whose application is either unknown or very remote. The second is composed of those general truths which still belong to pure theory, but lead, nevertheless, by a straight and short road to practical results. Methods of application and means of execution make up the third. Each of these different portions of science may be separately cultivated, although reason and experience show that none of them can prosper long, if it be absolutely cut off from the two others.

In America the purely practical part of science is admirably understood, and careful attention is paid to the theoretical portion which is immediately requisite to application. On this head the Americans always display a clear, free, original, and inventive power of mind. But hardly any one in the United States devotes himself to the essentially theoretical and abstract portion of human knowledge. In this respect the Americans carry to excess a tendency which is, I think, discernible, though in a less degree, amongst all democratic nations.

Nothing is more necessary to the culture of the higher sciences, or of the more elevated departments of science, than meditation; and nothing is less suited to meditation than the structure of democratic society. We do not find there, as amongst an aristocratic people, one class which clings to a state of repose because it is well off; and another which does not venture to stir because it despairs of improving its condition. Every one is actively in motion: some in quest of power, others of gain. In the midst of this universal tumult—this incessant conflict of jarring interests—this continual stride of men after fortune—where is that calm to be found which is necessary for the deeper combinations of the intellect? How can the mind dwell upon any single point, when everything whirls around it, and man himself is swept and beaten onwards by the heady current which rolls all things in its course? But the permanent agitation which subsists in the bosom of a peaceable and established democracy, must be distinguished from the tumultuous and revolutionary movements which almost always attend the birth and growth of democratic society. When a violent revolution occurs amongst a highly civilised people, it cannot fail to give a sudden impulse to their feelings and their opinions. This is more particularly true of democratic revolutions, which stir up all the classes of which a people is composed, and beget, at the same time, inordinate ambition in the breast of every member of the community. The French made most surprising advances in the exact sciences at the very time at which they were finishing the destruction of the remains of their former feudal society; yet this sudden fecundity is not to be attributed to democracy, but to the unexampled revolution which attended its growth. What happened at that period was a special incident, and it would be unwise to regard it as the test of a general principle.

Great revolutions are not more common amongst democratic nations than amongst others: I am even inclined to believe that they are less so. But there prevails amongst those populations a small distressing motion—a sort of incessant jostling of men—which annoys and disturbs the mind, without exciting or elevating it. Men who live in democratic communities not only seldom indulge in meditation, but they naturally entertain very little esteem for it. A democratic state of society and democratic institutions plunge the greater part of men in constant active life; and the habits of mind which are suited to an active life, are not always suited to a contemplative one. The man of action is frequently obliged to content himself with the best he can get, because he would never accomplish his purpose if he chose to carry every detail to perfection. He has perpetually occasion to rely on ideas which he has not had leisure to search to the bottom; for he is much more frequently aided by the opportunity of an idea than by its strict accuracy; and, in the long run, he risks less in making use of some false principles, than in spending his time in establishing all his principles on the basis of truth. The world is not led by long or learned demonstrations; a rapid glance at particular incidents, the daily study of the fleeting passions of the multitude, the accidents of the time, and the art of arming them to account, decide all its affairs.

In the ages in which active life is the condition of almost everyone, men are therefore generally led to attach an excessive value to the rapid bursts and superficial conceptions of the intellect; and, on the other hand, to depreciate below their true standard its slower and deeper labours. This opinion of the public influences the judgment of the men who cultivate the sciences; they are persuaded that they may succeed in those pursuits without meditation, or deterred from such pursuits as demand it.

There are several methods of studying the sciences. Amongst a multitude of men you will find a selfish, mercantile, and trading taste for the discoveries of the mind, which must not be confounded with that disinterested passion which is kindled in the heart of the few. A desire to utilise knowledge is one thing; the pure desire to know is another. I do not doubt that in a few minds and far between, an ardent, inexhaustible love of truth springs up, self-supported, and living in ceaseless fruition without ever attaining the satisfaction which it seeks. This ardent love it is—this proud, disinterested love of what is true—which raises men to the abstract sources of truth, to draw their mother-knowledge thence. If Pascal had had nothing in view but some large gain, or even if he had been stimulated by the love of fame alone, I cannot conceive that he would ever have been able to rally all the powers of his mind, as he did, for the better discovery of the most hidden things of the Creator. When I see him, as it were, tear his soul from the midst of all the cares of life to devote it wholly to these researches, and, prematurely snapping the links which bind the frame to life, die of old age before forty, I stand amazed, and I perceive that no ordinary cause is at work to produce efforts so extraordinary.

The future will prove whether these passions, at once so rare and so productive, come into being and into growth as easily in the midst of democratic as in aristocratic communities. For myself, I confess that I am slow to believe it. In aristocratic society, the class which gives the tone to opinion, and has the supreme guidance of affairs, being permanently and hereditarily placed above the multitude, naturally conceives a lofty idea of itself and of man. It loves to invent for him noble pleasures, to carve out splendid objects for his ambition. Aristocracies often commit very tyrannical and very inhuman actions; but they rarely entertain grovelling thoughts; and they show a kind of haughty contempt of little pleasures, even whilst they indulge in them. The effect is greatly to raise the general pitch of society. In aristocratic ages vast ideas are commonly entertained of the dignity, the power, and the greatness of man. These opinions exert their influence on those who cultivate the sciences, as well as on the rest of the community. They facilitate the natural impulse of the mind to the highest regions of thought, and they naturally prepare it to conceive a sublime—nay, almost a divine—love of truth. Men of science at such periods are consequently carried away by theory; and it even happens that they frequently conceive an inconsiderate contempt for the practical part of learning. 'Archimedes,' says Plutarch, 'was of so lofty a spirit, that he never condescended to write any treatise on the manner of constructing all these engines of offence and defence. And as he held this science of inventing and putting together engines, and all arts generally speaking which tended to any useful end in practice, to be vile, low, and mercenary, he spent his talents and his studious hours in writing of those things only whose beauty and subtilty had in them no admixture of necessity.' Such is the aristocratic aim of science; in democratic nations it cannot be the same.

The greater part of the men who constitute these nations are extremely eager in the pursuit of actual and physical gratification. As they are always dissatisfied with the position which they occupy, and are always free to leave it, they think of nothing but the means of changing their fortune, or of increasing it. To minds thus predisposed, every new method which leads by a shorter road to wealth, every machine which spares labour, every instrument which diminishes the cost of production, every discovery which facilitates pleasures or augments them, seems to be the grandest effort of the human intellect. It is chiefly from these motives that a democratic people addicts itself to scientific pursuits—that it understands, and that it respects them. In aristocratic ages, science is more particularly called upon to furnish gratification to the mind; in democracies, to the body. You may be sure that the more a nation is democratic, enlightened, and free, the greater will be the number of these interested promoters of scientific genius, and the more will discoveries immediately applicable to productive industry confer gain, fame, and even power on their authors. For in democracies the working class takes a part in public affairs; and public honours, as well as pecuniary remuneration, may be awarded to those who deserve them. In a community thus organised it may easily be conceived that the human mind may be led insensibly to the neglect of theory; and that it is urged, on the contrary, with unparalleled vehemence to the applications of science, or at least to that portion of theoretical science which is necessary to those who make such applications. In vain will some innate propensity raise the mind towards the loftier spheres of the intellect; interest draws it down to the middle zone. There it may develop all its energy and restless activity, there it may engender all its wonders. These very Americans, who have not discovered one of the general laws of mechanics, have introduced into navigation an engine which changes the aspect of the world.

Assuredly I do not contend that the democratic nations of our time are destined to witness the extinction of the transcendent luminaries of man's intelligence, nor even that no new lights will ever start into existence. At the age at which the world has now arrived, and amongst so many cultivated nations, perpetually excited by the fever of productive industry, the bonds which connect the different parts of science together cannot fail to strike the observation; and the taste for practical science itself, if it be enlightened, ought to lead men not to neglect theory. In the midst of such numberless attempted applications of so many experiments, repeated every day, it is almost impossible that general laws should not frequently be brought to light; so that great discoveries would be frequent, though great inventors be rare. I believe, moreover, in the high calling of scientific minds. If the democratic principle does not, on the one hand, induce men to cultivate science for its own sake, on the other it enormously increases the number of those who do cultivate it. Nor is it credible that, from amongst so great a multitude no speculative genius should from time to time arise, inflamed by the love of truth alone. Such a one, we may be sure, would dive into the deepest mysteries of nature, whatever be the spirit of his country or his age. He requires no assistance in his course—enough that he be not checked in it.

All that I mean to say is this:—permanent inequality of conditions leads men to confine themselves to the arrogant and sterile research of abstract truths; whilst the social condition and the institutions of democracy prepare them to seek the immediate and useful practical results of the sciences. This tendency is natural and inevitable: it is curious to be acquainted with it, and it may be necessary to point it out. If those who are called upon to guide the nations of our time clearly discerned from afar off these new tendencies, which will soon be irresistible, they would understand that, possessing education and freedom, men living in democratic ages cannot fail to improve the industrial part of science; and that henceforward all the efforts of the constituted authorities ought to be directed to support the highest branches of learning, and to foster the nobler passion for science itself. In the present age the human mind must be coerced into theoretical studies; it runs of its own accord to practical applications; and, instead of perpetually referring it to the minute examination of secondary effects, it is well to divert it from them sometimes, in order to raise it up to the contemplation of primary causes. Because the civilisation of ancient Rome perished in consequence of the invasion of the Barbarians, we are perhaps too apt to think that civilisation cannot perish in any other manner. If the light by which we are guided is ever extinguished, it will dwindle by degrees, and expire of itself. By dint of close adherence to mere applications, principles would be lost sight of; and when the principles were wholly forgotten, the methods derived from them would be ill pursued. New methods could no longer be invented, and men would continue to apply, without intelligence, and without art, scientific processes no longer understood.

When Europeans first arrived in China, three hundred years ago, they found that almost all the arts had reached a certain degree of perfection there; and they were surprised that a people which had attained this point should not have gone beyond it. At a later period they discovered some traces of the higher branches of science which were lost. The nation was absorbed in productive industry: the greater part of its scientific processes had been preserved, but science itself no longer existed there. This served to explain the strangely motionless state in which they found the minds of this people. The Chinese, in following the track of their forefathers, had forgotten the reasons by which the latter had been guided. They still used the formula, without asking for its meaning: they retained the instrument, but they no longer possessed the art of altering or renewing it. The Chinese, then, had lost the power of change; for them to improve was impossible. They were compelled, at all times and in all points, to imitate their predecessors, lest they should stray into utter darkness, by deviating for an instant from the path already laid down for them. The source of human knowledge was all but dry; and though the stream still ran on, it could neither swell its waters nor alter its channel. Notwithstanding this, China had subsisted peaceably for centuries. The invaders who had conquered the country assumed the manners of the inhabitants, and order prevailed there. A sort of physical prosperity was everywhere discernible: revolutions were rare, and war was, so to speak, unknown.

It is then a fallacy to flatter ourselves with the reflection that the Barbarians are still far from us; for if there be some nations which allow civilisation to be torn from their grasp, there are others who trample it themselves under their feet.

Chapter XI

Of the Spirit in Which the Americans Cultivate the Arts

It would be to waste the time of my readers and my own if I strove to demonstrate how the general mediocrity of fortunes, the absence of superfluous wealth, the universal desire of comfort, and the constant efforts by which everyone attempts to procure it, make the taste for the useful predominate over the love of the beautiful in the heart of man. Democratic nations, amongst which all these things exist, will therefore cultivate the arts which serve to render life easy, in preference to those whose object is to adorn it. They will habitually prefer the useful to the beautiful, and they will require that the beautiful should be useful. But I propose to go further; and after having pointed out this first feature, to sketch several others.

It commonly happens that in the ages of privilege the practice of almost all the arts becomes a privilege; and that every profession is a separate walk, upon which it is not allowable for every one to enter. Even when productive industry is free, the fixed character which belongs to aristocratic nations gradually segregates all the persons who practise the same art, till they form a distinct class, always composed of the same families, whose members are all known to each other, and amongst whom a public opinion of their own and a species of corporate pride soon spring up. In a class or guild of this kind, each artisan has not only his fortune to make, but his reputation to preserve. He is not exclusively swayed by his own interest, or even by that of his customer, but by that of the body to which he belongs; and the interest of that body is, that each artisan should produce the best possible workmanship. In aristocratic ages, the object of the arts is therefore to manufacture as well as possible—not with the greatest despatch, or at the lowest rate.

When, on the contrary, every profession is open to all—when a multitude of persons are constantly embracing and abandoning it—and when its several members are strangers to each other, indifferent, and from their numbers hardly seen amongst themselves; the social tie is destroyed, and each workman, standing alone, endeavours simply to gain the greatest possible quantity of money at the least possible cost. The will of the customer is then his only limit. But at the same time a corresponding revolution takes place in the customer also. In countries in which riches as well as power are concentrated and retained in the hands of the few, the use of the greater part of this world's goods belongs to a small number of individuals, who are always the same. Necessity, public opinion, or moderate desires exclude all others from the enjoyment of them. As this aristocratic class remains fixed at the pinnacle of greatness on which it stands, without diminution or increase, it is always acted upon by the same wants and affected by them in the same manner. The men of whom it is composed naturally derive from their superior and hereditary position a taste for what is extremely well-made and lasting. This affects the general way of thinking of the nation in relation to the arts. It often occurs, among such a people, that even the peasant will rather go without the object he covets, than procure it in a state of imperfection. In aristocracies, then, the handicraftsmen work for only a limited number of very fastidious customers: the profit they hope to make depends principally on the perfection of their workmanship.

Such is no longer the case when, all privileges being abolished, ranks are intermingled, and men are forever rising or sinking upon the ladder of society. Amongst a democratic people a number of citizens always exist whose patrimony is divided and decreasing. They have contracted, under more prosperous circumstances, certain wants, which remain after the means of satisfying such wants are gone; and they are anxiously looking out for some surreptitious method of providing for them. On the other hand, there are always in democracies a large number of men whose fortune is upon the increase, but whose desires grow much faster than their fortunes: and who gloat upon the gifts of wealth in anticipation, long before they have means to command them. Such men are eager to find some short cut to these gratifications, already almost within their reach. From the combination of these causes the result is, that in democracies there are always a multitude of individuals whose wants are above their means, and who are very willing to take up with imperfect satisfaction rather than abandon the object of their desires.

The artisan readily understands these passions, for he himself partakes in them: in an aristocracy he would seek to sell his workmanship at a high price to the few; he now conceives that the more expeditious way of getting rich is to sell them at a low price to all. But there are only two ways of lowering the price of commodities. The first is to discover some better, shorter, and more ingenious method of producing them: the second is to manufacture a larger quantity of goods, nearly similar, but of less value. Amongst a democratic population, all the intellectual faculties of the workman are directed to these two objects: he strives to invent methods which may enable him not only to work better, but quicker and cheaper; or, if he cannot succeed in that, to diminish the intrinsic qualities of the thing he makes, without rendering it wholly unfit for the use for which it is intended. When none but the wealthy had watches, they were almost all very good ones: few are now made which are worth much, but everybody has one in his pocket. Thus the democratic principle not only tends to direct the human mind to the useful arts, but it induces the artisan to produce with greater rapidity a quantity of imperfect commodities, and the consumer to content himself with these commodities.

Not that in democracies the arts are incapable of producing very commendable works, if such be required. This may occasionally be the case, if customers appear who are ready to pay for time and trouble. In this rivalry of every kind of industry—in the midst of this immense competition and these countless experiments, some excellent workmen are formed who reach the utmost limits of their craft. But they have rarely an opportunity of displaying what they can do; they are scrupulously sparing of their powers; they remain in a state of accomplished mediocrity, which condemns itself, and, though it be very well able to shoot beyond the mark before it, aims only at what it hits. In aristocracies, on the contrary, workmen always do all they can; and when they stop, it is because they have reached the limit of their attainments.

When I arrive in a country where I find some of the finest productions of the arts, I learn from this fact nothing of the social condition or of the political constitution of the country. But if I perceive that the productions of the arts are generally of an inferior quality, very abundant and very cheap, I am convinced that, amongst the people where this occurs, privilege is on the decline, and that ranks are beginning to intermingle, and will soon be confounded together.

The handicraftsmen of democratic ages endeavour not only to bring their useful productions within the reach of the whole community, but they strive to give to all their commodities attractive qualities which they do not in reality possess. In the confusion of all ranks everyone hopes to appear what he is not, and makes great exertions to succeed in this object. This sentiment indeed, which is but too natural to the heart of man, does not originate in the democratic principle; but that principle applies it to material objects. To mimic virtue is of every age; but the hypocrisy of luxury belongs more particularly to the ages of democracy.

To satisfy these new cravings of human vanity the arts have recourse to every species of imposture: and these devices sometimes go so far as to defeat their own purpose. Imitation diamonds are now made which may be easily mistaken for real ones; as soon as the art of fabricating false diamonds shall have reached so high a degree of perfection that they cannot be distinguished from real ones, it is probable that both one and the other will be abandoned, and become mere pebbles again.

This leads me to speak of those arts which are called the fine arts, by way of distinction. I do not believe that it is a necessary effect of a democratic social condition and of democratic institutions to diminish the number of men who cultivate the fine arts; but these causes exert a very powerful influence on the manner in which these arts are cultivated. Many of those who had already contracted a taste for the fine arts are impoverished: on the other hand, many of those who are not yet rich begin to conceive that taste, at least by imitation; and the number of consumers increases, but opulent and fastidious consumers become more scarce. Something analogous to what I have already pointed out in the useful arts then takes place in the fine arts; the productions of artists are more numerous, but the merit of each production is diminished. No longer able to soar to what is great, they cultivate what is pretty and elegant; and appearance is more attended to than reality. In aristocracies a few great pictures are produced; in democratic countries, a vast number of insignificant ones. In the former, statues are raised of bronze; in the latter, they are modelled in plaster.

When I arrived for the first time at New York, by that part of the Atlantic Ocean which is called the Narrows, I was surprised to perceive along the shore, at some distance from the city, a considerable number of little palaces of white marble, several of which were built after the models of ancient architecture. When I went the next day to inspect more closely the building which had particularly attracted my notice, I found that its walls were of whitewashed brick, and its columns of painted wood. All the edifices which I had admired the night before were of the same kind.

The social condition and the institutions of democracy impart, moreover, certain peculiar tendencies to all the imitative arts, which it is easy to point out. They frequently withdraw them from the delineation of the soul to fix them exclusively on that of the body: and they substitute the representation of motion and sensation for that of sentiment and thought: in a word, they put the Real in the place of the Ideal. I doubt whether Raphael studied the minutest intricacies of the mechanism of the human body as thoroughly as the draughtsmen of our own time. He did not attach the same importance to rigorous accuracy on this point as they do, because he aspired to surpass nature. He sought to make of man something which should be superior to man, and to embellish beauty's self. David and his scholars were, on the contrary, as good anatomists as they were good painters. They wonderfully depicted the models which they had before their eyes, but they rarely imagined anything beyond them: they followed nature with fidelity: whilst Raphael sought for something better than nature. They have left us an exact portraiture of man; but he discloses in his works a glimpse of the Divinity. This remark as to the manner of treating a subject is no less applicable to the choice of it. The painters of the Middle Ages generally sought far above themselves, and away from their own time, for mighty subjects, which left to their imagination an unbounded range. Our painters frequently employ their talents in the exact imitation of the details of private life, which they have always before their eyes; and they are forever copying trivial objects, the originals of which are only too abundant in nature.

Chapter XII

Why the Americans Raise Some Monuments So Insignificant, and Others So Important

I have just observed, that in democratic ages monuments of the arts tend to become more numerous and less important. I now hasten to point out the exception to this rule. In a democratic community individuals are very powerless; but the State which represents them all, and contains them all in its grasp, is very powerful. Nowhere do citizens appear so insignificant as in a democratic nation; nowhere does the nation itself appear greater, or does the mind more easily take in a wide general survey of it. In democratic communities the imagination is compressed when men consider themselves; it expands indefinitely when they think of the State. Hence it is that the same men who live on a small scale in narrow dwellings, frequently aspire to gigantic splendour in the erection of their public monuments.

The Americans traced out the circuit of an immense city on the site which they intended to make their capital, but which, up to the present time, is hardly more densely peopled than Pontoise, though, according to them, it will one day contain a million of inhabitants. They have already rooted up trees for ten miles round, lest they should interfere with the future citizens of this imaginary metropolis. They have erected a magnificent palace for Congress in the centre of the city, and have given it the pompous name of the Capitol. The several States of the Union are every day planning and erecting for themselves prodigious undertakings, which would astonish the engineers of the great European nations. Thus democracy not only leads men to a vast number of inconsiderable productions; it also leads them to raise some monuments on the largest scale: but between these two extremes there is a blank. A few scattered remains of enormous buildings can therefore teach us nothing of the social condition and the institutions of the people by whom they were raised.

I may add, though the remark leads me to step out of my subject, that they do not make us better acquainted with its greatness, its civilisation, and its real prosperity. Whensoever a power of any kind shall be able to make a whole people cooperate in a single undertaking, that power, with a little knowledge and a great deal of time, will succeed in obtaining something enormous from the cooperation of efforts so multiplied. But this does not lead to the conclusion that the people was very happy, very enlightened, or even very strong.

The Spaniards found the City of Mexico full of magnificent temples and vast palaces; but that did not prevent Cortes from conquering the Mexican Empire with six hundred foot soldiers and sixteen horses. If the Romans had been better acquainted with the laws of hydraulics, they would not have constructed all the aqueducts which surround the ruins of their cities—they would have made a better use of their power and their wealth. If they had invented the steam-engine, perhaps they would not have extended to the extremities of their empire those long artificial roads which are called Roman roads. These things are at once the splendid memorials of their ignorance and of their greatness. A people which should leave no other vestige of its track than a few leaden pipes in the earth and a few iron rods upon its surface, might have been more the master of Nature than the Romans.

Chapter XIII

Literary Characteristics of Democratic Ages

When a traveller goes into a bookseller's shop in the United States, and examines the American books upon the shelves, the number of works appears extremely great; whilst that of known authors appears, on the contrary, to be extremely small. He will first meet with a number of elementary treatises, destined to teach the rudiments of human knowledge. Most of these books are written in Europe; the Americans reprint them, adapting them to their own country. Next comes an enormous quantity of religious works, Bibles, sermons, edifying anecdotes, controversial divinity, and reports of charitable societies; lastly, appears the long catalogue of political pamphlets. In America, parties do not write books to combat each others' opinions, but pamphlets which are circulated for a day with incredible rapidity, and then expire. In the midst of all these obscure productions of the human brain are to be found the more remarkable works of that small number of authors, whose names are, or ought to be, known to Europeans.

Although America is perhaps in our days the civilised country in which literature is least attended to, a large number of persons are nevertheless to be found there who take an interest in the productions of the mind, and who make them, if not the study of their lives, at least the charm of their leisure hours. But England supplies these readers with the larger portion of the books which they require. Almost all important English books are republished in the United States. The literary genius of Great Britain still darts its rays into the recesses of the forests of the New World. There is hardly a pioneer's hut which does not contain a few odd volumes of Shakespeare. I remember that I read the feudal play of Henry V. for the first time in a log-house.

Not only do the Americans constantly draw upon the treasures of English literature, but it may be said with truth that they find the literature of England growing on their own soil. The larger part of that small number of men in the United States who are engaged in the composition of literary works are English in substance, and still more so in form. Thus they transport into the midst of democracy the ideas and literary fashions which are current amongst the aristocratic nation they have taken for their model. They paint with colours borrowed from foreign manners; and as they hardly ever represent the country they were born in as it really is, they are seldom popular there. The citizens of the United States are themselves so convinced that it is not for them that books are published, that before they can make up their minds upon the merit of one of their authors, they generally wait till his fame has been ratified in England, just as in pictures the author of an original is held to be entitled to judge of the merit of a copy. The inhabitants of the United States have then at present, properly speaking, no literature. The only authors whom I acknowledge as American are the journalists. They indeed are not great writers, but they speak the language of their countrymen, and make themselves heard by them. Other authors are aliens; they are to the Americans what the imitators of the Greeks and Romans were to us at the revival of learning—an object of curiosity, not of general sympathy. They amuse the mind, but they do not act upon the manners of the people.

I have already said that this state of things is very far from originating in democracy alone, and that the causes of it must be sought for in several peculiar circumstances independent of the democratic principle. If the Americans, retaining the same laws and social condition, had had a different origin, and had been transported into another country, I do not question that they would have had a literature. Even as they now are, I am convinced that they will ultimately have one; but its character will be different from that which marks the American literary productions of our time, and that character will be peculiarly its own. Nor is it impossible to trace this character beforehand.

I suppose an aristocratic people amongst whom letters are cultivated; the labours of the mind, as well as the affairs of state, are conducted by a ruling class in society. The literary as well as the political career is almost entirely confined to this class, or to those nearest to it in rank. These premises suffice to give me a key to all the rest. When a small number of the same men are engaged at the same time upon the same objects, they easily concert with one another, and agree upon certain leading rules which are to govern them each and all. If the object which attracts the attention of these men is literature, the productions of the mind will soon be subjected by them to precise canons, from which it will no longer be allowable to depart. If these men occupy a hereditary position in the country, they will be naturally inclined, not only to adopt a certain number of fixed rules for themselves, but to follow those which their forefathers laid down for their own guidance; their code will be at once strict and traditional. As they are not necessarily engrossed by the cares of daily life—as they have never been so, any more than their fathers were before them—they have learned to take an interest, for several generations back, in the labours of the mind. They have learned to understand literature as an art, to love it in the end for its own sake, and to feel a scholar-like satisfaction in seeing men conform to its rules. Nor is this all: the men of whom I speak began and will end their lives in easy or in affluent circumstances; hence they have naturally conceived a taste for choice gratifications, and a love of refined and delicate pleasures. Nay more, a kind of indolence of mind and heart, which they frequently contract in the midst of this long and peaceful enjoyment of so much welfare, leads them to put aside, even from their pleasures, whatever might be too startling or too acute. They had rather be amused than intensely excited; they wish to be interested, but not to be carried away.

Now let us fancy a great number of literary performances executed by the men, or for the men, whom I have just described, and we shall readily conceive a style of literature in which everything will be regular and pre-arranged. The slightest work will be carefully touched in its least details; art and labour will be conspicuous in everything; each kind of writing will have rules of its own, from which it will not be allowed to swerve, and which distinguish it from all others. Style will be thought of almost as much importance as thought; and the form will be no less considered than the matter: the diction will be polished, measured, and uniform. The tone of the mind will be always dignified, seldom very animated; and writers will care more to perfect what they produce than to multiply their productions. It will sometimes happen that the members of the literary class, always living amongst themselves and writing for themselves alone, will lose sight of the rest of the world, which will infect them with a false and laboured style; they will lay down minute literary rules for their exclusive use, which will insensibly lead them to deviate from common sense, and finally to transgress the bounds of nature. By dint of striving after a mode of parlance different from the vulgar, they will arrive at a sort of aristocratic jargon, which is hardly less remote from pure language than is the coarse dialect of the people. Such are the natural perils of literature amongst aristocracies. Every aristocracy which keeps itself entirely aloof from the people becomes impotent—a fact which is as true in literature as it is in politics.[1]

Let us now turn the picture and consider the other side of it; let us transport ourselves into the midst of a democracy, not unprepared by ancient traditions and present culture to partake in the pleasures of the mind. Ranks are there intermingled and confounded; knowledge and power are both infinitely subdivided, and, if I may use the expression, scattered on every side. Here then is a motley multitude, whose intellectual wants are to be supplied. These new votaries of the pleasures of the mind have not all received the same education; they do not possess the same degree of culture as their fathers, nor any resemblance to them—nay, they perpetually differ from themselves, for they live in a state of incessant change of place, feelings, and fortunes. The mind of each member of the community is therefore unattached to that of his fellow-citizens by tradition or by common habits; and they have never had the power, the inclination, nor the time to concert together. It is, however, from the bosom of this heterogeneous and agitated mass that authors spring; and from the same source their profits and their fame are distributed. I can without difficulty understand that, under these circumstances, I must expect to meet in the literature of such a people with but few of those strict conventional rules which are admitted by readers and by writers in aristocratic ages. If it should happen that the men of some one period were agreed upon any such rules, that would prove nothing for the following period; for amongst democratic nations each new generation is a new people. Amongst such nations, then, literature will not easily be subjected to strict rules, and it is impossible that any such rules should ever be permanent.

In democracies it is by no means the case that all the men who cultivate literature have received a literary education; and most of those who have some tinge of belles-lettres are either engaged in politics, or in a profession which only allows them to taste occasionally and by stealth the pleasures of the mind. These pleasures, therefore, do not constitute the principal charm of their lives; but they are considered as a transient and necessary recreation amidst the serious labours of life. Such man can never acquire a sufficiently intimate knowledge of the art of literature to appreciate its more delicate beauties; and the minor shades of expression must escape them. As the time they can devote to letters is very short, they seek to make the best use of the whole of it. They prefer books which may be easily procured, quickly read, and which require no learned researches to be understood. They ask for beauties, self-proffered and easily enjoyed; above all, they must have what is unexpected and new. Accustomed to the struggle, the crosses, and the monotony of practical life, they require rapid emotions, startling passages—truths or errors brilliant enough to rouse them up, and to plunge them at once, as if by violence, into the midst of a subject.

Why should I say more? or who does not understand what is about to follow, before I have expressed it? Taken as a whole, literature in democratic ages can never present, as it does in the periods of aristocracy, an aspect of order, regularity, science, and art; its form will, on the contrary, ordinarily be slighted, sometimes despised. Style will frequently be fantastic, incorrect, overburdened, and loose—almost always vehement and bold. Authors will aim at rapidity of execution, more than at perfection of detail. Small productions will be more common than bulky books; there will be more wit than erudition, more imagination than profundity; and literary performances will bear marks of an untutored and rude vigour of thought—frequently of great variety and singular fecundity. The object of authors will be to astonish rather than to please, and to stir the passions more than to charm the taste. Here and there, indeed, writers will doubtless occur who will choose a different track, and who will, if they are gifted with superior abilities, succeed in finding readers, in spite of their defects or their better qualities; but these exceptions will be rare, and even the authors who shall so depart from the received practice in the main subject of their works, will always relapse into it in some lesser details.

I have just depicted two extreme conditions: the transition by which a nation passes from the former to the latter is not sudden but gradual, and marked with shades of very various intensity. In the passage which conducts a lettered people from the one to the other, there is almost always a moment at which the literary genius of democratic nations has its confluence with that of aristocracies, and both seek to establish their joint sway over the human mind. Such epochs are transient, but very brilliant: they are fertile without exuberance, and animated without confusion. The French literature of the eighteenth century may serve as an example.

I should say more than I mean if I were to assert that the literature of a nation is always subordinate to its social condition and its political constitution. I am aware that, independently of these causes, there are several others which confer certain characteristics on literary productions; but these appear to me to be the chief. The relations which exist between the social and political condition of a people and the genius of its authors are always very numerous: whoever knows the one is never completely ignorant of the other.

Chapter XIV

The Trade of Literature

Democracy not only infuses a taste for letters among the trading classes, but introduces a trading spirit into literature. In aristocracies, readers are fastidious and few in number; in democracies, they are far more numerous and far less difficult to please. The consequence is, that among aristocratic nations, no one can hope to succeed without immense exertions, and that these exertions may bestow a great deal of fame, but can never earn much money; whilst among democratic nations, a writer may flatter himself that he will obtain at a cheap rate a meagre reputation and a large fortune. For this purpose he need not be admired; it is enough that he is liked. The ever-increasing crowd of readers, and their continual craving for something new, insure the sale of books which nobody much esteems.

In democratic periods the public frequently treat authors as kings do their courtiers; they enrich, and they despise them. What more is needed by the venal souls which are born in courts, or which are worthy to live there? Democratic literature is always infested with a tribe of writers who look upon letters as a mere trade: and for some few great authors who adorn it you may reckon thousands of idea-mongers.

Chapter XV

The Study of Greek and Latin Literature Peculiarly Useful in Democratic Communities

What was called the People in the most democratic republics of antiquity, was very unlike what we designate by that term. In Athens, all the citizens took part in public affairs; but there were only twenty thousand citizens to more than three hundred and fifty thousand inhabitants. All the rest were slaves, and discharged the greater part of those duties which belong at the present day to the lower or even to the middle classes. Athens, then, with her universal suffrage, was after all merely an aristocratic republic in which all the nobles had an equal right to the government. The struggle between the patricians and plebeians of Rome must be considered in the same light: it was simply an intestine feud between the elder and younger branches of the same family. All the citizens belonged, in fact, to the aristocracy, and partook of its character.

It is moreover to be remarked, that amongst the ancients books were always scarce and dear; and that very great difficulties impeded their publication and circulation. These circumstances concentrated literary tastes and habits amongst a small number of men, who formed a small literary aristocracy out of the choicer spirits of the great political aristocracy. Accordingly nothing goes to prove that literature was ever treated as a trade amongst the Greeks and Romans.

These peoples, which not only constituted aristocracies, but very polished and free nations, of course imparted to their literary productions the defects and the merits which characterise the literature of aristocratic ages. And indeed a very superficial survey of the literary remains of the ancients will suffice to convince us, that if those writers were sometimes deficient in variety, or fertility in their subjects, or in boldness, vivacity, or power of generalisation in their thoughts, they always displayed exquisite care and skill in their details. Nothing in their works seems to be done hastily or at random: every line is written for the eye of the connoisseur, and is shaped after some conception of ideal beauty. No literature places those fine qualities, in which the writers of democracies are naturally deficient, in bolder relief than that of the ancients; no literature, therefore, ought to be more studied in democratic ages. This study is better suited than any other to combat the literary defects inherent in those ages; as for their more praiseworthy literary qualities, they will spring up of their own accord, without its being necessary to learn to acquire them.

It is important that this point should be clearly understood. A particular study may be useful to the literature of a people, without being appropriate to its social and political wants. If men were to persist in teaching nothing but the literature of the dead languages in a community where everyone is habitually led to make vehement exertions to augment or to maintain his fortune, the result would be a very polished, but a very dangerous, race of citizens. For as their social and political condition would give them every day a sense of wants which their education would never teach them to supply, they would perturb the State, in the name of the Greeks and Romans, instead of enriching it by their productive industry.

It is evident that in democratic communities the interest of individuals, as well as the security of the commonwealth, demands that the education of the greater number should be scientific, commercial, and industrial, rather than literary. Greek and Latin should not be taught in all schools; but it is important that those who by their natural disposition or their fortune are destined to cultivate letters or prepared to relish them, should find schools where a complete knowledge of ancient literature may be acquired, and where the true scholar may be formed. A few excellent universities would do more towards the attainment of this object than a vast number of bad grammar-schools, where superfluous matters, badly learned, stand in the way of sound instruction in necessary studies.

All who aspire to literary excellence in democratic nations, ought frequently to refresh themselves at the springs of ancient literature: there is no more wholesome course for the mind. Not that I hold the literary productions of the ancients to be irreproachable; but I think that they have some especial merits, admirably calculated to counterbalance our peculiar defects. They are a prop on the side on which we are in most danger of falling.

Chapter XVI

The Effect of Democracy on Language

If the reader has rightly understood what I have already said on the subject of literature in general, he will have no difficulty in comprehending that species of influence which a democratic social condition and democratic institutions may exercise over language itself, which is the chief instrument of thought.

American authors may truly be said to live more in England than in their own country; since they constantly study the English writers, and take them every day for their models. But such is not the case with the bulk of the population, which is more immediately subjected to the peculiar causes acting upon the United States. It is not then to the written, but to the spoken language that attention must be paid, if we would detect the modifications which the idiom of an aristocratic people may undergo when it becomes the language of a democracy.

Englishmen of education, and more competent judges than I can be myself of the nicer shades of expression, have frequently assured me that the language of the educated classes in the United States is notably different from that of the educated classes in Great Britain. They complain not only that the Americans have brought into use a number of new words—the difference and the distance between the two countries might suffice to explain that much—but that these new words are more especially taken from the jargon of parties, the mechanical arts, or the language of trade. They assert, in addition to this, that old English words are often used by the Americans in new acceptations; and lastly, that the inhabitants of the United States frequently intermingle their phraseology in the strangest manner, and sometimes place words together which are always kept apart in the language of the mother-country. These remarks, which were made to me at various times by persons who appeared to be worthy of credit, led me to reflect upon the subject; and my reflections brought me, by theoretical reasoning, to the same point at which my informants had arrived by practical observation.

In aristocracies, language must naturally partake of that state of repose in which everything remains. Few new words are coined, because few new things are made; and even if new things were made, they would be designated by known words, whose meaning has been determined by tradition. If it happens that the human mind bestirs itself at length, or is roused by light breaking in from without, the novel expressions which are introduced are characterised by a degree of learning, intelligence, and philosophy, which shows that they do not originate in a democracy. After the fall of Constantinople had turned the tide of science and literature towards the west, the French language was almost immediately invaded by a multitude of new words, which had all Greek or Latin roots. An erudite neologism then sprang up in France which was confined to the educated classes, and which produced no sensible effect, or at least a very gradual one, upon the people. All the nations of Europe successively exhibited the same change. Milton alone introduced more than six hundred words into the English language, almost all derived from the Latin, the Greek, or the Hebrew. The constant agitation which prevails in a democratic community tends unceasingly, on the contrary, to change the character of the language, as it does the aspect of affairs. In the midst of this general stir and competition of minds, a great number of new ideas are formed, old ideas are lost, or reappear, or are subdivided into an infinite variety of minor shades. The consequence is, that many words must fall into desuetude, and others must be brought into use.

Democratic nations love change for its own sake; and this is seen in their language as much as in their politics. Even when they do not need to change words, they sometimes feel a wish to transform them. The genius of a democratic people is not only shown by the great number of words they bring into use, but also by the nature of the ideas these new words represent. Amongst such a people the majority lays down the law in language as well as in everything else; its prevailing spirit is as manifest in that as in other respects. But the majority is more engaged in business than in study—in political and commercial interests than in philosophical speculation or literary pursuits. Most of the words coined or adopted for its use will therefore bear the mark of these habits; they will mainly serve to express the wants of business, the passions of party, or the details of the public administration. In these departments the language will constantly spread, whilst on the other hand it will gradually lose ground in metaphysics and theology.

As to the source from which democratic nations are wont to derive their new expressions, and the manner in which they go to work to coin them, both may easily be described. Men living in democratic countries know but little of the language which was spoken at Athens and at Rome, and they do not care to dive into the lore of antiquity to find the expression they happen to want. If they have sometimes recourse to learned etymologies, vanity will induce them to search at the roots of the dead languages; but erudition does not naturally furnish them with its resources. The most ignorant, it sometimes happens, will use them most. The eminently democratic desire to get above their own sphere will often lead them to seek to dignify a vulgar profession by a Greek or Latin name. The lower the calling is, and the more remote from learning, the more pompous and erudite is its appellation. Thus the French rope-dancers have transformed themselves into '*Acrobates*' and '*Funambules*.'

In the absence of knowledge of the dead languages, democratic nations are apt to borrow words from living tongues; for their mutual intercourse becomes perpetual, and the inhabitants of different countries imitate each other the more readily as they grow more like each other every day.

But it is principally upon their own languages that democratic nations attempt to perpetrate innovations. From time to time they resume forgotten expressions in their vocabulary, which they restore to use; or they borrow from some particular class of the community a term peculiar to it, which they introduce with a figurative meaning into the language of daily life. Many expressions which originally belonged to the technical language of a profession or a party, are thus drawn into general circulation.

The most common expedient employed by democratic nations to make an innovation in language consists in giving some unwonted meaning to an expression already in use. This method is very simple, prompt, and convenient; no learning is required to use it aright, and ignorance itself rather facilitates the practice; but that practice is most dangerous to the language. When a democratic people doubles the meaning of a word in this way, they sometimes render the signification which it retains as ambiguous as that which it acquires. An author begins by a slight deflection of a known expression from its primitive meaning, and he adapts it, thus modified, as well as he can to his subject. A second writer twists the sense of the expression in another way; a third takes possession of it for another purpose; and as there is no common appeal to the sentence of a permanent tribunal which may definitely settle the signification of the word, it remains in an ambiguous condition. The consequence is that writers hardly ever appear to dwell upon a single thought, but they always seem to point their aim at a knot of ideas, leaving the reader to judge which of them has been hit. This is a deplorable consequence of democracy. I had rather that the language should be made hideous with words imported from the Chinese, the Tartars, or the Hurons, than that the meaning of a word in our own language should become determinate. Harmony and uniformity are only secondary beauties in composition; many of these things are conventional, and, strictly speaking, it is possible to forego them; but without clear phraseology there is no good language.

The principle of equality necessarily introduces several other changes into language. In aristocratic ages, when each nation tends to stand aloof from all others and likes to have distinct characteristics of its own, it often happens that several peoples which have a common origin become nevertheless estranged from each other, so that, without ceasing to understand die same language, they no longer all speak it in the same manner. In these ages each nation is divided into a certain number of classes, which see but little of each other, and do not intermingle. Each of these classes contracts, and invariably retains, habits of mind peculiar to itself, and adopts by choice certain words and certain terms, which afterwards pass from generation to generation, like their estates. The same idiom then comprises a language of the poor and a language of the rich—a language of the citizen and a language of the nobility—a learned language and a vulgar one. The deeper the divisions, and the more impassable the barriers of society become, the more must this be the case. I would lay a wager, that amongst the castes of India there are amazing variations of language, and that there is almost as much difference between the language of the pariah and that of the Brahmin as there is in their dress. When, on the contrary, men, being no longer restrained by ranks, meet on terms of constant intercourse—when castes are destroyed, and the classes of society are recruited and intermixed with each other, all the words of a language are mingled. Those which are unsuitable to the greater number perish; the remainder form a common store, whence everyone chooses pretty nearly at random. Almost all the different dialects which divided the idioms of European nations are manifestly declining; there is no *patois* in the New World, and it is disappearing every day from the old countries.

The influence of this revolution in social conditions is as much felt in style as it is in phraseology. Not only does everyone use the same words, but a habit springs up of using them without discrimination. The rules which style had set up are almost abolished: the line ceases to be drawn between expressions which seem by their very nature vulgar, and others which appear to be refined. Persons springing from different ranks of society carry the terms and expressions they are accustomed to use with them, into whatever circumstances they may pass; thus the origin of words is lost like the origin of individuals, and there is as much confusion in language as there is in society.

I am aware that in the classification of words there are rules which do not belong to one form of society any more than to another, but which are derived from the nature of things. Some expressions and phrases are vulgar, because the ideas they are meant to express are low in themselves; others are of a higher character, because the objects they are intended to designate are naturally elevated. No intermixture of ranks will ever efface these differences. But the principle of equality cannot fail to root out whatever is merely conventional and arbitrary in the forms of thought. Perhaps the necessary classification which I pointed out in the last sentence will always be less respected by a democratic people than by any other, because amongst such a people there are no men who are permanently disposed by education, culture, and leisure to study the natural laws of language, and who cause those laws to be respected by their own observance of them.

I shall not quit this topic without touching on a feature of democratic languages, which is perhaps more characteristic of them than any other. It has already been shown that democratic nations have a taste, and sometimes a passion, for general ideas, and that this arises from their peculiar merits and defects. This liking for general ideas is displayed in democratic languages by the continual use of generic terms or abstract expressions, and by the manner in which they are employed. This is the great merit and the great imperfection of these languages. Democratic nations are passionately addicted to generic terms or abstract expressions, because these modes of speech enlarge thought, and assist the operations of the mind by enabling it to include several objects in a small compass. A French democratic writer will be apt to say *capacités* in the abstract for men of capacity, and without particularizing the objects to which their capacity is applied: he will talk about *actualités* to designate in one word the things passing before his eyes at the instant; and he will comprehend under the term *éventualités* whatever may happen in the universe, dating from the moment at which he speaks. Democratic writers are perpetually coining words of this kind, in which they sublimate into further abstraction the abstract terms of the language. Nay, more, to render their mode of speech more succinct, they personify the subject of these abstract terms, and make it act like a real entity. Thus they would say in French, '*La force des choses veut que les capacités gouvement.*'

I cannot better illustrate what I mean than by my own example. I have frequently used the word EQUALITY in an absolute senses—nay, I have personified equality in several places; thus I have said that equality does such and such things, or refrains from doing others. It may be affirmed that the writers of the age of Louis XIV. would not have used these expressions: they would never have thought of using the word equality without applying it to some particular object; and they would rather have renounced the term altogether than have consented to make a living personage of it.

These abstract terms which abound in democratic languages, and which are used on every occasion without attaching them to any particular fact, enlarge and obscure the thoughts they are intended to convey; they render the mode of speech more succinct, and the idea contained in it less clear. But with regard to language, democratic nations prefer obscurity to labour. I know not indeed whether this loose style has not some secret charm for those who speak and write amongst these nations. As the men who live there are frequently left to the efforts of their individual powers of mind, they are almost always a prey to doubt; and as their situation in life is forever changing, they are never held fast to any of their opinions by the certain tenure of their fortunes. Men living in democratic countries are, then, apt to entertain unsettled ideas, and they require loose expressions to convey them. As they never know whether the idea they express today will be appropriate to the new position they may occupy tomorrow, they naturally acquire a liking for abstract terms. An abstract term is like a box with a false bottom: you may put in it what ideas you please, and take them out again without being observed.

Amongst all nations, generic and abstract terms form the basis of language. I do not, therefore, affect to expel these terms from democratic languages; I simply remark that men have an especial tendency, in the ages of democracy, to multiply words of this kind—to take them always by themselves in their most abstract acceptation, and to use them on all occasions, even when the nature of the discourse does not require them.

Chapter XVII

Of Some of the Sources of Poetry amongst Democratic Nations

Various different significations have been given to the word Poetry. It would weary my readers if I were to lead them into a discussion as to which of these definitions ought to be selected: I prefer telling them at once that which I have chosen. In my opinion, Poetry is the search and the delineation of the ideal. The poet is he who, by suppressing a part of what exists, by adding some imaginary touches to the picture, and by combining certain real circumstances, but which do not in fact concurrently happen, completes and extends the work of nature. Thus the object of poetry is not to represent what is true, but to adorn it, and to present to the mind some loftier imagery. Verse, regarded as the ideal beauty of language, may be eminently poetical; but verse does not, of itself, constitute poetry.

I now proceed to inquire whether, amongst the actions, the sentiments, and the opinions of democratic nations, there are any which lead to a conception of ideal beauty, and which may for this reason be considered as natural sources of poetry. It must in the first place, be acknowledged that the taste for ideal beauty, and the pleasure derived from the expression of it, are never so intense or so diffused amongst a democratic as amongst an aristocratic people. In aristocratic nations it sometimes happens that the body goes on to act as it were spontaneously, whilst the higher faculties are bound and burdened by repose. Amongst these nations the people will very often display poetic tastes, and sometimes allow their fancy to range beyond and above what surrounds them. But in democracies the love of physical gratification, the notion of bettering one's condition, the excitement of competition, the charm of anticipated success, are so many spurs to urge men onwards in the active professions they have embraced, without allowing them to deviate for an instant from the track.

The main stress of the faculties is to this point. The imagination is not extinct; but its chief function is to devise what may be useful, and to represent what is real.

The principle of equality not only diverts men from the description of ideal beauty—it also diminishes the number of objects to be described. Aristocracy, by maintaining society in a fixed position, is favourable to the solidify and duration of positive religions, as well as to the stability of political institutions. It not only keeps the human mind within a certain sphere of belief, but it predisposes the mind to adopt one faith rather than another. An aristocratic people will always be prone to place intermediate powers between God and man. In this respect it may be said that the aristocratic element is favourable to poetry. When the universe is peopled with supernatural creatures, not palpable to the senses but discovered by the mind, the imagination ranges freely, and poets, finding a thousand subjects to delineate, also find a countless audience to take an interest in their productions. In democratic ages it sometimes happens, on the contrary, that men are as much afloat in matters of belief as they are in their laws. Scepticism then draws the imagination of poets back to earth, and confines them to the real and visible world. Even when the principle of equality does not disturb religious belief, it tends to simplify it, and to divert attention from secondary agents, to fix it principally on the Supreme Power. Aristocracy naturally leads the human mind to the contemplation of the past, and fixes it there. Democracy, on the contrary, gives men a sort of instinctive distaste for what is ancient. In this respect aristocracy is far more favourable to poetry; for things commonly grow larger and more obscure as they are more remote; and for this two-fold reason they are better suited to the delineation of the ideal.

After having deprived poetry of the past, the principle of equality robs it in part of the present. Amongst aristocratic nations there are a certain number of privileged personages, whose situation is, as it were, without and above the condition of man; to these, power, wealth, fame, wit, refinement, and distinction in all things appear peculiarly to belong. The crowd never sees them very closely, or does not watch them in minute details; and little is needed to make the description of such men poetical. On the other hand, amongst the same people, you will meet with classes so ignorant, low, and enslaved, that they are no less fit objects for poetry from the excess of their rudeness and wretchedness, than the former are from their greatness and refinement. Besides, as the different classes of which an aristocratic community is composed are widely separated, and imperfectly acquainted with each other, the imagination may always represent them with some addition to, or some subtraction from, what they really are. In democratic communities, where men are all insignificant and very much alike, each man instantly sees all his fellows when he surveys himself. The poets of democratic ages can never, therefore, take any man in particular as the subject of a piece; for an object of slender importance, which is distinctly seen on all sides, will never lend itself to an ideal conception. Thus the principle of equality; in proportion as it has established itself in the world, has dried up most of the old springs of poetry. Let us now attempt to show what new ones it may disclose.

When scepticism had depopulated heaven, and the progress of equality had reduced each individual to smaller and better-known proportions, the poets, not yet aware of what they could substitute for the great themes which were departing together with the aristocracy, turned their eyes to inanimate nature. As they lost sight of gods and heroes, they set themselves to describe streams and mountains. Thence originated in the last century, that kind of poetry which has been called, by way of distinction, the descriptive. Some have thought that this sort of delineation, embellished with all the physical and inanimate objects which cover the earth, was the kind of poetry peculiar to democratic ages; but I believe this to be an error, and that it only belongs to a period of transition.

I am persuaded that in the end democracy diverts the imagination from all that is external to man, and fixes it on man alone. Democratic nations may amuse themselves for a while with considering the productions of nature; but they are only excited in reality by a survey of themselves. Here, and here alone, the true sources of poetry amongst such nations are to be found; and it may be believed that the poets who shall neglect to draw their inspirations hence, will lose all sway over the minds which they would enchant, and will be left in the end with none but unimpassioned spectators of their transports. I have shown how the ideas of progression and of the indefinite perfectibility of the human race belong to democratic ages. Democratic nations care but little for what has been, but they are haunted by visions of what will be; in this direction their unbounded imagination grows and dilates beyond all measure. Here then is the wildest range open to the genius of poets, which allows them to remove their performances to a sufficient distance from the eye. Democracy shuts the past against the poet, but opens the future before him. As all the citizens who compose a democratic community are nearly equal and alike, the poet cannot dwell upon any one of them; but the nation itself invites the exercise of his powers. The general similitude of individuals, which renders any one of them taken separately an improper subject of poetry, allows poets to include them all in the same imagery, and to take a general survey of the people itself. Democractic nations have a clearer perception than any others of their own aspect; and an aspect so imposing is admirably fitted to the delineation of the ideal.

I readily admit that the Americans have no poets; I cannot allow that they have no poetic ideas. In Europe people talk a great deal of the wilds of America, but the

Americans themselves never think about them: they are insensible to the wonders of inanimate nature, and they may be said not to perceive the mighty forests which surround them till they fall beneath the hatchet. Their eyes are fixed upon another sight: the American people views its own march across these wilds—drying swamps, turning the course of rivers, peopling solitudes, and subduing nature. This magnificent image of themselves does not meet the gaze of the Americans at intervals only; it may be said to haunt every one of them in his least as well as in his most important actions, and to be always flitting before his mind. Nothing conceivable is so petty, so insipid, so crowded with paltry interests, in one word so anti-poetic, as the life of a man in the United States. But amongst the thoughts which it suggests there is always one which is full of poetry, and that is the hidden nerve which gives vigour to the frame.

In aristocratic ages each people, as well as each individual, is prone to stand separate and aloof from all others. In democratic ages, the extreme fluctuations of men and the impatience of their desires keep them perpetually on the move; so that the inhabitants of different countries intermingle, see, listen to, and borrow from each other's stores. It is not only then the members of the same community who grow more alike; communities are themselves assimilated to one another, and the whole assemblage presents to the eye of the spectator one vast democracy, each citizen of which is a people. This displays the aspect of mankind for the first time in the broadest light. All that belongs to the existence of the human race taken as a whole, to its vicissitudes and to its future, becomes an abundant mine of poetry. The poets who lived in aristocratic ages have been eminently successful in their delineations of certain incidents in the life of a people or a man; but none of them ever ventured to include within his performances the destinies of mankind—a task which poets writing in democratic ages may attempt. At that same time at which every man, raising his eyes above his country, begins at length to discern mankind at large, the Divinity is more and more manifest to the human mind in full and entire majesty. If in democratic ages faith in positive religions be often shaken, and the belief in intermediate agents, by whatever name they are called, be overcast; on the other hand men are disposed to conceive a far broader idea of Providence itself, and its interference in human affairs assumes a new and more imposing appearance to their eyes. Looking at the human race as one great whole, they easily conceive that its destinies are regulated by the same design; and in the actions of every individual they are led to acknowledge a trace of that universal and eternal plan on which God rules our race. This consideration may be taken as another prolific source of poetry which is opened in democratic ages. Democratic poets will always appear trivial and frigid if they seek to invest gods, demons, or angels, with corporeal forms, and if they attempt to draw them down from heaven to dispute the supremacy of earth. But if they strive to connect the great events they commemorate with the general providential designs which govern the universe, and, without showing the finger of the Supreme Governor, reveal the thoughts of the Supreme Mind, their works will be admired and understood, for the imagination of their contemporaries takes this direction of its own accord.

It may be foreseen in the like manner that poets living in democratic ages will prefer the delineation of passions and ideas to that of persons and achievements. The language, the dress, and the daily actions of men in democracies are repugnant to ideal conceptions. These things are not poetical in themselves; and, if it were otherwise, they would cease to be so, because they are too familiar to all those to whom the poet would speak of them. This forces the poet constantly to search below the external surface which is palpable to the senses, in order to read the inner soul: and nothing lends itself more to the delineation of the Ideal than the scrutiny of the hidden depths in the immaterial nature of man. I need not to ramble over earth and sky to discover a wondrous object woven of contrasts, of greatness and littleness infinite, of intense gloom and of amazing brightness capable—at once of exciting pity, admiration, terror, contempt. I find that object in myself. Man springs out of nothing, crosses Time, and disappears forever in the bosom of God; he is seen but for a moment, staggering on the verge of the two abysses, and there he is lost. If man were wholly ignorant of himself, he would have no poetry in him; for it is impossible to describe what the mind does not conceive. If man clearly discerned his own nature, his imagination would remain idle, and would have nothing to add to the picture. But the nature of man is sufficiently disclosed for him to apprehend something of himself; and sufficiently obscure for all the rest to be plunged in thick darkness, in which he gropes for ever—and for ever in vain—to lay hold on some completer notion of his being.

Amongst a democratic people poetry will not be fed with legendary lays or the memorials of old traditions. The poet will not attempt to people the universe with supernatural beings in whom his readers and his own fancy have ceased to believe; nor will he present virtues and vices in the mask of frigid personification, which are better received under their own features. All these resources fail him; but Man remains, and the poet needs no more. The destinies of mankind—man himself, taken aloof from his age and his country, and standing in the presence of Nature and of God, with his passions, his doubts, his rare prosperities, and inconceivable wretchedness—will become the chief, if not the sole theme of poetry amongst these nations. Experience may confirm this assertion, if we consider the productions of the greatest poets who have appeared since the world has been turned to democracy. The authors of our age who have so admirably delineated the features of Faust, Childe Harold, Réné, and Jocelyn, did not seek to record the actions of an individual, but to enlarge and to throw light on some of the obscurer recesses of the human heart. Such are the poems of democracy. The principle of equality does not then destroy all the subjects of poetry: it renders them less numerous, but more vast.

Chapter XVIII

Of the Inflated Style of American Writers and Orators

I have frequently remarked that the Americans, who generally treat of business in clear, plain language, devoid of all ornament, and so extremely simple as to be often coarse, are apt to become inflated as soon as they attempt a more poetical diction. They then vent their pomposity from one end of a harangue to the other; and to hear them lavish imagery on every occasion, one might fancy that they never spoke of anything with simplicity. The English are more rarely given to a similar failing. The cause of this may be pointed out without much difficulty. In democratic communities each citizen is habitually engaged in the contemplation of a very puny object, namely himself. If he ever raises his looks higher, he then perceives nothing but the immense form of society at large, or the still more imposing aspect of mankind. His ideas are all either extremely minute and clear, or extremely general and vague: what lies between is an open void. When he has been drawn out of his own sphere, therefore, he always expects that some amazing object will be offered to his attention; and it is on these terms alone that he consents to tear himself for an instant from the petty complicated cares which form the charm and the excitement of his life. This appears to me sufficiently to explain why men in democracies, whose concerns are in general so paltry, call upon their poets for conceptions so vast and descriptions so unlimited.

The authors, on their part, do not fail to obey a propensity of which they themselves partake; they perpetually inflate their imaginations, and expanding them beyond all bounds, they not unfrequently abandon the great in order to reach the gigantic. By these means they hope to attract the observation of the multitude, and to fix it easily upon themselves: nor are their hopes disappointed; for as the multitude seeks for nothing in poetry but subjects of very vast dimensions, it has neither the time to measure with accuracy the proportions of all the subjects set before it, nor a taste sufficiently correct to perceive at once in what respect they are out of proportion. The author and the public at once vitiate one another.

We have just seen that amongst democratic nations, the sources of poetry are grand, but not abundant. They are soon exhausted: and poets, not finding the elements of the Ideal in what is real and true, abandon them entirely and create monsters. I do not fear that the poetry of democratic nations will prove too insipid, or that it will fly too near the ground; I rather apprehend that it will be forever losing itself in the clouds, and that it will range at last to purely imaginary regions, I fear that the productions of democratic poets may often be surcharged with immense and incoherent imagery, with exaggerated descriptions and strange creations; and that the fantastic beings of their brain may sometimes make us regret the world of reality.

Chapter XIX

Some Observations on the Drama amongst Democratic Nations

When the revolution which subverts the social and political state of an aristocratic people begins to penetrate into literature, it generally first manifests itself in the drama, and it always remains conspicuous there. The spectator of a dramatic piece is, to a certain extent, taken by surprise by the impression it conveys. He has no time to refer to his memory, or to consult those more able to judge than himself. It does not occur to him to resist the new literary tendencies which begin to be felt by him; he yields to them before he knows what they are. Authors are very prompt in discovering which way the taste of the public is thus secretly inclined. They shape their productions accordingly; and the literature of the stage, after having served to indicate the approaching literary revolution, speedily completes its accomplishment. If you would judge beforehand of the literature of a people which is lapsing into democracy, study its dramatic productions.

The literature of the stage, moreover, even amongst aristocratic nations, constitutes the most democratic part of their literature. No kind of literary gratification is so much within the reach of the multitude as that which is derived from theatrical representations. Neither preparation nor study is required to enjoy them: they lay hold on you in the midst of your prejudices and your ignorance. When the yet untutored love of the pleasures of the mind begins to affect a class of the community, it instantly draws them to the stage. The theatres of aristocratic nations have always been filled with spectators not belonging to the aristocracy. At the theatre alone the higher ranks mix with the middle and the lower classes; there alone do the former consent to listen to the opinion of the latter, or at least to allow them to give an opinion at all. At the theatre, men of cultivation and of literary attainments have always had more difficulty than elsewhere in making their taste prevail over that of the people, and in preventing themselves from being carried away by the latter. The pit has frequently made laws for the boxes.

If it be difficult for an aristocracy to prevent the people from getting the upper hand in the theatre, it will readily be understood that the people will be supreme there when democratic principles have crept into the laws and manners—when ranks are intermixed—when minds, as well as fortunes, are brought more nearly together—and when the upper class has lost, with its hereditary wealth, its power, its precedents, and its leisure. The tastes and propensities natural to democratic nations, in respect to literature, will therefore first be discernible in the drama, and it may be foreseen that they will break out there with vehemence. In written productions, the literary canons of aristocracy will be gently, gradually, and, so to speak, legally modified; at the theatre they will be riotously overthrown.

The drama brings out most of the good qualities, and almost all the defects, inherent in democratic literature. Democratic peoples hold erudition very cheap, and care but little for what occurred at Rome and Athens; they want to hear something which concerns themselves, and the delineation of the present age is what they demand.

When the heroes and the manners of antiquity are frequently brought upon the stage, and dramatic authors faithfully observe the rules of antiquated precedent, that is enough to warrant a conclusion that the democratic classes have not yet got the upper hand of the theatres. Racine makes a very humble apology in the preface to the 'Britannicus' for having disposed of Junia amongst the Vestals, who, according to Aulus Gellius, he says, 'admitted no one below six years of age nor above ten.' We may be sure that he would neither have accused himself of the offence, nor defended himself from censure, if he had written for our contemporaries. A fact of this kind not only illustrates the state of literature at the time when it occurred, but also that of society itself. A democratic stage does not prove that the nation is in a state of democracy, for, as we have just seen, even in aristocracies it may happen that democratic tastes affect the drama; but when the spirit of aristocracy reigns exclusively on the stage, the fact irrefragably demonstrates that the whole of society is aristocratic; and it may be boldly inferred that the same lettered and learned class which sways the dramatic writers commands the people and governs the country.

The refined tastes and the arrogant bearing of an aristocracy will rarely fail to lead it, when it manages the stage, to make a kind of selection in human nature. Some of the conditions of society claim its chief interest; and the scenes which delineate their manners are preferred upon the stage. Certain virtues, and even certain vices, are thought more particularly to deserve to figure there; and they are applauded whilst all others are excluded. Upon the stage, as well as elsewhere, an aristocratic audience will only meet personages of quality, and share the emotions of kings. The same thing applies to style: an aristocracy is apt to impose upon dramatic authors certain modes of expression which give the key in which everything is to be delivered. By these means the stage frequently comes to delineate only one side of man, or sometimes even to represent what is not to be met with in human nature at all—to rise above nature and to go beyond it.

In democratic communities the spectators have no such partialities, and they rarely display any such antipathies: they like to see upon the stage that medley of conditions, of feelings, and of opinions, which occurs before their eyes. The drama becomes more striking, more common, and more true. Sometimes, however, those who write for the stage in democracies also transgress the bounds of human nature—but it is on a different side from their predecessors. By seeking to represent in minute detail the little singularities of the moment and the peculiar characteristics of certain personages, they forget to portray the general features of the race.

When the democratic classes rule the stage, they introduce as much licence in the manner of treating subjects as in the choice of them. As the love of the drama is, of all literary tastes, that which is most natural to democratic nations, the number of authors and of spectators, as well as of theatrical representations, is constantly increasing amongst these communities. A multitude composed of elements so different, and scattered in so many different places, cannot acknowledge the same rules or submit to the same laws. No concurrence is possible amongst judges so numerous, who know not when they may meet again; and therefore each pronounces his own sentence on the piece. If the effect of democracy is generally to question the authority of all literary rules and conventions, on the stage it abolishes them altogether, and puts in their place nothing but the whim of each author and of each public.

The drama also displays in an especial manner the truth of what I have said before in speaking more generally of style and art in democratic literature. In reading the criticisms which were occasioned by the dramatic productions of the age of Louis XIV., one is surprised to remark the great stress which the public laid on the probability of the plot, and the importance which was attached to the perfect consistency of the characters, and to their doing nothing which could not be easily explained and understood. The value which was set upon the forms of language at that period, and the paltry strife about words with which dramatic authors were assailed, are no less surprising. It would seem that the men of the age of Louis XIV. attached very exaggerated importance to those details, which may be perceived in the study, but which escape attention on the stage. For, after all, the principal object of a dramatic piece is to be performed, and its chief merit is to affect the audience. But the audience and the readers in that age were the same: on quitting the theatre they called up the author for judgment to their own firesides. In democracies, dramatic pieces are listened to, but not read. Most of those who frequent the amusements of the stage do not go there to seek the pleasures of the mind, but the keen emotions of the heart. They do not expect to hear a fine literary work, but to see a play; and provided the author writes the language of his country correctly enough to be understood, and that his characters excite curiosity and awaken sympathy, the audience are satisfied. They ask no more of fiction, and immediately return to real life. Accuracy of style is therefore less required, because the attentive observance of its rules is less perceptible on the stage. As for the probability of the plot, it is incompatible with perpetual novelty, surprise, and rapidity of invention. It is therefore neglected, and the public excuses the neglect. You may be sure that if you succeed in bringing your audience into the presence of something that affects them, they will not care by what road you brought them there; and they will never reproach you for having excited their emotions in spite of dramatic rules.

The Americans very broadly display all the different propensities which I have here described when they go to the theatres; but it must be acknowledged that as yet a very small number of them go to theatres at all. Although playgoers and plays have prodigiously increased in the United States in the last forty years, the population indulges in this kind of amusement with the greatest reserve. This is attributable to peculiar causes, which the reader is already acquainted with, and of which a few words will suffice to remind him. The Puritans who founded the American republics were not only enemies to amusements, but they professed an especial abhorrence for the stage. They considered it as an abominable pastime; and as long as their principles prevailed with undivided sway, scenic performances were wholly unknown amongst them. These opinions of the first fathers of the colony have left very deep marks on the minds of their descendants. The extreme regularity of habits and the great strictness of manners which are observable in the United States, have as yet opposed additional obstacles to the growth of dramatic art. There are no dramatic subjects in a country which has witnessed no great political catastrophes, and in which love invariably leads by a straight and easy road to matrimony. People who spend every day in the week in making money, and the Sunday in going to church, have nothing to invite the Muse of Comedy.

A single fact suffices to show that the stage is not very popular in the United States. The Americans, whose laws allow of the utmost freedom and even licence of language in all other respects, have nevertheless subjected their dramatic authors to a sort of censorship. Theatrical performances can only take place by permission of the municipal authorities. This may serve to show how much communities are like individuals; they surrender themselves unscrupulously to their ruling passions, and afterwards take the greatest care not to yield too much to the vehemence of tastes which they do not possess.

No portion of literature is connected by closer or more numerous ties with the present condition of society than the drama. The drama of one period can never be suited to the following age, if in the interval an important revolution has changed the manners and the laws of the nation. The great authors of a preceding age may be read; but pieces written for a different public will not be followed. The dramatic authors of the past live only in books. The traditional taste of certain individuals, vanity, fashion, or the genius of an actor may sustain or resuscitate for a time the aristocratic drama amongst a democracy; but it will speedily fall away of itself—not overthrown, but abandoned.

Chapter XX

Characteristics of Historians in Democratic Ages

Historians who write in aristocratic ages are wont to refer all occurrences to the particular will or temper of certain individuals; and they are apt to attribute the most important revolutions to very slight accidents. They trace out the smallest causes with sagacity, and frequently leave the greatest unperceived. Historians who live in democratic ages exhibit precisely opposite characteristics. Most of them attribute hardly any influence to the individual over the destiny of the race, nor to citizens over the fate of a people; but, on the other hand, they assign great general causes to all petty incidents. These contrary tendencies explain each other.

When the historian of aristocratic ages surveys the theatre of the world, he at once perceives a very small number of prominent actors, who manage the whole piece. These great personages, who occupy the front of the stage, arrest the observation, and fix it on themselves; and whilst the historian is bent on penetrating the secret motives which make them speak and act, the rest escape his memory. The importance of the things which some men are seen to do, gives him an exaggerated estimate of the influence which one man may possess; and naturally leads him to think, that in order to explain the impulses of the multitude, it is necessary to refer them to the particular influence of some one individual.

When, on the contrary, all the citizens are independent of one another, and each of them is individually weak, no one is seen to exert a great, or still less a lasting power, over the community. At first sight, individuals appear to be absolutely devoid of any influence over it; and society would seem to advance alone by the free and voluntary concurrence of all the men who compose it. This naturally prompts the mind to search for that general reason which operates upon so many men's faculties at the same time, and turns them simultaneously in the same direction.

I am very well convinced that even amongst democratic nations, the genius, the vices, or the virtues of certain individuals retard or accelerate the natural current of a people's history: but causes of this secondary and fortuitous nature are infinitely more various, more concealed, more complex, less powerful, and consequently less easy to trace in periods of equality than in ages of aristocracy, when the task of the historian is simply to detach from the mass of general events the particular influences of one man or of a few men. In the former case the historian is soon wearied by the toil; his mind loses itself in this labyrinth; and, in his inability clearly to discern or conspicuously to point out the influence of individuals, he denies their existence. He prefers talking about the characteristics of race, the physical conformation of the country, or the genius of civilisation—which abridges his own labours, and satisfies his reader far better at less cost.

M. de Lafayette says somewhere in his Memoirs that the exaggerated system of general causes affords surprising consolations to second-rate statesmen. I will add, that its effects are not less consolatory to second-rate historians; it can always furnish a few mighty reasons to extricate them from the most difficult part of their work, and it indulges the indolence or incapacity of their minds, whilst it confers upon them the honours of deep thinking.

For myself, I am of opinion that at all times one great portion of the events of this world are attributable to general facts, and another to special influences. These two kinds of cause are always in operation: their proportion only varies. General facts serve to explain more things in democratic than in aristocratic ages, and fewer things are then assignable to special influences. At periods of aristocracy the reverse takes place: special influences are stronger, general causes weaker—unless indeed we consider as a general cause the fact itself of the inequality of conditions, which allows some individuals to baffle the natural tendencies of all the rest. The historians who seek to describe what occurs in democratic societies are right, therefore, in assigning much to general causes, and in devoting their chief attention to discover them; but they are wrong in wholly denying the special influence of individuals, because they cannot easily trace or follow it.

The historians who live in democratic ages are not only prone to assign a great cause to every incident, but they are also given to connect incidents together, so as to deduce a system from them. In aristocratic ages, as the attention of historians is constantly drawn to individuals, the connection of events escapes them; or rather, they do not believe in any such connection. To them the clue of history seems every instant crossed and broken by the step of man. In democratic ages, on the contrary, as the historian sees much more of actions than of actors, he may easily establish some kind of sequency and methodical order amongst the former. Ancient literature, which is so rich in fine historical compositions, does not contain a single great historical system, whilst the poorest of modern literatures abound with them. It would appear that the ancient historians did not make sufficient use of those general theories which our historical writers are ever ready to carry to excess.

Those who write in democratic ages have another more dangerous tendency. When the traces of individual action upon nations are lost, it often happens that the world goes on to move, though the moving agent is no longer discoverable. As it becomes extremely difficult to discern and to analyse the reasons which, acting separately on the volition of each member of the community, concur in the end to produce movement in the old mass, men are led to believe that this movement is involuntary, and that societies unconsciously obey some superior force ruling over them. But even when the general fact which governs the private volition of all individuals is supposed to be discovered upon the earth, the principle of human free-will is not secure. A cause sufficiently extensive to affect millions of men at once, and sufficiently strong to bend them all together in the same direction, may well seem irresistible: having seen that mankind do yield to it, the mind is close upon the inference that mankind cannot resist it.

Historians who live in democratic ages, then, not only deny that the few have any power of acting upon the destiny of a people, but they deprive the people themselves of the power of modifying their own condition, and they subject them either to an inflexible Providence, or to some blind necessity. According to them, each nation is indissolubly bound by its position, its origin, its precedents, and its character, to a certain lot which no efforts can ever change. They involve generation in generation, and thus, going back from age to age, and from necessity to necessity, up to the origin of the world, they forge a close and enormous chain, which girds and binds the human race. To their minds it is not enough to show what events have occurred: they would fain show that events could not have occurred otherwise. They take a nation arrived at a certain stage of its history, and they affirm that it could not but follow the track which brought it thither. It is easier to make such an assertion than to show by what means the nation might have adopted a better course.

In reading the historians of aristocratic ages, and especially those of antiquity, it would seem that, to be master of his lot, and to govern his fellow-creatures, man requires only to be master of himself. In perusing the historical volumes which our age has produced, it would seem that man is utterly powerless over himself and over all around him. The historians of antiquity taught how to command: those of our time teach only how to obey; in their writings the author often appears great, but humanity is always diminutive. If this doctrine of necessity, which is so attractive to those who write history in democratic ages, passes from authors to their readers, till it infects the whole mass of the community and gets possession of the public mind, it will soon paralyse the activity of modern society, and reduce Christians to the level of the Turks. I would moreover observe, that such principles are peculiarly dangerous at the period at which we are arrived. Our contemporaries are but too prone to doubt of the human freewill, because each of them feels himself confined on every side by his own weakness; but they are still willing to acknowledge the strength and independence of men united in society. Let not this principle be lost sight of; for the great object in our time is to raise the faculties of men, not to complete their prostration.

Chapter XXI

Of Parliamentary Eloquence in the United States

Amongst aristocratic nations all the members of the community are connected with and dependent upon each other; the graduated scale of different ranks acts as a tie, which keeps everyone in his proper place and the whole body in subordination. Something of the same kind always occurs in the political assemblies of these nations. Parties naturally range themselves under certain leaders, whom they obey by a sort of instinct, which is only the result of habits contracted elsewhere. They carry the manners of general society into the lesser assemblage.

In democratic countries it often happens that a great number of citizens are tending to the same point; but each one only moves thither, or at least flatters himself that he moves, of his own accord. Accustomed to regulate his doings by personal impulse alone, he does not willingly submit to dictation from without. This taste and habit of independence accompany him into the councils of the nation. If he consents to connect himself with other men in the prosecution of the same purpose, at least he chooses to remain free to contribute to the common success after his own fashion. Hence it is that in democratic countries parties are so impatient of control, and are never manageable except in moments of great public danger. Even then, the authority of leaders, which under such circumstances may be able to make men act or speak, hardly ever reaches the extent of making them keep silence.

Amongst aristocratic nations the members of political assemblies are at the same time members of the aristocracy. Each of them enjoys high established rank in his own right, and the position which he occupies in the assembly is often less important in his eyes than that which he fills in the country. This consoles him for playing no part in the discussion of public affairs, and restrains him from too eagerly attempting to play an insignificant one.

In America, it generally happens that a representative only becomes somebody from his position in the assembly. He is therefore perpetually haunted by a craving to acquire importance there, and he feels a petulant desire to be constantly obtruding his opinions upon the House. His own vanity is not the only stimulant which urges him on in this course, but that of his constituents, and the continual necessity of propitiating them. Amongst aristocratic nations a member of the legislature is rarely in strict dependence upon his constituents: he is frequently to them a sort of unavoidable representative; sometimes they are themselves strictly dependent upon him; and if at length they reject him, he may easily get elected elsewhere, or, retiring from public life, he may still enjoy the pleasures of splendid idleness. In a democratic country like the United States a Representative has hardly ever a lasting hold on the minds of his constituents. However small an electoral body may be, the fluctuations of democracy are constantly changing its aspect; it must, therefore, be courted unceasingly. He is never sure of his supporters, and, if they forsake him, he is left without a resource; for his natural position is not sufficiently elevated for him to be easily known to those not close to him; and, with the complete state of independence prevailing among the people, he cannot hope that his friends or the Government will send him down to be returned by an electoral body unacquainted with him. The seeds of his fortune are, therefore, sown in his own neighbourhood; from that nook of earth he must start, to raise himself to the command of a people and to influence the destinies of the world. Thus it is natural that in democratic countries the members of political assemblies think more of their constituents than of their party, whilst in aristocracies they think more of their party than of their constituents.

But what ought to be said to gratify constituents is not always what ought to be said in order to serve the party to which representatives profess to belong. The general interest of a party frequently demands that members belonging to it should not speak on great questions which they understand imperfectly; that they should speak but little on those minor questions which impede the great ones; lastly, and for the most part, that they should not speak at all. To keep silence is the most useful service that an indifferent spokesman can render to the commonwealth. Constituents, however, do not think so. The population of a district sends a representative to take a part in the government of a country, because they entertain a very lofty notion of his merits. As men appear greater in proportion to the littleness of the objects by which they are surrounded, it may be assumed that the opinion entertained of the delegate will be so much the higher as talents are more rare among his constituents. It will therefore frequently happen that the less constituents have to expect from their representative, the more they will anticipate from him; and, however incompetent he may be, they will not fail to call upon him for signal exertions, corresponding to the rank they have conferred upon him.

Independently of his position as a legislator of the State, electors also regard their representative as the natural patron of the constituency in the legislature; they almost consider him as the proxy of each of his supporters, and they flatter themselves that he will not be less zealous in defence of their private interests than of those of the country. Thus electors are well assured beforehand that the representative of their choice will be an orator; that he will speak often if he can, and that in case he is forced to refrain, he will strive at any rate to compress into his less frequent orations an inquiry into all the great questions of state, combined with a statement of all the petty grievances they have themselves to complain of; so that, though he be not able to come forward frequently, he should on each occasion prove what he is capable of doing; and that, instead of perpetually lavishing his powers, he should occasionally condense them in a small compass, so as to furnish a sort of complete and brilliant epitome of his constituents and of himself. On these terms they will vote for him at the next election. These conditions drive worthy men of humble abilities to despair, who, knowing their own powers, would never voluntarily have come forward. But thus urged on, the representative begins to speak, to the great alarm of his friends; and rushing imprudently into the midst of the most celebrated orators, he perplexes the debate and wearies the House.

All laws which tend to make the representative more dependent on the elector, not only affect the conduct of the legislators, as I have remarked elsewhere, but also their language. They exercise a simultaneous influence on affairs themselves, and on the manner in which affairs are discussed.

There is hardly a member of Congress who can make up his mind to go home without having despatched at least one speech to his constituents; nor who will endure any interruption until he has introduced into his harangue whatever useful suggestions may be made touching the four-and-twenty States of which the Union is composed, and especially the district which he represents. He therefore presents to the mind of his auditors a succession of great general truths (which he himself only comprehends, and expresses, confusedly), and of petty minutiæ, which he is but too able to discover and to point out. The consequence is that the debates of that great assembly are frequently vague and perplexed, and that they seem rather to drag their slow length along than to advance towards a distinct object. Some such state of things will, I believe, always arise in the public assemblies of democracies.

Propitious circumstances and good laws might succeed in drawing to the legislature of a democratic people men very superior to those who are returned by the Americans to Congress; but nothing will ever prevent the men of slender abilities who sit there from obtruding themselves with complacency, and in all ways, upon the public. The evil does not appear to me to be susceptible of entire cure, because it not only originates in the tactics of that Assembly, but in its constitution and in that of the country. The inhabitants of the United States seem themselves to consider the matter in this light; and they show their long experience of parliamentary life not by abstaining from making bad speeches, but by courageously submitting to hear them made. They are resigned to it, as to an evil which they know to be inevitable.

We have shown the petty side of political debates in democratic assemblies—let us now exhibit the more imposing one. The proceedings within the Parliament of England for the last one hundred and fifty years have never occasioned any great sensation out of that country; the opinions and feelings expressed by the speakers have never awakened much sympathy, even amongst the nations placed nearest to the great arena of British liberty; whereas Europe was excited by the very first debates which took place in the small colonial assemblies of America at the time of the revolution. This was attributable not only to particular and fortuitous circumstances, but to general and lasting causes. I can conceive nothing more admirable or more powerful than a great orator debating on great questions of state in a democratic assembly. As no particular class is ever represented there by men commissioned to defend its own interests, it is always to the whole nation, and in the name of the whole nation, that the orator speaks. This expands his thoughts, and heightens his power of language. As precedents have there but little weight—as there are no longer any privileges attached to certain property, nor any rights inherent in certain bodies or in certain individuals, the mind must have recourse to general truths derived from human nature to resolve the particular question under discussion. Hence the political debates of a democratic people, however small it may be, have a degree of breadth which frequently renders them attractive to mankind. All men are interested by them, because they treat of *man,* who is everywhere the same. Amongst the greatest aristocratic nations, on the contrary, the most general questions are almost always argued on some special grounds derived from the practice of a particular time, or the rights of a particular class; which interest that class alone, or at most the people amongst whom that class happens to exist. It is owing to this, as much as to the greatness of the French people, and the favourable disposition of the nations who listen to them, that the great effect which the French political debates sometimes produce in the world, must be attributed. The orators of France frequently speak to mankind, even when they are addressing their countrymen only.

DEMOCRACY IN AMERICA

SECOND PART

SECOND BOOK

Influence of Democracy on the Feelings of the Americans

Chapter I

Why Democratic Nations Show a More Ardent and Enduring Love of Equality Than of Liberty

The first and most intense passion which is engendered by the equality of conditions is, I need hardly say, the love of that same equality. My readers will therefore not be surprised that I speak of it before all others. Everybody has remarked that in our time, and especially in France, this passion for equality is every day gaining ground in the human heart. It has been said a hundred times that our contemporaries are far more ardently and tenaciously attached to equality than to freedom; but as I do not find that the causes of the fact have been sufficiently analysed, I shall endeavour to point them out.

It is possible to imagine an extreme point at which freedom and equality would meet and be confounded together. Let us suppose that all the members of the community take a part in the government, and that each of them has an equal right to take a part in it. As none is different from his fellows, none can exercise a tyrannical power: men will be perfectly free, because they will all be entirely equal; and they will all be perfectly equal, because they will be entirely free. To this ideal state democratic nations tend. Such is the completest form that equality can assume upon earth; but there are a thousand others which, without being equally perfect, are not less cherished by those nations.

The principle of equality may be established in civil society, without prevailing in the political world. Equal rights may exist of indulging in the same pleasures, of entering the same professions, of frequenting the same places—in a word, of living in the same manner and seeking wealth by the same means, although all men do not take an equal share in the government. A kind of equality may even be established in the political world, though there should be no political freedom there. A man may be the equal of all his countrymen save one, who is the master of all without distinction, and who selects equally from among them all the agents of his power. Several other combinations might be easily imagined, by which very great equality would be united to institutions more or less free, or even to institutions wholly without freedom. Although men cannot become absolutely equal unless they be entirely free, and consequently equality, pushed to its furthest extent, may be confounded with freedom, yet there is good reason for distinguishing the one from the other. The taste which men have for liberty, and that which they feel for equality, are, in fact, two different things; and I am not afraid to add that, amongst democratic nations, they are two unequal things.

Upon close inspection, it will be seen that there is in every age some peculiar and preponderating fact with which all others are connected; this fact almost always gives birth to some pregnant idea or some ruling passion, which attracts to itself, and bears away in its course, all the feelings and opinions of the time: it is like a great stream, towards which each of the surrounding rivulets seems to flow. Freedom has appeared in the world at different times and under various forms; it has not been exclusively bound to any social condition, and it is not confined to democracies. Freedom cannot, therefore, form the distinguishing characteristic of democratic ages. The peculiar and preponderating fact which marks those ages as its own is the equality of conditions; the ruling passion of men in those periods is the love of this equality. Ask not what singular charm the men of democratic ages find in being equal, or what special reasons they may have for clinging so tenaciously to equality rather than to the other advantages which society holds out to them: equality is the distinguishing characteristic of the age they live in; that, of itself, is enough to explain that they prefer it to all the rest.

But independently of this reason there are several others, which will at all times habitually lead men to prefer equality to freedom. If a people could ever succeed in destroying, or even in diminishing, the equality which prevails in its own body, this could only be accomplished by long and laborious efforts. Its social condition must be modified, its laws abolished, its opinions superseded, its habits changed, its manners corrupted. But political liberty is more easily lost; to neglect to hold it fast is to allow it to escape. Men therefore not only cling to equality because it is dear to them; they also adhere to it because they think it will last forever.

That political freedom may compromise in its excesses the tranquillity, the property, the lives of individuals, is obvious to the narrowest and most unthinking minds. But, on the contrary, none but attentive and clear-sighted men perceive the perils with which equality threatens us, and they commonly avoid pointing them out. They know that the calamities they apprehend are remote, and flatter themselves that they will only fall upon future generations, for which the present generation takes but little thought. The evils which freedom sometimes brings with it are immediate; they are apparent to all, and all are more or less affected by them. The evils which extreme equality may produce are slowly disclosed; they creep gradually into the social frame; they are only seen at intervals, and at the moment at which they become most violent habit already causes them to be no longer felt. The advantages which freedom brings are only shown by length of time; and it is always easy to mistake the cause in which they originate. The advantages of equality are instantaneous, and they may constantly be traced from their source. Political liberty bestows exalted pleasures, from time to time, upon a certain number of citizens. Equality every day confers a number of small enjoyments on every man. The charms of equality are every instant felt, and are within the reach of all; the noblest hearts are not insensible to them, and the most vulgar souls exult in them. The passion which equality engenders must therefore be at once strong and general. Men cannot enjoy political liberty unpurchased by some sacrifices, and they never obtain it without great exertions. But the pleasures of equality are self-proffered: each of the petty incidents of life seems to occasion them, and in order to taste them nothing is required but to live.

Democratic nations are at all times fond of equality, but there are certain epochs at which the passion they entertain for it swells to the height of fury. This occurs at the moment when the old social system, long menaced, completes its own destruction after a last intestine struggle, and when the barriers of rank are at length thrown down. At such times men pounce upon equality as their booty, and they cling to it as to some precious treasure which they fear to lose. The passion for equality penetrates on every side into men's hearts, expands there, and fills them, entirely. Tell them not that by this blind surrender of themselves to an exclusive passion they risk their dearest interests: they are deaf. Show them not freedom escaping from their grasp, whilst they are looking another way: they are blind—or rather, they can discern but one sole object to be desired in the universe.

What I have said is applicable to all democratic nations: what I am about to say concerns the French alone. Amongst most modern nations, and especially amongst all those of the continent of Europe, the taste and the idea of freedom only began to exist and to extend themselves at the time when social conditions were tending to equality, and as a consequence of that very equality. Absolute kings were the most efficient levellers of ranks amongst their subjects. Amongst these nations equality preceded freedom: equality was therefore a fact of some standing when freedom was still a novelty: the one had already created customs, opinions, and laws belonging to it, when the other, alone and for the first time, came into actual existence. Thus the latter was still only an affair of opinion and of taste, whilst the former had already crept into the habits of the people, possessed itself of their manners, and given a particular turn to the smallest actions of their lives. Can it be wondered that the men of our own time prefer the one to the other?

I think that democratic communities have a natural taste for freedom: left to themselves, they will seek it, cherish it, and view any privation of it with regret. But for equality, their passion is ardent, insatiable, incessant, invincible: they call for equality in freedom; and if they cannot obtain that, they still call for equality in slavery. They will endure poverty, servitude, barbarism—but they will not endure aristocracy. This is true at all times, and especially true in our own. All men and all powers seeking to cope with this irresistible passion, will be overthrown and destroyed by it. In our age, freedom cannot be established without it, and despotism itself cannot reign without its support.

Chapter II

Of Individualism[1] in Democratic Countries

I have shown how it is that in ages of equality every man seeks for his opinions within himself: I am now about to show how it is that, in the same ages, all his feelings are turned towards himself alone. *Individualism* is a novel expression, to which a novel idea has given birth. Our fathers were only acquainted with egotism. Egotism is a passionate and exaggerated love of self, which leads a man to connect everything with his own person, and to prefer himself to everything in the world. Individualism is a mature and calm feeling, which disposes each member of the community to sever himself from the mass of his fellow-creatures; and to draw apart with his family and his friends; so that, after he has thus formed a little circle of his own, he willingly leaves society at large to itself. Egotism originates in blind instinct: individualism proceeds from erroneous judgment more than from depraved feelings; it originates as much in the deficiencies of the mind as in the perversity of the heart. Egotism blights the germ of all virtue; individualism, at first, only saps the virtues of public life; but, in the long run, it attacks and destroys all others, and is at length absorbed in downright egotism. Egotism is a vice as old as the world, which does not belong to one form of society more than to another: individualism is of democratic origin, and it threatens to spread in the same ratio as the equality of conditions.

Amongst aristocratic nations, as families remain for centuries in the same condition, often on the same spot, all generations become as it were contemporaneous. A man almost always knows his forefathers, and respects them: he thinks he already sees his remote descendants, and he loves them. He willingly imposes duties on himself towards the former and the latter; and he will frequently sacrifice his personal gratifications to those who went before and to those who will come after him. Aristocratic institutions have, moreover, the effect of closely binding every man to several of his fellow-citizens. As the classes of an aristocratic people are strongly marked and permanent, each of them is regarded by its own members as a sort of lesser country, more tangible and more cherished than the country at large. As in aristocratic communities all the citizens occupy fixed positions, one above the other, the result is that each of them always sees a man above himself whose patronage is necessary to him, and below himself another man whose cooperation he may claim. Men living in aristocratic ages are therefore almost always closely attached to something placed out of their own sphere, and they are often disposed to forget themselves. It is true that in those ages the notion of human fellowship is faint, and that men seldom think of sacrificing themselves for mankind; but they often sacrifice themselves for other men. In democratic ages, on the contrary, when the duties of each individual to the race are much more clear, devoted service to any one man becomes more rare; the bond of human affection is extended, but it is relaxed.

Amongst democratic nations new families are constantly springing up, others are constantly falling away, and all that remain change their condition; the woof of time is every instant broken, and the track of generations effaced. Those who went before are soon forgotten; of those who will come after no one has any idea: the interest of man is confined to those in close propinquity to himself. As each class approximates to other classes, and intermingles with them, its members become indifferent and as strangers to one another. Aristocracy had made a chain of all the members of the community, from the peasant to the king: democracy breaks that chain, and severs every link of it. As social conditions become more equal, the number of persons increases who, although they are neither rich enough nor powerful enough to exercise any great influence over their fellow-creatures, have nevertheless acquired or retained sufficient education and fortune to satisfy their own wants. They owe nothing to any man, they expect nothing from any man; they acquire the habit of always considering themselves as standing alone, and they are apt to imagine that their whole destiny is in their own hands. Thus not only does democracy make every man forget his ancestors, but it hides his descendants, and separates his contemporaries from him; it throws him back forever upon himself alone, and threatens in the end to confine him entirely within the solitude of his own heart.

CHAPTER III

Individualism Stronger at the Close of a Democratic Revolution Than at Other Periods

The period when the construction of democratic society upon the ruins of an aristocracy has just been completed, is especially that at which this separation of men from one another, and the egotism resulting from it, most forcibly strike the observation. Democratic communities not only contain a large number of independent citizens, but they are constantly filled with men who, having entered but yesterday upon their independent condition, are intoxicated with their new power. They entertain a presumptuous confidence in their strength, and as they do not suppose that they can henceforward ever have occasion to claim the assistance of their fellow-creatures, they do not scruple to show that they care for nobody but themselves.

An aristocracy seldom yields without a protracted struggle, in the course of which implacable animosities are kindled between the different classes of society. These passions survive the victory, and traces of them may be observed in the midst of the democratic confusion which ensues. Those members of the community who were at the top of the late gradations of rank cannot immediately forget their former greatness; they will long regard themselves as aliens in the midst of the newly-composed society. They look upon all those whom this state of society has made their equals as oppressors, whose destiny can excite no sympathy; they have lost sight of their former equals, and feel no longer bound by a common interest to their fate: each of them, standing aloof, thinks that he is reduced to care for himself alone. Those, on the contrary, who were formerly at the foot of the social scale, and who have been brought up to the common level by a sudden revolution, cannot enjoy their newly-acquired independence without secret uneasiness; and if they meet with some of their former superiors on the same footing as themselves, they stand aloof from them with an expression of triumph and of fear. It is, then, commonly at the outset of democratic society that citizens are most disposed to live apart. Democracy leads men not to draw near to their fellow-creatures; but democratic revolutions lead them to shun each other, and perpetuate in a state of equality the animosities which the state of inequality engendered. The great advantage of the Americans is that they have arrived at a state of democracy without having to endure a democratic revolution; and that they are born equal, instead of becoming so.

Chapter IV

That the Americans Combat the Effects of Individualism by Free Institutions

Despotism, which is of a very timorous nature, is never more secure of continuance than when it can keep men asunder; and all its influence is commonly exerted for that purpose. No vice of the human heart is so acceptable to it as egotism: a despot easily forgives his subjects for not loving him, provided they do not love each other. He does not ask them to assist him in governing the State; it is enough that they do not aspire to govern it themselves. He stigmatizes as turbulent and unruly spirits those who would combine their exertions to promote the prosperity of the community, and, perverting the natural meaning of words, he applauds as good citizens those who have no sympathy for any but themselves. Thus the vices which despotism engenders are precisely those which equality fosters. These two things mutually and perniciously complete and assist each other. Equality places men side by side, unconnected by any common tie; despotism raises barriers to keep them asunder; the former predisposes them not to consider their fellow-creatures, the latter makes general indifference a sort of public virtue.

Despotism then, which is at all times dangerous, is more particularly to be feared in democratic ages. It is easy to see that in those same ages men stand most in need of freedom. When the members of a community are forced to attend to public affairs, they are necessarily drawn from the circle of their own interests, and snatched at times from self-observation. As soon as a man begins to treat of public affairs in public, he begins to perceive that he is not so independent of his fellowmen as he had at first imagined, and that, in order to obtain their support, he must often lend them his cooperation.

When the public is supreme, there is no man who does not feel the value of public goodwill, or who does not endeavour to court it by drawing to himself the esteem and affection of those amongst whom he is to live. Many of the passions which congeal and keep asunder human hearts, are then obliged to retire and hide below the surface. Pride must be dissembled; disdain dares not break out; egotism fears its own self. Under a free government, as most public offices are elective, the men whose elevated minds or aspiring hopes are too closely circumscribed in private life, constantly feel that they cannot do without the population which surrounds them. Men learn at such times to think of their fellowmen from ambitious motives; and they frequently find it, in a manner, their interest to forget themselves.

I may here be met by an objection derived from electioneering intrigues, the meannesses of candidates, and the calumnies of their opponents. These are opportunities for animosity which occur the oftener the more frequent elections become. Such evils are doubtless great, but they are transient; whereas the benefits which attend them remain. The desire of being elected may lead some men for a time to violent hostility; but this same desire leads all men in the long run mutually to support each other; and if it happens that an election accidentally severs two friends, the electoral system brings a multitude of citizens permanently together, who would always have remained unknown to each other. Freedom engenders private animosities, but despotism gives birth to general indifference.

The Americans have combated by free institutions the tendency of equality to keep men asunder, and they have subdued it. The legislators of America did not suppose that a general representation of the whole nation would suffice to ward off a disorder at once so natural to the frame of democratic society, and so fatal: they also thought that it would be well to infuse political life into each portion of the territory, in order to multiply to an infinite extent opportunities of acting in concert for all the members of the community, and to make them constantly feel their mutual dependence on each other. The plan was a wise one. The general affairs of a country only engage the attention of leading politicians, who assemble from time to time in the same places; and as they often lose sight of each other afterwards, no lasting ties are established between them. But if the object be to have the local affairs of a district conducted by the men who reside there, the same persons are always in contact, and they are, in a manner, forced to be acquainted, and to adapt themselves to one another.

It is difficult to draw a man out of his own circle to interest him in the destiny of the state, because he does not clearly understand what influence the destiny of the state can have upon his own lot. But if it be proposed to make a road cross the end of his estate, he will see at a glance that there is a connection between this small public affair and his greatest private affairs; and he will discover, without its being shown to him, the close tie which unites private to general interest. Thus, far more may be done by intrusting to the citizens the administration of minor affairs than by surrendering to them the control of important ones, towards interesting them in the public welfare, and convincing them that they constantly stand in need one of the other in order to provide for it. A brilliant achievement may win for you the favour of a people at one stroke; but to earn the love and respect of the population which surrounds you, a long succession of little services rendered and of obscure good deeds—a constant habit of kindness, and an established reputation for disinterestedness—will be required. Local freedom, then, which leads a great number of citizens to value the affection of their neighbours and of their kindred, perpetually brings men together, and forces them to help one another, in spite of the propensities which sever them.

In the United States the more opulent citizens take great care not to stand aloof from the people; on the contrary, they constantly keep on easy terms with the lower classes: they listen to them, they speak to them every day. They know that the rich in democracies always stand in need of the poor; and that in democratic ages you attach a poor man to you more by your manner than by benefits conferred. The magnitude of such benefits, which sets off the difference of conditions, causes a secret irritation to those who reap advantage from them; but the charm of simplicity of manners is almost irresistible: their affability carries men away, and even their want of polish is not always displeasing. This truth does not take root at once in the minds of the rich. They generally resist it as long as the democratic revolution lasts, and they do not acknowledge it immediately after that revolution is accomplished. They are very ready to do good to the people, but they still choose to keep them at arm's length; they think that is sufficient, but they are mistaken. They might spend fortunes thus without warming the hearts of the population around them;—that population does not ask them for the sacrifice of their money, but of their pride.

It would seem as if every imagination in the United States were upon the stretch to invent means of increasing the wealth and satisfying the wants of the public. The best-informed inhabitants of each district constantly use their information to discover new truths which may augment the general prosperity; and if they have made any such discoveries, they eagerly surrender them to the mass of the people.

When the vices and weaknesses, frequently exhibited by those who govern in America, are closely examined, the prosperity of the people occasions—but improperly occasions—surprise. Elected magistrates do not make the American democracy flourish; it flourishes because the magistrates are elective.

It would be unjust to suppose that the patriotism and the zeal which every American displays for the welfare of his fellow-citizens are wholly insincere. Although private interest directs the greater part of human actions in the United States as well as elsewhere, it does not regulate them all. I must say that I have often seen Americans make great and real sacrifices to the public welfare; and I have remarked a hundred instances in which they hardly ever failed to lend faithful support to each other. The free institutions which the inhabitants of the United States possess, and the political rights of which they make so much use, remind every citizen, and in a thousand ways, that he lives in society. They every instant impress upon his mind the notion that it is the duty, as well as the interest of men, to make themselves useful to their fellow-creatures; and as he sees no particular ground of animosity to them, since he is never either their master or their slave, his heart readily leans to the side of kindness. Men attend to the interests of the public, first by necessity, afterwards by choice: what was intentional becomes an instinct; and by dint of working for the good of one's fellow citizens, the habit and the taste for serving them is at length acquired.

Many people in France consider equality of conditions as one evil, and political freedom as a second. When they are obliged to yield to the former, they strive at least to escape from the latter. But I contend that in order to combat the evils which equality may produce, there is only one effectual remedy—namely, political freedom.

Chapter V

Of the Use Which the Americans Make of Public Associations in Civil Life

I do not propose to speak of those political associations—by the aid of which men endeavour to defend themselves against the despotic influence of a majority—or against the aggressions of regal power. That subject I have already treated. If each citizen did not learn, in proportion as he individually becomes more feeble, and consequently more incapable of preserving his freedom single-handed, to combine with his fellow-citizens for the purpose of defending it, it is clear that tyranny would unavoidably increase together with equality.

Those associations only which are formed in civil life, without reference to political objects, are here adverted to. The political associations which exist in the United States are only a single feature in the midst of the immense assemblage of associations in that country. Americans of all ages, all conditions, and all dispositions, constantly form associations. They have not only commercial and manufacturing companies, in which all take part, but associations of a thousand other kinds—religious, moral, serious, futile, extensive or restricted, enormous or diminutive. The Americans make associations to give entertainments, to found establishments for education, to build inns, to construct churches, to diffuse books, to send missionaries to the antipodes; and in this manner they found hospitals, prisons, and schools. If it be proposed to advance some truth, or to foster some feeling by the encouragement of a great example, they form a society. Wherever, at the head of some new undertaking, you see the Government in France, or a man of rank in England, in the United States you will be sure to find an association. I met with several kinds of associations in America, of which I confess I had no previous notion; and I have often admired the extreme skill with which the inhabitants of the United States succeed in proposing a common object to the exertions of a great many men, and in getting them voluntarily to pursue it. I have since travelled over England, whence the Americans have taken some of their laws and many of their customs; and it seemed to me that the principle of association was by no means so constantly or so adroitly used in that country. The English often perform great things singly; whereas the Americans form associations for the smallest undertakings. It is evident that the former people consider association as a powerful means of action, but the latter seem to regard it as the only means they have of acting.

Thus the most democratic country on the face of the earth is that in which men have in our time carried to the highest perfection the art of pursuing in common the object of their common desires, and have applied this new science to the greatest number of purposes. Is this the result of accident? or is there in reality any necessary connection between the principle of association and that of equality? Aristocratic communities always contain, amongst a multitude of persons who by themselves are powerless, a small number of powerful and wealthy citizens, each of whom can achieve great undertakings single-handed. In aristocratic societies men do not need to combine in order to act, because they are strongly held together. Every wealthy and powerful citizen constitutes the head of a permanent and compulsory association, composed of all those who are dependent upon him, or whom he makes subservient to the execution of his designs. Amongst democratic nations, on the contrary, all the citizens are independent and feeble; they can do hardly anything by themselves, and none of them can oblige his fellowmen to lend him their assistance. They all, therefore, fall into a state of incapacity, if they do not learn voluntarily to help each other. If men living in democratic countries had no right and no inclination to associate for political purposes, their independence would be in great jeopardy; but they might long preserve their wealth and their cultivation: whereas if they never acquired the habit of forming associations in ordinary life, civilisation itself would be endangered. A people amongst which individuals should lose the power of achieving great things single-handed, without acquiring the means of producing them by united exertions, would soon relapse into barbarism.

Unhappily, the same social condition which renders associations so necessary to democratic nations, renders their formation more difficult amongst those nations than amongst all others. When several members of an aristocracy agree to combine, they easily succeed in doing so; as each of them brings great strength to the partnership, the number of its members may be very limited; and when the members of an association are limited in number, they may easily become mutually acquainted, understand each other, and establish fixed regulations. The same opportunities do not occur amongst democratic nations, where the associated members must always be very numerous for their association to have any power.

I am aware that many of my countrymen are not in the least embarrassed by this difficulty. They contend that the more enfeebled and incompetent the citizens become, the more able and active the Government ought to be rendered, in order that society at large may execute what individuals can no longer accomplish. They believe this answers the whole difficulty, but I think they are mistaken. A Government might perform the part of some of the largest American companies; and several States, members of the Union, have already attempted it; but what political power could ever carry on the vast multitude of lesser undertakings which the American citizens perform every day, with the assistance of the principle of association? It is easy to foresee that the time is drawing near when man will be less and less able to produce, of himself alone, the commonest necessaries of life. The task of the governing power will therefore perpetually increase, and its very efforts will extend it every day. The more it stands in the place of associations, the more will individuals, losing the notion of combining together, require its assistance: these are causes and effects which unceasingly engender each other. Will the administration of the country ultimately assume the management of all the manufacturers, which no single citizen is able to carry on? And if a time at length arrives, when, in consequence of the extreme subdivision of landed property, the soil is split into an infinite number of parcels, so that it can only be cultivated by companies of husbandmen, will it be necessary that the head of the government should leave the helm of state to follow the plough? The morals and the intelligence of a democratic people would be as much endangered as its business and manufactures, if the government ever wholly usurped the place of private companies.

Feelings and opinions are recruited, the heart is enlarged, and the human mind is developed by no other means than by the reciprocal influence of men upon each other. I have shown that these influences are almost null in democratic countries; they must therefore be artificially created, and this can only be accomplished by associations.

When the members of an aristocratic community adopt a new opinion, or conceive a new sentiment, they give it a station, as it were, beside themselves, upon the lofty platform where they stand; and opinions or sentiments so conspicuous to the eyes of the multitude are easily introduced into the minds or hearts of all around. In democratic countries the governing power alone is naturally in a condition to act in this manner; but it is easy to see that its action is always inadequate, and often dangerous. A government can no more be competent to keep alive and to renew the circulation of opinions and feelings amongst a great people, than to manage all the speculations of productive industry. No sooner does a government attempt to go beyond its political sphere and to enter upon this new track, than it exercises, even unintentionally, an insupportable tyranny; for a government can only dictate strict rules, the opinions which it favours are rigidly enforced, and it is never easy to discriminate between its advice and its commands. Worse still will be the case if the government really believes itself interested in preventing all circulation of ideas; it will then stand motionless, and oppressed by the heaviness of voluntary torpor. Governments therefore should not be the only active powers: associations ought, in democratic nations, to stand in lieu of those powerful private individuals whom the equality of conditions has swept away.

As soon as several of the inhabitants of the United States have taken up an opinion or a feeling which they wish to promote in the world, they look out for mutual assistance; and as soon as they have found each other out, they combine. From that moment they are no longer isolated men, but a power seen from afar, whose actions serve for an example, and whose language is listened to. The first time I heard in the United States that a hundred thousand men had bound themselves publicly to abstain from spirituous liquors, it appeared to me more like a joke than a serious engagement; and I did not at once perceive why these temperate citizens could not content themselves with drinking water by their own firesides. I at last understood that three hundred thousand Americans, alarmed by the progress of drunkenness around them, had made up their minds to patronise temperance. They acted just in the same way as a man of high rank who should dress very plainly, in order to inspire the humbler orders with a contempt of luxury. It is probable that if these hundred thousand men had lived in France, each of them would singly have memorialised the government to watch the public-houses all over the kingdom.

Nothing, in my opinion, is more deserving of our attention than the intellectual and moral associations of America. The political and industrial associations of that country strike us forcibly; but the others elude our observation, or if we discover them, we understand them imperfectly, because we have hardly ever seen anything of the kind. It must, however, be acknowledged that they are as necessary to the American people as the former, and perhaps more so. In democratic countries the science of association is the mother of science; the progress of all the rest depends upon the progress it has made. Amongst the laws which rule human societies there is one which seems to be more precise and clear than all others. If men are to remain civilised, or to become so, the art of associating together must grow and improve in the same ratio in which the equality of conditions is increased.

Chapter VI

Of the Relation between Public Associations and Newspapers

When men are no longer united amongst themselves by firm and lasting ties, it is impossible to obtain the cooperation of any great number of them, unless you can persuade every man whose concurrence you require that this private interest obliges him voluntarily to unite his exertions to the exertions of all the rest. This can only be habitually and conveniently effected by means of a newspaper; nothing but a newspaper can drop the same thought into a thousand minds at the same moment. A newspaper is an adviser who does not require to be sought, but who comes of his own accord, and talks to you briefly every day of the common weal, without distracting you from your private affairs.

Newspapers therefore become more necessary in proportion as men become more equal, and individualism more to be feared. To suppose that they only serve to protect freedom would be to diminish their importance: they maintain civilisation. I shall not deny that in democratic countries newspapers frequently lead the citizens to launch together in very ill-digested schemes; but if there were no newspapers there would be no common activity. The evil which they produce is therefore much less than that which they cure.

The effect of a newspaper is not only to suggest the same purpose to a great number of persons, but also to furnish means for executing in common the designs which they may have singly conceived. The principal citizens who inhabit an aristocratic country discern each other from afar; and if they wish to unite their forces, they move towards each other, drawing a multitude of men after them. It frequently happens, on the contrary, in democratic countries, that a great number of men who wish or who want to combine cannot accomplish it, because as they are very insignificant and lost amidst the crowd, they cannot see, and know not where to find, one another. A newspaper then takes up the notion or the feeling which had occurred simultaneously, but singly, to each of them. All are then immediately guided towards this beacon; and these wandering minds, which had long sought each other in darkness, at length meet and unite.

The newspaper brought them together, and the newspaper is still necessary to keep them united. In order that an association amongst a democratic people should have any power, it must be a numerous body. The persons of whom it is composed are therefore scattered over a wide extent, and each of them is detained in the place of his domicile by the narrowness of his income, or by the small unremitting exertions by which he earns it. Means then must be found to converse every day without seeing each other, and to take steps in common without having met. Thus hardly any democratic association can do without newspapers. There is consequently a necessary connection between public associations and newspapers: newspapers make associations, and associations make newspapers; and if it has been correctly advanced that associations will increase in number as the conditions of men become more equal, it is not less certain that the number of newspapers increases in proportion to that of associations. Thus it is in America that we find at the same time the greatest number of associations and of newspapers.

This connection between the number of newspapers and that of associations leads us to the discovery of a further connection between the state of the periodical press and the form of the administration in a country; and shows that the number of newspapers must diminish or increase amongst a democratic people, in proportion as its administration is more or less centralised. For amongst democratic nations the exercise of local powers cannot be intrusted to the principal members of the community as in aristocracies. Those powers must either be abolished, or placed in the hands of very large numbers of men, who then in fact constitute an association permanently established by law for the purpose of administering the affairs of a certain extent of territory; and they require a journal, to bring to them every day, in the midst of their own minor concerns, some intelligence of the state of their public weal. The more numerous local powers are, the greater is the number of men in whom they are vested by law; and as this want is hourly felt, the more profusely do newspapers abound.

The extraordinary subdivision of administrative power has much more to do with the enormous number of American newspapers than the great political freedom of the country and the absolute liberty of the press. If all the inhabitants of the Union had the suffrage—but a suffrage which should only extend to the choice of their legislators in Congress—they would require but few newspapers, because they would only have to act together on a few very important but very rare occasions. But within the pale of the great association of the nation, lesser associations have been established by law in every country, every city, and indeed in every village, for the purposes of local administration. The laws of the country thus compel every American to cooperate every day of his life with some of his fellow-citizens for a common purpose, and each one of them requires a newspaper to inform him what all the others are doing.

I am of opinion that a democratic people,[1] without any national representative assemblies, but with a great number of small local powers, would have in the end more newspapers than another people governed by a centralised administration and an elective legislation. What best explains to me the enormous circulation of the daily press in the United States, is that amongst the Americans I find the utmost national freedom combined with local freedom of every kind. There is a prevailing opinion in France and England that the circulation of newspapers would be indefinitely increased by removing the taxes which have been laid upon the press. This is a very exaggerated estimate of the effects of such a reform. Newspapers increase in numbers, not according to their cheapness, but according to the more or less frequent want which a great number of men may feel for intercommunication and combination.

In like manner I should attribute the increasing influence of the daily press to causes more general than those by which it is commonly explained. A newspaper can only subsist on the condition of publishing sentiments or principles common to a large number of men. A newspaper therefore always represents an association which is composed of its habitual readers. This association may be more or less defined, more or less restricted, more or less numerous; but the fact that the newspaper keeps alive, is a proof that at least the germ of such an association exists in the minds of its readers.

This leads me to a last reflection, with which I shall conclude this chapter. The more equal the conditions of men become, and the less strong men individually are, the more easily do they give way to the current of the multitude, and the more difficult is it for them to adhere by themselves to an opinion which the multitude discard. A newspaper represents an association; it may be said to address each of its readers in the name of all the others, and to exert its influence over them in proportion to their individual weakness. The power of the newspaper press must therefore increase as the social conditions of men become more equal.

Chapter VII

Connection of Civil and Political Associations

There is only one country on the face of the earth where the citizens enjoy unlimited freedom of association for political purposes. This same country is the only one in the world where the continual exercise of the right of association has been introduced into civil life, and where all the advantages which civilisation can confer are procured by means of it. In all the countries where political associations are prohibited, civil associations are rare. It is hardly probable that this is the result of accident; but the inference should rather be, that there is a natural, and perhaps a necessary, connection between these two kinds of associations. Certain men happen to have a common interest in some concern—either a commercial undertaking is to be managed, or some speculation in manufactures to be tried; they meet, they combine, and thus by degrees they become familiar with the principle of association. The greater is the multiplicity of small affairs, the more do men, even without knowing it, acquire facility in prosecuting great undertakings in common. Civil associations, therefore, facilitate political association: but, on the other hand, political association singularly strengthens and improves associations for civil purposes. In civil life every man may, strictly speaking, fancy that he can provide for his own wants; in politics, he can fancy no such thing. When a people, then, have any knowledge of public life, the notion of association, and the wish to coalesce, present themselves every day to the minds of the whole community: whatever natural repugnance may restrain men from acting in concert, they will always be ready to combine for the sake of a party. Thus political life makes the love and practice of association more general; it imparts a desire of union, and teaches the means of combination to numbers of men who would have always lived apart.

Politics not only give birth to numerous associations, but to associations of great extent. In civil life it seldom happens that any one interest draws a very large number of men to act in concert; much skill is required to bring such an interest into existence: but in politics opportunities present themselves every day. Now it is solely in great associations that the general value of the principle of association is displayed. Citizens who are individually powerless, do not very clearly anticipate the strength which they may acquire by uniting together; it must be shown to them in order to be understood. Hence it is often easier to collect a multitude for a public purpose than a few persons; a thousand citizens do not see what interest they have in combining together—ten thousand will be perfectly aware of it. In politics men combine for great undertakings; and the use they make of the principle of association in important affairs practically teaches them that it is their interest to help each other in those of less moment. A political association draws a number of individuals at the same time out of their own circle: however they may be naturally kept asunder by age, mind, and fortune, it places them nearer together and brings them into contact. Once met, they can always meet again.

Men can embark in few civil partnerships without risking a portion of their possessions; this is the case with all manufacturing and trading companies. When men are as yet but little versed in the art of association, and are unacquainted with its principal rules, they are afraid, when first they combine in this manner, of buying their experience dear. They therefore prefer depriving themselves of a powerful instrument of success to running the risks which attend the use of it. They are, however, less reluctant to join political associations, which appear to them to be without danger, because they adventure no money in them. But they cannot belong to these associations for any length of time without finding out how order is maintained amongst a large number of men, and by what contrivance they are made to advance, harmoniously and methodically, to the same object. Thus they learn to surrender their own will to that of all the rest, and to make their own exertions subordinate to the common impulse—things which it is not less necessary to know in civil than in political associations. Political associations may therefore be considered as large free schools, where all the members of the community go to learn the general theory of association.

But even if political association did not directly contribute to the progress of civil association, to destroy the former would be to impair the latter. When citizens can only meet in public for certain purposes, they regard such meetings as a strange proceeding of rare occurrence, and they rarely think at all about it. When they are allowed to meet freely for all purposes, they ultimately look upon public association as the universal, or in a manner the sole means, which men can employ to accomplish the different purposes they may have in view. Every new want instantly revives the notion. The art of association then becomes, as I have said before, the mother of action, studied and applied by all.

When some kinds of associations are prohibited and others allowed, it is difficult to distinguish the former from the latter beforehand. In this state of doubt men abstain from them altogether, and a sort of public opinion passes current, which tends to cause any association whatsoever to be regarded as a bold and almost an illicit enterprise.[1]

It is therefore chimerical to suppose that the spirit of association, when it is repressed on some one point, will nevertheless display the same vigour on all others; and that if men be allowed to prosecute certain undertakings in common, that is quite enough for them eagerly to set about them. When the members of a community are allowed and accustomed to combine for all purposes, they will combine as readily for the lesser as for the more important ones; but if they are only allowed to combine for small affairs, they will be neither inclined nor able to effect it. It is in vain that you will leave them entirely free to prosecute their business on joint-stock account: they will hardly care to avail themselves of the rights you have granted to them; and, after having exhausted your strength in vain efforts to put down prohibited associations, you will be surprised that you cannot persuade men to form the associations you encourage.

I do not say that there can be no civil associations in a country where political association is prohibited; for men can never live in society without embarking in some common undertakings: but I maintain that in such a country civil associations will always be few in number, feebly planned, unskilfully managed, that they will never form any vast designs, or that they will fail in the execution of them.

This naturally leads me to think that freedom of association in political matters is not so dangerous to public tranquillity as is supposed; and that possibly, after having agitated society for some time, it may strengthen the State in the end. In democratic countries political associations are, so to speak, the only powerful persons who aspire to rule the State. Accordingly, the governments of our time look upon associations of this kind just as sovereigns in the Middle Ages regarded the great vassals of the crown: they entertain a sort of instinctive abhorrence of them, and they combat them on all occasions. They bear, on the contrary, a natural goodwill to civil associations, because they readily discover that, instead of directing the minds of the community to public affairs, these institutions serve to divert them from such reflections; and that, by engaging them more and more in the pursuit of objects which cannot be attained without public tranquillity, they deter them from revolutions. But these governments do not attend to the fact that political associations tend amazingly to multiply and facilitate those of a civil character, and that in avoiding a dangerous evil they deprive themselves of an efficacious remedy.

When you see the Americans freely and constantly forming associations for the purpose of promoting some political principle, of raising one man to the head of affairs, or of wresting power from another, you have some difficulty in understanding that men so independent do not constantly fall into the abuse of freedom. If, on the other hand, you survey the infinite number of trading companies which are in operation in the United States, and perceive that the Americans are on every side unceasingly engaged in the execution of important and difficult plans, which the slightest revolution would throw into confusion, you will readily comprehend why people so well employed are by no means tempted to perturb the State, nor to destroy that public tranquillity by which they all profit.

Is it enough to observe these things separately, or should we not discover the hidden tie which connects them? In their political associations, the Americans of all conditions, minds, and ages, daily acquire a general taste for association, and grow accustomed to the use of it. There they meet together in large numbers, they converse, they listen to each other, and they are mutually stimulated to all sorts of undertakings. They afterwards transfer to civil life the notions they have thus acquired, and make them subservient to a thousand purposes. Thus it is by the enjoyment of a dangerous freedom that the Americans learn the art of rendering the dangers of freedom less formidable.

If a certain moment in the existence of a nation be selected, it is easy to prove that political associations perturb the State, and paralyse productive industry; but take the whole life of a people, and it may perhaps be easy to demonstrate that freedom of association in political matters is favourable to the prosperity and even to the tranquillity of the community.

I said in the former part of this work, 'The unrestrained liberty of political association cannot be entirely assimilated to the liberty of the press. The one is at the same time less necessary and more dangerous than the other. A nation may confine it within certain limits without ceasing to be mistress of itself; and it may sometimes be obliged to do so in order to maintain its own authority.' And further on I added: 'It cannot be denied that the unrestrained liberty of association for political purposes is the last degree of liberty which a people is fit for. If it does not throw them into anarchy, it perpetually brings them, as it were, to the verge of it.' Thus I do not think that a nation is always at liberty to invest its citizens with an absolute right of association for political purposes; and I doubt whether, in any country or in any age, it be wise to set no limits to freedom of association. A certain nation, it is said, could not maintain tranquillity in the community, cause the laws to be respected, or establish a lasting government, if the right of association were not confined within narrow limits. These blessings are doubtless invaluable, and I can imagine that, to acquire or to preserve them, a nation may impose upon itself severe temporary restrictions: but still it is well that the nation should know at what price these blessings are purchased. I can understand that it may be advisable to cut off a man's arm in order to save his life; but it would be ridiculous to assert that he will be as dexterous as he was before he lost it.

Chapter VIII

The Americans Combat Individualism by the Principle of Interest Rightly Understood

When the world was managed by a few rich and powerful individuals, these persons loved to entertain a lofty idea of the duties of man. They were fond of professing that it is praiseworthy to forget oneself, and that good should be done without hope of reward, as it is by the Deity himself. Such were the standard opinions of that time in morals. I doubt whether men were more virtuous in aristocratic ages than in others; but they were incessantly talking of the beauties of virtue, and its utility was only studied in secret. But since the imagination takes less lofty flights and every man's thoughts are centred in himself, moralists are alarmed by this idea of self-sacrifice, and they no longer venture to present it to the human mind. They therefore content themselves with inquiring whether the personal advantage of each member of the community does not consist in working for the good of all; and when they have hit upon some point on which private interest and public interest meet and amalgamate, they are eager to bring it into notice. Observations of this kind are gradually multiplied: what was only a single remark becomes a general principle; and it is held as a truth that man serves himself in serving his fellow-creatures, and that his private interest is to do good.

I have already shown, in several parts of this work, by what means the inhabitants of the United States almost always manage to combine their own advantage with that of their fellow-citizens: my present purpose is to point out the general rule which enables them to do so. In the United States hardly anybody talks of the beauty of virtue; but they maintain that virtue is useful, and prove it every day. The American moralists do not profess that men ought to sacrifice themselves for their fellow-creatures *because* it is noble to make such sacrifices; but they boldly aver that such sacrifices are as necessary to him who imposes them upon himself as to him for whose sake they are made. They have found out that in their country and their age man is brought home to himself by an irresistible force; and losing all hope of stopping that force, they turn all their thoughts to the direction of it. They therefore do not deny that every man may follow his own interest; but they endeavour to prove that it is the interest of every man to be virtuous. I shall not here enter into the reasons they allege, which would divert me from my subject: suffice it to say that they have convinced their fellow-countrymen.

Montaigne said long ago: 'Were I not to follow the straight road for its straightness, I should follow it for having found by experience that in the end it is commonly the happiest and most useful track.' The doctrine of interest rightly understood is not, then, new, but amongst the Americans of our time it finds universal acceptance: it has become popular there; you may trace it at the bottom of all their actions, you will remark it in all they say. It is as often to be met with on the lips of the poor man as of the rich. In Europe the principle of interest is much grosser than it is in America, but at the same time it is less common, and especially it is less avowed; amongst us, men still constantly feign great abnegation which they no longer feel. The Americans, on the contrary, are fond of explaining almost all the actions of their lives by the principle of interest rightly understood; they show with complacency how an enlightened regard for themselves constantly prompts them to assist each other, and inclines them willingly to sacrifice a portion of their time and property to the welfare of the State. In this respect I think they frequently fail to do themselves justice; for in the United States, as well as elsewhere, people are sometimes seen to give way to those disinterested and spontaneous impulses which are natural to man; but the Americans seldom allow that they yield to emotions of this kind; they are more anxious to do honour to their philosophy than to themselves.

I might here pause, without attempting to pass a judgment on what I have described. The extreme difficulty of the subject would be my excuse, but I shall not avail myself of it; and I had rather that my readers, clearly perceiving my object, should refuse to follow me than that I should leave them in suspense. The principle of interest rightly understood is not a lofty one, but it is clear and sure. It does not aim at mighty objects, but it attains without excessive exertion all those at which it aims. As it lies within the reach of all capacities, everyone can without difficulty apprehend and retain it. By its admirable conformity to human weaknesses, it easily obtains great dominion; nor is that dominion precarious, since the principle checks one personal interest by another, and uses, to direct the passions, the very same instrument which excites them. The principle of interest rightly understood produces no great acts of self-sacrifice, but it suggests daily small acts of self-denial. By itself it cannot suffice to make a man virtuous, but it disciplines a number of citizens in habits of regularity, temperance, moderation, foresight, self-command; and, if it does not lead men straight to virtue by the will, it gradually draws them in that direction by their habits. If the principle of interest rightly understood were to sway the whole moral world, extraordinary virtues would doubtless be more rare; but I think that gross depravity would then also be less common. The principle of interest rightly understood perhaps prevents some men from rising far above the level of mankind; but a great number of other men, who were falling far below it, are caught and restrained by it. Observe some few individuals, they are lowered by it; survey mankind, it is raised. I am not afraid to say that the principle of interest, rightly understood, appears to me the best suited of all philosophical theories to the wants of the men of our time, and that I regard it as their chief remaining security against themselves. Towards it, therefore, the minds of the moralists of our age should turn; even should they judge it to be incomplete, it must nevertheless be adopted as necessary.

I do not think upon the whole that there is more egotism amongst us than in America; the only difference is, that there it is enlightened—here it is not. Every American will sacrifice a portion of his private interests to preserve the rest; we would fain preserve the whole, and oftentimes the whole is lost. Everybody I see about me seems bent on teaching his contemporaries, by precept and example, that what is useful is never wrong. Will nobody undertake to make them understand how what is right may be useful? No power upon earth can prevent the increasing equality of conditions from inclining the human mind to seek out what is useful, or from leading every member of the community to be wrapped up in himself. It must therefore be expected that personal interest will become more than ever the principal, if not the sole, spring of men's actions; but it remains to be seen how each man will understand his personal interest. If the members of a community, as they become more equal, become more ignorant and coarse, it is difficult to foresee to what pitch of stupid excesses their egotism may lead them; and no one can foretell into what disgrace and wretchedness they would plunge themselves, lest they should have to sacrifice something of their own well-being to the prosperity of their fellow-creatures. I do not think that the system of interest, as it is professed in America, is, in all its parts, self-evident; but it contains a great number of truths so evident that men, if they are but educated, cannot fail to see them. Educate, then, at any rate; for the age of implicit self-sacrifice and instinctive virtues is already flitting far away from us, and the time is fast approaching when freedom, public peace, and social order itself will not be able to exist without education.

Chapter IX

That the Americans Apply the Principle of Interest Rightly Understood to Religious Matters

If the principle of interest rightly understood had nothing but the present world in view, it would be very insufficient; for there are many sacrifices which can only find their recompense in another; and whatever ingenuity may be put forth to demonstrate the utility of virtue, it will never be an easy task to make that man live aright who has no thoughts of dying. It is therefore necessary to ascertain whether the principle of interest rightly understood is easily compatible with religious belief. The philosophers who inculcate this system of morals tell men, that to be happy in this life they must watch their own passions and steadily control their excess; that lasting happiness can only be secured by renouncing a thousand transient gratifications; and that a man must perpetually triumph over himself, in order to secure his own advantage. The founders of almost all religions have held the same language. The track they point out to man is the same, only that the goal is more remote; instead of placing in this world the reward of the sacrifices they impose, they transport it to another. Nevertheless I cannot believe that all those who practise virtue from religious motives are only actuated by the hope of a recompense. I have known zealous Christians who constantly forgot themselves, to work with greater ardour for the happiness of their fellow-men; and I have heard them declare that all they did was only to earn the blessings of a future state. I cannot but think that they deceive themselves; I respect them too much to believe them.

Christianity indeed teaches that a man must prefer his neighbour to himself, in order to gain eternal life; but Christianity also teaches that men ought to benefit their fellow-creatures for the love of God. A sublime expression! Man, searching by his intellect into the Divine conception, and seeing that order is the purpose of God, freely combines to prosecute the great design; and whilst he sacrifices his personal interests to this consummate order of all created things, expects no other recompense than the pleasure of contemplating it. I do not believe that interest is the sole motive of religious men: but I believe that interest is the principal means which religions themselves employ to govern men, and I do not question that this way they strike into the multitude and become popular. It is not easy clearly to perceive why the principle of interest rightly understood should keep aloof from religious opinions; and it seems to me more easy to show why it should draw men to them. Let it be supposed that, in order to obtain happiness in this world, a man combats his instinct on all occasions and deliberately calculates every action of his life; that, instead of yielding blindly to the impetuosity of first desires, he has learned the art of resisting them, and that he has accustomed himself to sacrifice without an effort the pleasure of a moment to the lasting interest of his whole life. If such a man believes in the religion which he professes, it will cost him but little to submit to the restrictions it may impose. Reason herself counsels him to obey, and habit has prepared him to endure them. If he should have conceived any doubts as to the object of his hopes, still he will not easily allow himself to be stopped by them; and he will decide that it is wise to risk some of the advantages of this world, in order to preserve his rights to the great inheritance promised him in another. 'To be mistaken in believing that the Christian religion is true,' says Pascal, 'is no great loss to anyone; but how dreadful to be mistaken in believing it to be false!'

The Americans do not affect a brutal indifference to a future state; they affect no puerile pride in despising perils which they hope to escape from. They therefore profess their religion without shame and without weakness; but there generally is, even in their zeal, something so indescribably tranquil, methodical, and deliberate, that it would seem as if the head, far more than the heart, brought them to the foot of the altar. The Americans not only follow their religion from interest, but they often place in this world the interest which makes them follow it. In the middle ages the clergy spoke of nothing but a future state; they hardly cared to prove that a sincere Christian may be a happy man here below. But the American preachers are constantly referring to the earth; and it is only with great difficulty that they can divert their attention from it. To touch their congregations, they always show them how favourable religious opinions are to freedom and public tranquillity; and it is often difficult to ascertain from their discourses whether the principal object of religion is to procure eternal felicity in the other world, or prosperity in this.

Chapter X

Of the Taste for Physical Well-Being in America

In America the passion for physical well-being is not always exclusive, but it is general; and if all do not feel it in the same manner, yet it is felt by all. Carefully to satisfy all, even the least wants of the body, and to provide the little conveniences of life, is uppermost in every mind. Something of an analogous character is more and more apparent in Europe. Amongst the causes which produce these similar consequences in both hemispheres, several are so connected with my subject as to deserve notice.

When riches are hereditarily fixed in families, there are a great number of men who enjoy the comforts of life without feeling an exclusive taste for those comforts. The heart of man is not so much caught by the undisturbed possession of anything valuable as by the desire, as yet imperfectly satisfied, of possessing it, and by the incessant dread of losing it. In aristocratic communities, the wealthy, never having experienced a condition different from their own, entertain no fear of changing it; the existence of such conditions hardly occurs to them. The comforts of life are not to them the end of life, but simply a way of living; they regard them as existence itself—enjoyed, but scarcely thought of. As the natural and instinctive taste which all men feel for being well off is thus satisfied without trouble and without apprehension, their faculties are turned elsewhere, and cling to more arduous and more lofty undertakings, which excite and engross their minds. Hence it is that, in the midst of physical gratifications, the members of an aristocracy often display a haughty contempt of these very enjoyments, and exhibit singular powers of endurance under the privation of them. All the revolutions which have ever shaken or destroyed aristocracies, have shown how easily men accustomed to superfluous luxuries can do without the necessaries of life; whereas men who have toiled to acquire a competency can hardly live after they have lost it.

If I turn my observation from the upper to the lower classes, I find analogous effects produced by opposite causes. Amongst a nation where aristocracy predominates in society, and keeps it stationary, the people in the end get as much accustomed to poverty as the rich to their opulence. The latter bestow no anxiety on their physical comforts, because they enjoy them without an effort; the former do not think of things which they despair of obtaining, and which they hardly know enough of to desire them. In communities of this kind, the imagination of the poor is driven to seek another world; the miseries of real life inclose it around, but it escapes from their control, and flies to seek its pleasures far beyond. When, on the contrary, the distinctions of ranks are confounded together and privileges are destroyed—when hereditary property is subdivided, and education and freedom widely diffused, the desire of acquiring the comforts of the world haunts the imagination of the poor, and the dread of losing them that of the rich. Many scanty fortunes spring up; those who possess them have a sufficient share of physical gratifications to conceive a taste for these pleasures—not enough to satisfy it. They never procure them without exertion, and they never indulge in them without apprehension. They are therefore always straining to pursue or to retain gratifications so delightful, so imperfect, so fugitive.

If I were to inquire what passion is most natural to men who are stimulated and circumscribed by the obscurity of their birth or the mediocrity of their fortune, I could discover none more peculiarly appropriate to their condition than this love of physical prosperity. The passion for physical comforts is essentially a passion of the middle classes: with those classes it grows and spreads, with them it preponderates. From them it mounts into the higher orders of society, and descends into the mass of the people. I never met in America with any citizen so poor as not to cast a glance of hope and envy on the enjoyments of the rich, or whose imagination did not possess itself by anticipation of those good things which fate still obstinately withheld from him. On the other hand, I never perceived amongst the wealthier inhabitants of the United States that proud contempt of physical gratifications which is sometimes to be met with even in the most opulent and dissolute aristocracies. Most of these wealthy persons were once poor; they have felt the sting of want; they were long a prey to adverse fortunes; and now that the victory is won, the passions which accompanied the contest have survived it: their minds are, as it were, intoxicated by the small enjoyments which they have pursued for forty years. Not but that in the United States, as elsewhere, there are a certain number of wealthy persons who, having come into their property by inheritance, possess, without exertion, an opulence they have not earned. But even these men are not less devotedly attached to the pleasures of material life. The love of well-being is now become the predominant taste of the nation; the great current of man's passions runs in that channel, and sweeps everything along in its course.

Chapter XI

Peculiar Effects of the Love of Physical Gratifications in Democratic Ages

It may be supposed, from what has just been said, that the love of physical gratifications must constantly urge the Americans to irregularities in morals, disturb the peace of families, and threaten the security of society at large. Such is not the case: the passion for physical gratifications produces in democracies effects very different from those which it occasions in aristocratic nations. It sometimes happens that, wearied with public affairs and sated with opulence, amidst the ruin of religious belief and the decline of the State, the heart of an aristocracy may by degrees be seduced to the pursuit of sensual enjoyments only. At other times the power of the monarch or the weakness of the people, without stripping the nobility of their fortune, compels them to stand aloof from the administration of affairs, and whilst the road to mighty enterprise is closed, abandons them to the inquietude of their own desires; they then fall back heavily upon themselves, and seek in the pleasures of the body oblivion of their former greatness. When the members of an aristocratic body are thus exclusively devoted to the pursuit of physical gratifications, they commonly concentrate in that direction all the energy which they derive from their long experience of power. Such men are not satisfied with the pursuit of comfort; they require sumptuous depravity and splendid corruption. The worship they pay the senses is a gorgeous one; and they seem to vie with each other in the art of degrading their own natures. The stronger, the more famous, and the more free an aristocracy has been, the more depraved will it then become; and however brilliant may have been the lustre of its virtues, I dare predict that they will always be surpassed by the splendour of its vices.

The taste for physical gratifications leads a democratic people into no such excesses. The love of well-being is there displayed as a tenacious, exclusive, universal passion; but its range is confined. To build enormous palaces, to conquer or to mimic nature, to ransack the world in order to gratify the passions of a man, is not thought of: but to add a few roods of land to your field, to plant an orchard, to enlarge a dwelling, to be always making life more comfortable and convenient, to avoid trouble, and to satisfy the smallest wants without effort and almost without cost. These are small objects, but the soul clings to them; it dwells upon them closely and day by day, till they at last shut out the rest of the world, and sometimes intervene between itself and Heaven.

This, it may be said, can only be applicable to those members of the community who are in humble circumstances; wealthier individuals will display tastes akin to those which belonged to them in aristocratic ages. I contest the proposition: in point of physical gratifications, the most opulent members of a democracy will not display tastes very different from those of the people; whether it be that, springing from the people, they really share those tastes, or that they esteem it a duty to submit to them. In democratic society the sensuality of the public has taken a moderate and tranquil course, to which all are bound to conform: it is as difficult to depart from the common rule by one's vices as by one's virtues. Rich men who live amidst democratic nations are therefore more intent on providing for their smallest wants than for their extraordinary enjoyments; they gratify a number of petty desires, without indulging in any great irregularities of passion: thus they are more apt to become enervated than debauched.

The especial taste which the men of democratic ages entertain for physical enjoyments is not naturally opposed to the principles of public order; nay, it often stands in need of order that it may be gratified. Nor is it adverse to regularity of morals, for good morals contribute to public tranquillity and are favourable to industry. It may even be frequently combined with a species of religious morality: men wish to be as well off as they can in this world, without foregoing their chance of another. Some physical gratifications cannot be indulged in without crime; from such they strictly abstain. The enjoyment of others is sanctioned by religion and morality; to these the heart, the imagination, and life itself are unreservedly given up; till, in snatching at these lesser gifts, men lose sight of those more precious possessions which constitute the glory and the greatness of mankind. The reproach I address to the principle of equality, is not that it leads men away in the pursuit of forbidden enjoyments, but that it absorbs them wholly in quest of those which are allowed. By these means, a kind of virtuous materialism may ultimately be established in the world, which would not corrupt, but enervate the soul, and noiselessly unbend its springs of action.

Chapter XII

Causes of Fanatical Enthusiasm in Some Americans

Although the desire of acquiring the good things of this world is the prevailing passion of the American people, certain momentary outbreaks occur, when their souls seem suddenly to burst the bonds of matter by which they are restrained, and to soar impetuously towards Heaven. In all the States of the Union, but especially in the half-peopled country of the Far West, wandering preachers may be met with who hawk about the word of God from place to place. Whole families—old men, women, and children—cross rough passes and untrodden wilds, coming from a great distance, to join a camp-meeting, where they totally forget for several days and nights, in listening to these discourses, the cares of business and even the most urgent wants of the body. Here and there, in the midst of American society, you meet with men, full of a fanatical and almost wild enthusiasm, which hardly exists in Europe. From time to time strange sects arise, which endeavor to strike out extraordinary paths to eternal happiness. Religious insanity is very common in the United States.

Nor ought these facts to surprise us. It was not man who implanted in himself the taste for what is infinite and the love of what is immortal: those lofty instincts are not the offspring of his capricious will; their steadfast foundation is fixed in human nature, and they exist in spite of his efforts. He may cross and distort them—destroy them he cannot. The soul has wants which must be satisfied; and whatever pains be taken to divert it from itself, it soon grows weary, restless, and disquieted amidst the enjoyments of sense. If ever the faculties of the great majority of mankind were exclusively bent upon the pursuit of material objects, it might be anticipated that an amazing reaction would take place in the souls of some men. They would drift at large in the world of spirits, for fear of remaining shackled by the close bondage of the body.

It is not then wonderful if, in the midst of a community whose thoughts tend earthward, a small number of individuals are to be found who turn their looks to Heaven. I should be surprised if mysticism did not soon make some advance amongst a people solely engaged in promoting its own worldly welfare. It is said that the deserts of the Thebaid were peopled by the persecutions of the emperors and the massacres of the Circus; I should rather say that it was by the luxuries of Rome and the Epicurean philosophy of Greece. If their social condition, their present circumstances, and their laws did not confine the minds of the Americans so closely to the pursuit of worldly welfare, it is probable that they would display more reserve and more experience whenever their attention is turned to things immaterial, and that they would check themselves without difficulty. But they feel imprisoned within bounds which they will apparently never be allowed to pass. As soon as they have passed these bounds, their minds know not where to fix themselves, and they often rush unrestrained beyond the range of common sense.

Chapter XIII

Causes of the Restless Spirit of the Americans in the Midst of Their Prosperity

In certain remote corners of the Old World you may still sometimes stumble upon a small district which seems to have been forgotten amidst the general tumult, and to have remained stationary whilst everything around it was in motion. The inhabitants are for the most part extremely ignorant and poor; they take no part in the business of the country, and they are frequency oppressed by the government; yet their countenances are generally placid, and their spirits light. In America I saw the freest and most enlightened men, placed in the happiest circumstances which the world affords: it seemed to me as if a cloud habitually hung upon their brow, and I thought them serious and almost sad even in their pleasures. The chief reason of this contrast is that the former do not think of the ills they endure—the latter are forever brooding over advantages they do not possess. It is strange to see with what feverish ardour the Americans pursue their own welfare; and to watch the vague dread that constantly torments them lest they should not have chosen the shortest path which may lead to it. A native of the United States clings to this world's goods as if he were certain never to die; and he is so hasty in grasping at all within his reach, that one would suppose he was constantly afraid of not living long enough to enjoy them. He clutches everything, he holds nothing fast, but soon loosens his grasp to pursue fresh gratifications.

In the United States a man builds a house to spend his latter years in it, and he sells it before the roof is on: he plants a garden, and lets it just as the trees are coming into bearing: he brings a field into tillage, and leaves other men to gather the crops: he embraces a profession, and gives it up: he settles in a place, which he soon afterwards leaves, to carry his changeable longings elsewhere. If his private affairs leave him any leisure, he instantly plunges into the vortex of politics; and if at the end of a year of unremitting labour he finds he has a few days' vacation, his eager curiosity whirls him over the vast extent of the United States, and he will travel fifteen hundred miles in a few days, to shake off his happiness. Death at length overtakes him, but it is before he is weary of his bootless chase of that complete felicity which is for ever on the wing.

At first sight there is something surprising in this strange unrest of so many happy men, restless in the midst of abundance. The spectacle itself is however as old as the world; the novelty is to see a whole people furnish an exemplification of it. Their taste for physical gratifications must be regarded as the original source of that secret inquietude which the actions of the Americans betray, and of that inconstancy of which they afford fresh examples every day. He who has set his heart exclusively upon the pursuit of worldly welfare is always in a hurry, for he has but a limited time at his disposal to reach it, to grasp it, and to enjoy it. The recollection of the brevity of life is a constant spur to him. Besides the good things which he possesses, he every instant fancies a thousand others which death will prevent him from trying if he does not try them soon. This thought fills him with anxiety, fear, and regret, and keeps his mind in ceaseless trepidation, which leads him perpetually to change his plans and his abode. If in addition to the taste for physical well-being a social condition be superadded, in which the laws and customs make no condition permanent, here is a great additional stimulant to this restlessness of temper. Men will then be seen continually to change their track, for fear of missing the shortest cut to happiness. It may readily be conceived that if men, passionately bent upon physical gratifications, desire eagerly, they are also easily discouraged: as their ultimate object is to enjoy, the means to reach that object must be prompt and easy, or the trouble of acquiring the gratification would be greater than the gratification itself. Their prevailing frame of mind then is at once ardent and relaxed, violent and enervated. Death is often less dreaded than perseverance in continuous efforts to one end.

The equality of conditions leads by a still straighter road to several of the effects which I have here described. When all the privileges of birth and fortune are abolished, when all professions are accessible to all, and a man's own energies may place him at the top of any one of them, an easy and unbounded career seems open to his ambition, and he will readily persuade himself that he is born to no vulgar destinies. But this is an erroneous notion, which is corrected by daily experience. The same equality which allows every citizen to conceive these lofty hopes, renders all the citizens less able to realise them: it circumscribes their powers on every side, whilst it gives freer scope to their desires. Not only are they themselves powerless, but they are met at every step by immense obstacles, which they did not at first perceive. They have swept away the privileges of some of their fellow-creatures which stood in their way, but they have opened the door to universal competition: the barrier has changed its shape rather than its position. When men are nearly alike, and all follow the same track, it is very difficult for any one individual to walk quick and cleave a way through the dense throng which surrounds and presses him. This constant strife between the propensities springing from the equality of conditions and the means it supplies to satisfy them, harasses and wearies the mind.

It is possible to conceive men arrived at a degree of freedom which should completely content them; they would then enjoy their independence without anxiety and without impatience. But men will never establish any equality with which they can be contented. Whatever efforts a people may make, they will never succeed in reducing all the conditions of society to a perfect level; and even if they unhappily attained that absolute and complete depression, the inequality of minds would still remain, which, coming directly from the hand of God, will forever escape the laws of man. However democratic then the social state and the political constitution of a people may be, it is certain that every member of the community will always find out several points about him which command his own position; and we may foresee that his looks will be doggedly fixed in that direction. When inequality of conditions is the common law of society, the most marked inequalities do not strike the eye: when everything is nearly on the same level, the slightest are marked enough to hurt it. Hence the desire of equality always becomes more insatiable in proportion as equality is more complete.

Amongst democratic nations men easily attain a certain equality of conditions: they can never attain the equality they desire. It perpetually retires from before them, yet without hiding itself from their sight, and in retiring draws them on. At every moment they think they are about to grasp it; it escapes at every moment from their hold. They are near enough to see its charms, but too far off to enjoy them; and before they have fully tasted its delights they die. To these causes must be attributed that strange melancholy which oftentimes will haunt the inhabitants of democratic countries in the midst of their abundance, and that disgust at life which sometimes seizes upon them in the midst of calm and easy circumstances. Complaints are made in France that the number of suicides increases; in America suicide is rare, but insanity is said to be more common than anywhere else. These are all different symptoms of the same disease. The Americans do not put an end to their lives, however disquieted they may be, because their religion forbids it; and amongst them materialism may be said hardly to exist, notwithstanding the general passion for physical gratification. The will resists—reason frequently gives way.

In democratic ages enjoyments are more intense than in the ages of aristocracy, and especially the number of those who partake in them is larger: but, on the other hand, it must be admitted that man's hopes and his desires are oftener blasted, the soul is more stricken and perturbed, and care itself more keen.

Chapter XIV

Taste for Physical Gratifications United in America to Love of Freedom and Attention to Public Affairs

When a democratic state turns to absolute monarchy, the activity which was before directed to public and to private affairs is all at once centred upon the latter: the immediate consequence is, for some time, great physical prosperity; but this impulse soon slackens, and the amount of productive industry is checked. I know not if a single trading or manufacturing people can be cited, from the Tyrians down to the Florentines and the English, who were not a free people also. There is therefore a close bond and necessary relation between these two elements—freedom and productive industry. This proposition is generally true of all nations, but especially of democratic nations. I have already shown that men who live in ages of equality continually require to form associations in order to procure the things they covet; and, on the other hand, I have shown how great political freedom improves and diffuses the art of association. Freedom, in these ages, is therefore especially favourable to the production of wealth; nor is it difficult to perceive that despotism is especially adverse to the same result. The nature of despotic power in democratic ages is not to be fierce or cruel, but minute and meddling. Despotism of this kind, though it does not trample on humanity, is directly opposed to the genius of commerce and the pursuits of industry.

Thus the men of democratic ages require to be free in order more readily to procure those physical enjoyments for which they are always longing. It sometimes happens, however, that the excessive taste they conceive for these same enjoyments abandons them to the first master who appears. The passion for worldly welfare then defeats itself, and, without perceiving it, throws the object of their desires to a greater distance.

There is, indeed, a most dangerous passage in the history of a democratic people. When the taste for physical gratifications amongst such a people has grown more rapidly than their education and their experience of free institutions, the time will come when men are carried away, and lose all self-restraint, at the sight of the new possessions they are about to lay hold upon. In their intense and exclusive anxiety to make a fortune, they lose sight of the close connection which exists between the private fortune of each of them and the prosperity of all. It is not necessary to do violence to such a people in order to strip them of the rights they enjoy; they themselves willingly loosen their hold. The discharge of political duties appears to them to be a troublesome annoyance, which diverts them from their occupations and business. If they be required to elect representatives, to support the Government by personal service, to meet on public business, they have no time—they cannot waste their precious time in useless engagements: such idle amusements are unsuited to serious men who are engaged with the more important interests of life. These people think they are following the principle of self-interest, but the idea they entertain of that principle is a very rude one; and the better to look after what they call their business, they neglect their chief business, which is to remain their own masters.

As the citizens who work do not care to attend to public business, and as the class which might devote its leisure to these duties has ceased to exist, the place of the Government is, as it were, unfilled. If at that critical moment some able and ambitious man grasps the supreme power, he will find the road to every kind of usurpation open before him. If he does but attend for some time to the material prosperity of the country, no more will be demanded of him. Above all he must insure public tranquillity: men who are possessed by the passion of physical gratification generally find out that the turmoil of freedom disturbs their welfare, before they discover how freedom itself serves to promote it. If the slightest rumour of public commotion intrudes into the petty pleasures of private life, they are aroused and alarmed by it. The fear of anarchy perpetually haunts them, and they are always ready to fling away their freedom at the first disturbance.

I readily admit that public tranquillity is a great good; but at the same time I cannot forget that all nations have been enslaved by being kept in good order. Certainly it is not to be inferred that nations ought to despise public tranquillity; but that state ought not to content them. A nation which asks nothing of its government but the maintenance of order is already a slave at heart—the slave of its own well-being, awaiting but the hand that will bind it. By such a nation the despotism of faction is not less to be dreaded than the despotism of an individual. When the bulk of the community is engrossed by private concerns, the smallest parties need not despair of getting the upper hand in public affairs. At such times it is not rare to see upon the great stage of the world, as we see at our theatres, a multitude represented by a few players, who alone speak in the name of an absent or inattentive crowd: they alone are in action whilst all are stationary; they regulate everything by their own caprice; they change the laws, and tyrannise at will over the manners of the country; and then men wonder to see into how small a number of weak and worthless hands a great people may fall.

Hitherto the Americans have fortunately escaped all the perils which I have just pointed out; and in this respect they are really deserving of admiration. Perhaps there is no country in the world where fewer idle men are to be met with than in America, or where all who work are more eager to promote their own welfare. But if the passion of the Americans for physical gratifications is vehement, at least it is not discriminating; and reason, though unable to restrain it, still directs its course. An American attends to his private concerns as if he were alone in the world, and the next minute he gives himself up to the common weal as if he had forgotten them. At one time he seems animated by the most selfish cupidity, at another by the most lively patriotism. The human heart cannot be thus divided. The inhabitants of the United States alternately display so strong and so similar a passion for their own welfare and for their freedom, that it may be supposed that these passions are united and mingled in some part of their character. And indeed the Americans believe their freedom to be the best instrument and surest safeguard of their welfare: they are attached to the one by the other. They by no means think that they are not called upon to take a part in the public weal; they believe, on the contrary, that their chief business is to secure for themselves a government which will allow them to acquire the things they covet, and which will not debar them from the peaceful enjoyment of those possessions which they have acquired.

Chapter XV

That Religious Belief Sometimes Turns the Thoughts of the Americans to Immaterial Pleasures

In the United States, on the seventh day of every week, the trading and working life of the nation seems suspended; all noises cease; a deep tranquillity, say rather the solemn calm of meditation, succeeds the turmoil of the week, and the soul resumes possession and contemplation of itself. Upon this day the marts of traffic are deserted; every member of the community, accompanied by his children, goes to church, where he listens to strange language which would seem unsuited to his ear. He is told of the countless evils caused by pride and covetousness: he is reminded of the necessity of checking his desires, of the finer pleasures which belong to virtue alone, and of the true happiness which attends it. On his return home, he does not turn to the ledgers of his calling, but he opens the book of Holy Scripture; there he meets with sublime or affecting descriptions of the greatness and goodness of the Creator, of the infinite magnificence of the handiwork of God, of the lofty destinies of man, of his duties, and of his immortal privileges. Thus it is that the American at times steals an hour from himself; and laying aside for a while the petty passions which agitate his life, and the ephemeral interests which engross it, he strays at once into an ideal world, where all is great, eternal, and pure.

I have endeavoured to point out in another part of this work the causes to which the maintenance of the political institutions of the Americans is attributable; and religion appeared to be one of the most prominent amongst them. I am now treating of the Americans in an individual capacity, and I again observe that religion is not less useful to each citizen than to the whole State. The Americans show, by their practice, that they feel the high necessity of imparting morality to democratic communities by means of religion. What they think of themselves in this respect is a truth of which every democratic nation ought to be thoroughly persuaded.

I do not doubt that the social and political constitution of a people predisposes them to adopt a certain belief and certain tastes, which afterwards flourish without difficulty amongst them; whilst the same causes may divert a people from certain opinions and propensities, without any voluntary effort, and, as it were, without any distinct consciousness, on their part. The whole art of the legislator is correctly to discern beforehand these natural inclinations of communities of men, in order to know whether they should be assisted, or whether it may not be necessary to check them. For the duties incumbent on the legislator differ at different times; the goal towards which the human race ought ever to be tending is alone stationary; the means of reaching it are perpetually to be varied.

If I had been born in an aristocratic age, in the midst of a nation where the hereditary wealth of some, and the irremediable penury of others, should equally divert men from the idea of bettering their condition, and hold the soul as it were in a state of torpor fixed on the contemplation of another world, I should then wish that it were possible for me to rouse that people to a sense of their wants; I should seek to discover more rapid and more easy means for satisfying the fresh desires which I might have awakened; and, directing the most strenuous efforts of the human mind to physical pursuits, I should endeavour to stimulate it to promote the well-being of man. If it happened that some men were immoderately incited to the pursuit of riches, and displayed an excessive liking for physical gratifications, I should not be alarmed; these peculiar symptoms would soon be absorbed in the general aspect of the people.

The attention of the legislators of democracies is called to other cares. Give democratic nations education and freedom, and leave them alone. They will soon learn to draw from this world all the benefits which it can afford; they will improve each of the useful arts, and will day by day render life more comfortable, more convenient, and more easy. Their social condition naturally urges them in this direction; I do not fear that they will slacken their course.

But whilst man takes delight in this honest and lawful pursuit of his well-being, it is to be apprehended that he may in the end lose the use of his sublimest faculties; and that whilst he is busied in improving all around him, he may at length degrade himself. Here, and here only, does the peril lie. It should therefore be the unceasing object of the legislators of democracies, and of all the virtuous and enlightened men who live there, to raise the souls of their fellow-citizens, and keep them lifted up towards Heaven. It is necessary that all who feel an interest in the future destinies of democratic society should unite, and that all should make joint and continual efforts to diffuse the love of the infinite, a sense of greatness, and a love of pleasures not of earth. If amongst the opinions of a democratic people any of those pernicious theories exist which tend to inculcate that all perishes with the body, let men by whom such theories are professed be marked as the natural foes of such a people.

The Materialists are offensive to me in many respects; their doctrines I hold to be pernicious, and I am disgusted at their arrogance. If their system could be of any utility to man, it would seem to be by giving him a modest opinion of himself. But these reasoners show that it is not so; and when they think they have said enough to establish that they are brutes, they show themselves as proud as if they had demonstrated that they are gods. Materialism is, amongst all nations, a dangerous disease of the human mind; but it is more especially to be dreaded amongst a democratic people, because it readily amalgamates with that vice which is most familiar to the heart under such circumstances. Democracy encourages a taste for physical gratification: this taste, if it become excessive, soon disposes men to believe that all is matter only; and materialism, in turn, hurries them back with mad impatience to these same delights: such is the fatal circle within which democratic nations are driven round. It were well that they should see the danger and hold back.

Most religions are only general, simple, and practical means of teaching men the doctrine of the immortality of the soul. That is the greatest benefit which a democratic people derives, from its belief, and hence belief is more necessary to such a people than to all others. When therefore any religion has struck its roots deep into a democracy, beware lest you disturb them; but rather watch it carefully, as the most precious bequest of aristocratic ages. Seek not to supersede the old religious opinions of men by new ones; lest in the passage from one faith to another, the soul being left for a while stripped of all belief, the love of physical gratifications should grow upon it and fill it wholly.

The doctrine of metempsychosis is assuredly not more rational than that of materialism; nevertheless if it were absolutely necessary that a democracy should choose one of the two, I should not hesitate to decide that the community would run less risk of being brutalized by believing that the soul of man will pass into the carcass of a hog, than by believing that the soul of man is nothing at all. The belief in a supersensual and immortal principle, united for a time to matter, is so indispensable to man's greatness, that its effects are striking even when it is not united to the doctrine of future reward and punishment; and when it holds no more than that after death the Divine principle contained in man is absorbed in the Deity, or transferred to animate the frame of some other creature. Men holding so imperfect a belief will still consider the body as the secondary and inferior portion of their nature, and they will despise it even whilst they yield to its influence; whereas they have a natural esteem and secret admiration for the immaterial part of man, even though they sometimes refuse to submit to its dominion. That is enough to give a lofty cast to their opinions and their tastes, and to bid them tend with no interested motive, and as it were by impulse, to pure feelings and elevated thoughts.

It is not certain that Socrates and his followers had very fixed opinions as to what would befall man hereafter; but the sole point of belief on which they were determined—that the soul has nothing in common with the body, and survives it—was enough to give the Platonic philosophy that sublime aspiration by which it is distinguished. It is clear from the works of Plato, that many philosophical writers, his predecessors or contemporaries, professed materialism. These writers have not reached us, or have reached us in mere fragments. The same thing has happened in almost all ages; the greater part of the most famous minds in literature adhere to the doctrines of a supersensual philosophy. The instinct and the taste of the human race maintain those doctrines; they save them oftentimes in spite of men themselves, and raise the names of their defenders above the tide of time. It must not then be supposed that at any period or under any political condition, the passion for physical gratifications, and the opinions which are superinduced by that passion, can ever content a whole people. The heart of man is of a larger mould: it can at once comprise a taste for the possessions of earth and the love of those of Heaven: at times it may seem to cling devotedly to the one, but it will never be long without thinking of the other.

If it be easy to see that it is more particularly important in democratic ages that spiritual opinions should prevail, it is not easy to say by what means those who govern democratic nations may make them predominate. I am no believer in the prosperity, any more than in the durability, of official philosophies; and as to state religions, I have always held, that if they be sometimes of momentary service to the interests of political power, they always, sooner or later, become fatal to the church. Nor do I think with those who assert, that to raise religion in the eyes of the people, and to make them do honour to her spiritual doctrines, it is desirable indirectly to give her ministers a political influence which the laws deny them. I am so much alive to the almost inevitable dangers which beset religious belief whenever the clergy take part in public affairs, and I am so convinced that Christianity must be maintained at any cost in the bosom of modern democracies, that I had rather shut up the priesthood within the sanctuary than allow them to step beyond it.

What means then remain in the hands of constituted authorities to bring men back to spiritual opinions, or to hold them fast to the religion by which those opinions are suggested? My answer will do me harm in the eyes of politicians. I believe that the sole effectual means which governments can employ in order to have the doctrine of the immortality of the soul duly respected, is ever to act as if they believed in it themselves; and I think that it is only by scrupulous conformity to religious morality in great affairs that they can hope to teach the community at large to know, to love, and to observe it in the lesser concerns of life.

Chapter XVI

That Excessive Care of Worldly Welfare May Impair That Welfare

There is a closer tie than is commonly supposed between the improvement of the soul and the amelioration of what belongs to the body. Man may leave these two things apart, and consider each of them alternately; but he cannot sever them entirely without at last losing sight of one and of the other. The beasts have the same senses as ourselves, and very nearly the same appetites. We have no sensual passions which are not common to our race and theirs, and which are not to be found, at least in the germ, in a dog as well as in a man. Whence is it then that the animals can only provide for their first and lowest wants, whereas we can infinitely vary and endlessly increase our enjoyments?

We are superior to the beasts in this, that we use our souls to find out those material benefits to which they are only led by instinct. In man, the angel teaches the brute the art of contenting its desires. It is because man is capable of rising above the things of the body, and of contemning life itself, of which the beasts have not the least notion, that he can multiply these same things of the body to a degree which inferior races are equally unable to conceive. Whatever elevates, enlarges, and expands the soul, renders it more capable of succeeding in those very undertakings which concern it not. Whatever, on the other hand, enervates or lowers it, weakens it for all purposes, the chiefest, as well as the least, and threatens to render it almost equally impotent for the one and for the other. Hence the soul must remain great and strong, though it were only to devote its strength and greatness from time to time to the service of the body. If men were ever to content themselves with material objects, it is probable that they would lose by degrees the art of producing them; and they would enjoy them in the end, like the brutes, without discernment and without improvement.

Chapter XVII

That in Times Marked by Equality of Conditions and Sceptical Opinions, It Is Important to Remove to a Distance the Objects of Human Actions

In the ages of faith the final end of life is placed beyond life. The men of those ages therefore naturally, and in a manner involuntarily, accustom themselves to fix their gaze for a long course of years on some immovable object, towards which they are constantly tending; and they learn by insensible degrees to repress a multitude of petty passing desires, in order to be the better able to content that great and lasting desire which possesses them. When these same men engage in the affairs of this world, the same habits may be traced in their conduct. They are apt to set up some general and certain aim and end to their actions here below, towards which all their efforts are directed: they do not turn from day to day to chase some novel object of desire, but they have settled designs which they are never weary of pursuing. This explains why religious nations have so often achieved such lasting results: for whilst they were thinking only of the other world, they had found out the great secret of success in this. Religions give men a general habit of conducting themselves with a view to futurity: in this respect they are not less useful to happiness in this life than to felicity hereafter; and this is one of their chief political characteristics.

But in proportion as the light of faith grows dim, the range of man's sight is circumscribed, as if the end and aim of human actions appeared every day to be more within his reach. When men have once allowed themselves to think no more of what is to befall them after life, they readily lapse into that complete and brutal indifference to futurity, which is but too conformable to some propensities of mankind. As soon as they have lost the habit of placing their chief hopes upon remote events, they naturally seek to gratify without delay their smallest desires; and no sooner do they despair of living for ever, than they are disposed to act as if they were to exist but for a single day. In sceptical ages it is always therefore to be feared that men may perpetually give way to their daily casual desires; and that, wholly renouncing whatever cannot be acquired without protracted effort, they may establish nothing great, permanent, and calm.

If the social condition of a people, under these circumstances, becomes democratic, the danger which I here point out is thereby increased. When everyone is constantly striving to change his position—when an immense field for competition is thrown open to all—when wealth is amassed or dissipated in the shortest possible space of time amidst the turmoil of democracy, visions of sudden and easy fortunes—of great possessions easily won and lost—of chance, under all its forms—haunt the mind. The instability of society itself fosters the natural instability of man's desires. In the midst of these perpetual fluctuations of his lot, the present grows upon his mind, until it conceals futurity from his sight, and his looks go no further than the morrow.

In those countries in which unhappily irreligion and democracy coexist, the most important duty of philosophers and of those in power is to be always striving to place the objects of human actions far beyond man's immediate range. Circumscribed by the character of his country and his age, the moralist must learn to vindicate his principles in that position. He must constantly endeavour to show his contemporaries, that, even in the midst of the perpetual commotion around them, it is easier than they think to conceive and to execute protracted undertakings. He must teach them that, although the aspect of mankind may have changed, the methods by which men may provide for their prosperity in this world are still the same; and that amongst democratic nations, as well as elsewhere, it is only by resisting a thousand petty selfish passions of the hour that the general and unquenchable passion for happiness can be satisfied.

The task of those in power is not less clearly marked out. At all times it is important that those who govern nations should act with a view to the future: but this is even more necessary in democratic and sceptical ages than in any others. By acting thus, the leading men of democracies not only make public affairs prosperous, but they also teach private individuals, by their example, the art of managing private concerns. Above all they must strive as much as possible to banish chance from the sphere of politics. The sudden and undeserved promotion of a courtier produces only a transient impression in an aristocratic country, because the aggregate institutions and opinions of the nation habitually compel men to advance slowly in tracks which they cannot get out of. But nothing is more pernicious than similar instances of favour exhibited to the eyes of a democratic people: they give the last impulse to the public mind in a direction where everything hurries it onwards. At times of scepticism and equality more especially, the favour of the people or of the prince, which chance may confer or chance withhold, ought never to stand in lieu of attainments or services. It is desirable that every advancement should there appear to be the result of some effort; so that no greatness should be of too easy acquirement, and that ambition should be obliged to fix its gaze long upon an object before it is gratified. Governments must apply themselves to restore to men that love of the future with which religion and the state of society no longer inspire them; and, without saying so, they must practically teach the community day by day that wealth, fame, and power are the rewards of labour—that great success stands at the utmost range of long desires, and that nothing lasting is obtained but what is obtained by toil. When men have accustomed themselves to foresee from afar what is likely to befall them in the world and to feed upon hopes, they can hardly confine their minds within the precise circumference of life, and they are ready to break the boundary and cast their looks beyond. I do not doubt that, by training the members of a community to think of their future condition in this world, they would be gradually and unconsciously brought nearer to religious convictions. Thus the means which allow men, up to a certain point to go without religion, are perhaps after all the only means we still possess for bringing mankind back by a long and roundabout path to a state of faith.

Chapter XVIII

That amongst the Americans All Honest Callings Are Honourable

Amongst a democratic people, where there is no hereditary wealth, every man works to earn a living, or has worked, or is born of parents who have worked. The notion of labour is therefore presented to the mind on every side as the necessary, natural, and honest condition of human existence. Not only is labour not dishonourable amongst such a people, but it is held in honour: the prejudice is not against it, but in its favour. In the United States a wealthy man thinks that he owes it to public opinion to devote his leisure to some kind of industrial or commercial pursuit, or to public business. He would think himself in bad repute if he employed his life solely in living. It is for the purpose of escaping this obligation to work, that so many rich Americans come to Europe, where they find some scattered remains of aristocratic society, amongst which idleness is still held in honour.

Equality of conditions not only ennobles the notion of labour in men's estimation, but it raises the notion of labour as a source of profit. In aristocracies it is not exactly labour that is despised, but labour with a view to profit. Labour is honorific in itself, when it is undertaken at the sole bidding of ambition or of virtue. Yet in aristocratic society it constantly happens that he who works for honour is not insensible to the attractions of profit. But these two desires only intermingle in the innermost depths of his soul: he carefully hides from every eye the point at which they join; he would fain conceal it from himself. In aristocratic countries there are few public officers who do not affect to serve their country without interested motives. Their salary is an incident of which they think but little, and of which they always affect not to think at all. Thus the notion of profit is kept distinct from that of labour; however they may be united in point of fact, they are not thought of together.

In democratic communities these two notions are, on the contrary, always palpably united. As the desire of well-being is universal—as fortunes are slender or fluctuating—as everyone wants either to increase his own resources, or to provide fresh ones for his progeny, men clearly see that it is profit which, if not wholly, at least partially, leads them to work. Even those who are principally actuated by the love of fame are necessarily made familiar with the thought that they are not exclusively actuated by that motive; and they discover that the desire of getting a living is mingled in their minds with the desire of making life illustrious.

As soon as, on the one hand, labour is held by the whole community to be an honourable necessity of man's condition, and, on the other, as soon as labour is always ostensibly performed, wholly or in part, for the purpose of earning remuneration, the immense interval which separated different callings in aristocratic societies disappears. If all are not alike, all at least have one feature in common. No profession exists in which men do not work for money; and the remuneration which is common to them all gives them all an air of resemblance. This serves to explain the opinions which the Americans entertain with respect to different callings. In America no one is degraded because he works, for everyone about him works also; nor is anyone humiliated by the notion of receiving pay, for the President of the United States also works for pay. He is paid for commanding, other men for obeying orders. In the United States professions are more or less laborious, more or less profitable; but they are never either high or low: every honest calling is honourable.

Chapter XIX

That Almost All the Americans Follow Industrial Callings

Agriculture is, perhaps, of all the useful arts that which improves most slowly amongst democratic nations. Frequently, indeed, it would seem to be stationary, because other arts are making rapid strides towards perfection. On the other hand, almost all the tastes and habits which the equality of condition engenders naturally lead men to commercial and industrial occupations.

Suppose an active, enlightened, and free man, enjoying a competency, but full of desires: he is too poor to live in idleness; he is rich enough to feel himself protected from the immediate fear of want, and he thinks how he can better his condition. This man has conceived a taste for physical gratifications, which thousands of his fellow-men indulge in around him; he has himself begun to enjoy these pleasures, and he is eager to increase his means of satisfying these tastes more completely. But life is slipping away, time is urgent—to what is he to turn? The cultivation of the ground promises an almost certain result to his exertions, but a slow one; men are not enriched by it without patience and toil. Agriculture is therefore only suited to those who have already large superfluous wealth, or to those whose penury bids them only seek a bare subsistence. The choice of such a man as we have supposed is soon made; he sells his plot of ground, leaves his dwelling, and embarks in some hazardous but lucrative calling. Democratic communities abound in men of this kind; and in proportion as the equality of conditions becomes greater, their multitude increases. Thus democracy not only swells the number of workingmen, but it leads men to prefer one kind of labour to another; and whilst it diverts them from agriculture, it encourages their taste for commerce and manufactures.[1]

This spirit may be observed even amongst the richest members of the community. In democratic countries, however opulent a man is supposed to be, he is almost always discontented with his fortune, because he finds that he is less rich than his father was, and he fears that his sons will be less rich than himself. Most rich men in democracies are therefore constantly haunted by the desire of obtaining wealth, and they naturally turn their attention to trade and manufactures, which appear to offer the readiest and most powerful means of success. In this respect they share the instincts of the poor, without feeling the same necessities; say rather, they feel the most imperious of all necessities, that of not sinking in the world.

In aristocracies the rich are at the same time those who govern. The attention which they unceasingly devote to important public affairs diverts them from the lesser cares which trade and manufactures demand. If the will of an individual happens, nevertheless, to turn his attention to business, the will of the body to which he belongs will immediately debar him from pursuing it; for however men may declaim against the rule of numbers, they cannot wholly escape their sway; and even amongst those aristocratic bodies which most obstinately refuse to acknowledge the rights of the majority of the nation, a private majority is formed which governs the rest.[2]

In democratic countries, where money does not lead those who possess it to political power, but often removes them from it, the rich do not know how to spend their leisure. They are driven into active life by the inquietude and the greatness of their desires, by the extent of their resources, and by the taste for what is extraordinary, which is almost always felt by those who rise, by whatsoever means, above the crowd. Trade is the only road open to them. In democracies nothing is more great or more brilliant than commerce: it attracts the attention of the public, and fills the imagination of the multitude; all energetic passions are directed towards it. Neither their own prejudices, nor those of anybody else, can prevent the rich from devoting themselves to it. The wealthy members of democracies never form a body which has manners and regulations of its own; the opinions peculiar to their class do not restrain them, and the common opinions of their country urge them on. Moreover, as all the large fortunes which are to be met with in a democratic community are of commercial growth, many generations must succeed each other before their possessors can have entirely laid aside their habits of business.

Circumscribed within the narrow space which politics leave them, rich men in democracies eagerly embark in commercial enterprise: there they can extend and employ their natural advantages; and indeed it is even by the boldness and the magnitude of their industrial speculations that we may measure the slight esteem in which productive industry would have been held by them, if they had been born amidst an aristocracy.

A similar observation is likewise applicable to all men living in democracies, whether they be poor or rich. Those who live in the midst of democratic fluctuations have always before their eyes the phantom of chance; and they end by liking all undertakings in which chance plays a part. They are therefore all led to engage in commerce, not only for the sake of the profit it holds out to them, but for the love of the constant excitement occasioned by that pursuit.

The United States of America have only been emancipated for half a century [in 1840] from the state of colonial dependence in which they stood to Great Britain; the number of large fortunes there is small, and capital is still scarce. Yet no people in the world has made such rapid progress in trade and manufactures as the Americans: they constitute at the present day the second maritime nation in the world; and although their manufactures have to struggle with almost insurmountable natural impediments, they are not prevented from making great and daily advances. In the United States the greatest undertakings and speculations are executed without difficulty, because the whole population is engaged in productive industry, and because the poorest as well as the most opulent members of the commonwealth are ready to combine their efforts for these purposes. The consequence is, that a stranger is constantly amazed by the immense public works executed by a nation which contains, so to speak, no rich men. The Americans arrived but as yesterday on the territory which they inhabit, and they have already changed the whole, order of nature for their own advantage. They have joined the Hudson to the Mississippi, and made the Atlantic Ocean communicate with the Gulf of Mexico, across a continent of more than five hundred leagues in extent which separates the two seas. The longest railroads which have been constructed up to the present time are in America. But what most astonishes me in the United States, is not so much the marvellous grandeur of some undertakings, as the innumerable multitude of small ones. Almost all the farmers of the United States combine some trade with agriculture; most of them make agriculture itself a trade. It seldom happens that an American farmer settles for good upon the land which he occupies: especially in the districts of the far West he brings land into tillage in order to sell it again, and not to farm it: he builds a farmhouse on the speculation that, as the state of the country will soon be changed by the increase of population, a good price will be gotten for it. Every year a swarm of the inhabitants of the North arrive in the Southern States, and settle in the parts where the cotton-plant and the sugar-cane grow. These men cultivate the soil in order to make it produce in a few years enough to enrich them; and they already look forward to the time when they may return home to enjoy the competency thus acquired. Thus the Americans carry their business-like qualities into agriculture; and their trading passions are displayed in that as in their other pursuits.

The Americans make immense progress in productive industry, because they all devote themselves to it at once; and for this same reason they are exposed to very unexpected and formidable embarrassments. As they are all engaged in commerce, their commercial affairs are affected by such various and complex causes that it is impossible to foresee what difficulties may arise. As they are all more or less engaged in productive industry, at the least shock given to business all private fortunes are put in jeopardy at the same time, and the State is shaken. I believe that the return of these commercial panics is an endemic disease of the democratic nations of our age. It may be rendered less dangerous, but it cannot be cured; because it does not originate in accidental circumstances, but in the temperament of these nations.

Chapter XX

That Aristocracy May Be Engendered by Manufactures

I have shown that democracy is favourable to the growth of manufactures, and that it increases without limit the numbers of the manufacturing classes: we shall now see by what side-road manufacturers may possibly in their turn bring men back to aristocracy. It is acknowledged that when a workman is engaged every day upon the same detail, the whole commodity is produced with greater ease, promptitude, and economy. It is likewise acknowledged that the cost of the production of manufactured goods is diminished by the extent of the establishment in which they are made, and by the amount of capital employed or of credit. These truths had long been imperfectly discerned, but in our time they have been demonstrated. They have been already applied to many very important kinds of manufactures, and the humblest will gradually be governed by them. I know of nothing in politics which deserves to fix the attention of the legislator more closely than these two new axioms of the science of manufactures.

When a workman is unceasingly and exclusively engaged in the fabrication of one thing, he ultimately does his work with singular dexterity; but at the same time he loses the general faculty of applying his mind to the direction of the work. He every day becomes more adroit and less industrious; so that it may be said of him, that in proportion as the workman improves the man is degraded. What can be expected of a man who has spent twenty years of his life in making heads for pins? and to what can that mighty human intelligence, which has so often stirred the world, be applied in him, except it be to investigate the best method of making pins' heads? When a workman has spent a considerable portion of his existence in this manner, his thoughts are forever set upon the object of his daily toil; his body has contracted certain fixed habits, which it can never shake off: in a word, he no longer belongs to himself, but to the calling which he has chosen. It is in vain that laws and manners have been at the pains to level all barriers round such a man, and to open to him on every side a thousand different paths to fortune; a theory of manufactures more powerful than manners and laws binds him to a craft, and frequently to a spot, which he cannot leave: it assigns to him a certain place in society, beyond which he cannot go: in the midst of universal movement it has rendered him stationary.

In proportion as the principle of the division of labour is more extensively applied, the workman becomes more weak, more narrow-minded, and more dependent. The art advances, the artisan recedes. On the other hand, in proportion as it becomes more manifest that the productions of manufactures are by so much the cheaper and better as the manufacture is larger and the amount of capital employed more considerable, wealthy and educated men come forward to embark in manufactures which were heretofore abandoned to poor or ignorant handicraftsmen. The magnitude of the efforts required, and the importance of the results to be obtained, attract them. Thus at the very time at which the science of manufactures lowers the class of workmen, it raises the class of masters.

Whereas the workman concentrates his faculties more and more upon the study of a single detail, the master surveys a more extensive whole, and the mind of the latter is enlarged in proportion as that of the former is narrowed. In a short time the one will require nothing but physical strength without intelligence; the other stands in need of science, and almost of genius, to insure success. This man resembles more and more the administrator of a vast empire—that man, a brute. The master and the workman have then here no similarity, and their differences increase every day. They are only connected as the two rings at the extremities of a long chain. Each of them fills the station which is made for him, and out of which he does not get: the one is continually, closely, and necessarily dependent upon the other, and seems as much born to obey as that other is to command. What is this but aristocracy?

As the conditions of men constituting the nation become more and more equal, the demand for manufactured commodities becomes more general and more extensive; and the cheapness which places these objects within the reach of slender fortunes becomes a great element of success. Hence there are every day more men of great opulence and education who devote their wealth and knowledge to manufactures; and who seek, by opening large establishments, and by a strict division of labour, to meet the fresh demands which are made on all sides. Thus, in proportion as the mass of the nation turns to democracy, that particular class which is engaged in manufactures becomes more aristocratic. Men grow more alike in the one—more different in the other; and inequality increases in the less numerous class in the same ratio in which it decreases in the community. Hence it would appear, on searching to the bottom, that aristocracy should naturally spring out of the bosom of democracy.

But this kind of aristocracy by no means resembles those kinds which preceded it. It will be observed at once, that as it applies exclusively to manufactures and to some manufacturing callings, it is a monstrous exception in the general aspect of society. The small aristocratic societies which are formed by some manufacturers in the midst of the immense democracy of our age, contain, like the great aristocratic societies of former ages, some men who are very opulent, and a multitude who are wretchedly poor. The poor have few means of escaping from their condition and becoming rich; but the rich are constantly becoming poor, or they give up business when they have realised a fortune. Thus the elements of which the class of the poor is composed are fixed; but the elements of which the class of the rich is composed are not so. To say the truth, though there are rich men, the class of rich men does not exist; for these rich individuals have no feelings or purposes in common, no mutual traditions or mutual hopes; there are therefore members, but no body.

Not only are the rich not compactly united amongst themselves, but there is no real bond between them and the poor. Their relative position is not a permanent one; they are constantly drawn together or separated by their interests. The workman is generally dependent on the master, but not on any particular master; these two men meet in the factory, but know not each other elsewhere; and whilst they come into contact on one point, they stand very wide apart on all others. The manufacturer asks nothing of the workman but his labour; the workman expects nothing from him but his wages. The one contracts no obligation to protect, nor the other to defend; and they are not permanently connected either by habit or by duty. The aristocracy created by business rarely settles in the midst of the manufacturing population which it directs; the object is not to govern that population, but to use it. An aristocracy thus constituted can have no great hold upon those whom it employs; and even if it succeed in retaining them at one moment, they escape the next; it knows not how to will, and it cannot act. The territorial aristocracy of former ages was either bound by law, or thought itself bound by usage, to come to the relief of its serving-men, and to succour their distresses. But the manufacturing aristocracy of our age first impoverishes and debases the men who serve it, and then abandons them to be supported by the charity of the public. This is a natural consequence of what has been said before. Between the workmen and the master there are frequent relations, but no real partnership.

I am of opinion, upon the whole, that the manufacturing aristocracy which is growing up under our eyes is one of the harshest which ever existed in the world; but at the same time it is one of the most confined and least dangerous. Nevertheless the friends of democracy should keep their eyes anxiously fixed in this direction; for if ever a permanent inequality of conditions and aristocracy again penetrate into the world, it may be predicted that this is the channel by which they will enter.

DEMOCRACY IN AMERICA

SECOND PART

THIRD BOOK

Influence of Democracy on Manners, Properly So Called

Chapter I

That Manners Are Softened as Social Conditions Become More Equal

We perceive that for several ages social conditions have tended to equality, and we discover that in the course of the same period the manners of society have been softened. Are these two things merely contemporaneous, or does any secret link exist between them, so that the one cannot go on without making the other advance? Several causes may concur to render the manners of a people less rude; but, of all these causes, the most powerful appears to me to be the equality of conditions. Equality of conditions and growing civility in manners are, then, in my eyes, not only contemporaneous occurrences, but correlative facts. When the fabulists seek to interest us in the actions of beasts, they invest them with human notions and passions; the poets who sing of spirits and angels do the same; there is no wretchedness so deep, nor any happiness so pure, as to fill the human mind and touch the heart, unless we are ourselves held up to our own eyes under other features.

This is strictly applicable to the subject upon which we are at present engaged. When all men are irrevocably marshalled in an aristocratic community, according to their professions, their property, and their birth, the members of each class, considering themselves as children of the same family, cherish a constant and lively sympathy towards each other, which can never be felt in an equal degree by the citizens of a democracy. But the same feeling does not exist between the several classes towards each other. Amongst an aristocratic people each caste has its own opinions, feelings, rights, manners, and modes of living. Thus the men of whom each caste is composed do not resemble the mass of their fellow-citizens; they do not think or feel in the same manner, and they scarcely believe that they belong to the same human race.

They cannot, therefore, thoroughly understand what others feel, nor judge of others by themselves. Yet they are sometimes eager to lend each other mutual aid; but this is not contrary to my previous observation. These aristocratic institutions, which made the beings of one and the same race so different, nevertheless bound them to each other by close political ties. Although the serf had no natural interest in the fate of nobles, he did not the less think himself obliged to devote his person to the service of that noble who happened to be his lord; and although the noble held himself to be of a different nature from that of his serfs, he nevertheless held that his duty and his honour constrained him to defend, at the risk of his own life, those who dwelt upon his domains.

It is evident that these mutual obligations did not originate in the law of nature, but in the law of society; and that the claim of social duty was more stringent than that of mere humanity. These services were not supposed to be due from man to man, but to the vassal or to the lord. Feudal institutions awakened a lively sympathy for the sufferings of certain men, but none at all for the miseries of mankind. They infused generosity rather than mildness into the manners of the time, and although they prompted men to great acts of self-devotion, they engendered no real sympathies; for real sympathies can only exist between those who are alike; and in aristocratic ages men acknowledge none but the members of their own caste to be like themselves.

When the chroniclers of the Middle Ages, who all belonged to the aristocracy by birth or education, relate the tragical end of a noble, their grief flows apace; whereas they tell you at a breath, and without wincing, of massacres and tortures inflicted on the common sort of people. Not that these writers felt habitual hatred or systematic disdain for the people; war between the several classes of the community was not yet declared. They were impelled by an instinct rather than by a passion; as they had formed no clear notion of a poor man's sufferings, they cared but little for his fate. The same feelings animated the lower orders whenever the feudal tie was broken. The same ages which witnessed so many heroic acts of self-devotion on the part of vassals for their lords, were stained with atrocious barbarities, exercised from time to time by the lower classes on the higher. It must not be supposed that this mutual insensibility arose solely from the absence of public order and education; for traces of it are to be found in the following centuries, which became tranquil and enlightened whilst they remained aristocratic. In 1675 the lower classes in Brittany revolted at the imposition of a new tax. These disturbances were put down with unexampled atrocity. Observe the language in which Madame de Sévigné, a witness of these horrors, relates them to her daughter:—

'AUX ROCHERS,' 30 OCTOBRE, 1675.

'Mon Dieu, ma fille, que votre lettre d'Aix est plaisante! Au moins relisez vos lettres avant que de les envoyer; laissez-vous surpendre à leur agrément, et consolez-vous par ce plaisir de la peine que vous avez d'en tant écrire. Vous avez done baisé toute la Provence? il n'y aurait pas satisfaction à baiser toute la Bretagne, à moins qu'on n'aimat à sentir le vin. . . . Voulez-vous savoir des nouvelles de Rennes? On a fait une taxe de cent mille écus sur le bourgeois; et si on ne trouve point cette somme dans vingt-quatre heures, elle sera doublée et exigible par les soldats. On a chassé et banni toute une grand rue, et défendu de les recueillir sous peine de la vie; de sorte qu'on voyait tous ces misérables, veillards, femmes accouchées, enfans, errer en pleurs au sortir de cette ville sans savoir où aller. On roua avant-hier un violon, qui avait commencé la danse et la pillerie du papier timbré; il a été écartelé après sa mort, et ses quatre quartiers exposés aux quatre coins de la ville. On a pris soixante bourgeois, et on commence demain les punitions. Cette province est un bel exemple pour les autres, et surtout de respecter les gouverneurs et les gouvernantes, et de ne point jeter de pierres dans leur jardin.[1]

'Madame de Tarente était hier dans ces bois par un temps enchanté: il n'est question ni de chambre ni de collation; elle entre par la barrière et s'en retourne de mêame. . . .'

In another letter she adds:—

'Vous me parlez bien plaisamment de nos misères; nous ne sommes plus si roués; un en huit jours, pour entretenir la justice. Il est vrai que la penderie me paraît maintenant un refraîchissement. J'ai une tout autre idée de la justice, depuis que je suis en ce pays. Vos galériens me paraissent une société d'honnêates gens qui se sont retirès du monde pour mener une vie douce.'

It would be a mistake to suppose that Madame de Sévigné, who wrote these lines, was a selfish or cruel person; she was passionately attached to her children, and very ready to sympathise in the sorrows of her friends; nay, her letters show that she treated her vassals and servants with kindness and indulgence. But Madame de Sévigné had no clear notion of suffering in anyone who was not a person of quality.

In our time the harshest man writing to the most insensible person of his acquaintance would not venture wantonly to indulge in the cruel jocularity which I have quoted; and even if his own manners allowed him to do so, the manners of society at large would forbid it. Whence does this arise? Have we more sensibility than our forefathers? I know not that we have; but I am sure that our insensibility is extended to a far greater range of objects. When all the ranks of a community are nearly equal, as all men think and feel in nearly the same manner, each of them may judge in a moment of the sensations of all the others; he casts a rapid glance upon himself, and that is enough. There is no wretchedness into which he cannot readily enter, and a secret instinct reveals to him its extent. It signifies not that strangers or foes be the sufferers; imagination puts him in their place; something like a personal feeling is mingled with his pity, and makes himself suffer whilst the body of his fellow-creature is in torture. In democratic ages men rarely sacrifice themselves for one another; but they display general compassion for the members of the human race. They inflict no useless ills; and they are happy to relieve the griefs of others, when they can do so without much hurting themselves; they are not disinterested, but they are humane.

Although the Americans have, in a manner, reduced egotism to a social and philosophical theory, they are nevertheless extremely open to compassion. In no country is criminal justice administered with more mildness than in the United States. Whilst the English seem disposed carefully to retain the bloody traces of the dark ages in their penal legislation, the Americans have almost expunged capital punishment from their codes. North America is, I think, the only one country upon earth in which the life of no one citizen has been taken for a political offence in the course of the last fifty years. The circumstance which conclusively shows that this singular mildness of the Americans arises chiefly from their social condition, is the manner in which they treat their slaves. Perhaps there is not, upon the whole, a single European colony in the New World in which the physical condition of the blacks is less severe than in the United States; yet the slaves still endure horrid sufferings there, and are constantly exposed to barbarous punishments. It is easy to perceive that the lot of these unhappy beings inspires their masters with but little compassion, and that they look upon slavery, not only as an institution which is profitable to them, but as an evil which does not affect them. Thus the same man who is full of humanity towards his fellow-creatures when they are at the same time his equals, becomes insensible to their afflictions as soon as that equality ceases. His mildness should therefore be attributed to the equality of conditions, rather than to civilisation and education.

What I have here remarked of individuals is, to a certain extent, applicable to nations. When each nation has its distinct opinions, belief, laws, and customs, it looks upon itself as the whole of mankind, and is moved by no sorrows but its own. Should war break out between two nations animated by this feeling, it is sure to be waged with great cruelty. At the time of their highest culture, the Romans slaughtered the generals of their enemies, after having dragged them in triumph behind a car; and they flung their prisoners to the beasts of the Circus for the amusement of the people. Cicero, who declaimed so vehemently at the notion of crucifying a

Roman citizen, had not a word to say against these horrible abuses of victory. It is evident that in his eyes a barbarian did not belong to the same human race as a Roman. On the contrary, in proportion as nations become more like each other, they become reciprocally more compassionate, and the law of nations is mitigated.

Chapter II

That Democracy Renders the Habitual Intercourse of the Americans Simple and Easy

Democracy does not attach men strongly to each other; but it places their habitual intercourse upon an easier footing. If two Englishmen chance to meet at the Antipodes, where they are surrounded by strangers whose language and manners are almost unknown to them, they will first stare at each other with much curiosity and a kind of secret uneasiness; they will then turn away, or, if one accosts the other, they will take care only to converse with a constrained and absent air upon very unimportant subjects. Yet there is no enmity between these men; they have never seen each other before, and each believes the other to be a respectable person. Why then should they stand so cautiously apart? We must go back to England to learn the reason.

When it is birth alone, independent of wealth, which classes men in society, every one knows exactly what his own position is upon the social scale; he does not seek to rise, he does not fear to sink. In a community thus organised, men of different castes communicate very little with each other; but if accident brings them together, they are ready to converse without hoping or fearing to lose their own position. Their intercourse is not upon a footing of equality, but it is not constrained. When monied aristocracy succeeds to aristocracy of birth, the case is altered. The privileges of some are still extremely great, but the possibility of acquiring those privileges is open to all: whence it follows that those who possess them are constantly haunted by the apprehension of losing them, or of other men's sharing them; those who do not yet enjoy them long to possess them at any cost, or, if they fail to appear at least to possess them—which is not impossible. As the social importance of men is no longer ostensibly and permanently fixed by blood, and is infinitely varied by wealth, ranks still exist, but it is not easy clearly to distinguish at a glance those who respectively belong to them. Secret hostilities then arise in the community; one set of men endeavour by innumerable artifices to penetrate, or to appear to penetrate, amongst those who are above them; another set are constantly in arms against these usurpers of their rights; or rather the same individual does both at once, and whilst he seeks to raise himself into a higher circle, he is always on the defensive against the intrusion of those below him.

Such is the condition of England at the present time; and I am of opinion that the peculiarity before adverted to is principally to be attributed to this cause. As aristocratic pride is still extremely great amongst the English, and as the limits of aristocracy are ill-defined, everybody lives in constant dread lest advantage should be taken of his familiarity. Unable to judge at once of the social position of those he meets, an Englishman prudently avoids all contact with them. Men are afraid lest some slight service rendered should draw them into an unsuitable acquaintance; they dread civilities, and they avoid the obtrusive gratitude of a stranger quite as much as his hatred. Many people attribute these singular anti-social propensities, and the reserved and taciturn bearing of the English, to purely physical causes. I may admit that there is something of it in their race, but much more of it is attributable to their social condition, as is proved by the contrast of the Americans.

In America, where the privileges of birth never existed, and where riches confer no peculiar rights on their possessors, men unacquainted with each other are very ready to frequent the same places, and find neither peril nor advantage in the free interchange of their thoughts. If they meet by accident, they neither seek nor avoid intercourse; their manner is therefore natural, frank, and open: it is easy to see that they hardly expect or apprehend anything from each other, and that they do not care to display, any more than to conceal, their position in the world. If their demeanour is often cold and serious, it is never haughty or constrained; and if they do not converse, it is because they are not in a humour to talk, not because they think it their interest to be silent. In a foreign country two Americans are at once friends, simply because they are Americans. They are repulsed by no prejudice; they are attracted by their common country. For two Englishmen the same blood is not enough; they must be brought together by the same rank. The Americans remark this unsociable mood of the English as much as the French do, and they are not less astonished by it. Yet the Americans are connected with England by their origin, their religion, their language, and partially by their manners; they only differ in their social condition. It may therefore be inferred that the reserve of the English proceeds from the constitution of their country much more than from that of its inhabitants.

Chapter III

Why the Americans Show So Little Sensitiveness in Their Own Country, and Are So Sensitive in Europe

The temper of the Americans is vindictive, like that of all serious and reflecting nations. They hardly ever forget an offence, but it is not easy to offend them; and their resentment is as slow to kindle as it is to abate. In aristocratic communities where a small number of persons manage everything, the outward intercourse of men is subject to settled conventional rules. Every one then thinks he knows exactly what marks of respect or of condescension he ought to display, and none are presumed to be ignorant of the science of etiquette. These usages of the first class in society afterwards serve as a model to all the others; besides which each of the latter lays down a code of its own, to which, all its members are bound to conform. Thus the rules of politeness form a complex system of legislation, which it is difficult to be perfectly master of, but from which it is dangerous for anyone to deviate; so that men are constantly exposed involuntarily to inflict or to receive bitter affronts. But as the distinctions of rank are obliterated, as men differing in education and in birth meet and mingle in the same places of resort, it is almost impossible to agree upon the rules of good breeding. As its laws are uncertain, to disobey them is not a crime, even in the eyes of those who know what they are; men attach more importance to intentions than to forms, and they grow less civil, but at the same time less quarrelsome. There are many little attentions which an American does not care about; he thinks they are not due to him, or he presumes that they are not known to be due: he therefore either does not perceive a rudeness or he forgives it; his manners become less courteous, and his character more plain and masculine.

The mutual indulgence which the Americans display, and the manly confidence with which they treat each other, also result from another deeper and more general cause, which I have already adverted to in the preceding chapter. In the United States the distinctions of rank in civil society are slight, in political society they are null; an American, therefore, does not think himself bound to pay particular attentions to any of his fellow-citizens, nor does he require such attentions from them towards himself. As he does not see that it is his interest eagerly to seek the company of any of his countrymen, he is slow to fancy that his own company is declined: despising no one on account of his station, he does not imagine that anyone can despise him for that cause; and until he has clearly perceived an insult, he does not suppose that an affront was intended. The social condition of the Americans naturally accustoms them not to take offence in small matters; and, on the other hand, the democratic freedom which they enjoy transfuses this same mildness of temper into the character of the nation. The political institutions of the United States constantly bring citizens of all ranks into contact, and compel them to pursue great undertakings in concert. People thus engaged have scarcely time to attend to the details of etiquette, and they are besides too strongly interested in living harmoniously for them to stick at such things. They therefore soon acquire a habit of considering the feelings and opinions of those whom they meet more than their manners, and they do not allow themselves to be annoyed by trifles.

I have often remarked in the United States that it is not easy to make a man understand that his presence may be dispensed with; hints will not always suffice to shake him off. I contradict an American at every word he says, to show him that his conversation bores me; he instantly labours with fresh pertinacity to convince me; I preserve a dogged silence, and he thinks I am meditating deeply on the truths which he is uttering; at last I rush from his company, and he supposes that some urgent business hurries me elsewhere. This man will never understand that he wearies me to extinction unless I tell him so: and the only way to get rid of him is to make him my enemy for life.

It appears surprising at first sight that the same man transported to Europe suddenly becomes so sensitive and captious, that I often find it as difficult to avoid offending him here as it was to put him out of countenance. These two opposite effects proceed from the same cause. Democratic institutions generally give men a lofty notion of their country and of themselves. An American leaves his country with a heart swollen with pride; on arriving in Europe he at once finds out that we are not so engrossed by the United States and the great people which inhabits them as he had supposed, and this begins to annoy him. He has been informed that the conditions of society are not equal in our part of the globe, and he observes that among the nations of Europe the traces of rank are not wholly obliterated; that wealth and birth still retain some indeterminate privileges, which force themselves upon his notice whilst they elude definition. He is therefore profoundly ignorant of the place which he ought to occupy in this half-ruined scale of classes, which are sufficiently distinct to hate and despise each other, yet sufficiently alike for him to be always confounding them. He is afraid of ranging himself too high—still more is he afraid of being ranged too low; this twofold peril keeps his mind constantly on the stretch, and embarrasses all he says and does. He learns from tradition that in Europe ceremonial observances were infinitely varied according to different ranks; this recollection of former times completes his perplexity, and he is the more afraid of not obtaining those marks of respect which are due to him, as he does not exactly know in what they consist. He is like a man surrounded by traps: society is not a recreation for him, but a serious toil: he weighs your least actions, interrogates your looks, and scrutinises all you say, lest there should be some hidden allusion to affront him. I doubt whether there was ever a provincial man of quality so punctilious in breeding as he is: he endeavours to attend to the slightest rules of etiquette, and does not allow one of them to be waived towards himself: he is full of scruples and at the same time of pretensions; he wishes to do enough, but fears to do too much; and as he does not very well know the limits of the one or of the other, he keeps up a haughty and embarrassed air of reserve.

But this is not all: here is yet another double of the human heart. An American is forever talking of the admirable equality which prevails in the United States; aloud he makes it the boast of his country, but in secret he deplores it for himself; and he aspires to show that, for his part, he is an exception to the general state of things which he vaunts. There is hardly an American to be met with who does not claim some remote kindred with the first founders of the colonies; and as for the scions of the noble families of England, America seemed to me to be covered with them. When an opulent American arrives in Europe, his first care is to surround himself with all the luxuries of wealth: he is so afraid of being taken for the plain citizen of a democracy, that he adopts a hundred distorted ways of bringing some new instance of his wealth before you every day. His house will be in the most fashionable part of the town: he will always be surrounded by a host of servants. I have heard an American complain, that in the best houses of Paris the society was rather mixed; the taste which prevails there was not pure enough for him; and he ventured to hint that, in his opinion, there was a want of elegance of manner; he could not accustom himself to see wit concealed under such unpretending forms.

These contrasts ought not to surprise us. If the vestiges of former aristocratic distinctions were not so completely effaced in the United States, the Americans would be less simple and less tolerant in their own country—they would require less, and be less fond of borrowed manners in ours.

CHAPTER IV

Consequences of the Three Preceding Chapters

When men feel a natural compassion for their mutual sufferings—when they are brought together by easy and frequent intercourse, and no sensitive feelings keep them asunder, it may readily be supposed that they will lend assistance to one another whenever it is needed. When an American asks for the cooperation of his fellow-citizens it is seldom refused, and I have often seen it afforded spontaneously and with great goodwill. If an accident happens on the highway, everybody hastens to help the sufferer; if some great and sudden calamity befalls a family, the purses of a thousand strangers are at once willingly opened, and small but numerous donations pour in to relieve their distress. It often happens amongst the most civilised nations of the globe, that a poor wretch is as friendless in the midst of a crowd as the savage in his wilds: this is hardly ever the case in the United States. The Americans, who are always cold and often coarse in their manners, seldom show insensibility; and if they do not proffer services eagerly, yet they do not refuse to render them.

All this is not in contradiction to what I have said before on the subject of individualism. The two things are so far from combating each other, that I can see how they agree. Equality of conditions, whilst it makes men feel their independence, shows them their own weakness: they are free, but exposed to a thousand accidents; and experience soon teaches them that, although they do not habitually require the assistance of others, a time almost always comes when they cannot do without it. We constantly see in Europe that men of the same profession are ever ready to assist each other; they are all exposed to the same ills, and that is enough to teach them to seek mutual preservatives, however hard-hearted and selfish they may otherwise be.

When one of them falls into danger, from which the others may save him by a slight transient sacrifice or a sudden effort, they do not fail to make the attempt. Not that they are deeply interested in his fate; for if, by chance, their exertions are unavailing, they immediately forget the object of them, and return to their own business; but a sort of tacit and almost involuntary agreement has been passed between them, by which each one owes to the others a temporary support which he may claim for himself in turn. Extend to a people the remark here applied to a class, and you will understand my meaning. A similar covenant exists in fact between all the citizens of a democracy: they all feel themselves subject to the same weakness and the same dangers; and their interest, as well as their sympathy, makes it a rule with them to lend each other mutual assistance when required. The more equal social conditions become, the more do men display this reciprocal disposition to oblige each other. In democracies no great benefits are conferred, but good offices are constantly rendered: a man seldom displays self-devotion, but all men are ready to be of service to one another.

CHAPTER V

How Democracy Affects the Relation of Masters and Servants

An American who had travelled for a long time in Europe once said to me, 'The English treat their servants with a stiffness and imperiousness of manner which surprise us; but on the other hand the French sometimes treat their attendants with a degree of familiarity or of politeness which we cannot conceive. It looks as if they were afraid to give orders: the posture of the superior and the inferior is ill-maintained.'—The remark was a just one, and I have often made it myself. I have always considered England as the country in the world where, in our time, the bond of domestic service is drawn most tightly, and France as the country where it is most relaxed. Nowhere have I seen masters stand so high or so low as in these two countries. Between these two extremes the Americans are to be placed. Such is the fact as it appears upon the surface of things: to discover the causes of that fact, it is necessary to search the matter thoroughly.

No communities have ever yet existed in which social conditions have been so equal that there were neither rich nor poor, and consequently neither masters nor servants. Democracy does not prevent the existence of these two classes, but it changes their dispositions and modifies their mutual relations. Amongst aristocratic nations servants form a distinct class, not more variously composed than that of masters. A settled order is soon established; in the former as well as in the latter class a scale is formed, with numerous distinctions or marked gradations of rank, and generations succeed each other thus without any change of position. These two communities are superposed one above the other, always distinct, but regulated by analogous principles. This aristocratic constitution does not exert a less powerful influence on the notions and manners of servants than on those of masters; and, although the effects are different, the same cause may easily be traced. Both classes constitute small communities in the heart of the nation, and certain permanent notions of right and wrong are ultimately engendered amongst them. The different acts of human life are viewed by one particular and unchanging light. In the society of servants, as in that of masters, men exercise a great influence over each other: they acknowledge settled rules, and in the absence of law they are guided by a sort of public opinion: their habits are settled, and their conduct is placed under a certain control.

These men, whose destiny is to obey, certainly do not understand fame, virtue, honesty, and honour in the same manner as their masters; but they have a pride, a virtue, and an honesty pertaining to their condition; and they have a notion, if I may use the expression, of a sort of servile honour.[1] Because a class is mean, it must not be supposed that all who belong to it are mean-hearted; to think so would be a great mistake. However lowly it may be, he who is foremost there, and who has no notion of quitting it, occupies an aristocratic position which inspires him with lofty feelings, pride, and self-respect, that fit him for the higher virtues and actions above the common. Amongst aristocratic nations it was by no means rare to find men of noble and vigorous minds in the service of the great, who felt not the servitude they bore, and who submitted to the will of their masters without any fear of their displeasure. But this was hardly ever the case amongst the inferior ranks of domestic servants. It may be imagined that he who occupies the lowest stage of the order of menials stands very low indeed. The French created a word on purpose to designate the servants of the aristocracy—they called them 'lacqueys.' This word 'lacquey' served as the strongest expression, when all others were exhausted, to designate human meanness. Under the old French monarchy, to denote by a single expression a low-spirited contemptible fellow, it was usual to say that he had the '*soul of a lacquey*'; the term was enough to convey all that was intended.

The permanent inequality of conditions not only gives servants certain peculiar virtues and vices, but it places them in a peculiar relation with respect to their masters. Amongst aristocratic nations the poor man is familiarised from his childhood with the notion of being commanded: to whichever side he turns his eyes the graduated structure of society and the aspect of obedience meet his view. Hence in those countries the master readily obtains prompt, complete, respectful, and easy obedience from his servants, because they revere in him not only their master but the class of masters. He weighs down their will by the whole weight of the aristocracy. He orders their actions—to a certain extent he even directs their thoughts. In aristocracies the master often exercises, even without being aware of it, an amazing sway over the opinions, the habits, and the manners of those who obey him, and his influence extends even further than his authority.

In aristocratic communities there are not only hereditary families of servants as well as of masters, but the same families of servants adhere for several generations to the same families of masters (like two parallel lines which neither meet nor separate); and this considerably modifies the mutual relations of these two classes of persons. Thus, although in aristocratic society the master and servant have no natural resemblance—although, on the contrary, they are placed at an immense distance on the scale of human beings by their fortune, education, and opinions—yet time ultimately binds them together. They are connected by a long series of common reminiscences, and however different they may be, they grow alike; whilst in democracies, where they are naturally almost alike, they always remain strangers to each other. Amongst an aristocratic people the master gets to look upon his servants as an inferior and secondary part of himself, and he often takes an interest in their lot by a last stretch of egotism.

Servants, on their part, are not averse to regard themselves in the same light; and they sometimes identify themselves with the person of the master, so that they become an appendage to him in their own eyes as well as in his. In aristocracies a servant fills a subordinate position which he cannot get out of; above him is another man, holding a superior rank which he cannot lose. On one side are obscurity, poverty, obedience for life; on the other, and also for life, fame, wealth, and command. The two conditions are always distinct and always in propinquity; the tie that connects them is as lasting as they are themselves. In this predicament the servant ultimately detaches his notion of interest from his own person; he deserts himself as it were, or rather he transports himself into the character of his master, and thus assumes an imaginary personality. He complacently invests himself with the wealth of those who command him; he shares their fame, exalts himself by their rank, and feeds his mind with borrowed greatness, to which he attaches more importance than those who fully and really possess it. There is something touching, and at the same time ridiculous, in this strange confusion of two different states of being. These passions of masters, when they pass into the souls of menials, assume the natural dimensions of the place they occupy—they are contracted and lowered. What was pride in the former becomes puerile vanity and paltry ostentation in the latter. The servants of a great man are commonly most punctilious as to the marks of respect due to him, and they attach more importance to his slightest privileges than he does himself. In France a few of these old servants of the aristocracy are still to be met with here and there; they have survived their race, which will soon disappear with them altogether. In the United States I never saw anyone at all like them. The Americans are not only unacquainted with the kind of man, but it is hardly possible to make them understand that such ever existed. It is scarcely less difficult for them to conceive it, than for us to form a correct notion of what a slave was amongst the Romans, or a serf in the Middle Ages. All these men were in fact, though in different degrees, results of the same cause: they are all retiring from our sight, and disappearing in the obscurity of the past, together with the social condition to which they owed their origin.

Equality of conditions turns servants and masters into new beings, and places them in new relative positions. When social conditions are nearly equal, men are constantly changing their situations in life: there is still a class of menials and a class of masters, but these classes are not always composed of the same individuals, stall less of the same families; and those who command are not more secure of perpetuity than those who obey. As servants do not form a separate people, they have no habits, prejudices, or manners peculiar to themselves; they are not remarkable for any particular turn of mind or moods of feeling. They know no vices or virtues of their condition, but they partake of the education, the opinions, the feelings, the virtues, and the vices of their contemporaries; and they are honest men or scoundrels in the same way as their masters are. The conditions of servants are not less equal than those of masters. As no marked ranks or fixed subordination are to be found amongst them, they will not display either the meanness or the greatness which characterise the aristocracy of menials as well as all other aristocracies. I never saw a man in the United States who reminded me of that class of confidential servants of which we still retain a reminiscence in Europe, neither did I ever meet with such a thing as a lacquey: all traces of the one and of the other have disappeared.

In democracies servants are not only equal amongst themselves, but it may be said that they are in some sort the equals of their masters. This requires explanation in order to be rightly understood. At any moment a servant may become a master, and he aspires to rise to that condition: the servant is therefore not a different man from the master. Why then has the former a right to command, and what compels the latter to obey?—the free and temporary consent of both their wills. Neither of them is by nature inferior to the other; they only become so for a time by covenant. Within the terms of this covenant, the one is a servant, the other a master; beyond it they are two citizens of the commonwealth—two men. I beg the reader particularly to observe that this is not only the notion which servants themselves entertain of their own condition; domestic service is looked upon by masters in the same light; and the precise limits of authority and obedience are as clearly settled in the mind of the one as in that of the other.

When the greater part of the community have long attained a condition nearly alike, and when equality is an old and acknowledged fact, the public mind, which is never affected by exceptions, assigns certain general limits to the value of man, above or below which no man can long remain placed. It is in vain that wealth and poverty, authority and obedience, accidentally interpose great distances between two men; public opinion, founded upon the usual order of things, draws them to a common level, and creates a species of imaginary equality between them, in spite of the real inequality of their conditions. This all-powerful opinion penetrates at length even into the hearts of those whose interest might arm them to resist it; it affects their judgment whilst it subdues their will. In their inmost convictions the master and the servant no longer perceive any deep-seated difference between them, and they neither hope nor fear to meet with any such at any time. They are therefore neither subject to disdain nor to anger, and they discern in each other neither humility nor pride. The master holds the contract of service to be the only source of his power, and the servant regards it as the only cause of his obedience. They do not quarrel about their reciprocal situations, but each knows his own and keeps it.

In the French army the common soldier is taken from nearly the same classes as the officer, and may hold the same commissions; out of the ranks he considers himself entirely equal to his military superiors, and in point of fact he is so; but when under arms he does not hesitate to obey, and his obedience is not the less prompt, precise, and ready, for being voluntary and defined. This example may give a notion of what takes place between masters and servants in democratic communities.

It would be preposterous to suppose that those warm and deep-seated affections, which are sometimes kindled in the domestic service of aristocracy, will ever spring up between these two men, or that they will exhibit strong instances of self-sacrifice. In aristocracies masters and servants live apart, and frequently their only intercourse is through a third person; yet they commonly stand firmly by one another. In democratic countries the master and the servant are close together; they are in daily personal contact, but their minds do not intermingle; they have common occupations, hardly ever common interests. Amongst such a people the servant always considers himself as a sojourner in the dwelling of his masters. He knew nothing of their forefathers,—he will see nothing of their descendants,—he has nothing lasting to expect from their hand. Why then should he confound his life with theirs, and whence should so strange a surrender of himself proceed? The reciprocal position of the two men is changed,—their mutual relations must be so too.

I would fain illustrate all these reflections by the example of the Americans; but for this purpose the distinctions of persons and places must be accurately traced. In the south of the Union, slavery exists; all that I have just said is consequently inapplicable there. In the north, the majority of servants are either freed-men or the children of freed-men; these persons occupy a contested position in the public estimation; by the laws they are brought up to the level of their masters,—by the manners of the country they are obstinately detruded from it. They do not themselves clearly know their proper place, and they are almost always either insolent or craven. But in the Northern States, especially in New England, there are a certain number of whites, who agree, for wages, to yield a temporary obedience to the will of their fellow-citizens. I have heard that these servants commonly perform the duties of their situation with punctuality and intelligence; and that without thinking themselves naturally inferior to the person who orders them, they submit without reluctance to obey him. They appear to me to carry into service some of those manly habits which independence and equality engender. Having once selected a hard way of life, they do not seek to escape from it by indirect means; and they have sufficient respect for themselves, not to refuse to their master that obedience which they have freely promised. On their part, masters require nothing of their servants but the faithful and rigorous performance of the covenant: they do not ask for marks of respect, they do not claim their love or devoted attachment; it is enough that, as servants, they are exact and honest. It would not then be true to assert that, in democratic society, the relation of servants and masters is disorganised: it is organised on another footing; the rule is different, but there is a rule.

It is not my purpose to inquire whether the new state of things which I have just described is inferior to that which preceded it, or simply different. Enough for me that it is fixed and determined: for what is most important to meet with among men is not any given ordering, but order. But what shall I say of those sad and troubled times at which equality is established in the midst of the tumult of revolution,—when democracy, after having been introduced into the state of society, still struggles with difficulty against the prejudices and manners of the country? The laws, and partially public opinion, already declare that no natural or permanent inferiority exists between the servant and the master. But this new belief has not yet reached the innermost convictions of the latter, or rather his heart rejects it; in the secret persuasion of his mind the master thinks that he belongs to a peculiar and superior race; he dares not say so, but he shudders whilst he allows himself to be dragged to the same level. His authority over his servants becomes timid and at the same time harsh: he has already ceased to entertain for them the feelings of patronising kindness which long uncontested power always engenders, and he is surprised that, being changed himself, his servant changes also. He wants his attendants to form regular and permanent habits, in a condition of domestic service which is only temporary: he requires that they should appear contented with and proud of a servile condition, which they will one day shake off—that they should sacrifice themselves to a man who can neither protect nor ruin them,—and in short that they should contract an indissoluble engagement to a being like themselves, and one who will last no longer than they will.

Amongst aristocratic nations it often happens that the condition of domestic service does not degrade the character of those who enter upon it, because they neither know nor imagine any other; and the amazing inequality which is manifest between them and their master appears to be the necessary and unavoidable consequence of some hidden law of Providence. In democracies the condition of domestic service does not degrade the character of those who enter upon it, because it is freely chosen, and adopted for a time only; because it is not stigmatised by public opinion, and creates no permanent inequality between the servant and the master. But whilst the transition from one social condition to another is going on, there is almost always a time when men's minds fluctuate between the aristocratic notion of subjection and the democratic notion of obedience. Obedience then loses its moral importance in the eyes of him who obeys; he no longer considers it as a species of divine obligation, and he does not yet view it under its purely human aspect; it has to him no character of sanctity or of justice, and he submits to it as to a degrading but profitable condition. At that moment a confused and imperfect phantom of equality haunts the minds of servants; they do not at once perceive whether the equality to which they are entitled is to be found within or without the pale of domestic service; and they rebel in their hearts against a subordination to which they have subjected themselves, and from which they derive actual profit. They consent to serve, and they blush to obey; they like the advantages of service, but not the master; or rather, they are not sure that they ought not themselves to be masters, and they are inclined to consider him who orders them as an unjust usurper of their own rights. Then it is that the dwelling of every citizen offers a spectacle somewhat analogous to the gloomy aspect of political society. A secret and intestine warfare is going on there between powers, ever rivals and suspicious of one another: the master is ill-natured and weak, the servant ill-natured and intractable; the one constantly attempts to evade by unfair restrictions his obligation to protect and to remunerate,—the other his obligation to obey. The reins of domestic government dangle between them, to be snatched at by one or the other. The lines which divide authority from oppression, liberty from license, and right from might, are to their eyes so jumbled together and confused, that no one knows exactly what he is, or what he may be, or what he ought to be. Such a condition is not democracy, but revolution.

Chapter VI

That Democratic Institutions and Manners Tend to Raise Rents and Shorten the Terms of Leases

What has been said of servants and masters is applicable, to a certain extent, to land-owners and farming tenants; but this subject deserves to be considered by itself. In America there are, properly speaking, no tenant farmers; every man owns the ground he tills. It must be admitted that democratic laws tend greatly to increase the number of land-owners, and to diminish that of farming tenants. Yet what takes place in the United States is much less attributable to the institutions of the country than to the country itself. In America land is cheap, and any one may easily become a land-owner; its returns are small, and its produce cannot well be divided between a land-owner and a farmer. America therefore stands alone in this as well as in many other respects, and it would be a mistake to take it as an example.

I believe that in democratic as well as in aristocratic countries there will be land-owners and tenants, but the connection existing between them will be of a different kind. In aristocracies the hire of a farm is paid to the landlord, not only in rent, but in respect, regard, and duty; in democracies the whole is paid in cash. When estates are divided and passed from hand to hand, and the permanent connection which existed between families and the soil is dissolved, the land-owner and the tenant are only casually brought into contact. They meet for a moment to settle the conditions of the agreement, and then lose sight of each other; they are two strangers brought together by a common interest, and who keenly talk over a matter of business, the sole object of which is to make money.

In proportion as property is subdivided and wealth distributed over the country, the community is filled with people whose former opulence is declining, and with others whose fortunes are of recent growth and whose wants increase more rapidly than their resources. For all such persons the smallest pecuniary profit is a matter of importance, and none of them feel disposed to waive any of their claims, or to lose any portion of their income. As ranks are intermingled, and as very large as well as very scanty fortunes become more rare, every day brings the social condition of the land-owner nearer to that of the farmer; the one has not naturally any uncontested superiority over the other; between two men who are equal, and not at ease in their circumstances, the contract of hire is exclusively an affair of money. A man whose estate extends over a whole district, and who owns a hundred farms, is well aware of the importance of gaining at the same time the affections of some thousands of men; this object appears to call for his exertions, and to attain it he will readily make considerable sacrifices. But he who owns a hundred acres is insensible to similar considerations, and he cares but little to win the private regard of his tenant.

An aristocracy does not expire like a man in a single day; the aristocratic principle is slowly undermined in men's opinion, before it is attacked in their laws. Long before open war is declared against it, the tie which had hitherto united the higher classes to the lower may be seen to be gradually relaxed. Indifference and contempt are betrayed by one class, jealousy and hatred by the others; the intercourse between rich and poor becomes less frequent and less kind, and rents are raised. This is not the consequence of a democratic revolution, but its certain harbinger; for an aristocracy which has lost the affections of the people, once and for ever, is like a tree dead at the root, which is the more easily torn up by the winds the higher its branches have spread.

In the course of the last fifty years the rents of farms have amazingly increased, not only in France but throughout the greater part of Europe. The remarkable improvements which have taken place in agriculture and manufactures within the same period do not suffice in my opinion to explain this fact; recourse must be had to another cause more powerful and more concealed. I believe that cause is to be found in the democratic institutions which several European nations have adopted, and in the democratic passions which more or less agitate all the rest. I have frequently heard great English land-owners congratulate themselves that, at the present day, they derive a much larger income from their estates than their fathers did. They have perhaps good reasons to be glad; but most assuredly they know not what they are glad of. They think they are making a clear gain, when it is in reality only an exchange; their influence is what they are parting with for cash; and what they gain in money will ere long be lost in power.

There is yet another sign by which it is easy to know that a great democratic revolution is going on or approaching. In the middle ages almost all lands were leased for lives, or for very long terms; the domestic economy of that period shows that leases for ninety-nine years were more frequent then than leases for twelve years are now. Men then believed that families were immortal; men's conditions seemed settled for ever, and the whole of society appeared to be so fixed, that it was not supposed that anything would ever be stirred or shaken in its structure. In ages of equality, the human mind takes a different bent; the prevailing notion is that nothing abides, and man is haunted by the thought of mutability. Under this impression the land-owner and the tenant himself are instinctively averse to protracted terms of obligation; they are afraid of being tied up tomorrow by the contract which benefits them today. They have vague anticipations of some sudden and unforeseen change in their conditions; they mistrust themselves; they fear lest their taste should change, and lest they should lament that they cannot rid themselves of what they coveted; nor are such fears unfounded, for in democratic ages that which is most fluctuating amidst the fluctuation of all around is the heart of man.

Chapter VII

Influence of Democracy on Wages

Most of the remarks which I have already made in speaking of servants and masters, may be applied to masters and workmen. As the gradations of the social scale come to be less observed, whilst the great sink the humble rise, and as poverty as well as opulence ceases to be hereditary, the distance both in reality and in opinion, which heretofore separated the workman from the master is lessened every day. The workman conceives a more lofty opinion of his rights, of his future, of himself; he is filled with new ambition and with new desires, he is harassed by new wants. Every instant he views with longing eyes the profits of his employer; and in order to share them, he strives to dispose of his labour at a higher rate, and he generally succeeds at length in the attempt. In democratic countries, as well as elsewhere, most of the branches of productive industry are carried on at a small cost, by men little removed by their wealth or education above the level of those whom they employ. These manufacturing speculators are extremely numerous; their interests differ; they cannot therefore easily concert or combine their exertions. On the other hand the workmen have almost always some sure resources, which enable them to refuse to work when they cannot get what they conceive to be the fair price of their labour. In the constant struggle for wages which is going on between these two classes, their strength is divided, and success alternates from one to the other. It is even probable that in the end the interest of the working class must prevail; for the high wages which they have already obtained make them every day less dependent on their masters; and as they grow more independent, they have greater facilities for obtaining a further increase of wages.

I shall take for example that branch of productive industry which is still at the present day the most generally followed in France, and in almost all the countries of the world—I mean the cultivation of the soil. In France most of those who labour for hire in agriculture, are themselves owners of certain plots of ground, which just enable them to subsist without working for anyone else. When these labourers come to offer their services to a neighbouring land-owner or farmer, if he refuses them a certain rate of wages, they retire to their own small property and await another opportunity.

I think that, upon the whole, it may be asserted that a slow and gradual rise of wages is one of the general laws of democratic communities. In proportion as social conditions become more equal, wages rise; and as wages are higher, social conditions become more equal. But a great and gloomy exception occurs in our own time. I have shown in a preceding chapter that aristocracy, expelled from political society, has taken refuge in certain departments of productive industry, and has established its sway there under another form; this powerfully affects the rate of wages. As a large capital is required to embark in the great manufacturing speculations to which I allude, the number of persons who enter upon them is exceedingly limited: as their number is small, they can easily concert together, and fix the rate of wages as they please. Their workmen on the contrary are exceedingly numerous, and the number of them is always increasing; for, from time to time, an extraordinary run of business takes place, during which wages are inordinately high, and they attract the surrounding population to the factories. But, when once men have embraced that line of life, we have already seen that they cannot quit it again, because they soon contract habits of body and mind which unfit them for any other sort of toil. These men have generally but little education and industry, with but few resources; they stand therefore almost at the mercy of the master. When competition, or other fortuitous circumstances, lessen his profits, he can reduce the wages of his workmen almost at pleasure, and make from them what he loses by the chances of business. Should the workmen strike, the master, who is a rich man, can very well wait without being ruined until necessity brings them back to him; but they must work day by day or they die, for their only property is in their hands. They have long been impoverished by oppression, and the poorer they become the more easily may they be oppressed: they can never escape from this fatal circle of cause and consequence. It is not then surprising that wages, after having sometimes suddenly risen, are permanently lowered in this branch of industry; whereas in other callings the price of labour, which generally increases but little, is nevertheless constantly augmented.

This state of dependence and wretchedness, in which a part of the manufacturing population of our time lives, forms an exception to the general rule, contrary to the state of all the rest of the community; but, for this very reason, no circumstance is more important or more deserving of the especial consideration of the legislator; for when the whole of society is in motion, it is difficult to keep any one class stationary; and when the greater number of men are opening new paths to fortune, it is no less difficult to make the few support in peace their wants and their desires.

Chapter VIII

Influence of Democracy on Kindred

I have just examined the changes which the equality of conditions produces in the mutual relations of the several members of the community amongst democratic nations, and amongst the Americans in particular. I would now go deeper, and inquire into the closer ties of kindred: my object here is not to seek for new truths, but to show in what manner facts already known are connected with my subject.

It has been universally remarked, that in our time the several members of a family stand upon an entirely new footing towards each other; that the distance which formerly separated a father from his sons has been lessened; and that paternal authority, if not destroyed, is at least impaired. Something analogous to this, but even more striking, may be observed in the United States. In America the family, in the Roman and aristocratic signification of the word, does not exist. All that remains of it are a few vestiges in the first years of childhood, when the father exercises, without opposition, that absolute domestic authority, which the feebleness of his children renders necessary, and which their interest, as well as his own incontestable superiority, warrants. But as soon as the young American approaches manhood, the ties of filial obedience are relaxed day by day: master of his thoughts, he is soon master of his conduct. In America there is, strictly speaking, no adolescence: at the close of boyhood the man appears, and begins to trace out his own path. It would be an error to suppose that this is preceded by a domestic struggle, in which the son has obtained by a sort of moral violence the liberty that his father refused him. The same habits, the same principles which impel the one to assert his independence, predispose the other to consider the use of that independence as an incontestable right. The former does not exhibit any of those rancorous or irregular passions which disturb men long after they have shaken off an established authority; the latter feels none of that bitter and angry regret which is apt to survive a bygone power. The father foresees the limits of his authority long beforehand, and when the time arrives he surrenders it without a struggle: the son looks forward to the exact period at which he will be his own master; and he enters upon his freedom without precipitation and without effort, as a possession which is his own and which no one seeks to wrest from him.[1]

It may perhaps not be without utility to show how these changes which take place in family relations, are closely connected with the social and political revolution which is approaching its consummation under our own observation. There are certain great social principles, which a people either introduces everywhere, or tolerates nowhere. In countries which are aristocratically constituted with all the gradations of rank, the Government never makes a direct appeal to the mass of the governed: as men are united together, it is enough to lead the foremost, the rest will follow. This is equally applicable to the family, as to all aristocracies which have a head. Amongst aristocratic nations, social institutions recognize, in truth, no one in the family but the father; children are received by society at his hands; society governs him, he governs them. Thus the parent has not only a natural right, but he acquires a political right, to command them: he is the author and the support of his family; but he is also its constituted ruler. In democracies, where the government picks out every individual singly from the mass, to make him subservient to the general laws of the community, no such intermediate person is required: a father is there, in the eye of the law, only a member of the community, older and richer than his sons.

When most of the conditions of life are extremely unequal, and the inequality of these conditions is permanent, the notion of a superior grows upon the imaginations of men: if the law invested him with no privileges, custom and public opinion would concede them. When, on the contrary, men differ but little from each other, and do not always remain in dissimilar conditions of life, the general notion of a superior becomes weaker and less distinct: it is vain for legislation to strive to place him who obeys very much beneath him who commands; the manners of the time bring the two men nearer to one another, and draw them daily towards the same level. Although the legislation of an aristocratic people should grant no peculiar privileges to the heads of families; I shall not be the less convinced that their power is more respected and more extensive than in a democracy; for I know that, whatsoever the laws may be, superiors always appear higher and inferiors lower in aristocracies than amongst democratic nations.

When men live more for the remembrance of what has been than for the care of what is, and when they are more given to attend to what their ancestors thought than to think themselves, the father is the natural and necessary tie between the past and the present—the link by which the ends of these two chains are connected. In aristocracies, then, the father is not only the civil head of the family, but the oracle of its traditions, the expounder of its customs, the arbiter of its manners. He is listened to with deference, he is addressed with respect, and the love which is felt for him is always tempered with fear. When the condition of society becomes democratic, and men adopt as their general principle that it is good and lawful to judge of all things for oneself, using former points of belief not as a rule of faith but simply as a means of information, the power which the opinions of a father exercise over those of his sons diminishes as well as his legal power.

Perhaps the subdivision of estates which democracy brings with it contributes more than anything else to change the relations existing between a father and his children. When the property of the father of a family is scanty, his son and himself constantly live in the same place, and share the same occupations: habit and necessity bring them together, and force them to hold constant communication: the inevitable consequence is a sort of familiar intimacy, which renders authority less absolute, and which can ill be reconciled with the external forms of respect. Now in democratic countries the class of those who are possessed of small fortunes is precisely that which gives strength to the notions, and a particular direction to the manners, of the community. That class makes its opinions preponderate as universally as its will, and even those who are most inclined to resist its commands are carried away in the end by its example. I have known eager opponents of democracy who allowed their children to address them with perfect colloquial equality.

Thus, at the same time that the power of aristocracy is declining, the austere, the conventional, and the legal part of parental authority vanishes, and a species of equality prevails around the domestic hearth. I know not, upon the whole, whether society loses by the change, but I am inclined to believe that man individually is a gainer by it. I think that, in proportion as manners and laws become more democratic, the relation of father and son becomes more intimate and more affectionate; rules and authority are less talked of; confidence and tenderness are oftentimes increased, and it would seem that the natural bond is drawn closer in proportion as the social bond is loosened. In a democratic family the father exercises no other power than that with which men love to invest the affection and the experience of age; his orders would perhaps be disobeyed, but his advice is for the most part authoritative. Though he be not hedged in with ceremonial respect, his sons at least accost him with confidence; no settled form of speech is appropriated to the mode of addressing him, but they speak to him constantly, and are ready to consult him day by day; the master and the constituted ruler have vanished—the father remains. Nothing more is needed, in order to judge of the difference between the two states of society in this respect, than to peruse the family correspondence of aristocratic ages. The style is always correct, ceremonious, stiff, and so cold that the natural warmth of the heart can hardly be felt in the language. The language, on the contrary, addressed by a son to his father in democratic countries is always marked by mingled freedom, familiarity and affection, which at once show that new relations have sprung up in the bosom of the family.

A similar revolution takes place in the mutual relations of children. In aristocratic families, as well as in aristocratic society, every place is marked out beforehand. Not only does the father occupy a separate rank, in which he enjoys extensive privileges, but even the children are not equal amongst themselves. The age and sex of each irrevocably determine his rank, and secure to him certain privileges: most of these distinctions are abolished or diminished by democracy. In aristocratic families the eldest son, inheriting the greater part of the property, and almost all the rights of the family, becomes the chief, and, to a certain extent, the master, of his brothers. Greatness and power are for him—for them, mediocrity and dependence. Nevertheless it would be wrong to suppose that, amongst aristocratic nations, the privileges of the eldest son are advantageous to himself alone, or that they excite nothing but envy and hatred in those around him. The eldest son commonly endeavours to procure wealth and power for his brothers, because the general splendour of the house is reflected back on him who represents it; the younger sons seek to back the elder brother in all his undertakings, because the greatness and power of the head of the family better enable him to provide for all its branches. The different members of an aristocratic family are therefore very closely bound together; their interests are connected, their minds agree, but their hearts are seldom in harmony.

Democracy also binds brothers to each other, but by very different means. Under democratic laws all the children are perfectly equal, and consequently independent; nothing brings them forcibly together, but nothing keeps them apart; and as they have the same origin, as they are trained under the same roof, as they are treated with the same care, and as no peculiar privilege distinguishes or divides them, the affectionate and youthful intimacy of early years easily springs up between them. Scarcely any opportunities occur to break the tie thus formed at the outset of life; for their brotherhood brings them daily together, without embarrassing them. It is not, then, by interest, but by common associations and by the free sympathy of opinion and of taste, that democracy unites brothers to each other. It divides their inheritance, but it allows their hearts and minds to mingle together. Such is the charm of these democratic manners, that even the partisans of aristocracy are caught by it; and after having experienced it for some time, they are by no means tempted to revert to the respectful and frigid observance of aristocratic families. They would be glad to retain the domestic habits of democracy, if they might throw off its social conditions and its laws; but these elements are indissolubly united, and it is impossible to enjoy the former without enduring the latter.

The remarks I have made on filial love and fraternal affection are applicable to all the passions which emanate spontaneously from human nature itself. If a certain mode of thought or feeling is the result of some peculiar condition of life, when that condition is altered nothing whatever remains of the thought or feeling. Thus a law may bind two members of the community very closely to one another; but that law being abolished, they stand asunder. Nothing was more strict than the tie which united the vassal to the lord under the feudal system; at the present day the two men know not each other; the fear, the gratitude, and the affection which formerly connected them have vanished, and not a vestige of the tie remains. Such, however, is not the case with those feelings which are natural to mankind. Whenever a law attempts to tutor these feelings in any particular manner, it seldom fails to weaken them; by attempting to add to their intensity, it robs them of some of their elements, for they are never stronger than when left to themselves.

Democracy, which destroys or obscures almost all the old conventional rules of society, and which prevents men from readily assenting to new ones, entirely effaces most of the feelings to which these conventional rules have given rise; but it only modifies some others, and frequently imparts to them a degree of energy and sweetness unknown before. Perhaps it is not impossible to condense into a single proposition the whole meaning of this chapter, and of several others that preceded it. Democracy loosens social ties, but it draws the ties of nature more tight; it brings kindred more closely together, whilst it places the various members of the community more widely apart.

Chapter IX

Education of Young Women in the United States

No free communities ever existed without morals; and, as I observed in the former part of this work, morals are the work of woman. Consequently, whatever affects the condition of women, their habits and their opinions, has great political importance in my eyes. Amongst almost all Protestant nations young women are far more the mistresses of their own actions than they are in Catholic countries. This independence is still greater in Protestant countries like England, which have retained or acquired the right of self-government; the spirit of freedom is then infused into the domestic circle by political habits and by religious opinions. In the United States the doctrines of Protestantism are combined with great political freedom and a most democratic state of society; and nowhere are young women surrendered so early or so completely to their own guidance. Long before an American girl arrives at the age of marriage, her emancipation from maternal control begins; she has scarcely ceased to be a child when she already thinks for herself, speaks with freedom, and acts on her own impulse. The great scene of the world is constantly open to her view: far from seeking concealment, it is every day disclosed to her more completely, and she is taught to survey it with a firm and calm gaze. Thus the vices and dangers of society are early revealed to her; as she sees them clearly, she views them without, illusions, and braves them without fear; for she is full of reliance on her own strength, and her reliance seems to be shared by all who are about her. An American girl scarcely ever displays that virginal bloom in the midst of young desires, or that innocent and ingenuous grace which usually attends the European woman in the transition from girlhood to youth. It is rarely that an American woman at any age displays childish timidity or ignorance. Like the young women of Europe, she seeks to please, but she knows precisely the cost of pleasing. If she does not abandon herself to evil, at least she knows that it exists; and she is remarkable rather for purity of manners than for chastity of mind. I have been frequently surprised, and almost frightened, at the singular address and happy boldness with which young women in America contrive to manage their thoughts and their language amidst all the difficulties of stimulating conversation; a philosopher would have stumbled at every step along the narrow path which they trod without accidents and without effort. It is easy indeed to perceive that, even amidst the independence of early youth, an American woman is always mistress of herself; she indulges in all permitted pleasures, without yielding herself up to any of them; and her reason never allows the reins of self-guidance to drop, though it often seems to hold them loosely.

In France, where remnants of every age are still so strangely mingled in the opinions and tastes of the people, women commonly receive a reserved, retired, and almost cloistral education, as they did in aristocratic times; and then they are suddenly abandoned, without a guide and without assistance, in the midst of all the irregularities inseparable from democratic society. The Americans are more consistent. They have found out that in a democracy the independence of individuals cannot fail to be very great, youth premature, tastes ill-restrained, customs fleeting, public opinion often unseeded and powerless, paternal authority weak, and marital authority contested. Under these circumstances, believing that they had little chance of repressing in woman the most vehement passions of the human heart, they held that the surer way was to teach her the art of combating those passions for herself. As they could not prevent her virtue from being exposed to frequent danger, they determined that she should know how best to defend it; and more reliance was placed on the free vigour of her will than on safeguards which have been shaken or overthrown. Instead then of inculcating mistrust of herself, they constantly seek to enhance their confidence in her own strength of character. As it is neither possible nor desirable to keep a young woman in perpetual or complete ignorance, they hasten to give her a precocious knowledge on all subjects. Far from hiding the corruptions of the world from her, they prefer that she should see them at once and train herself to shun them; and they hold it of more importance to protect her conduct than to be overscrupulous of her innocence.

Although the Americans are a very religious people, they do not rely on religion alone to defend the virtue of woman; they seek to arm her reason also. In this they have followed the same method as in several other respects; they first make the most vigorous efforts to bring individual independence to exercise a proper control over itself, and they do not call in the aid of religion until they have reached the utmost limits of human strength. I am aware that an education of this kind is not without danger; I am sensible that it tends to invigorate the judgment at the expense of the imagination, and to make cold and virtuous women instead of affectionate wives and agreeable companions to man. Society may be more tranquil and better regulated, but domestic life has often fewer charms. These, however, are secondary evils, which may be braved for the sake of higher interests. At the stage at which we are now arrived the time for choosing is no longer within our control; a democratic education is indispensable to protect women from the dangers with which democratic institutions and manners surround them.

Chapter X

The Young Woman in the Character of a Wife

In America the independence of woman is irrevocably lost in the bonds of matrimony: if an unmarried woman is less constrained there than elsewhere, a wife is subjected to stricter obligations. The former makes her father's house an abode of freedom and of pleasure; the latter lives in the home of her husband as if it were a cloister. Yet these two different conditions of life are perhaps not so contrary as may be supposed, and it is natural that the American women should pass through the one to arrive at the other.

Religious peoples and trading nations entertain peculiarly serious notions of marriage: the former consider the regularity of woman's life as the best pledge and most certain sign of the purity of her morals; the latter regard it as the highest security for the order and prosperity of the household. The Americans are at the same time a puritanical people and a commercial nation: their religious opinions, as well as their trading habits, consequently lead them to require much abnegation on the part of woman, and a constant sacrifice of her pleasures to her duties which is seldom demanded of her in Europe. Thus in the United States the inexorable opinion of the public carefully circumscribes woman within the narrow circle of domestic interest and duties, and forbids her to step beyond it.

Upon her entrance into the world a young American woman finds these notions firmly established; she sees the rules which are derived from them; she is not slow to perceive that she cannot depart for an instant from the established usages of her contemporaries, without putting in jeopardy her peace of mind, her honour, nay even her social existence; and she finds the energy required for such an act of submission in the firmness of her understanding and in the virile habits which her education has given her. It may be said that she has learned by the use of her independence to surrender it without a struggle and without a murmur when the time comes for making the sacrifice. But no American woman falls into the toils of matrimony as into a snare held out to her simplicity and ignorance. She has been taught beforehand what is expected of her, and voluntarily and freely does she enter upon this engagement. She supports her new condition with courage, because she chose it. As in America paternal discipline is very relaxed and the conjugal tie very strict, a young woman does not contract the latter without considerable circumspection and apprehension. Precocious marriages are rare. Thus American women do not marry until their understandings are exercised and ripened; whereas in other countries most women generally only begin to exercise and to ripen their understandings after marriage.

I by no means suppose, however, that the great change which takes place in all the habits of women in the United States, as soon as they are married, ought solely to be attributed to the constraint of public opinion: it is frequently imposed upon themselves by the sole effort of their own will. When the time for choosing a husband is arrived, that cold and stern reasoning power which has been educated and invigorated by the free observation of the world, teaches an American woman that a spirit of levity and independence in the bonds of marriage is a constant subject of annoyance, not of pleasure; it tells her that the amusements of the girl cannot become the recreations of the wife, and that the sources of a married woman's happiness are in the home of her husband. As she clearly discerns beforehand the only road which can lead to domestic happiness, she enters upon it at once, and follows it to the end without seeking to turn back.

The same strength of purpose which the young wives of America display, in bending themselves at once and without repining to the austere duties of their new condition, is no less manifest in all the great trials of their lives. In no country in the world are private fortunes more precarious than in the United States. It is not uncommon for the same man, in the course of his life, to rise and sink again through all the grades which lead from opulence to poverty. American women support these vicissitudes with calm and unquenchable energy: it would seem that their desires contract, as easily as they expand, with their fortunes.[1]

The greater part of the adventurers who migrate every year to people the western wilds, belong, as I observed in the former part of this work, to the old Anglo-American race of the Northern States. Many of these men, who rush so boldly onwards in pursuit of wealth, were already in the enjoyment of a competency in their own part of the country. They take their wives along with them, and make them share the countless perils and privations which always attend the commencement of these expeditions. I have often met, even on the verge of the wilderness, with young women, who after having been brought up amidst all the comforts of the large towns of New England, had passed, almost without any intermediate stage, from the wealthy abode of their parents to a comfortless hovel in a forest. Fever, solitude, and a tedious life had not broken the springs of their courage. Their features were impaired and faded, but their looks were firm: they appeared to be at once sad and resolute. I do not doubt that these young American women had amassed, in the education of their early years, that inward strength which they displayed under these circumstances. The early culture of the girl may still therefore be traced, in the United States, under the aspect of marriage: her part is changed, her habits are different, but her character is the same.

Chapter XI

That the Equality of Conditions Contributes to the Maintenance of Good Morals in America

Some philosophers and historians have said, or have hinted, that the strictness of female morality was increased or diminished simply by the distance of a country from the equator. This solution of the difficulty was an easy one; and nothing was required but a globe and a pair of compasses to settle in an instant one of the most difficult problems in the condition of mankind. But I am not aware that this principle of the materialists is supported by facts. The same nations have been chaste or dissolute at different periods of their history; the strictness or the laxity of their morals depended therefore on some variable cause, not only on the natural qualities of their country, which were invariable. I do not deny that in certain climates the passions which are occasioned by the mutual attraction of the sexes are peculiarly intense; but I am of opinion that this natural intensity may always be excited or restrained by the condition of society and by political institutions.

Although the travellers who have visited North America differ on a great number of points, they all agree in remarking that morals are far more strict there than elsewhere. It is evident that on this point the Americans are very superior to their progenitors the English. A superficial glance at the two nations will establish the fact. In England, as in all other countries of Europe, public malice is constantly attacking the frailties of women. Philosophers and statesmen are heard to deplore that morals are not sufficiently strict, and the literary productions of the country constantly lead one to suppose so. In America all books, novels not excepted, suppose women to be chaste, and no one thinks of relating affairs of gallantry. No doubt this great regularity of American morals originates partly in the country, in the race of the people, and in their religion: but all these causes, which operate elsewhere, do not suffice to account for it; recourse must be had to some special reason. This reason appears to me to be the principle of equality and the institutions derived from it. Equality of conditions does not of itself engender regularity of morals, but it unquestionably facilitates and increases it.[1]

Amongst aristocratic nations birth and fortune frequently make two such different beings of man and woman, that they can never be united to each other. Their passions draw them together, but the condition of society, and the notions suggested by it, prevent them from contracting a permanent and ostensible tie. The necessary consequence is a great number of transient and clandestine connections. Nature secretly avenges herself for the constraint imposed upon her by the laws of man. This is not so much the case when the equality of conditions has swept away all the imaginary, or the real, barriers which separated man from woman. No girl then believes that she cannot become the wife of the man who loves her; and this renders all breaches of morality before marriage very uncommon: for, whatever be the credulity of the passions, a woman will hardly be able to persuade herself that she is beloved, when her lover is perfectly free to marry her and does not.

The same cause operates, though more indirectly, on married life. Nothing better serves to justify an illicit passion, either to the minds of those who have conceived it or to the world which looks on, than compulsory or accidental marriages.[2] In a country in which a woman is always free to exercise her power of choosing, and in which education has prepared her to choose rightly, public opinion is inexorable to her faults. The rigour of the Americans arises in part from this cause. They consider marriages as a covenant which is often onerous, but every condition of which the parties are strictly bound to fulfil, because they knew all those conditions beforehand, and were perfectly free not to have contracted them.

The very circumstances which render matrimonial fidelity more obligatory also render it more easy. In aristocratic countries the object of marriage is rather to unite property than persons; hence the husband is sometimes at school and the wife at nurse when they are betrothed. It cannot be wondered at if the conjugal tie which holds the fortunes of the pair united allows their hearts to rove; this is the natural result of the nature of the contract. When, on the contrary, a man always chooses a wife for himself, without any external coercion or even guidance, it is generally a conformity of tastes and opinions which brings a man and a woman together, and this same conformity keeps and fixes them in close habits of intimacy.

Our forefathers had conceived a very strange notion on the subject of marriage: as they had remarked that the small number of love-matches which occurred in their time almost always turned out ill, they resolutely inferred that it was exceedingly dangerous to listen to the dictates of the heart on the subject. Accident appeared to them to be a better guide than choice. Yet it was not very difficult to perceive that the examples which they witnessed did in fact prove nothing at all. For in the first place, if democratic nations leave a woman at liberty to choose her husband, they take care to give her mind sufficient knowledge, and her will sufficient strength, to make so important a choice: whereas the young women who, amongst aristocratic nations, furtively elope from the authority of their parents to throw themselves of their own accord into the arms of men whom they have had neither time to know, nor ability to judge of, are totally without those securities. It is not surprising that they make a bad use of their freedom of action the first time they avail themselves of it; nor that they fall into such cruel mistakes, when, not having received a democratic education, they choose to marry in conformity to democratic customs. But this is not all. When a man and woman are bent upon marriage in spite of the differences of an aristocratic state of society, the difficulties to be overcome are enormous. Having broken or relaxed the bonds of filial obedience, they have then to emancipate themselves by a final effort from the sway of custom and the tyranny of opinion; and when at length they have succeeded in this arduous task, they stand estranged from their natural friends and kinsmen: the prejudice they have crossed separates them from all, and places them in a situation which soon breaks their courage and sours their hearts. If, then, a couple married in this manner are first unhappy and afterwards criminal, it ought not to be attributed to the freedom of their choice, but rather to their living in a community in which this freedom of choice is not admitted.

Moreover it should not be forgotten that the same effort which makes a man violently shake off a prevailing error, commonly impels him beyond the bounds of reason; that, to dare to declare war, in however just a cause, against the opinion of one's age and country, a violent and adventurous spirit is required, and that men of this character seldom arrive at happiness or virtue, whatever be the path they follow. And this, it may be observed by the way, is the reason why in the most necessary and righteous revolutions, it is so rare to meet with virtuous or moderate revolutionary characters. There is then no just ground for surprise if a man, who in an age of aristocracy chooses to consult nothing but his own opinion and his own taste in the choice of a wife, soon finds that infractions of morality and domestic wretchedness invade his household: but when this same line of action is in the natural and ordinary course of things, when it is sanctioned by parental authority and backed by public opinion, it cannot be doubted that the internal peace of families will be increased by it, and conjugal fidelity more rigidly observed.

Almost all men in democracies are engaged in public or professional life; and on the other hand the limited extent of common incomes obliges a wife to confine herself to the house, in order to watch in person and very closely over the details of domestic economy. All these distinct and compulsory occupations are so many natural barriers, which, by keeping the two sexes asunder, render the solicitations of the one less frequent and less ardent—the resistance of the other more easy.

Not indeed that the equality of conditions can ever succeed in making men chaste, but it may impart a less dangerous character to their breaches of morality. As no one has then either sufficient time or opportunity to assail a virtue armed in self-defence, there will be at the same time a great number of courtesans and a great number of virtuous women. This state of things causes lamentable cases of individual hardship, but it does not prevent the body of society from being strong and alert: it does not destroy family ties, or enervate the morals of the nation. Society is endangered not by the great profligacy of a few, but by laxity of morals amongst all. In the eyes of a legislator, prostitution is less to be dreaded than intrigue.

The tumultuous and constantly harassed life which equality makes men lead, not only distracts them from the passion of love, by denying them time to indulge in it, but it diverts them from it by another more secret but more certain road. All men who live in democratic ages more or less contract the ways of thinking of the manufacturing and trading classes; their minds take a serious, deliberate, and positive turn; they are apt to relinquish the ideal, in order to pursue some visible and proximate object, which appears to be the natural and necessary aim of their desires. Thus the principle of equality does not destroy the imagination, but lowers its flight to the level of the earth. No men are less addicted to reverie than the citizens of a democracy; and few of them are ever known to give way to those idle and solitary meditations which commonly precede and produce the great emotions of the heart. It is true they attach great importance to procuring for themselves that sort of deep, regular, and quiet affection which constitutes the charm and safeguard of life, but they are not apt to run after those violent and capricious sources of excitement which disturb and abridge it.

I am aware that all this is only applicable in its full extent to America, and cannot at present be extended to Europe. In the course of the last half-century, whilst laws and customs have impelled several European nations with unexampled force towards democracy, we have not had occasion to observe that the relations of man and woman have become more orderly or more chaste. In some places the very reverse may be detected: some classes are more strict—the general morality of the people appears to be more lax. I do not hesitate to make the remark, for I am as little disposed to flatter my contemporaries as to malign them. This fact must distress, but it ought not to surprise us. The propitious influence which a democratic state of society may exercise upon orderly habits, is one of those tendencies which can only be discovered after a time. If the equality of conditions is favourable to purity of morals, the social commotion by which conditions are rendered equal is adverse to it. In the last fifty years, during which France has been undergoing this transformation, that country has rarely had freedom, always disturbance. Amidst this universal confusion of notions and this general stir of opinions—amidst this incoherent mixture of the just and unjust, of truth and falsehood, of right and might—public virtue has become doubtful, and private morality wavering. But all revolutions, whatever may have been their object or their agents, have at first produced similar consequences; even those which have in the end drawn the bonds of morality more tightly began by loosening them. The violations of morality which the French frequently witness do not appear to me to have a permanent character; and this is already betokened by some curious signs of the times.

Nothing is more wretchedly corrupt than an aristocracy which retains its wealth when it has lost its power, and which still enjoys a vast deal of leisure after it is reduced to mere vulgar pastimes. The energetic passions and great conceptions which animated it heretofore, leave it then; and nothing remains to it but a host of petty consuming vices, which cling about it like worms upon a carcass. No one denies that the French aristocracy of the last century was extremely dissolute; whereas established habits and ancient belief still preserved some respect for morality amongst the other classes of society. Nor will it be contested that at the present day the remnants of that same aristocracy exhibit a certain severity of morals; whilst laxity of morals appears to have spread amongst the middle and lower ranks. So that the same families which were most profligate fifty years ago are now-a-days the most exemplary, and democracy seems only to have strengthened the morality of the aristocratic classes. The French Revolution, by dividing the fortunes of the nobility, by forcing them to attend assiduously to their affairs and to their families, by making them live under the same roof with their children, and in short by giving a more rational and serious turn to their minds, has imparted to them, almost without their being aware of it, a reverence for religious belief, a love of order, of tranquil pleasures, of domestic endearments, and of comfort; whereas the rest of the nation, which had naturally these same tastes, was carried away into excesses by the effort which was required to overthrow the laws and political habits of the country. The old French aristocracy has undergone the consequences of the revolution, but it neither felt the revolutionary passions nor shared in the anarchical excitement which produced that crisis; it may easily be conceived that this aristocracy feels the salutary influence of the revolution in its manners, before those who achieve it. It may therefore be said, though at first it seems paradoxical, that, at the present day, the most antidemocratic classes of the nation principally exhibit the kind of morality which may reasonably be anticipated from democracy. I cannot but think that when we shall have obtained all the effects of this democratic revolution, after having got rid of the tumult it has caused, the observations which are now only applicable to the few will gradually become true of the whole community.

Chapter XII

How the Americans Understand the Equality of the Sexes

I have shown how democracy destroys or modifies the different inequalities which originate in society; but is this all? or does it not ultimately affect that great inequality of man and woman which has seemed, up to the present day, to be eternally based in human nature? I believe that the social changes which bring nearer to the same level the father and son, the master and servant, and superiors and inferiors generally speaking, will raise woman and make her more and more the equal of man. But here, more than ever, I feel the necessity of making myself clearly understood; for there is no subject on which the coarse and lawless fancies of our age have taken a freer range.

There are people in Europe who, confounding together the different characteristics of the sexes, would make of man and woman beings not only equal but alike. They would give to both the same functions, impose on both the same duties, and grant to both the same rights; they would mix them in all things—their occupations, their pleasures, their business. It may readily be conceived, that by thus attempting to make one sex equal to the other, both are degraded; and from so preposterous a medley of the works of nature nothing could ever result but weak men and disorderly women.

It is not thus that the Americans understand that species of democratic equality which may be established between the sexes. They admit, that as nature has appointed such wide differences between the physical and moral constitution of man and woman, her manifest design was to give a distinct employment to their various faculties; and they hold that improvement does not consist in making beings so dissimilar do pretty nearly the same things, but in getting each of them to fulfil their respective tasks in the best possible manner. The Americans have applied to the sexes the great principle of political economy which governs the manufactures of our age, by carefully dividing the duties of man from those of woman, in order that the great work of society may be the better carried on.

In no country has such constant care been taken as in America to trace two clearly distinct lines of action for the two sexes, and to make them keep pace one with the other, but in two pathways which are always different. American women never manage the outward concerns of the family, or conduct a business, or take a part in political life; nor are they, on the other hand, ever compelled to perform the rough labour of the fields, or to make any of those laborious exertions which demand the exertion of physical strength. No families are so poor as to form an exception to this rule. If on the one hand an American woman cannot escape from the quiet circle of domestic employments, on the other hand she is never forced to go beyond it. Hence it is that the women of America, who often exhibit a masculine strength of understanding and a manly energy, generally preserve great delicacy of personal appearance and always retain the manners of women, although they sometimes show that they have the hearts and minds of men.

Nor have the Americans ever supposed that one consequence of democratic principles is the subversion of marital power, of the confusion of the natural authorities in families. They hold that every association must have a head in order to accomplish its object, and that the natural head of the conjugal association is man. They do not therefore deny him the right of directing his partner; and they maintain, that in the smaller association of husband and wife, as well as in the great social community, the object of democracy is to regulate and legalise the powers which are necessary, not to subvert all power. This opinion is not peculiar to one sex, and contested by the other: I never observed that the women of America consider conjugal authority as a fortunate usurpation of their rights, nor that they thought themselves degraded by submitting to it. It appeared to me, on the contrary, that they attach a sort of pride to the voluntary surrender of their own will, and make it their boast to bend themselves to the yoke, not to shake it off. Such at least is the feeling expressed by the most virtuous of their sex; the others are silent; and in the United States it is not the practice for a guilty wife to clamour for the rights of women, whilst she is trampling on her holiest duties.

It has often been remarked that in Europe a certain degree of contempt lurks even in the flattery which men lavish upon women: although a European frequently affects to be the slave of woman, it may be seen that he never sincerely thinks her his equal. In the United States men seldom compliment women, but they daily show

how much they esteem them. They constantly display an entire confidence in the understanding of a wife, and a profound respect for her freedom; they have decided that her mind is just as fitted as that of a man to discover the plain truth, and her heart as firm to embrace it; and they have never sought to place her virtue, any more than his, under the shelter of prejudice, ignorance, and fear. It would seem that in Europe, where man so easily submits to the despotic sway of women, they are nevertheless curtailed of some of the greatest qualities of the human species, and considered as seductive but imperfect beings; and (what may well provoke astonishment) women ultimately look upon themselves in the same light, and almost consider it as a privilege that they are entitled to show themselves futile, feeble, and timid. The women of America claim no such privileges.

Again, it may be said that in our morals we have reserved strange immunities to man; so that there is, as it were, one virtue for his use, and another for the guidance of his partner; and that, according to the opinion of the public, the very same act may be punished alternately as a crime or only as a fault. The Americans know not this iniquitous division of duties and rights; amongst them the seducer is as much dishonoured as his victim. It is true that the Americans rarely lavish upon women those eager attentions which are commonly paid them in Europe; but their conduct to women always implies that they suppose them to be virtuous and refined; and such is the respect entertained for the moral freedom of the sex, that in the presence of a woman the most guarded language is used, lest her ear should be offended by an expression. In America a young unmarried woman may, alone and without fear, undertake a long journey.

The legislators of the United States, who have mitigated almost all the penalties of criminal law, still make rape a capital offence, and no crime is visited with more inexorable severity by public opinion. This may be accounted for; as the Americans can conceive nothing more precious than a woman's honour, and nothing which ought so much to be respected as her independence, they hold that no punishment is too severe for the man who deprives her of them against her will. In France, where the same offence is visited with far milder penalties, it is frequency difficult to get a verdict from a jury against the prisoner. Is this a consequence of contempt of decency or contempt of women? I cannot but believe that it is a contempt of one and of the other.

Thus the Americans do not think that man and woman have either the duty or the right to perform the same offices, but they show an equal regard for both their respective parts; and though their lot is different, they consider both of them as beings of equal value. They do not give to the courage of woman the same form or the same direction as to that of man; but they never doubt her courage: and if they hold that man and his partner ought not always to exercise their intellect and understanding in the same manner, they at least believe the understanding of the one to be as sound as that of the other, and her intellect to be as clear. Thus, then, whilst they have allowed the social inferiority of woman to subsist, they have done all they could to raise her morally and intellectually to the level of man; and in this respect they appear to me to have excellently understood the true principle of democratic improvement. As for myself, I do not hesitate to avow that, although the women of the United States are confined within the narrow circle of domestic life, and their situation is in some respects one of extreme dependence, I have nowhere seen woman occupying a loftier position; and if I were asked, now that I am drawing to the close of this work, in which I have spoken of so many important things done by the Americans, to what the singular prosperity and growing strength of that people ought mainly to be attributed, I should reply,—to the superiority of their women.

Chapter XIII

That the Principle of Equality Naturally Divides the Americans into a Number of Small Private Circles

It may probably be supposed that the final consequence and necessary effect of democratic institutions is to confound together all the members of the community in private as well as in public life, and to compel them all to live in common; but this would be to ascribe a very coarse and oppressive form to the equality which originates in democracy. No state of society or laws can render men so much alike, but that education, fortune, and tastes will interpose some differences between them; and, though different men may sometimes find it their interest to combine for the same purposes, they will never make it their pleasure. They will therefore always tend to evade the provisions of legislation, whatever they may be; and departing in some one respect from the circle within which they were to be bounded, they will set up, close by the great political community, small private circles, united together by the similitude of their conditions, habits, and manners.

In the United States the citizens have no sort of pre-eminence over each other; they owe each other no mutual obedience or respect; they all meet for the administration of justice, for the government of the State, and in general to treat of the affairs which concern their common welfare; but I never heard that attempts have been made to bring them all to follow the same diversions, or to amuse themselves promiscuously in the same places of recreation. The Americans, who mingle so readily in their political assemblies and courts of justice, are wont on the contrary carefully to separate into small distinct circles, in order to indulge by themselves in the enjoyments of private life. Each of them is willing to acknowledge all his fellow-citizens as his equals, but he will only receive a very limited number of them amongst his friends or his guests. This appears to me to be very natural. In proportion as the circle of public society is extended, it may be anticipated that the sphere of private intercourse will be contracted; far from supposing that the members of modern society will ultimately live in common, I am afraid that they may end by forming nothing but small coteries.

Amongst aristocratic nations the different classes are like vast chambers, out of which it is impossible to get, into which it is impossible to enter. These classes have no communication with each other, but within their pale men necessarily live in daily contact; even though they would not naturally suit, the general conformity of a similar condition brings them nearer together. But when neither law nor custom professes to establish frequent and habitual relations between certain men, their intercourse originates in the accidental analogy of opinions and tastes; hence private society is infinitely varied. In democracies, where the members of the community never differ much from each other, and naturally stand in such propinquity that they may all at any time be confounded in one general mass, numerous artificial and arbitrary distinctions spring up, by means of which every man hopes to keep himself aloof, lest he should be carried away in the crowd against his will. This can never fail to be the case; for human institutions may be changed, but not man: whatever may be the general endeavour of a community to render its members equal and alike, the personal pride of individuals will always seek to rise above the line, and to form somewhere an inequality to their own advantage.

In aristocracies men are separated from each other by lofty stationary barriers; in democracies they are divided by a number of small and almost invisible threads, which are constantly broken or moved from place to place. Thus, whatever may be the progress of equality, in democratic nations a great number of small private communities will always be formed within the general pale of political society; but none of them will bear any resemblance in its manners to the highest class in aristocracies.

Chapter XIV

Some Reflections on American Manners

Nothing seems at first sight less important than the outward form of human actions, yet there is nothing upon which men set more store: they grow used to everything except to living in a society which has not their own manners. The influence of the social and political state of a country upon manners is therefore deserving of serious examination. Manners are, generally, the product of the very basis of the character of a people, but they are also sometimes the result of an arbitrary convention between certain men; thus they are at once natural and acquired. When certain men perceive that they are the foremost persons in society, without contestation and without effort—when they are constantly engaged on large objects, leaving the more minute details to others—and when they live in the enjoyment of wealth which they did not amass and which they do not fear to lose, it may be supposed that they feel a kind of haughty disdain of the petty interests and practical cares of life, and that their thoughts assume a natural greatness, which their language and their manners denote. In democratic countries manners are generally devoid of dignity, because private life is there extremely petty in its character; and they are frequently low, because the mind has few opportunities of rising above the engrossing cares of domestic interests. True dignity in manners consists in always taking one's proper station, neither too high nor too low; and this is as much within the reach of a peasant as of a prince. In democracies all stations appear doubtful; hence it is that the manners of democracies, though often full of arrogance, are commonly wanting in dignity, and, moreover, they are never either well-disciplined or accomplished.

The men who live in democracies are too fluctuating for a certain number of them ever to succeed in laying down a code of good breeding, and in forcing people to follow it. Every man therefore behaves after his own fashion, and there is always a certain incoherence in the manners of such times, because they are moulded upon the feelings and notions of each individual, rather than upon an ideal model proposed for general imitation. This, however, is much more perceptible at the time when an aristocracy has just been overthrown than after it has long been destroyed. New political institutions and new social elements then bring to the same places of resort, and frequently compel to live in common, men whose education and habits are still amazingly dissimilar, and this renders the motley composition of society peculiarly visible. The existence of a former strict code of good breeding is still remembered, but what it contained or where it is to be found is already forgotten. Men have lost the common law of manners, and they have not yet made up their minds to do without it; but everyone endeavours to make to himself some sort of arbitrary and variable rule, from the remnant of former usages; so that manners have neither the regularity and the dignity which they often display amongst aristocratic nations, nor the simplicity and freedom which they sometimes assume in democracies; they are at once constrained and without constraint.

This, however, is not the normal state of things. When the equality of conditions is long established and complete, as all men entertain nearly the same notions and do nearly the same things, they do not require to agree or to copy from one another in order to speak or act in the same manner: their manners are constantly characterised by a number of lesser diversities, but not by any great differences. They are never perfectly alike, because they do not copy from the same pattern; they are never very unlike, because their social condition is the same. At first sight a traveller would observe that the manners of all the Americans are exactly similar; it is only upon close examination that the peculiarities in which they differ may be detected.

The English make game of the manners of the Americans; but it is singular that most of the writers who have drawn these ludicrous delineations belonged themselves to the middle classes in England, to whom the same delineations are exceedingly applicable: so that these pitiless censors for the most part furnish an example of the very thing they blame in the United States; they do not perceive that they are deriding themselves, to the great amusement of the aristocracy of their own country.

Nothing is more prejudicial to democracy than its outward forms of behaviour: many men would willingly endure its vices, who cannot support its manners. I cannot, however, admit that there is nothing commendable in the manners of a democratic people. Amongst aristocratic nations, all who live within reach of the first class in society commonly strain to be like it, which gives rise to ridiculous and insipid imitations. As a democratic people does not possess any models of high breeding, at least it escapes the daily necessity of seeing wretched copies of them. In democracies manners are never so refined as amongst aristocratic nations, but on the other hand they are never so coarse. Neither the coarse oaths of the populace, nor the elegant and choice expressions of the nobility are to be heard there: the manners of such a people are often vulgar, but they are neither brutal nor mean. I have already observed that in democracies no such thing as a regular code of good breeding can be laid down; this has some inconveniences and some advantages. In aristocracies the rules of propriety impose the same demeanour on everyone; they make all the members of the same class appear alike, in spite of their private inclinations; they adorn and they conceal the natural man. Amongst a democratic people manners are neither so tutored nor so uniform, but they are frequently more sincere. They form, as it were, a light and loosely-woven veil, through which the real feelings and private opinions of each individual are easily discernible. The form and the substance of human actions often, therefore, stand in closer relation; and if the great picture of human life be less embellished, it is more true. Thus it may be said, in one sense, that the effect of democracy is not exactly to give men any particular manners, but to prevent them from having manners at all.

The feelings, the passions, the virtues, and the vices of an aristocracy may sometimes reappear in a democracy, but not its manners; they are lost, and vanish forever, as soon as the democratic revolution is completed. It would seem that nothing is more lasting than the manners of an aristocratic class, for they are preserved by that class for some time after it has lost its wealth and its power—nor so fleeting, for no sooner have they disappeared than not a trace of them is to be found; and it is scarcely possible to say what they have been as soon as they have ceased to be. A change in the state of society works this miracle, and a few generations suffice to consummate it. The principal characteristics of aristocracy are handed down by history after an aristocracy is destroyed, but the light and exquisite touches of manners are effaced from men's memories almost immediately after its fall. Men can no longer conceive what these manners were when they have ceased to witness them; they are gone, and their departure was unseen, unfelt; for in order to feel that refined enjoyment which is derived from choice and distinguished manners, habit and education must have prepared the heart, and the taste for them is lost almost as easily as the practice of them. Thus not only a democratic people cannot have aristocratic manners, but they neither comprehend nor desire them; and as they never have thought of them, it is to their minds as if such things had never been. Too much importance should not be attached to this loss, but it may well be regretted.

I am aware that it has not unfrequently happened that the same men have had very high-bred manners and very low-born feelings: the interior of courts has sufficiently shown what imposing externals may conceal the meanest hearts. But though the manners of aristocracy did not constitute virtue, they sometimes embellish virtue itself. It was no ordinary sight to see a numerous and powerful class of men, whose every outward action seemed constantly to be dictated by a natural elevation of thought and feeling, by delicacy and regularity of taste, and by urbanity of manners. Those manners threw a pleasing illusory charm over human nature; and though the picture was often a false one, it could not be viewed without a noble satisfaction.

Chapter XV

Of the Gravity of the Americans, and Why It Does Not Prevent Them from Often Committing Inconsiderate Actions

Men who live in democratic countries do not value the simple, turbulent, or coarse diversions in which the people indulge in aristocratic communities: such diversions are thought by them to be puerile or insipid. Nor have they a greater inclination for the intellectual and refined amusements of the aristocratic classes. They want something productive and substantial in their pleasures; they want to mix actual fruition with their joy. In aristocratic communities the people readily give themselves up to bursts of tumultuous and boisterous gaiety, which shake off at once the recollection of their privations: the natives of democracies are not fond of being thus violently broken in upon, and they never lose sight of their own selves without regret. They prefer to these frivolous delights those more serious and silent amusements which are like business, and which do not drive business wholly from their minds. An American, instead of going in a leisure hour to dance merrily at some place of public resort, as the fellows of his calling continue to do throughout the greater part of Europe, shuts himself up at home to drink. He thus enjoys two pleasures; he can go on thinking of his business, and he can get drunk decently by his own fireside.

I thought that the English constituted the most serious nation on the face of the earth, but I have since seen the Americans and have changed my opinion. I do not mean to say that temperament has not a great deal to do with the character of the inhabitants of the United States, but I think that their political institutions are a still more influential cause. I believe the seriousness of the Americans arises partly from their pride. In democratic countries even poor men entertain a lofty notion of their personal importance: they look upon themselves with complacency, and are apt to suppose that others are looking at them, too. With this disposition they watch their language and their actions with care, and do not lay themselves open so as to. betray their deficiencies; to preserve their dignity they think it necessary to retain their gravity.

But I detect another more deep-seated and powerful cause which instinctively produces amongst the Americans this astonishing gravity. Under a despotism communities give way at times to bursts of vehement joy; but they are generally gloomy and moody, because they are afraid. Under absolute monarchies tempered by the customs and manners of the country, their spirits are often cheerful and even, because as they have some freedom and a good deal of security, they are exempted from the most important cares of life; but all free peoples are serious, because their minds are habitually absorbed by the contemplation of some dangerous or difficult purpose. This is more especially the case amongst those free nations which form democratic communities. Then there are in all classes a very large number of men constantly occupied with the serious affairs of the government; and those whose thoughts are not engaged in the direction of the commonwealth are wholly engrossed by the acquisition of a private fortune. Amongst such a people a serious demeanour ceases to be peculiar to certain men, and becomes a habit of the nation.

We are told of small democracies in the days of antiquity, in which the citizens met upon the public places with garlands of roses, and spent almost all their time in dancing and theatrical amusements. I do not believe in such republics any more than in that of Plato; or, if the things we read of really happened, I do not hesitate to affirm that these supposed democracies were composed of very different elements from ours, and that they had nothing in common with the latter except their name. But it must not be supposed that, in the midst of all their toils, the people who live in democracies think themselves to be pitied; the contrary is remarked to be the case. No men are fonder of their own condition. Life would have no relish for them if they were delivered from the anxieties which harass them, and they show more attachment to their cares than aristocratic nations to their pleasures.

I am next led to inquire how it is that these same democratic nations, which are so serious, sometimes act in so inconsiderate a manner. The Americans, who almost always preserve a staid demeanour and a frigid air, nevertheless frequently allow themselves to be borne away, far beyond the bound of reason, by a sudden passion or a hasty opinion, and they sometimes gravely commit strange absurdities. This contrast ought not to surprise us. There is one sort of ignorance which originates in extreme publicity. In despotic states men know not how to act, because they are told nothing; in democratic nations they often act at random, because nothing is to be left untold.

The former do not know—the latter forget; and the chief features of each picture are lost to them in a bewilderment of details.

It is astonishing what imprudent language a public man may sometimes use in free countries, and especially in democratic states, without being compromised; whereas in absolute monarchies a few words dropped by accident are enough to unmask him forever, and ruin him without hope of redemption. This is explained by what goes before. When a man speaks in the midst of a great crowd, many of his words are not heard, or are forthwith obliterated from the memories of those who hear them; but amidst the silence of a mute and motionless throng the slightest whisper strikes the ear.

In democracies men are never stationary; a thousand chances waft them to and fro, and their life is always the sport of unforeseen or (so to speak) extemporaneous circumstances. Thus they are often obliged to do things which they have imperfectly learned, to say things they imperfectly understand, and to devote themselves to work for which they are unprepared by long apprenticeship. In aristocracies every man has one sole object which he unceasingly pursues, but amongst democratic nations the existence of man is more complex; the same mind will almost always embrace several objects at the same time, and these objects are frequently wholly foreign to each other: as it cannot know them all well, the mind is readily satisfied with imperfect notions of each.

When the inhabitant of democracies is not urged by his wants, he is so at least by his desires; for of all the possessions which he sees around him, none are wholly beyond his reach. He therefore does everything in a hurry, he is always satisfied with 'pretty well,' and never pauses more than an instant to consider what he has been doing. His curiosity is at once insatiable and cheaply satisfied; for he cares more to know a great deal quickly than to know anything well: he has no time and but little taste to search things to the bottom.

Thus then democratic peoples are grave, because their social and political condition constantly leads them to engage in serious occupations; and they act inconsiderately, because they give but little time and attention to each of these occupations. The habit of inattention must be considered as the greatest bane of the democratic character.

Chapter XVI

Why the National Vanity of the Americans Is More Restless and Captious Than That of the English

All free nations are vain-glorious, but national pride is not displayed by all in the same manner. The Americans in their intercourse with strangers appear impatient of the smallest censure and insatiable of praise. The most slender eulogium is acceptable to them; the most exalted seldom contents them; they unceasingly harass you to extort praise, and if you resist their entreaties they fall to praising themselves. It would seem as if, doubting their own merit, they wished to have it constantly exhibited before their eyes. Their vanity is not only greedy, but restless and jealous; it will grant nothing, whilst it demands everything, but is ready to beg and to quarrel at the same time. If I say to an American that the country he lives in is a fine one, 'Ay,' he replies, 'there is not its fellow in the world.' If I applaud the freedom which its inhabitants enjoy, he answers, 'Freedom is a fine thing, but few nations are worthy to enjoy it.' If I remark the purity of morals which distinguishes the United States, 'I can imagine,' says he, 'that a stranger, who has been struck by the corruption of all other nations, is astonished at the difference.' At length I leave him to the contemplation of himself; but he returns to the charge, and does not desist till he has got me to repeat all I had just been saying. It is impossible to conceive a more troublesome or more garrulous patriotism; it wearies even those who are disposed to respect it.[1]

Such is not the case with the English. An Englishman calmly enjoys the real or imaginary advantages which in his opinion his country possesses. If he grants nothing to other nations, neither does he solicit anything for his own. The censure of foreigners does not affect him, and their praise hardly flatters him; his position with regard to the rest of the world is one of disdainful and ignorant reserve: his pride requires no sustenance, it nourishes itself. It is remarkable that two nations, so recently sprung from the same stock, should be so opposite to one another in their manner of feeling and conversing.

In aristocratic countries the great possess immense privileges, upon which their pride rests, without seeking to rely upon the lesser advantages which accrue to them. As these privileges came to them by inheritance, they regard them in some sort as a portion of themselves, or at least as a natural right inherent in their own persons. They therefore entertain a calm sense of their superiority; they do not dream of vaunting privileges which everyone perceives and no one contests, and these things are not sufficiently new to them to be made topics of conversation. They stand unmoved in their solitary greatness, well assured that they are seen of all the world without any effort to show themselves off, and that no one will attempt to drive them from that position. When an aristocracy carries on the public affairs, its national pride naturally assumes this reserved, indifferent, and haughty form, which is imitated by all the other classes of the nation.

When, on the contrary, social conditions differ but little, the slightest privileges are of some importance; as every man sees around himself a million of people enjoying precisely similar or analogous advantages, his pride becomes craving and jealous, he clings to mere trifles, and doggedly defends them. In democracies, as the conditions of life are very fluctuating, men have almost always recently acquired the advantages which they possess; the consequence is that they feel extreme pleasure in exhibiting them, to show others and convince themselves that they really enjoy them. As at any instant these same advantages may be lost, their possessors are constantly on the alert, and make a point of showing that they still retain them. Men living in democracies love their country just as they love themselves, and they transfer the habits of their private vanity to their vanity as a nation. The restless and insatiable vanity of a democratic people originates so entirely in the equality and precariousness of social conditions, that the members of the haughtiest nobility display the very same passion in those lesser portions of their existence in which there is anything fluctuating or contested. An aristocratic class always differs greatly from the other classes of the nation, by the extent and perpetuity of its privileges; but it often happens that the only differences between the members who belong to it consist in small transient advantages, which may any day be lost or acquired.

The members of a powerful aristocracy, collected in a capital or a court, have been known to contest with virulence those frivolous privileges which depend on the caprice of fashion or the will of their master. These persons then displayed towards each other precisely the same puerile jealousies which animate the men of democracies, the same eagerness to snatch the smallest advantages which their equals contested, and the same desire to parade ostentatiously those of which they were in possession. If national pride ever entered into the minds of courtiers, I do not question that they would display it in the same manner as the members of a democratic community.

CHAPTER XVII

That the Aspect of Society in the United States Is at Once Excited and Monotonous

It would seem that nothing can be more adapted to stimulate and to feed curiosity than the aspect of the United States. Fortunes, opinions, and laws are there in ceaseless variation: it is as if immutable Nature herself were mutable, such are the changes worked upon her by the hand of man. Yet in the end the sight of this excited community becomes monotonous, and after having watched the moving pageant for a time the spectator is tired of it. Amongst aristocratic nations every man is pretty nearly stationary in his own sphere; but men are astonishingly unlike each other—their passions, their notions, their habits, and their tastes are essentially different: nothing changes, but everything differs. In democracies, on the contrary, all men are alike and do things pretty nearly alike. It is true that they are subject to great and frequent vicissitudes; but as the same events of good or adverse fortune are continually recurring, the name of the actors only is changed, the piece is always the same. The aspect of American society is animated, because men and things are always changing; but it is monotonous, because all these changes are alike.

Men living in democratic ages have many passions, but most of their passions either end in the love of riches or proceed from it. The cause of this is, not that their souls are narrower, but that the importance of money is really greater at such times. When all the members of a community are independent of or indifferent to each other, the cooperation of each of them can only be obtained by paying for it: this infinitely multiplies the purposes to which wealth may be applied, and increases its value. When the reverence which belonged to what is old has vanished, birth, condition, and profession no longer distinguish men, or scarcely distinguish them at all: hardly anything but money remains to create strongly marked differences between them, and to raise some of them above the common level. The distinction originating in wealth is increased by the disappearance and diminution of all other distinctions. Amongst aristocratic nations money only reaches to a few points on the vast circle of man's desires—in democracies it seems to lead to all. The love of wealth is therefore to be traced, either as a principal or an accessory motive, at the bottom of all that the Americans do: this gives to all their passions a sort of family likeness, and soon renders the survey of them exceedingly wearisome. This perpetual recurrence of the same passion is monotonous; the peculiar methods by which this passion seeks its own gratification are no less so.

In an orderly and constituted democracy like the United States, where men cannot enrich themselves by war, by public office, or by political confiscation, the love of wealth mainly drives them into business and manufactures. Although these pursuits often bring about great commotions and disasters, they cannot prosper without strictly regular habits and a long routine of petty uniform acts. The stronger the passion is, the more regular are these habits, and the more uniform are these acts. It may be said that it is the vehemence of their desires which makes the Americans so methodical; it perturbs their minds, but it disciplines their lives.

The remark I here apply to America may indeed be addressed to almost all our contemporaries. Variety is disappearing from the human race; the same ways of acting, thinking, and feeling are to be met with all over the world. This is not only because nations work more upon each other, and are more faithful in their mutual imitation; but as the men of each country relinquish more and more the peculiar opinions and feelings of a caste, a profession, or a family, they simultaneously arrive at something nearer to the constitution of man, which is everywhere the same. Thus they become more alike, even without having imitated each other. Like travellers scattered about some large wood, which is intersected by paths converging to one point, if all of them keep their eyes fixed upon that point and advance towards it, they insensibly draw nearer together—though they seek not, though they see not, though they know not each other; and they will be surprised at length to find themselves all collected on the same spot. All the nations which take, not any particular man, but man himself, as the object of their researches and their imitations, are tending in the end to a similar state of society, like these travellers converging to the central plot of the forest.

Chapter XVIII

Of Honour[1] in the United States and in Democratic Communities

It would seem that men employ two very distinct methods in the public estimation of the actions of their fellowmen; at one time they judge them by those simple notions of right and wrong which are diffused all over the world; at another they refer their decision to a few very special notions which belong exclusively to some particular age and country. It often happens that these two rules differ; they sometimes conflict: but they are never either entirely identified or entirely annulled by one another. Honour, at the periods of its greatest power, sways the will more than the belief of men; and even whilst they yield without hesitation and without a murmur to its dictates, they feel notwithstanding, by a dim but mighty instinct, the existence of a more general, more ancient, and more holy law, which they sometimes disobey although they cease not to acknowledge it. Some actions have been held to be at the same time virtuous and dishonourable—a refusal to fight a duel is a case in point.

I think these peculiarities may be otherwise explained than by the mere caprices of certain individuals and nations, as has hitherto been the customary mode of reasoning on the subject. Mankind is subject to general and lasting wants that have engendered moral laws, to the neglect of which men have ever and in all places attached the notion of censure and shame: to infringe them was *to do ill—to do well* was to conform to them. Within the bosom of this vast association of the human race, lesser associations have been formed which are called nations; and amidst these nations further subdivisions have assumed the names of classes or castes. Each of these associations forms, as it were, a separate species of the human race; and though it has no essential difference from the mass of mankind, to a certain extent it stands apart and has certain wants peculiar to itself. To these special wants must be attributed the modifications which affect in various degrees and in different countries the mode of considering human actions, and the estimate which ought to be formed of them. It is the general and permanent interest of mankind that men should not kill each other: but it may happen to be the peculiar and temporary interest of a people or a class to justify, or even to honour, homicide.

Honour is simply that peculiar rule, founded upon a peculiar state of society, by the application of which a people or a class allot praise or blame. Nothing is more unproductive to the mind than an abstract idea; I therefore hasten to call in the aid of facts and examples to illustrate my meaning.

I select the most extraordinary kind of honour which was ever known in the world, and that which we are best acquainted with, viz., aristocratic honour springing out of feudal society. I shall explain it by means of the principle already laid down, and I shall explain the principle by means of the illustration. I am not here led to inquire when and how the aristocracy of the middle ages came into existence, why it was so deeply severed from the remainder of the nation, or what founded and consolidated its power. I take its existence as an established fact, and I am endeavouring to account for the peculiar view which it took of the greater part of human actions. The first thing that strikes me is, that in the feudal world actions were not always praised or blamed with reference to their intrinsic worth, but that they were sometimes appreciated exclusively with reference to the person who was the actor or the object of them, which is repugnant to the general conscience of mankind. Thus some of the actions which were indifferent on the part of a man in humble life, dishonoured a noble; others changed their whole character according as the person aggrieved by them belonged or did not belong to the aristocracy. When these different notions first arose, the nobility formed a distinct body amidst the people, which it commanded from the inaccessible heights where it was ensconced. To maintain this peculiar position, which constituted its strength, it not only required political privileges, but it required a standard of right and wrong for its own especial use. That some particular virtue or vice belonged to the nobility rather than to the humble classes—that certain actions were guiltless when they affected the villain, which were criminal when they touched the noble—these were often arbitrary matters; but that honour or shame should be attached to a man's actions according to his condition, was a result of the internal constitution of an aristocratic community. This has been actually the case in all the countries which have had an aristocracy; as long as a trace of the principle remains, these peculiarities will still exist; to debauch a woman of colour scarcely injures the reputation of an American—to marry her dishonours him.

In some cases feudal honour enjoined revenge, and stigmatised the forgiveness of insults; in others it imperiously commanded men to conquer their own passions, and imposed forgetfulness of self. It did not make humanity or kindness its law, but it extolled generosity; it set more store on liberality than on benevolence; it allowed men to enrich themselves by gambling or by war, but not by labour; it preferred great crimes to small earnings; cupidity was less distasteful to it than avarice; violence it often sanctioned, but cunning and treachery it invariably reprobated as contemptible. These fantastical notions did not proceed exclusively from the caprices of those who entertained them. A class which has succeeded in placing itself at the head of and above all others, and which makes perpetual exertions to maintain this lofty position, must especially honour those virtues which are conspicuous for their dignity and splendour, and which may be easily combined with pride and the love of power. Such men would not hesitate to invert the natural order of the conscience in order to give those virtues precedence before all others. It may even be conceived that some of the more bold and brilliant vices would readily be set above the quiet, unpretending virtues. The very existence of such a class in society renders these things unavoidable.

The nobles of the middle ages placed military courage foremost amongst virtues, and in lieu of many of them. This was again a peculiar opinion which arose necessarily from the peculiarity of the state of society. Feudal aristocracy existed by war and for war; its power had been founded by arms, and by arms that power was maintained; it therefore required nothing more than military courage, and that quality was naturally exalted above all others; whatever denoted it, even at the expense of reason and humanity, was therefore approved and frequently enjoined by the manners of the time. Such was the main principle; the caprice of man was only to be traced in minuter details. That a man should regard a tap on the cheek as an unbearable insult, and should be obliged to kill in single combat the person who struck him thus lightly, is an arbitrary rule; but that a noble could not tranquilly receive an insult, and was dishonoured if he allowed himself to take a blow without fighting, were direct consequences of the fundamental principles and the wants of a military aristocracy.

Thus it was true to a certain extent to assert that the laws of honour were capricious; but these caprices of honour were always confined within certain necessary limits. The peculiar rule, which was called honour by our forefathers, is so far from being an arbitrary law in my eyes, that I would readily engage to ascribe its most incoherent and fantastical injunctions to a small number of fixed and invariable wants inherent in feudal society.

If I were to trace the notion of feudal honour into the domain of politics, I should not find it more difficult to explain its dictates. The state of society and the political institutions of the middle ages were such, that the supreme power of the nation never governed the community directly. That power did not exist in the eyes of the people: every man looked up to a certain individual whom he was bound to obey; by that intermediate personage he was connected with all the others. Thus in feudal society the whole system of the commonwealth rested upon the sentiment of fidelity to the person of the lord: to destroy that sentiment was to open the sluices of anarchy. Fidelity to a political superior was, moreover, a sentiment of which all the members of the aristocracy had constant opportunities of estimating the importance; for every one of them was a vassal as well as a lord, and had to command as well as to obey. To remain faithful to the lord, to sacrifice oneself for him if called upon, to share his good or evil fortunes, to stand by him in his undertakings whatever they might be—such were the first injunctions of feudal honour in relation to the political institutions of those times. The treachery of a vassal was branded with extraordinary severity by public opinion, and a name of peculiar infamy was invented for the offence which was called *felony*.

On the contrary, few traces are to be found in the middle ages of the passion which constituted the life of the nations of antiquity—I mean patriotism—the word itself is not of very ancient date in the language.[2] Feudal institutions concealed the country at large from men's sight, and rendered the love of it less necessary. The nation was forgotten in the passions which attached men to persons. Hence it was no part of the strict law of feudal honour to remain faithful to one's country. Not indeed that the love of their country did not exist in the hearts of our forefathers; but it constituted a dim and feeble instinct, which has grown more clear and strong in proportion as aristocratic classes have been abolished, and the supreme power of the nation centralised. This may be clearly seen from the contrary judgments which European nations have passed upon the various events of their histories, according to the generations by which such judgments have been formed. The circumstance which most dishonoured the Constable de Bourbon in the eyes of his contemporaries was that he bore arms against his king: that which most dishonours him in our eyes, is that he made war against his country; we brand him as deeply as our forefathers did, but for different reasons.

I have chosen the honour of feudal times by way of illustration of my meaning, because its characteristics are more distinctly marked and more familiar to us than those of any other period; but I might have taken an example elsewhere, and I should have reached the same conclusion by a different road. Although we are less perfectly acquainted with the Romans than with our own ancestors, yet we know that certain peculiar notions of glory and disgrace obtained amongst them, which were not solely derived from the general principles of right and wrong. Many human actions were judged differently, according as they affected a Roman citizen or a stranger, a freeman or a slave; certain vices were blazoned abroad, certain virtues were extolled above all others. 'In that age,' says Plutarch in the life of Coriolanus, 'martial prowess was more honoured and prized in Rome than all the other virtues, insomuch that it was called *virtus,* the name of virtue itself, by applying the name of the kind to this particular species; so that virtue in Latin was as much as to say valour.' Can anyone fail to recognise the peculiar want of that singular community which was formed for the conquest of the world?

Any nation would furnish us with similar grounds of observation; for, as I have already remarked, whenever men collect together as a distinct community, the notion of honour instantly grows up amongst them; that is to say, a system of opinions peculiar to themselves as to what is blamable or commendable; and these peculiar rules always originate in the special habits and special interests of the community. This is applicable to a certain extent to democratic communities as well as to others, as we shall now proceed to prove by the example of the Americans.[3] Some loose notions of the old aristocratic honour of Europe are still to be found scattered amongst the opinions of the Americans; but these traditional opinions are few in number, they have but little root in the country, and but little power. They are like a religion which has still some temples left standing, though men have ceased to believe in it. But amidst these half-obliterated notions of exotic honour, some new opinions have sprung up, which constitute what may be termed in our days American honour. I have shown how the Americans are constantly driven to engage in commerce and industry. Their origin, their social condition, their political institutions, and even the spot they inhabit, urge them irresistibly in this direction. Their present condition is then that of an almost exclusively manufacturing and commercial association, placed in the midst of a new and boundless country, which their principal object is to explore for purposes of profit. This is the characteristic which most peculiarly distinguishes the American people from all others at the present time. All those quiet virtues which tend to give a regular movement to the community, and to encourage business, will therefore be held in peculiar honour by that people, and to neglect those virtues will be to incur public contempt. All the more turbulent virtues, which often dazzle, but more frequently disturb society, will on the contrary occupy a subordinate rank in the estimation of this same people: they may be neglected without forfeiting the esteem of the community—to acquire them would perhaps be to run a risk of losing it.

The Americans make a no less arbitrary classification of men's vices. There are certain propensities which appear censurable to the general reason and the universal conscience of mankind, but which happen to agree with the peculiar and temporary wants of the American community: these propensities are lightly reproved, sometimes even encouraged; for instance, the love of wealth and the secondary propensities connected with it may be more particularly cited. To clear, to till, and to transform the vast uninhabited continent which is his domain, the American requires the daily support of an energetic passion; that passion can only be the love of wealth; the passion for wealth is therefore not reprobated in America, and provided it does not go beyond the bounds assigned to it for public security, it is held in honour. The American lauds as a noble and praiseworthy ambition what our own forefathers in the middle ages stigmatised as servile cupidity, just as he treats as a blind and barbarous frenzy that ardour of conquest and martial temper which bore them to battle. In the United States fortunes are lost and regained without difficulty; the country is boundless, and its resources inexhaustible. The people have all the wants and cravings of a growing creature; and whatever be their efforts, they are always surrounded by more than they can appropriate. It is not the ruin of a few individuals which may be soon repaired, but the inactivity and sloth of the community at large which would be fatal to such a people. Boldness of enterprise is the foremost cause of its rapid progress, its strength, and its greatness. Commercial business is there like a vast lottery, by which a small number of men continually lose, but the State is always a gainer; such a people ought therefore to encourage and do honour to boldness in commercial speculations. But any bold speculation risks the fortune of the speculator and of all those who put their trust in him. The Americans, who make a virtue of commercial temerity, have no right in any case to brand with disgrace those who practise it. Hence arises the strange indulgence which is shown to bankrupts in the United States; their honour does not suffer by such an accident. In this respect the Americans differ, not only from the nations of Europe, but from all the commercial nations of our time, and accordingly they resemble none of them in their position or their wants.

In America all those vices which tend to impair the purity of morals, and to destroy the conjugal tie, are treated with a degree of severity which is unknown in the rest of the world. At first sight this seems strangely at variance with the tolerance shown there on other subjects, and one is surprised to meet with a morality so relaxed and so austere amongst the self-same people. But these things are less incoherent than they seem to be. Public opinion in the United States very gently represses that love of wealth which promotes the commercial greatness and the prosperity of the nation, and it especially condemns that laxity of morals which diverts the human mind from the pursuit of well-being, and disturbs the internal order of domestic life which is so necessary to success in business. To earn the esteem of their countrymen, the Americans are therefore constrained to adapt themselves to orderly habits—and it may be said in this sense that they make it a matter of honour to live chastely.

On one point American honour accords with the notions of honour acknowledged in Europe; it places courage as the highest virtue, and treats it as the greatest of the moral necessities of man; but the notion of courage itself assumes a different aspect. In the United States martial valour is but little prized; the courage which is best known and most esteemed is that which emboldens men to brave the dangers of the ocean, in order to arrive earlier in port—to support the privations of the wilderness without complaint, and solitude more cruel than privations—the courage which renders them almost insensible to the loss of a fortune laboriously acquired, and instantly prompts to fresh exertions to make another. Courage of this kind is peculiarly necessary to the maintenance and prosperity of the American communities, and it is held by them in peculiar honour and estimation; to betray a want of it is to incur certain disgrace.

I have yet another characteristic point which may serve to place the idea of this chapter in stronger relief. In a democratic society like that of the United States, where fortunes are scanty and insecure, everybody works, and work opens a way to everything: this has changed the point of honour quite round, and has turned it against idleness. I have sometimes met in America with young men of wealth, personally disinclined to all laborious exertion, but who had been compelled to embrace a profession. Their disposition and their fortune allowed them to remain without employment; public opinion forbade it, too imperiously to be disobeyed. In the European countries, on the contrary, where aristocracy is still struggling with the flood which overwhelms it, I have often seen men, constantly spurred on by their wants and desires, remain in idleness, in order not to lose the esteem of their equals; and I have known them submit to ennui and privations rather than to work. No one can fail to perceive that these opposite obligations are two different rules of conduct, both nevertheless originating in the notion of honour.

What our forefathers designated as honour absolutely was in reality only one of its forms; they gave a generic name to what was only a species. Honour therefore is to be found in democratic as well as in aristocratic ages, but it will not be difficult to show that it assumes a different aspect in the former. Not only are its injunctions different, but we shall shortly see that they are less numerous, less precise, and that its dictates are less rigorously obeyed. The position of a caste is always much more peculiar than that of a people. Nothing is so much out of the way of the world as a small community invariably composed of the same families (as was for instance the aristocracy of the middle ages), whose object is to concentrate and to retain, exclusively and hereditarily, education, wealth, and power amongst its own members. But the more out of the way the position of a community happens to be, the more numerous are its special wants, and the more extensive are its notions of honour corresponding to those wants. The rules of honour will therefore always be less numerous amongst a people not divided into castes than amongst any other. If ever any nations are constituted in which it may even be difficult to find any peculiar classes of society, the notion of honour will be confined to a small number of precepts, which will be more and more in accordance with the moral laws adopted by the mass of mankind. Thus the laws of honour will be less peculiar and less multifarious amongst a democratic people than in an aristocracy. They will also be more obscure; and this is a necessary consequence of what goes before; for as the distinguishing marks of honour are less numerous and less peculiar, it must often be difficult to distinguish them. To this other reasons may be added. Amongst the aristocratic nations of the middle ages, generation succeeded generation in vain; each family was like a never-dying, ever stationary man, and the state of opinions was hardly more changeable than that of conditions. Everyone then had always the same objects before his eyes, which he contemplated from the same point; his eyes gradually detected the smallest details, and his discernment could not fail to become in the end clear and accurate. Thus not only had the men of feudal times very extraordinary opinions in matters of honour, but each of those opinions was present to their minds under a clear and precise form.

This can never be the case in America, where all men are in constant motion; and where society, transformed daily by its own operations, changes its opinions together with its wants. In such a country men have glimpses of the rules of honour, but they have seldom time to fix attention upon them.

But even if society were motionless, it would still be difficult to determine the meaning which ought to be attached to the word honour. In the middle ages, as each class had its own honour, the same opinion was never received at the same time by a large number of men; and this rendered it possible to give it a determined and accurate form, which was the more easy, as all those by whom it was received, having a perfectly identical and most peculiar position, were naturally disposed to agree upon the points of a law which was made for themselves alone. Thus the code of honour became a complete and detailed system, in which everything was anticipated and provided for beforehand, and a fixed and always palpable standard was applied to human actions. Amongst a democratic nation, like the Americans, in which ranks are identified, and the whole of society forms one single mass, composed of elements which are all analogous though not entirely similar, it is impossible ever to agree beforehand on what shall or shall not be allowed by the laws of honour. Amongst that people, indeed, some national wants do exist which give rise to opinions common to the whole nation on points of honour; but these opinions never occur at the same time, in the same manner, or with the same intensity to the minds of the whole community; the law of honour exists, but it has no organs to promulgate it.

The confusion is far greater still in a democratic country like France, where the different classes of which the former fabric of society was composed, being brought together but not yet mingled, import day by day into each other's circles various and sometimes conflicting notions of honour—where every man, at his own will and pleasure, forsakes one portion of his forefathers' creed, and retains another; so that, amidst so many arbitrary measures, no common rule can ever be established, and it is almost impossible to predict which actions will be held in honour and which will be thought disgraceful. Such times are wretched, but they are of short duration.

As honour, amongst democratic nations, is imperfectly defined, its influence is of course less powerful; for it is difficult to apply with certainty and firmness a law which is not distinctly known. Public opinion, the natural and supreme interpreter of the laws of honour, not clearly discerning to which side censure or approval ought to lean, can only pronounce a hesitating judgment. Sometimes the opinion of the public may contradict itself; more frequently it does not act, and lets things pass.

The weakness of the sense of honour in democracies also arises from several other causes. In aristocratic countries, the same notions of honour are always entertained by only a few persons, always limited in number, often separated from the rest of their fellow-citizens. Honour is easily mingled and identified in their minds with the idea of all that distinguishes their own position; it appears to them as the chief characteristic of their own rank; they apply its different rules with all the warmth of personal interest, and they feel (if I may use the expression) a passion for complying with its dictates. This truth is extremely obvious in the old black-letter law books on the subject of trial by battel. The nobles, in their disputes, were bound to use the lance and sword; whereas the villains used only sticks amongst themselves, 'inasmuch as,' to use the words of the old books, 'villains have no honour.' This did not mean, as it may be imagined at the present day, that these people were contemptible; but simply that their actions were not to be judged by the same rules which were applied to the actions of the aristocracy.

It is surprising, at first sight, that when the sense of honour is most predominant, its injunctions are usually most strange; so that the further it is removed from common reason the better it is obeyed; whence it has sometimes been inferred that the laws of honour were strengthened by their own extravagance. The two things indeed originate from the same source, but the one is not derived from the other. Honour becomes fantastical in proportion to the peculiarity of the wants which it denotes, and the paucity of the men by whom those wants are felt; and it is because it denotes wants of this kind that its influence is great. Thus the notion of honour is not the stronger for being fantastical, but it is fantastical and strong from the self-same cause.

Further, amongst aristocratic nations each rank is different, but all ranks are fixed; every man occupies a place in his own sphere which he cannot relinquish, and he lives there amidst other men who are bound by the same ties. Amongst these nations no man can either hope or fear to escape being seen; no man is placed so low but that he has a stage of his own, and none can avoid censure or applause by his obscurity. In democratic states on the contrary, where all the members of the community are mingled in the same crowd and in constant agitation, public opinion has no hold on men; they disappear at every instant, and elude its power. Consequently the dictates of honour will be there less imperious and less stringent; for honour acts solely for the public eye—differing in this respect from mere virtue, which lives upon itself contented with its own approval.

If the reader has distinctly apprehended all that goes before, he will understand that there is a close and necessary relation between the inequality of social conditions and what has here been styled honour—a relation which, if I am not mistaken, had not before been clearly pointed out. I shall therefore make one more attempt to illustrate it satisfactorily. Suppose a nation stands apart from the rest of mankind: independently of certain general wants inherent in the human race, it will also have wants and interests peculiar to itself: certain opinions of censure or approbation forthwith arise in the community, which are peculiar to itself, and which are styled honour by the members of that community. Now suppose that in this same nation a caste arises, which, in its turn, stands apart from all the other classes, and contracts certain peculiar wants, which give rise in their turn to special opinions. The honour of this caste, composed of a medley of the peculiar notions of the nation, and the still more peculiar notions of the caste, will be as remote as it is possible to conceive from the simple and general opinions of men.

Having reached this extreme point of the argument, I now return. When ranks are commingled and privileges abolished, the men of whom a nation is composed being once more equal and alike, their interests and wants become identical, and all the peculiar notions which each caste styled honour successively disappear: the notion of honour no longer proceeds from any other source than the wants peculiar to the nation at large, and it denotes the individual character of that nation to the world. Lastly, if it be allowable to suppose that all the races of mankind should be commingled, and that all the peoples of earth should ultimately come to have the same interests, the same wants, undistinguished from each other by any characteristic peculiarities, no conventional value whatever would then be attached to men's actions; they would all be regarded by all in the same light; the general necessities of mankind, revealed by conscience to every man, would become the common standard. The simple and general notions of right and wrong only would then be recognised in the world, to which, by a natural and necessary tie, the idea of censure or approbation would be attached. Thus, to comprise all my meaning in a single proposition, the dissimilarities and inequalities of men gave rise to the notion of honour; that notion is weakened in proportion as these differences are obliterated, and with them it would disappear.

Chapter XIX

Why So Many Ambitious Men and So Little Lofty Ambition Are to Be Found in the United States

The first thing which strikes a traveller in the United States is the innumerable multitude of those who seek to throw off their original condition; and the second is the rarity of lofty ambition to be observed in the midst of the universally ambitious stir of society. No Americans are devoid of a yearning desire to rise; but hardly any appear to entertain hopes of great magnitude, or to drive at very lofty aims. All are constantly seeking to acquire property, power, and reputation—few contemplate these things upon a great scale; and this is the more surprising, as nothing is to be discerned in the manners or laws of America to limit desire, or to prevent it from spreading its impulses in every direction. It seems difficult to attribute this singular state of things to the equality of social conditions; for at the instant when that same equality was established in France, the flight of ambition became unbounded. Nevertheless, I think that the principal cause which may be assigned to this fact is to be found in the social condition and democratic manners of the Americans.

All revolutions enlarge the ambition of men: this proposition is more peculiarly true of those revolutions which overthrow an aristocracy. When the former barriers which kept back the multitude from fame and power are suddenly thrown down, a violent and universal rise takes place towards that eminence so long coveted and at length to be enjoyed. In this first burst of triumph nothing seems impossible to any one: not only are desires boundless, but the power of satisfying them seems almost boundless, too. Amidst the general and sudden renewal of laws and customs, in this vast confusion of all men and all ordinances, the various members of the community rise and sink again with excessive rapidity; and power passes so quickly from hand to hand that none need despair of catching it in turn. It must be recollected, moreover, that the people who destroy an aristocracy have lived under its laws; they have witnessed its splendour, and they have unconsciously imbibed the feelings and notions which it entertained. Thus at the moment when an aristocracy is dissolved, its spirit still pervades the mass of the community, and its tendencies are retained long after it has been defeated. Ambition is therefore always extremely great as long as a democratic revolution lasts, and it will remain so for some time after the revolution is consummated. The reminiscence of the extraordinary events which men have witnessed is not obliterated from their memory in a day. The passions which a revolution has roused do not disappear at its close. A sense of instability remains in the midst of reestablished order: a notion of easy success survives the strange vicissitudes which gave it birth; desires still remain extremely enlarged, when the means of satisfying them are diminished day by day. The taste for large fortunes subsists, though large fortunes are rare: and on every side we trace the ravages of inordinate and hapless ambition kindled in hearts which they consume in secret and in vain.

At length, however, the last vestiges of the struggle are effaced; the remains of aristocracy completely disappear; the great events by which its fall was attended are forgotten; peace succeeds to war, and the sway of order is restored in the new realm; desires are again adapted to the means by which they may be fulfilled; the wants, the opinions, and the feelings of men cohere once more; the level of the community is permanently determined, and democratic society established. A democratic nation, arrived at this permanent and regular state of things, will present a very different spectacle from that which we have just described; and we may readily conclude that, if ambition becomes great whilst the conditions of society are growing equal, it loses that quality when they have grown so. As wealth is subdivided and knowledge diffused, no one is entirely destitute of education or of property; the privileges and disqualifications of caste being abolished, and men having shattered the bonds which held them fixed, the notion of advancement suggests itself to every mind, the desire to rise swells in every heart, and all men want to mount above their station: ambition is the universal feeling.

But if the equality of conditions gives some resources to all the members of the community, it also prevents any of them from having resources of great extent, which necessarily circumscribes their desires within somewhat narrow limits. Thus amongst democratic nations ambition is ardent and continual, but its aim is not habitually lofty; and life is generally spent in eagerly coveting small objects which are within reach. What chiefly diverts the men of democracies from lofty ambition is not the scantiness of their fortunes, but the vehemence of the exertions they daily make to improve them. They strain their faculties to the utmost to achieve paltry results, and this cannot fail speedily to limit their discernment and to circumscribe their powers. They might be much poorer and still be greater. The small number of opulent citizens who are to be found amidst a democracy do not constitute an exception to this rule. A man who raises himself by degrees to wealth and power, contracts, in the course of this protracted labour, habits of prudence and restraint which he cannot afterwards shake off. A man cannot enlarge his mind as he would his house. The same observation is applicable to the sons of such a man; they are born, it is true, in a lofty position, but their parents were humble; they have grown up amidst feelings and notions which they cannot afterwards easily get rid of; and it may be presumed that they will inherit the propensities of their father as well as his wealth. It may happen, on the contrary, that the poorest scion of a powerful aristocracy may display vast ambition, because the traditional opinions of his race and the general spirit of his order still buoy him up for some time above his fortune.

Another thing which prevents the men of democratic periods from easily indulging in the pursuit of lofty objects, is the lapse of time which they foresee must take place before they can be ready to approach them. 'It is a great advantage,' says Pascal, 'to be a man of quality, since it brings one man as forward at eighteen or twenty as another man would be at fifty, which is a clear gain of thirty years.' Those thirty years are commonly wanting to the ambitious characters of democracies. The principle of equality, which allows every man to arrive at everything, prevents all men from rapid advancement.

In a democratic society, as well as elsewhere, there are only a certain number of great fortunes to be made; and as the paths which lead to them are indiscriminately open to all, the progress of all must necessarily be slackened. As the candidates appear to be nearly alike, and as it is difficult to make a selection without infringing the principle of equality, which is the supreme law of democratic societies, the first idea which suggests itself is to make them all advance at the same rate and submit to the same probation. Thus in proportion as men become more alike, and the principle of equality is more peaceably and deeply infused into the institutions and manners of the country, the rules of advancement become more inflexible, advancement itself slower, the difficulty of arriving quickly at a certain height far greater. From hatred of privilege and from the embarrassment of choosing, all men are at last constrained, whatever may be their standard, to pass the same ordeal; all are indiscriminately subjected to a multitude of petty preliminary exercises, in which their youth is wasted and their imagination quenched, so that they despair of ever fully attaining what is held out to them; and when at length they are in a condition to perform any extraordinary acts, the taste for such things has forsaken them.

In China, where the equality of conditions is exceedingly great and very ancient, no man passes from one public office to another without undergoing a probationary trial. This probation occurs afresh at every stage of his career; and the notion is now so rooted in the manners of the people that I remember to have read a Chinese novel, in which the hero, after numberless crosses, succeeds at length in touching the heart of his mistress by taking honours. A lofty ambition breathes with difficulty in such an atmosphere.

The remark I apply to politics extends to everything; equality everywhere produces the same effects; where the laws of a country do not regulate and retard the advancement of men by positive enactment, competition attains the same end. In a well-established democratic community great and rapid elevation is therefore rare; it forms an exception to the common rule; and it is the singularity of such occurrences that makes men forget how rarely they happen. Men living in democracies ultimately discover these things; they find out at last that the laws of their country open a boundless field of action before them, but that no one can hope to hasten across it. Between them and the final object of their desires, they perceive a multitude of small intermediate impediments, which must be slowly surmounted: this prospect wearies and discourages their ambition at once. They therefore give up hopes so doubtful and remote, to search nearer to themselves for less lofty and more easy enjoyments. Their horizon is not bounded by the laws but narrowed by themselves.

I have remarked that lofty ambitions are more rare in the ages of democracy than in times of aristocracy: I may add that when, in spite of these natural obstacles, they do spring into existence, their character is different. In aristocracies the career of ambition is often wide, but its boundaries are determined. In democracies ambition commonly ranges in a narrower field, but if once it gets beyond that, hardly any limits can be assigned to it. As men are individually weak—as they live asunder, and in constant motion—as precedents are of little authority and laws but of short duration, resistance to novelty is languid, and the fabric of society never appears perfectly erect or firmly consolidated. So that, when once an ambitious man has the power in his grasp, there is nothing he may not dare; and when it is gone from him, he meditates the overthrow of the State to regain it. This gives to great political ambition a character of revolutionary violence, which it seldom exhibits to an equal degree in aristocratic communities. The common aspect of democratic nations will present a great number of small and very rational objects of ambition, from amongst which a few ill-controlled desires of a larger growth will at intervals break out: but no such a thing as ambition conceived and contrived on a vast scale is to be met with there.

I have shown elsewhere by what secret influence the principle of equality makes the passion for physical gratifications and the exclusive love of the present predominate in the human heart: these different propensities mingle with the sentiment of ambition, and tinge it, as it were, with their hues. I believe that ambitious men in democracies are less engrossed than any others with the interests and the judgment of posterity; the present moment alone engages and absorbs them. They are more apt to complete a number of undertakings with rapidity than to raise lasting monuments of their achievements; and they care much more for success than for fame. What they most ask of men is obedience—what they most covet is empire. Their manners have in almost all cases remained below the height of their station; the consequence is that they frequently carry very low tastes into their extraordinary fortunes, and that they seem to have acquired the supreme power only to minister to their coarse or paltry pleasures.

I think that in our time it is very necessary to cleanse, to regulate, and to adapt the feeling of ambition, but that it would be extremely dangerous to seek to impoverish and to repress it over-much. We should attempt to lay down certain extreme limits, which it should never be allowed to outstep; but its range within those established limits should not be too much checked. I confess that I apprehend much less for democratic society from the boldness than from the mediocrity of desires. What appears to me most to be dreaded is that, in the midst of the small incessant occupations of private life, ambition should lose its vigour and its greatness—that the passions of man should abate, but at the same time be lowered, so that the march of society should every day become more tranquil and less aspiring. I think then that the leaders of modern society would be wrong to seek to lull the community by a state of too uniform and too peaceful happiness; and that it is well to expose it from time to time to matters of difficulty and danger, in order to raise ambition and to give it a field of action. Moralists are constantly complaining that the ruling vice of the present time is pride. This is true in one sense, for indeed no one thinks that he is not better than his neighbour, or consents to obey his superior: but it is extremely false in another; for the same man who cannot endure subordination or equality, has so contemptible an opinion of himself that he thinks he is only born to indulge in vulgar pleasures. He willingly takes up with low desires, without daring to embark in lofty enterprises, of which he scarcely dreams. Thus, far from thinking that humility ought to be preached to our contemporaries, I would have endeavours made to give them a more enlarged idea of themselves and of their kind. Humility is unwholesome to them; what they most want is, in my opinion, pride. I would willingly exchange several of our small virtues for this one vice.

Chapter XX

The Trade of Place-Hunting in Certain Democratic Countries

In the United States as soon as a man has acquired some education and pecuniary resources, he either endeavours to get rich by commerce or industry, or he buys land in the bush and turns pioneer. All that he asks of the State is not to be disturbed in his toil, and to be secure of his earnings. Amongst the greater part of European nations, when a man begins to feel his strength and to extend his desires, the first thing that occurs to him is to get some public employment. These opposite effects, originating in the same cause, deserve our passing notice.

When public employments are few in number, ill-paid and precarious, whilst the different lines of business are numerous and lucrative, it is to business, and not to official duties, that the new and eager desires engendered by the principle of equality turn from every side. But if, whilst the ranks of society are becoming more equal, the education of the people remains incomplete, or their spirit the reverse of bold—if commerce and industry, checked in their growth, afford only slow and arduous means of making a fortune—the various members of the community, despairing of ameliorating their own condition, rush to the head of the State and demand its assistance. To relieve their own necessities at the cost of the public treasury, appears to them to be the easiest and most open, if not the only, way they have to rise above a condition which no longer contents them; place-hunting becomes the most generally followed of all trades. This must especially be the case, in those great centralised monarchies in which the number of paid offices is immense, and the tenure of them tolerably secure, so that no one despairs of obtaining a place, and of enjoying it as undisturbedly as a hereditary fortune.

I shall not remark that the universal and inordinate desire for place is a great social evil; that it destroys the spirit of independence in the citizen, and diffuses a venal and servile humour throughout the frame of society; that it stifles the manlier virtues: nor shall I be at the pains to demonstrate that this kind of traffic only creates an unproductive activity, which agitates the country without adding to its resources: all these things are obvious. But I would observe, that a government which encourages this tendency risks its own tranquillity, and places its very existence in great jeopardy. I am aware that at a time like our own, when the love and respect which formerly clung to authority are seen gradually to decline, it may appear necessary to those in power to lay a closer hold on every man by his own interest, and it may seem convenient to use his own passions to keep him in order and in silence; but this cannot be so long, and what may appear to be a source of strength for a certain time will assuredly become in the end a great cause of embarrassment and weakness.

Amongst democratic nations, as well as elsewhere, the number of official appointments has in the end some limits; but amongst those nations, the number of aspirants is unlimited; it perpetually increases, with a gradual and irresistible rise in proportion as social conditions become more equal, and is only checked by the limits of the population. Thus, when public employments afford the only outlet for ambition, the government necessarily meets with a permanent opposition at last; for it is tasked to satisfy with limited means unlimited desires. It is very certain that of all people in the world the most difficult to restrain and to manage are a people of solicitants. Whatever endeavours are made by rulers, such a people can never be contented; and it is always to be apprehended that they will ultimately overturn the constitution of the country, and change the aspect of the State, for the sole purpose of making a clearance of places. The sovereigns of the present age, who strive to fix upon themselves alone all those novel desires which are aroused by equality, and to satisfy them, will repent in the end, if I am not mistaken, that they ever embarked in this policy: they will one day discover that they have hazarded their own power, by making it so necessary; and that the more safe and honest course would have been to teach their subjects the art of providing for themselves.[1]

Chapter XXI

Why Great Revolutions Will Become More Rare

A people which has existed for centuries under a system of castes and classes can only arrive at a democratic state of society by passing through a long series of more or less critical transformations, accomplished by violent efforts, and after numerous vicissitudes; in the course of which, property, opinions, and power are rapidly transferred from one hand to another. Even after this great revolution is consummated, the revolutionary habits engendered by it may long be traced, and it will be followed by deep commotion. As all this takes place at the very time at which social conditions are becoming more equal, it is inferred that some concealed relation and secret tie exist between the principle of equality itself and revolution, insomuch that the one cannot exist without giving rise to the other.

On this point reasoning may seem to lead to the same result as experience. Amongst a people whose ranks are nearly equal, no ostensible bond connects men together, or keeps them settled in their station. None of them have either a permanent right or power to command—none are forced by their condition to obey; but every man, finding himself possessed of some education and some resources, may choose his own path and proceed apart from all his fellowmen. The same causes which make the members of the community independent of each other, continually impel them to new and restless desires, and constantly spur them onwards. It therefore seems natural that, in a democratic community, men, things, and opinions should be forever changing their form and place, and that democratic ages should be times of rapid and incessant transformation.

But is this really the case? does the equality of social conditions habitually and permanently lead men to revolution? does that state of society contain some perturbing principle which prevents the community from ever subsiding into calm, and disposes the citizens to alter incessantly their laws, their principles, and their manners? I do not believe it; and as the subject is important, I beg for the reader's close attention. Almost all the revolutions which have changed the aspect of nations have been made to consolidate or to destroy social inequality. Remove the secondary causes which have produced the great convulsions of the world, and you will almost always find the principle of inequality at the bottom. Either the poor have attempted to plunder the rich, or the rich to enslave the poor. If then a state of society can ever be founded in which every man shall have something to keep, and little to take from others, much will have been done for the peace of the world. I am aware that amongst a great democratic people there will always be some members of the community in great poverty, and others in great opulence; but the poor, instead of forming the immense majority of the nation, as is always the case in aristocratic communities, are comparatively few in number, and the laws do not bind them together by the ties of irremediable and hereditary penury. The wealthy, on their side, are scarce and powerless; they have no privileges which attract public observation; even their wealth, as it is no longer incorporated and bound up with the soil, is impalpable, and as it were invisible. As there is no longer a race of poor men, so there is no longer a race of rich men; the latter spring up daily from the multitude, and relapse into it again. Hence they do not form a distinct class, which may be easily marked out and plundered; and, moreover, as they are connected with the mass of their fellow-citizens by a thousand secret ties, the people cannot assail them without inflicting an injury upon itself. Between these two extremes of democratic communities stand an innumerable multitude of men almost alike, who, without being exactly either rich or poor, are possessed of sufficient property to desire the maintenance of order, yet not enough to excite envy. Such men are the natural enemies of violent commotions: their stillness keeps all beneath them and above them still, and secures the balance of the fabric of society. Not indeed that even these men are contented with what they have gotten, or that they feel a natural abhorrence for a revolution in which they might share the spoil without sharing the calamity; on the contrary, they desire, with unexampled ardour, to get rich, but the difficulty is to know from whom riches can be taken. The same state of society which constantly prompts desires, restrains these desires within necessary limits: it gives men more liberty of changing and less interest in change.

Not only are the men of democracies not naturally desirous of revolutions, but they are afraid of them. All revolutions more or less threaten the tenure of property: but most of those who live in democratic countries are possessed of property—not only are they possessed of property, but they live in the condition of men who set the greatest store upon their property. If we attentively consider each of the classes of which society is composed, it is easy to see that the passions engendered by property are keenest and most tenacious amongst the middle classes. The poor often care but little for what they possess, because they suffer much more from the want of what they have not, than they enjoy the little they have. The rich have many other passions besides that of riches to satisfy; and, besides, the long and arduous enjoyment of a great fortune sometimes makes them in the end insensible to its charms. But the men who have a competency, alike removed from opulence and from penury, attach an enormous value to their possessions. As they are still almost within the reach of poverty, they see its privations near at hand, and dread them; between poverty and themselves there is nothing but a scanty fortune, upon which they immediately fix their apprehensions and their hopes. Every day increases the interest they take in it, by the constant cares which it occasions; and they are the more attached to it by their continual exertions to increase the amount. The notion of surrendering the smallest part of it is insupportable to them, and they consider its total loss as the worst of misfortunes. Now these eager and apprehensive men of small property constitute the class which is constantly increased by the equality of conditions. Hence, in democratic communities, the majority of the people do not clearly see what they have to gain by a revolution, but they continually and in a thousand ways feel that they might lose by one.

I have shown in another part of this work that the equality of conditions naturally urges men to embark in commercial and industrial pursuits, and that it tends to increase and to distribute real property: I have also pointed out the means by which it inspires every man with an eager and constant desire to increase his welfare. Nothing is more opposed to revolutionary passions than these things. It may happen that the final result of a revolution is favourable to commerce and manufactures; but its first consequence will almost always be the ruin of manufactures and mercantile men, because it must always change at once the general principles of consumption, and temporarily upset the existing proportion between supply and demand. I know of nothing more opposite to revolutionary manners than commercial manners. Commerce is naturally adverse to all the violent passions; it loves to temporize, takes delight in compromise, and studiously avoids irritation. It is patient, insinuating, flexible, and never has recourse to extreme measures until obliged by the most absolute necessity. Commerce renders men independent of each other, gives them a lofty notion of their personal importance, leads them to seek to conduct their own affairs, and teaches how to conduct them well; it therefore prepares men for freedom, but preserves them from revolutions. In a revolution the owners of personal property have more to fear than all others; for on the one hand their property is often easy to seize, and on the other it may totally disappear at any moment—a subject of alarm to which the owners of real property are less exposed, since, although they may lose the income of their estates, they may hope to preserve the land itself through the greatest vicissitudes. Hence the former are much more alarmed at the symptoms of revolutionary commotion than the latter. Thus nations are less disposed to make revolutions in proportion as personal property is augmented and distributed amongst them, and as the number of those possessing it increases. Moreover, whatever profession men may embrace, and whatever species of property they may possess, one characteristic is common to them all. No one is fully contented with his present fortune—all are perpetually striving in a thousand ways to improve it. Consider any one of them at any period of his life, and he will be found engaged with some new project for the purpose of increasing what he has; talk not to him of the interests and the rights of mankind, this small domestic concern absorbs for the time all his thoughts, and inclines him to defer political excitement to some other season. This not only prevents men from making revolutions, but deters men from desiring them. Violent political passions have but little hold on those who have devoted all their faculties to the pursuit of their well-being. The ardour which they display in small matters calms their zeal for momentous undertakings.

From time to time indeed, enterprising and ambitious men will arise in democratic communities, whose unbounded aspirations cannot be contented by following the beaten track. Such men like revolutions and hail their approach; but they have great difficulty in bringing them about, unless unwonted events come to their assistance. No man can struggle with advantage against the spirit of his age and country; and, however powerful he may be supposed to be, he will find it difficult to make his contemporaries share in feelings and opinions which are repugnant to all their feelings and desires.

It is a mistake to believe that, when once the equality of conditions has become the old and uncontested state of society, and has imparted its characteristics to the manners of a nation, men will easily allow themselves to be thrust into perilous risks by an imprudent leader or a bold innovator. Not indeed that they will resist him openly, by well-contrived schemes, or even by a premeditated plan of resistance. They will not struggle energetically against him, sometimes they will even applaud him—but they do not follow him. To his vehemence they secretly oppose their inertia,—to his revolutionary tendencies their conservative interests,—their homely tastes to his adventurous passions,—their good sense to the flights of his genius,—to his poetry their prose. With immense exertion he raises them for an instant, but they speedily escape from him, and fall back, as it were, by their own weight. He strains himself to rouse the indifferent and distracted multitude, and finds at last that he is reduced to impotence, not because he is conquered, but because he is alone.

I do not assert that men living in democratic communities are naturally stationary; I think, on the contrary, that a perpetual stir prevails in the bosom of those societies, and that rest is unknown there; but I think that men bestir themselves within certain limits beyond which they hardly ever go. They are forever varying, altering, and restoring secondary matters; but they carefully abstain from touching what is fundamental. They love change, but they dread revolutions. Although the Americans are constantly modifying or abrogating some of their laws, they by no means display revolutionary passions. It may be easily seen, from the promptitude with which they check and calm themselves when public excitement begins to grow alarming, and at the very moment when passions seem most roused, that they dread a revolution as the worst of misfortunes, and that every one of them is inwardly resolved to make great sacrifices to avoid such a catastrophe. In no country in the world is the love of property more active and more anxious than in the United States; nowhere does the majority display less inclination for those principles which threaten to alter, in whatever manner, the laws of property. I have often remarked that theories which are of a revolutionary nature, since they cannot be put in practice without a complete and sometimes a sudden change in the state of property and persons, are much less favourably viewed in the United States than in the great monarchical countries of Europe: if some men profess them, the bulk of the people reject them with instinctive abhorrence. I do not hesitate to say that most of the maxims commonly called democratic in France would be proscribed by the democracy of the United States. This may easily be understood: in America men have the opinions and passions of democracy, in Europe we have still the passions and opinions of revolution. If ever America undergoes great revolutions, they will be brought about by the presence of the black race on the soil of the United States,—that is to say, they will owe their origin, not to the equality, but to the inequality, of conditions.

When social conditions are equal, every man is apt to live apart, centred in himself and forgetful of the public. If the rulers of democratic nations were either to neglect to correct this fatal tendency, or to encourage it from a notion that it weans men from political passions and thus wards off revolutions, they might eventually produce the evil they seek to avoid, and a time might come when the inordinate passions of a few men, aided by the unintelligent selfishness or the pusillanimity of the greater number, would ultimately compel society to pass through strange vicissitudes. In democratic communities revolutions are seldom desired except by a minority; but a minority may sometimes effect them. I do not assert that democratic nations are secure from revolutions; I merely say that the state of society in those nations does not lead to revolutions, but rather wards them off. A democratic people left to itself will not easily embark in great hazards; it is only led to revolutions unawares; it may sometimes undergo them, but it does not make them; and I will add that, when such a people has been allowed to acquire sufficient knowledge and experience, it will not suffer them to be made. I am well aware that in this respect public institutions may themselves do much; they may encourage or repress the tendencies which originate in the state of society. I therefore do not maintain, I repeat, that a people is secure from revolutions simply because conditions are equal in the community; but I think that, whatever the institutions of such a people may be, great revolutions will always be far less violent and less frequent than is supposed; and I can easily discern a state of polity, which, when combined with the principle of equality, would render society more stationary than it has ever been in our western part of the world.

The observations I have here made on events may also be applied in part to opinions. Two things are surprising in the United States,—the mutability of the greater part of human actions, and the singular stability of certain principles. Men are in constant motion; the mind of man appears almost unmoved. When once an opinion has spread over the country and struck root there, it would seem that no power on earth is strong enough to eradicate it. In the United States, general principles in religion, philosophy, morality, and even politics, do not vary, or at least are only modified by a hidden and often an imperceptible process: even the grossest prejudices are obliterated with incredible slowness, amidst the continual friction of men and things.

I hear it said that it is in the nature and the habits of democracies to be constantly changing their opinions and feelings. This may be true of small democratic nations, like those of the ancient world, in which the whole community could be assembled in a public place and then excited at will by an orator. But I saw nothing of the kind amongst the great democratic people which dwells upon the opposite shores of the Atlantic Ocean. What struck me in the United States was the difficulty in shaking the majority in an opinion once conceived, or of drawing it off from a leader once adopted. Neither speaking nor writing can accomplish it; nothing but experience will avail, and even experience must be repeated. This is surprising at first sight, but a more attentive investigation explains the fact. I do not think that it is as easy as is supposed to uproot the prejudices of a democratic people—to change its belief—to supersede principles once established, by new principles in religion, politics, and morals—in a word, to make great and frequent changes in men's minds. Not that the human mind is there at rest,—it is in constant agitation; but it is engaged in infinitely varying the consequences of known principles, and in seeking for new consequences, rather than in seeking for new principles. Its motion is one of rapid circumvolution, rather than of straightforward impulse by rapid and direct effort; it extends its orbit by small continual and hasty movements, but it does not suddenly alter its position.

Men who are equal in rights, in education, in fortune, or, to comprise all in one word, in their social condition, have necessarily wants, habits, and tastes which are hardly dissimilar. As they look at objects under the same aspect, their minds naturally tend to analogous conclusions; and, though each of them may deviate from his contemporaries and form opinions of his own, they will involuntarily and unconsciously concur in a certain number of received opinions. The more attentively I consider the effects of equality upon the mind, the more am I persuaded that the intellectual anarchy which we witness about us is not, as many men suppose, the natural state of democratic nations. I think it is rather to be regarded as an accident peculiar to their youth, and that it only breaks out at that period of transition when men have already snapped the former ties which bound them together, but are still amazingly different in origin, education, and manners; so that, having retained opinions, propensities and tastes of great diversity, nothing any longer prevents men from avowing them openly. The leading opinions of men become similar in proportion as their conditions assimilate; such appears to me to be the general and permanent law,—the rest is casual and transient.

I believe that it will rarely happen to any man amongst a democratic community, suddenly to frame a system of notions very remote from that which his contemporaries have adopted; and if some such innovator appeared, I apprehend that he would have great difficulty in finding listeners, still more in finding believers. When the conditions of men are almost equal, they do not easily allow themselves to be persuaded by each other. As they all live in close intercourse, as they have learned the same things together, and as they lead the same life, they are not naturally disposed to take one of themselves for a guide, and to follow him implicitly. Men seldom take the opinion of their equal, or of a man like themselves, upon trust. Not only is confidence in the superior attainments of certain individuals weakened amongst democratic nations, as I have elsewhere remarked, but the general notion of the intellectual superiority which any man whatsoever may acquire in relation to the rest of the community is soon overshadowed. As men grow more like each other, the doctrine of the equality of the intellect gradually infuses itself into their opinions; and it becomes more difficult for any innovator to acquire or to exert much influence over the minds of a people. In such communities sudden intellectual revolutions will therefore be rare; for, if we read aright the history of the world, we shall find that great and rapid changes in human opinions have been produced far less by the force of reasoning than by the authority of a name. Observe, too, that as the men who live in democratic societies are not connected with each other by any tie, each of them must be convinced individually; whilst in aristocratic society it is enough to convince a few,—the rest follow. If Luther had lived in an age of equality, and had not had princes and potentates for his audience, he would perhaps have found it more difficult to change the aspect of Europe. Not indeed that the men of democracies are naturally strongly persuaded of the certainty of their opinions, or are unwavering in belief; they frequently entertain doubts which no one, in their eyes, can remove. It sometimes happens at such times that the human mind would willingly change its position; but as nothing urges or guides it forwards, it oscillates to and fro without progressive motion.[1]

Even when the reliance of a democratic people has been won, it is still no easy matter to gain their attention. It is extremely difficult to obtain a hearing from men living in democracies, unless it be to speak to them of themselves. They do not attend to the things said to them, because they are always fully engrossed with the things they are doing. For indeed few men are idle in democratic nations; life is passed in the midst of noise and excitement, and men are so engaged in acting that little remains to them for thinking. I would especially remark that they are not only employed, but that they are passionately devoted to their employments. They are always in action, and each of their actions absorbs their faculties: the zeal which they display in business puts out the enthusiasm they might otherwise entertain for ideas. I think that it is extremely difficult to excite the enthusiasm of a democratic people for any theory which has not a palpable, direct, and immediate connexion with the daily occupations of life: therefore they will not easily forsake their old opinions; for it is enthusiasm which flings the minds of men out of the beaten track, and effects the great revolutions of the intellect as well as the great revolutions of the political world. Thus democratic nations have neither time nor taste to go in search of novel opinions. Even when those they possess become doubtful, they still retain them, because it would take too much time and inquiry to change them,—they retain them, not as certain, but as established.

There are yet other and more cogent reasons which prevent any great change from being easily effected in the principles of a democratic people. I have already adverted to them at the commencement of this part of my work. If the influence of individuals is weak and hardly perceptible amongst such a people, the power exercised by the mass upon the mind of each individual is extremely great,—I have already shown for what reasons. I would now observe that it is wrong to suppose that this depends solely upon the form of government, and that the majority would lose its intellectual supremacy if it were to lose its political power. In aristocracies men have often much greatness and strength of their own: when they find themselves at variance with the greater number of their fellow-countrymen, they withdraw to their own circle, where they support and console themselves. Such is not the case in a democratic country; there public favour seems as necessary as the air we breathe, and to live at variance with the multitude is, as it were, not to live. The multitude requires no laws to coerce those who think not like itself: public disapprobation is enough; a sense of their loneliness and impotence overtakes them and drives them to despair.

Whenever social conditions are equal, public opinion presses with enormous weight upon the mind of each individual; it surrounds, directs, and oppresses him; and this arises from the very constitution of society, much more than from its political laws. As men grow more alike, each man feels himself weaker in regard to all the rest; as he discerns nothing by which he is considerably raised above them, or distinguished from them, he mistrusts himself as soon as they assail him. Not only does he mistrust his strength, but he even doubts of his right; and he is very near acknowledging that he is in the wrong, when the greater number of his countrymen assert that he is so. The majority do not need to constrain him,—they convince him. In whatever way then the powers of a democratic community may be organized and balanced, it will always be extremely difficult to believe what the bulk of the people reject, or to profess what they condemn.

This circumstance is extraordinarily favourable to the stability of opinions. When an opinion has taken root amongst a democratic people, and established itself in the minds of the bulk of the community, it afterwards subsists by itself and is maintained without effort, because no one attacks it. Those who at first rejected it as false, ultimately receive it as the general impression; and those who still dispute it in their hearts, conceal their dissent; they are careful not to engage in a dangerous and useless conflict. It is true, that when the majority of a democratic people change their opinions, they may suddenly and arbitrarily effect strange revolutions in men's minds; but their opinions do not change without much difficulty, and it is almost as difficult to show that they are changed.

Time, events, or the unaided individual action of the mind, will sometimes undermine or destroy an opinion, without any outward sign of the change. It has not been openly assailed, no conspiracy has been formed to make war on it, but its followers one by one noiselessly secede,—day by day a few of them abandon it, until last it is only professed by a minority. In this state it will still continue to prevail.

As its enemies remain mute, or only interchange their thoughts by stealth, they are themselves unaware for a long period that a great revolution has actually been effected; and in this state of uncertainly they take no steps,—they observe each other and are silent. The majority have ceased to believe what they believed before; but they still affect to believe, and this empty phantom of public opinion in strong enough to chill innovators, and to keep them silent and at a respectful distance. We live at a time which has witnessed the most rapid changes of opinion in the minds of men; nevertheless it may be that the leading opinions of society will ere long be more settled than they have been for several centuries in our history: that time is not yet come, but it may perhaps be approaching. As I examine more closely the natural wants and tendencies of democratic nations, I grow persuaded that if ever social equality is generally and permanently established in the world, great intellectual and political revolutions will become more difficult and less frequent than is supposed. Because the men of democracies appear always excited, uncertain, eager, changeable in their wills and in their positions, it is imagined that they are suddenly to abrogate their laws, to adopt new opinions, and to assume new manners. But if the principle of equality predisposes men to change, it also suggests to them certain interests and tastes which cannot be satisfied without a settled order of things; equality urges them on, but at the same time it holds them back; it spurs them, but fastens them to earth;—it kindles their desires, but limits their powers. This, however, is not perceived at first; the passions which tend to sever the citizens of a democracy are obvious enough; but the hidden force which restrains and unites them is not discernible at a glance.

Amidst the ruins which surround me, shall I dare to say that revolutions are not what I most fear for coming generations? If men continue to shut themselves more closely within the narrow circle of domestic interests and to live upon that kind of excitement, it is to be apprehended that they may ultimately become inaccessible to those great and powerful public emotions which perturb nations,—but which enlarge them and recruit them. When property becomes so fluctuating, and the love of property so restless and so ardent, I cannot but fear that men may arrive at such a state as to regard every new theory as a peril, every innovation as an irksome toil, every social improvement as a stepping-stone to revolution, and so refuse to move altogether for fear of being moved too far. I dread, and I confess it, lest they should at last so entirely give way to a cowardly love of present enjoyment, as to lose sight of the interests of their future selves and of those of their descendants; and to prefer to glide along the easy current of life, rather than to make, when it is necessary, a strong and sudden effort to a higher purpose. It is believed by some that modern society will be ever changing its aspect; for myself, I fear that it will ultimately be too invariably fixed in the same institutions, the same prejudices, the same manners, so that mankind will be stopped and circumscribed; that the mind will swing backwards and forwards for ever, without begetting fresh ideas; that man will waste his strength in bootless and solitary trifling; and, though in continual motion, that humanity will cease to advance.

Chapter XXII

Why Democratic Nations Are Naturally Desirous of Peace, and Democratic Armies of War

The same interests, the same fears, the same passions which deter democratic nations from revolutions, deter them also from war; the spirit of military glory and the spirit of revolution are weakened at the same time and by the same causes. The ever-increasing numbers of men of property,—lovers of peace, the growth of personal wealth which war so rapidly consumes, the mildness of manners, the gentleness of heart, those tendencies to pity which are engendered by the equality of conditions, that coolness of understanding which renders men comparatively insensible to the violent and poetical excitement of arms,—all these causes concur to quench the military spirit. I think it may be admitted as a general and constant rule, that, amongst civilised nations, the warlike passions will become more rare and less intense in proportion as social conditions shall be more equal. War is nevertheless an occurrence to which all nations are subject, democratic nations as well as others. Whatever taste they may have for peace, they must hold themselves in readiness to repel aggression, or in other words they must have an army.

Fortune, which has conferred so many peculiar benefits upon the inhabitants of the United States, has placed them in the midst of a wilderness, where they have, so to speak, no neighbours: a few thousand soldiers are sufficient for their wants; but this is peculiar to America, not to democracy. The equality of conditions, and the manners as well as the institutions resulting from it, do not exempt a democratic people from the necessity of standing armies, and their armies always exercise a powerful influence over their fate. It is therefore of singular importance to inquire what are the natural propensities of the men of whom these armies are composed.

Amongst aristocratic nations, especially amongst those in which birth is the only source of rank, the same inequality exists in the army as in the nation; the officer is noble, the soldier is a serf; the one is naturally called upon to command, the other to obey. In aristocratic armies, the private soldier's ambition is therefore circumscribed within very narrow limits. Nor has the ambition of the officer an unlimited range. An aristocratic body not only forms a part of the scale of ranks in the nation, but it contains a scale of ranks within itself: the members of whom it is composed are placed one above another, in a particular and unvarying manner. Thus one man is born to the command of a regiment, another to that of a company; when once they have reached the utmost object of their hopes, they stop of their own accord, and remain contented with their lot. There is, besides, a strong cause, which, in aristocracies, weakens the officer's desire of promotion. Amongst aristocratic nations, an officer, independently of his rank in the army, also occupies an elevated rank in society; the former is almost always in his eyes only an appendage to the latter. A nobleman who embraces the profession of arms follows it less from motives of ambition than from a sense of the duties imposed on him by his birth. He enters the army in order to find an honourable employment for the idle years of his youth, and to be able to bring back to his home and his peers some honourable recollections of military life; but his principal object is not to obtain by that profession either property, distinction, or power, for he possesses these advantages in his own right, and enjoys them without leaving his home.

In democratic armies all the soldiers may become officers, which makes the desire of promotion general, and immeasurably extends the bounds of military ambition. The officer, on his part, sees nothing which naturally and necessarily stops him at one grade more than at another; and each grade has immense importance in his eyes, because his rank in society almost always depends on his rank in the army. Amongst democratic nations it often happens that an officer has no property but his pay, and no distinction but that of military honours: consequently as often as his duties change, his fortune changes, and he becomes, as it were, a new man. What was only an appendage to his position in aristocratic armies, has thus become the main point, the basis of his whole condition. Under the old French monarchy officers were always called by their titles of nobility; they are now always called by the title of their military rank. This little change in the forms of language suffices to show that a great revolution has taken place in the constitution of society and in that of the army. In democratic armies the desire of advancement is almost universal: it is ardent, tenacious, perpetual; it is strengthened by all other desires, and only extinguished with life itself. But it is easy to see, that of all armies in the world, those in which advancement must be slowest in time of peace are the armies of democratic countries. As the number of commissions is naturally limited, whilst the number of competitors is almost unlimited, and as the strict law of equality is over all alike, none can make rapid progress,—many can make no progress at all. Thus the desire of advancement is greater, and the opportunities of advancement fewer, there than elsewhere. All the ambitious spirits of a democratic army are consequently ardently desirous of war, because war makes vacancies, and warrants the violation of that law of seniority which is the sole privilege natural to democracy.

We thus arrive at this singular consequence, that of all armies those most ardently desirous of war are democratic armies, and of all nations those most fond of peace are democratic nations: and, what makes these facts still more extraordinary, is that these contrary effects are produced at the same time by the principle of equality.

All the members of the community, being alike, constantly harbour the wish, and discover the possibility, of changing their condition and improving their welfare: this makes them fond of peace, which is favourable to industry, and allows every man to pursue his own little undertakings to their completion. On the other hand, this same equality makes soldiers dream of fields of battle, by increasing the value of military honours in the eyes of those who follow the profession of arms, and by rendering those honours accessible to all. In either case the inquietude of the heart is the same, the taste for enjoyment as insatiable, the ambition of success as great—the means of gratifying it are alone different.

These opposite tendencies of the nation and the army expose democratic communities to great dangers. When a military spirit forsakes a people, the profession of arms immediately ceases to be held in honour, and military men fall to the lowest rank of the public servants: they are little esteemed, and no longer understood. The reverse of what takes place in aristocratic ages then occurs; the men who enter the army are no longer those of the highest, but of the lowest rank. Military ambition is only indulged in when no other is possible. Hence arises a circle of cause and consequence from which it is difficult to escape: the best part of the nation shuns the military profession because that profession is not honoured, and the profession is not honoured because the best part of the nation has ceased to follow it. It is then no matter of surprise that democratic armies are often restless, ill-tempered, and dissatisfied with their lot, although their physical condition is commonly far better, and their discipline less strict than in other countries. The soldier feels that he occupies an inferior position, and his wounded pride either stimulates his taste for hostilities which would render his services necessary, or gives him a turn for revolutions, during which he may hope to win by force of arms the political influence and personal importance now denied him. The composition of democratic armies makes this last-mentioned danger much to be feared. In democratic communities almost every man has some property to preserve; but democratic armies are generally led by men without property, most of whom have little to lose in civil broils. The bulk of the nation is naturally much more afraid of revolutions than in the ages of aristocracy, but the leaders of the army much less so.

Moreover, as amongst democratic nations (to repeat what I have just remarked) the wealthiest, the best educated, and the most able men seldom adopt the military profession, the army, taken collectively, eventually forms a small nation by itself, where the mind is less enlarged, and habits are more rude than in the nation at large. Now, this small uncivilised nation has arms in its possession, and alone knows how to use them: for, indeed, the pacific temper of the community increases the danger to which a democratic people is exposed from the military and turbulent spirit of the army. Nothing is so dangerous as an army amidst an unwarlike nation; the excessive love of the whole community for quiet continually puts its constitution at the mercy of the soldiery.

It may therefore be asserted, generally speaking, that if democratic nations are naturally prone to peace from their interests and their propensities, they are constantly drawn to war and revolutions by their armies. Military revolutions, which are scarcely ever to be apprehended in aristocracies, are always to be dreaded amongst democratic nations. These perils must be reckoned amongst the most formidable which beset their future fate, and the attention of statesmen should be sedulously applied to find a remedy for the evil.

When a nation perceives that it is inwardly affected by the restless ambition of its army, the first thought which occurs is to give this inconvenient ambition an object by going to war. I speak no ill of war: war almost always enlarges the mind of a people, and raises their character. In some cases it is the only check to the excessive growth of certain propensities which naturally spring out of the equality of conditions, and it must be considered as a necessary corrective to certain inveterate diseases to which democratic communities are liable. War has great advantages, but we must not flatter ourselves that it can diminish the danger I have just pointed out. That peril is only suspended by it, to return more fiercely when the war is over; for armies are much more impatient of peace after having tasted military exploits. War could only be a remedy for a people which should always be athirst for military glory. I foresee that all the military rulers who may rise up in great democratic nations, will find it easier to conquer with their armies, than to make their armies live at peace after conquest. There are two things which a democratic people will always find very difficult—to begin a war, and to end it.

Again, if war has some peculiar advantages for democratic nations, on the other hand it exposes them to certain dangers which aristocracies have no cause to dread to an equal extent. I shall only point out two of these. Although war gratifies the army, it embarrasses and often exasperates that countless multitude of men whose minor passions every day require peace in order to be satisfied. Thus there is some risk of its causing, under another form, the disturbance it is intended to prevent. No protracted war can fail to endanger the freedom of a democratic country. Not indeed that after every victory it is to be apprehended that the victorious generals will possess themselves by force of the supreme power, after the manner of Sylla and Cæsar: the danger is of another kind. War does not always give over democratic communities to military government, but it must invariably and immeasurably increase the powers of civil government; it must almost compulsorily concentrate the direction of all men and the management of all things in the hands of the administration. If it lead not to despotism by sudden violence, it prepares men for it more gently by their habits. All those who seek to destroy the liberties of a democratic nation ought to know that war is the surest and the shortest means to accomplish it. This is the first axiom of the science.

One remedy, which appears to be obvious when the ambition of Soldiers and officers becomes the subject of alarm, is to augment the number of commissions to be distributed by increasing the army. This affords temporary relief, but it plunges the country into deeper difficulties at some future period. To increase the army may produce a lasting effect in an aristocratic community, because military ambition is there confined to one class of men, and the ambition of each individual stops, as it were, at a certain limit; so that it may be possible to satisfy all who feel its influence. But nothing is gained by increasing the army amongst a democratic people, because the number of aspirants always rises in exactly the same ratio as the army itself. Those whose claims have been satisfied by the creation of new commissions are instantly succeeded by a fresh multitude beyond all power of satisfaction; and even those who were but now satisfied soon begin to crave more advancement; for the same excitement prevails in the ranks of the army as in the civil classes of democratic society, and what men want is not to reach a certain grade, but to have constant promotion. Though these wants may not be very vast, they are perpetually recurring. Thus a democratic nation, by augmenting its army, only allays for a time the ambition of the military profession, which soon becomes even more formidable, because the number of those who feel it is increased. I am of opinion that a restless and turbulent spirit is an evil inherent in the very constitution of democratic armies, and beyond hope of cure. The legislators of democracies must not expect to devise any military organisation capable by its influence of calming and restraining the military profession: their efforts would exhaust their powers, before the object is attained.

The remedy for the vices of the army is not to be found in the army itself, but in the country. Democratic nations are naturally afraid of disturbance and of despotism; the object is to turn these natural instincts into well-digested, deliberate, and lasting tastes. When men have at last learned to make a peaceful and profitable use of freedom, and have felt its blessings,—then they have conceived a manly love of order, and have freely submitted themselves to discipline,—these same men, if they follow the profession of arms, bring into it, unconsciously and almost against their will, these same habits and manners. The general spirit of the nation being infused into the spirit peculiar to the army, tempers the opinions and desires engendered by military life, or represses them by the mighty force of public opinion. Teach but the citizens to be educated, orderly, firm, and free, the soldiers will be disciplined and obedient. Any law which, in repressing the turbulent spirit of the army, should tend to diminish the spirit of freedom in the nation, and to overshadow the notion of law and right, would defeat its object: it would do much more to favour, than to defeat, the establishment of military tyranny.

After all, and in spite of all precautions, a large army amidst a democratic people will always be a source of great danger; the most effectual means of diminishing that danger would be to reduce the army, but this is a remedy which all nations have it not in their power to use.

Chapter XXIII

Which Is the Most Warlike and Most Revolutionary Class in Democratic Armies

It is a part of the essence of a democratic army to be very numerous in proportion to the people to which it belongs, as I shall hereafter show. On the other hand, men living in democratic times seldom choose a military life. Democratic nations are therefore soon led to give up the system of voluntary recruiting for that of compulsory enlistment. The necessity of their social condition compels them to resort to the latter means, and it may easily be foreseen that they will all eventually adopt it. When military service is compulsory, the burden is indiscriminately and equally borne by the whole community. This is another necessary consequence of the social condition of these nations, and of their notions. The government may do almost whatever it pleases, provided it appeals to the whole community at once: it is the unequal distribution of the weight, not the weight itself, which commonly occasions resistance. But as military service is common to all the citizens, the evident consequence is that each of them remains but for a few years on active duty. Thus it is in the nature of things that the soldier in democracies only passes through the army, whilst among most aristocratic nations the military profession is one which the soldier adopts, or which is imposed upon him, for life.

This has important consequences. Amongst the soldiers of a democratic army, some acquire a taste for military life, but the majority being enlisted against their will, and ever ready to go back to their homes, do not consider themselves as seriously engaged in the military profession, and are always thinking of quitting it. Such men do not contract the wants, and only half partake in the passions, which that mode of life engenders. They adapt themselves to their military duties, but their minds are still attached to the interests and the duties which engaged them in civil life. They do not therefore imbibe the spirit of the army,—or rather, they infuse the spirit of the community at large into the army, and retain it there. Amongst democratic nations the private soldiers remain most like civilians: upon them the habits of the nation have the firmest hold, and public opinion most influence. It is by the instrumentality of the private soldiers especially that it may be possible to infuse into a democratic army the love of freedom and the respect of rights, if these principles have once been successfully inculcated on the people at large. The reverse happens amongst aristocratic nations, where the soldiery have eventually nothing in common with their fellow-citizens, and where they live amongst them as strangers, and often as enemies. In aristocratic armies the officers are the conservative element, because the officers alone have retained a strict connexion with civil society, and never forego their purpose of resuming their place in it sooner or later: in democratic armies the private soldiers stand in this position, and from the same cause.

It often happens, on the contrary, that in these same democratic armies the officers contract tastes and wants wholly distinct from those of the nation—a fact which may be thus accounted for. Amongst democratic nations, the man who becomes an officer severs all the ties which bound him to civil life; he leaves it forever; he has no interest to resume it. His true country is the army, since he owes all he has to the rank he has attained in it; he therefore follows the fortunes of the army, rises or sinks with it, and henceforward directs all his hopes to that quarter only. As the wants of an officer are distinct from those of the country, he may perhaps ardently desire war, or labour to bring about a revolution at the very moment when the nation is most desirous of stability and peace. There are, nevertheless, some causes which allay this restless and warlike spirit. Though ambition is universal and continual amongst democratic nations, we have seen that it is seldom great. A man who, being born in the lower classes of the community, has risen from the ranks to be an officer, has already taken a prodigious step. He has gained a footing in a sphere above that which he filled in civil life, and he has acquired rights which most democratic nations will ever consider as inalienable.[1] He is willing to pause after so great an effort, and to enjoy what he has won. The fear of risking what he has already obtained damps the desire of acquiring what he has not got. Having conquered the first and greatest impediment which opposed his advancement, he resigns himself with less impatience to the slowness of his progress. His ambition will be more and more cooled in proportion as the increasing distinction of his rank teaches him that he has more to put in jeopardy. If I am not mistaken, the least warlike, and also the least revolutionary part, of a democratic army, will always be its chief commanders.

But the remarks I have just made on officers and soldiers are not applicable to a numerous class which in all armies fills the intermediate space between them,—I mean the class of non-commissioned officers. This class of non-commissioned officers which have never acted a part in history until the present century, is henceforward destined, I think, to play one of some importance. Like the officers, non-commissioned officers have broken, in their minds, all the ties which bound them to civil life; like the former, they devote themselves permanently to the service, and perhaps make it even more exclusively the object of all their desires: but non-commissioned officers are men who have not yet reached a firm and lofty post at which they may pause and breathe more freely, ere they can attain further promotion. By the very nature of his duties, which is invariable, a non-commissioned officer is doomed to lead an obscure, confined, comfortless, and precarious existence; as yet he sees nothing of military life but its dangers; he knows nothing but its privations and its discipline—more difficult to support than dangers: he suffers the more from his present miseries, from knowing that the constitution of society and of the army allow him to rise above them; he may, indeed, at any time obtain his commission, and enter at once upon command, honours, independence, rights, and enjoyments. Not only does this object of his hopes appear to him of immense importance, but he is never sure of reaching it till it is actually his own; the grade he fills is by no means irrevocable; he is always entirely abandoned to the arbitrary pleasure of his commanding officer, for this is imperiously required by the necessity of discipline: a slight fault, a whim, may always deprive him in an instant of the fruits of many years of toil and endeavour; until he has reached the grade to which he aspires he has accomplished nothing; not till he reaches that grade does his career seem to begin. A desperate ambition cannot fail to be kindled in a man thus incessantly goaded on by his youth, his wants, his passions, the spirit of his age, his hopes, and his fears. Non-commissioned officers are therefore bent on war,—on war always, and at any cost; but if war be denied them, then they desire revolutions to suspend the authority of established regulations, and to enable them, aided by the general confusion and the political passions of the time, to get rid of their superior officers and to take their places. Nor is it impossible for them to bring about such a crisis, because their common origin and habits give them much influence over the soldiers, however different may be their passions and their desires.

It would be an error to suppose that these various characteristics of officers, non-commissioned officers, and men, belong to any particular time or country; they will always occur at all times, and amongst all democratic nations. In every democratic army the non-commissioned officers will be the worst representatives of the pacific and orderly spirit of the country, and the private soldiers will be the best. The latter will carry with them into military life the strength or weakness of the manners of the nation; they will display a faithful reflection of the community: if that community is ignorant and weak, they will allow themselves to be drawn by their leaders into disturbances, either unconsciously or against their will; if it is enlightened and energetic, the community will itself keep them within the bounds of order.

Chapter XXIV

Causes Which Render Democratic Armies Weaker Than Other Armies at the Outset of a Campaign, and More Formidable in Protracted Warfare

Any army is in danger of being conquered at the outset of a campaign, after a long peace; any army which has long been engaged in warfare has strong chances of victory: this truth is peculiarly applicable to democratic armies. In aristocracies the military profession, being a privileged career, is held in honour even in time of peace. Men of great talents, great attainments, and great ambition embrace it; the army is in all respects on a level with the nation, and frequently above it. We have seen, on the contrary, that amongst a democratic people the choicer minds of the nation are gradually drawn away from the military profession, to seek by other paths, distinction, power, and especially wealth. After a long peace,—and in democratic ages the periods of peace are long,—the army is always inferior to the country itself. In this state it is called into active service; and until war has altered it, there is danger for the country as well as for the army.

I have shown that in democratic armies, and in time of peace, the rule of seniority is the supreme and inflexible law of advancement. This is not only a consequence, as I have before observed, of the constitution of these armies, but of the constitution of the people, and it will always occur. Again, as amongst these nations the officer derives his position in the country solely from his position in the army, and as he draws all the distinction and the competency he enjoys from the same source, he does not retire from his profession, or is not superannuated, till towards the extreme close of life. The consequence of these two causes is, that when a democratic people goes to war after a long interval of peace all the leading officers of the army are old men. I speak not only of the generals, but of the non-commissioned officers, who have most of them been stationary, or have only advanced step by step. It may be remarked with surprise, that in a democratic army after a long peace all the soldiers are mere boys, and all the superior officers in declining years; so that the former are wanting in experience, the latter in vigour. This is a strong element of defeat, for the first condition of successful generalship is youth: I should not have ventured to say so if the greatest captain of modern times had not made the observation.

These two causes do not act in the same manner upon aristocratic armies: as men are promoted in them by right of birth much more than by right of seniority, there are in all ranks a certain number of young men, who bring to their profession all the early vigour of body and mind. Again, as the men who seek for military honours amongst an aristocratic people, enjoy a settled position in civil society, they seldom continue in the army until old-age overtakes them. After having devoted the most vigorous years of youth to the career of arms, they voluntarily retire, and spend at home the remainder of their maturer years.

A long peace not only fills democratic armies with elderly officers, but it also gives to all the officers habits both of body and mind which render them unfit for actual service. The man who has long lived amidst the calm and lukewarm atmosphere of democratic manners can at first ill adapt himself to the harder toils and sterner duties of warfare; and if he has not absolutely lost the taste for arms, at least he has assumed a mode of life which unfits him for conquest.

Amongst aristocratic nations, the ease of civil life exercises less influence on the manners of the army, because amongst those nations the aristocracy commands the army: and an aristocracy, however plunged in luxurious pleasures, has always many other passions besides that of its own well-being, and to satisfy those passions more thoroughly its well-being will be readily sacrificed.[1]

I have shown that in democratic armies, in time of peace, promotion is extremely slow. The officers at first support this state of things with impatience, they grow excited, restless, exasperated, but in the end most of them make up their minds to it. Those who have the largest share of ambition and of resources quit the army; others, adapting their tastes and their desires to their scanty fortunes, ultimately look upon the military profession in a civil point of view. The quality they value most in it is the competency and security which attend it: their whole notion of the future rests upon the certainty of this little provision, and all they require is peaceably to enjoy it. Thus not only does a long peace fill an army with old men, but it frequently imparts the views of old men to those who are still in the prime of life.

I have also shown that amongst democratic nations in time of peace the military profession is held in little honour and indifferently followed. This want of public favour is a heavy discouragement to the army; it weighs down the minds of the troops, and when war breaks out at last, they cannot immediately resume their spring and vigour. No similar cause of moral weakness occurs in aristocratic armies: there the officers are never lowered either in their own eyes or in those of their countrymen, because, independently of their military greatness, they are personally great. But even if the influence of peace operated on the two kinds of armies in the same manner, the results would still be different. When the officers of an aristocratic army have lost their warlike spirit and the desire of raising themselves by service, they still retain a certain respect for the honour of their class, and an old habit of being foremost to set an example. But when the officers of a democratic army have no longer the love of war and the ambition of arms, nothing whatever remains to them.

I am therefore of opinion that, when a democratic people engages in a war after a long peace, it incurs much more risk of defeat than any other nation; but it ought not easily to be cast down by its reverses, for the chances of success for such an army are increased by the duration of the war. When a war has at length, by its long continuance, roused the whole community from their peaceful occupations and ruined their minor undertakings, the same passions which made them attach so much importance to the maintenance of peace will be turned to arms. War, after it has destroyed all modes of speculation, becomes itself the great and sole speculation, to which all the ardent and ambitious desires which equality engenders are exclusively directed. Hence it is that the selfsame democratic nations which are so reluctant to engage in hostilities, sometimes perform prodigious achievements when once they have taken the field. As the war attracts more and more of public attention, and is seen to create high reputations and great fortunes in a short space of time, the choicest spirits of the nation enter the military profession: all the enterprising, proud, and martial minds, no longer of the aristocracy solely, but of the whole country, are drawn in this direction. As the number of competitors for military honours is immense, and war drives every man to his proper level, great generals are always sure to spring up. A long war produces upon a democratic army the same effects that a revolution produces upon a people; it breaks through regulations, and allows extraordinary men to rise above the common level. Those officers whose bodies and minds have grown old in peace, are removed, or superannuated, or they die. In their stead a host of young men are pressing on, whose frames are already hardened, whose desires are extended and inflamed by active service. They are bent on advancement at all hazards, and perpetual advancement; they are followed by others with the same passions and desires, and after these are others yet unlimited by aught but the size of the army. The principle of equality opens the door of ambition to all, and death provides chances for ambition. Death is constantly thinning the ranks, making vacancies, closing and opening the career of arms.

There is moreover a secret connexion between the military character and the character of democracies, which war brings to light. The men of democracies are naturally passionately eager to acquire what they covet, and to enjoy it on easy conditions. They for the most part worship chance, and are much less afraid of death than of difficulty. This is the spirit which they bring to commerce and manufactures; and this same spirit, carried with them to the field of battle, induces them willingly to expose their lives in order to secure in a moment the rewards of victory. No kind of greatness is more pleasing to the imagination of a democratic people than military greatness,—a greatness of vivid and sudden lustre, obtained without toil, by nothing but the risk of life. Thus, whilst the interests and the tastes of the members of a democratic community divert them from war, their habits of mind fit them for carrying on war well; they soon make good soldiers, when they are roused from their business and their enjoyments. If peace is peculiarly hurtful to democratic armies, war secures to them advantages which no other armies ever possess; and these advantages, however little felt at first, cannot fail in the end to give them the victory. An aristocratic nation, which in a contest with a democratic people does not succeed in ruining the latter at the outset of the war, always runs a great risk of being conquered by it.

Chapter XXV

Of Discipline in Democratic Armies

It is a very general opinion, especially in aristocratic countries, that the great social equality which prevails in democracies ultimately renders the private soldier independent of the officer, and thus destroys the bond of discipline. This is a mistake, for there are two kinds of discipline, which it is important not to confound. When the officer is noble and the soldier a serf—one rich, the other poor—the former educated and strong, the latter ignorant and weak,—the strictest bond of obedience may easily be established between the two men. The soldier is broken in to military discipline, as it were, before he enters the army; or rather, military discipline is nothing but an enhancement of social servitude. In aristocratic armies the soldier will soon become insensible to everything but the orders of his superior officers; he acts without reflection, triumphs without enthusiasm, and dies without complaint: in this state he is no longer a man, but he is still a most formidable animal trained for war.

A democratic people must despair of ever obtaining from soldiers that blind, minute, submissive, and invariable obedience which an aristocratic people may impose on them without difficulty. The state of society does not prepare them for it, and the nation might be in danger of losing its natural advantages if it sought artificially to acquire advantages of this particular kind. Amongst democratic communities, military discipline ought not to attempt to annihilate the free spring of the faculties; all that can be done by discipline is to direct it; the obedience thus inculcated is less exact, but it is more eager and more intelligent. It has its root in the will of him who obeys: it rests not only on his instinct, but on his reason; and consequently it will often spontaneously become more strict as danger requires it. The discipline of an aristocratic army is apt to be relaxed in war, because that discipline is founded upon habits, and war disturbs those habits. The discipline of a democratic army on the contrary is strengthened in sight of the enemy, because every soldier then clearly perceives that he must be silent and obedient in order to conquer.

The nations which have performed the greatest warlike achievements knew no other discipline than that which I speak of. Amongst the ancients none were admitted into the armies but freemen and citizens, who differed but little from one another, and were accustomed to treat each other as equals. In this respect it may be said that the armies of antiquity were democratic, although they came out of the bosom of aristocracy; the consequence was that in those armies a sort of fraternal familiarity prevailed between the officers and the men. Plutarch's lives of great commanders furnish convincing instances of the fact: the soldiers were in the constant habit of freely addressing their general, and the general listened to and answered whatever the soldiers had to say: they were kept in order by language and by example, far more than by constraint or punishment; the general was as much their companion as their chief. I know not whether the soldiers of Greece and Rome ever carried the minutiae of military discipline to the same degree of perfection as the Russians have done; but this did not prevent Alexander from conquering Asia—and Rome, the world.

Chapter XXVI

Some Considerations on War in Democratic Communities

When the principle of equality is in growth, not only amongst a single nation, but amongst several neighboring nations at the same time, as is now the case in Europe, the inhabitants of these different countries, notwithstanding the dissimilarity of language, of customs, and of laws, nevertheless resemble each other in their equal dread of war and their common love of peace.[1] It is in vain that ambition or anger puts arms in the hands of princes; they are appeased in spite of themselves by a species of general apathy, and good-will, which makes the sword drop from their grasp, and wars become more rare. As the spread of equality, taking place in several countries at once, simultaneously impels their various inhabitants to follow manufactures and commerce, not only do their tastes grow alike, but their interests are so mixed and entangled with one another that no nation can inflict evils on other nations without those evils falling back upon itself; and all nations ultimately regard war as a calamity, almost as severe to the conqueror as to the conquered. Thus, on the one hand, it is extremely difficult in democratic ages to draw nations into hostilities; but on the other hand, it is almost impossible that any two of them should go to war without embroiling the rest. The interests of all are so interlaced, their opinions and their wants so much alike, that none can remain quiet when the others stir. Wars therefore become more rare, but when they break out they spread over a larger field. Neighbouring democratic nations not only become alike in some respects, but they eventually grow to resemble each other in almost all.[2] This similitude of nations has consequences of great importance in relation to war.

If I inquire why it is that the Helvetic Confederacy made the greatest and most powerful nations of Europe tremble in the fifteenth century, whilst at the present day the power of that country is exactly proportioned to its population, I perceive that the Swiss are become like all the surrounding communities, and those surrounding communities like the Swiss: so that as numerical strength now forms the only difference between them, victory necessarily attends the largest army. Thus one of the consequences of the democratic revolution which is going on in Europe is to make numerical strength preponderate on all fields of battle, and to constrain all small nations to incorporate themselves with large States, or at least to adopt the policy of the latter. As numbers are the determining cause of victory, each people ought of course to strive by all the means in its power to bring the greatest possible number of men into the field. When it was possible to enlist a kind of troops superior to all others, such as the Swiss infantry or the French horse of the sixteenth century, it was not thought necessary to raise very large armies; but the case is altered when one soldier is as efficient as another.

The same cause which begets this new want also supplies means of satisfying it; for, as I have already observed, when men are all alike, they are all weak, and the supreme power of the State is naturally much stronger amongst democratic nations than elsewhere. Hence, whilst these nations are desirous of enrolling the whole male population in the ranks of the army, they have the power of effecting this object: the consequence is, that in democratic ages armies seem to grow larger in proportion as the love of war declines. In the same ages, too, the manner of carrying on war is likewise altered by the same causes. Machiavelli observes in 'The Prince,' 'that it is much more difficult to subdue a people which has a prince and his barons for its leaders, than a nation which is commanded by a prince and his slaves.' To avoid offence, let us read public functionaries for slaves, and this important truth will be strictly applicable to our own time.

A great aristocratic people cannot either conquer its neighbours, or be conquered by them, without great difficulty. It cannot conquer them, because all its forces can never be collected and held together for a considerable period: it cannot be conquered, because an enemy meets at every step small centres of resistance by which invasion is arrested. War against an aristocracy may be compared to war in a mountainous country; the defeated party has constant opportunities of rallying its forces to make a stand in a new position. Exactly the reverse occurs amongst democratic nations: they easily bring their whole disposable force into the field, and when the nation is wealthy and populous it soon becomes victorious; but if ever it is conquered, and its territory invaded, it has few resources at command; and if the enemy takes the capital, the nation is lost. This may very well be explained: as each member of the community is individually isolated and extremely powerless, no one of the whole body can either defend himself or present a rallying point to others. Nothing is strong in a democratic country except the State; as the military strength of the State is destroyed by the destruction of the army, and its civil power paralysed by the capture of the chief city, all that remains is only a multitude without strength or government, unable to resist the organised power by which it is assailed. I am aware that this danger may be lessened by the creation of provincial liberties, and consequently of provincial powers, but this remedy will always be insufficient. For after such a catastrophe, not only is the population unable to carry on hostilities, but it may be apprehended that they will not be inclined to attempt it.

In accordance with the law of nations adopted in civilised countries, the object of wars is not to seize the property of private individuals, but simply to get possession of political power. The destruction of private property is only occasionally resorted to for the purpose of attaining the latter object. When an aristocratic country is invaded after the defeat of its army, the nobles, although they are at the same time the wealthiest members of the community, will continue to defend themselves individually rather than submit; for if the conqueror remained master of the country, he would deprive them of their political power, to which they cling even more closely than to their property. They therefore prefer fighting to subjection, which is to them the greatest of all misfortunes; and they readily carry the people along with them because the people has long been used to follow and obey them, and besides has but little to risk in the war. Amongst a nation in which equality of conditions prevails, each citizen, on the contrary, has but slender share of political power, and often has no share at all; on the other hand, all are independent, and all have something to lose; so that they are much less afraid of being conquered, and much more afraid of war, than an aristocratic people. It will always be extremely difficult to decide a democratic population to take up arms, when hostilities have reached its own territory. Hence the necessity of giving to such a people the rights and the political character which may impart to every citizen some of those interests that cause the nobles to act for the public welfare in aristocratic countries.

It should never be forgotten by the princes and other leaders of democratic nations, that nothing but the passion and the habit of freedom can maintain an advantageous contest with the passion and the habit of physical well-being. I can conceive nothing better prepared for subjection, in case of defeat, than a democratic people without free institutions.

Formerly it was customary to take the field with a small body of troops, to fight in small engagements, and to make long, regular sieges: modern tactics consist in fighting decisive battles, and, as soon as a line of march is open before the army, in rushing upon the capital city, in order to terminate the war at a single blow. Napoleon, it is said, was the inventor of this new system; but the invention of such a system did not depend on any individual man, whoever he might be. The mode in which Napoleon carried on war was suggested to him by the state of society in his time; that mode was successful, because it was eminently adapted to that state of society, and because he was the first to employ it. Napoleon was the first commander who marched at the head of an army from capital to capital, but the road was opened for him by the ruin of feudal society. It may fairly be believed that, if that extraordinary man had been born three hundred years ago, he would not have derived the same results from his method of warfare, or, rather, that he would have had a different method.

I shall add but a few words on civil wars, for fear of exhausting the patience of the reader. Most of the remarks which I have made respecting foreign wars are applicable *à fortiori* to civil wars. Men living in democracies are not naturally prone to the military character; they sometimes assume it, when they have been dragged by compulsion to the field; but to rise in a body and voluntarily to expose themselves to the horrors of war, and especially of civil war, is a course which the men of democracies are not apt to adopt. None but the most adventurous members of the community consent to run into such risks; the bulk of the population remains motionless. But even if the population were inclined to act, considerable obstacles would stand in their way; for they can resort to no old and well-established influence which they are willing to obey—no well-known leaders to rally the discontented, as well as to discipline and to lead them—no political powers subordinate to the supreme power of the nation, which afford an effectual support to the resistance directed against the government. In democratic countries the moral power of the majority is immense, and the physical resources which it has at its command are out of all proportion to the physical resources which may be combined against it. Therefore the party which occupies the seat of the majority, which speaks in its name and wields its power, triumphs instantaneously and irresistibly over all private resistance; it does not even give such opposition time to exist, but nips it in the bud. Those who in such nations seek to effect a revolution by force of arms have no other resource than suddenly to seize upon the whole engine of government as it stands, which can better be done by a single blow than by a war; for as soon as there is a regular war, the party which represents the State is always certain to conquer. The only case in which a civil war could arise is, if the army should divide itself into two factions, the one raising the standard of rebellion, the other remaining true to its allegiance. An army constitutes a small community, very closely united together, endowed with great powers of vitality, and able to supply its own wants for some time. Such a war might be bloody, but it could not be long; for either the rebellious army would gain over the government by the sole display of its resources, or by its first victory, and then the war would be over; or the struggle would take place, and then that portion of the army which should not be supported by the organised powers of the State would speedily either disband itself or be destroyed. It may therefore be admitted as a general truth, that in ages of equality civil wars will become much less frequent and less protracted.[3]

Second Part

Fourth Book

Influence of Democratic Opinions and Sentiments on Political Society

I should imperfectly fulfil the purpose of this book, if, after having shown what opinions and sentiments are suggested by the principle of equality, I did not point out, ere I conclude, the general influence which these same opinions and sentiments may exercise upon the government of human societies. To succeed in this object I shall frequently have to retrace my steps; but I trust the reader will not refuse to follow me through paths already known to him, which may lead to some new truth.

Chapter I

That Equality Naturally Gives Men a Taste for Free Institutions

The principle of equality, which makes men independent of each other, gives them a habit and a taste for following, in their private actions, no other guide but their own will. This complete independence, which they constantly enjoy towards their equals and in the intercourse of private life, tends to make them look upon all authority with a jealous eye, and speedily suggests to them the notion and the love of political freedom. Men living at such times have a natural bias to free institutions. Take any one of them at a venture, and search if you can his most deep-seated instincts; you will find that of all governments he will soonest conceive and most highly value that government, whose head he has himself elected, and whose administration he may control. Of all the political effects produced by the equality of conditions, this love of independence is the first to strike the observing, and to alarm the timid; nor can it be said that their alarm is wholly misplaced, for anarchy has a more formidable aspect in democratic countries than elsewhere. As the citizens have no direct influence on each other, as soon as the supreme power of the nation fails, which kept them all in their several stations, it would seem that disorder must instantly reach its utmost pitch, and that, every man drawing aside in a different direction, the fabric of society must at once crumble away.

I am however persuaded that anarchy is not the principal evil which democratic ages have to fear, but the least. For the principle of equality begets two tendencies; the one leads men straight to independence, and may suddenly drive them into anarchy; the other conducts them by a longer, more secret, but more certain road, to servitude. Nations readily discern the former tendency, and are prepared to resist it; they are led away by the latter, without perceiving its drift; hence it is peculiarly important to point it out. For myself, I am so far from urging as a reproach to the principle of equality that it renders men untractable, that this very circumstance principally calls forth my approbation. I admire to see how it deposits in the mind and heart of man the dim conception and instinctive love of political independence, thus preparing the remedy for the evil which it engenders; it is on this very account that I am attached to it.

Chapter II

That the Notions of Democratic Nations on Government Are Naturally Favourable to the Concentration of Power

The notion of secondary powers, placed between the sovereign and his subjects, occurred naturally to the imagination of aristocratic nations, because those communities contained individuals or families raised above the common level, and apparently destined to command by their birth, their education, and their wealth. This same notion is naturally wanting in the minds of men in democratic ages, for converse reasons: it can only be introduced artificially, it can only be kept there with difficulty; whereas they conceive, as it were without thinking upon the subject, the notion of a sole and central power which governs the whole community by its direct influence. Moreover in politics, as well as in philosophy and in religion, the intellect of democratic nations is peculiarly open to simple and general notions. Complicated systems are repugnant to it, and its favourite conception is that of a great nation composed of citizens all resembling the same pattern, and all governed by a single power.

The very next notion to that of a sole and central power, which presents itself to the minds of men in the ages of equality, is the notion of uniformity of legislation. As every man sees that he differs but little from those about him, he cannot understand why a rule which is applicable to one man should not be equally applicable to all others. Hence the slightest privileges are repugnant to his reason; the faintest dissimilarities in the political institutions of the same people offend him, and uniformity of legislation appears to him to be the first condition of good government. I find, on the contrary, that this same notion of a uniform rule, equally binding on all the members of the community, was almost unknown to the human mind in aristocratic ages; it was either never entertained, or it was rejected. These contrary tendencies of opinion ultimately turn on either side to such blind instincts and such ungovernable habits that they still direct the actions of men, in spite of particular exceptions. Notwithstanding the immense variety of conditions in the middle ages, a certain number of persons existed at that period in precisely similar circumstances; but this did not prevent the laws then in force from assigning to each of them distinct duties and different rights. On the contrary, at the present time all the powers of government are exerted to impose the same customs and the same laws on populations which have as yet but few points of resemblance. As the conditions of men become equal amongst a people, individuals seem of less importance, and society of greater dimensions; or rather, every citizen, being assimilated to all the rest, is lost in the crowd, and nothing stands conspicuous but the great and imposing image of the people at large. This naturally gives the men of democratic periods a lofty opinion of the privileges of society, and a very humble notion of the rights of individuals; they are ready to admit that the interests of the former are everything, and those of the latter nothing. They are willing to acknowledge that the power which represents the community has far more information and wisdom than any of the members of that community; and that it is the duty, as well as the right, of that power to guide as well as govern each private citizen.

If we closely scrutinise our contemporaries, and penetrate to the root of their political opinions, we shall detect some of the notions which I have just pointed out, and we shall perhaps be surprised to find so much accordance between men who are so often at variance. The Americans hold, that in every state the supreme power ought to emanate from the people; but when once that power is constituted, they can conceive, as it were, no limits to it, and they are ready to admit that it has the right to do whatever it pleases. They have not the slightest notion of peculiar privileges granted to cities, families or persons: their minds appear never to have foreseen that it might be possible not to apply with strict uniformity the same laws to every part, and to all the inhabitants. These same opinions are more and more diffused in Europe; they even insinuate themselves amongst those nations which most vehemently reject the principle of the sovereignty of the people. Such nations assign a different origin to the supreme power, but they ascribe to that power the same characteristics. Amongst them all, the idea of intermediate powers is weakened and obliterated: the idea of rights inherent in certain individuals is rapidly disappearing from the minds of men; the idea of the omnipotence and sole authority of society at large rises to fill its place. These ideas take root and spread in proportion as social conditions become more equal, and men more alike; they are engendered by equality, and in turn they hasten the progress of equality.

In France, where the revolution of which I am speaking has gone further than in any other European country, these opinions have got complete hold of the public mind. If we listen attentively to the language of the various parties in France, we shall find that there is not one which has not adopted them. Most of these parties censure the conduct of the government, but they all hold that the government ought perpetually to act and interfere in everything that is done. Even those which are most at variance are nevertheless agreed upon this head. The unity, the ubiquity, the omnipotence of the supreme power, and the uniformity of its rules, constitute the principal characteristics of all the political systems which have been put forward in our age. They recur even in the wildest visions of political regeneration: the human mind pursues them in its dreams. If these notions spontaneously arise in the minds of private individuals, they suggest themselves still more forcibly to the minds of princes. Whilst the ancient fabric of European society is altered and dissolved, sovereigns acquire new conceptions of their opportunities and their duties; they learn for the first time that the central power which they represent may and ought to administer by its own agency, and on a uniform plan, all the concerns of the whole community. This opinion, which, I will venture to say, was never conceived before our time by the monarchs of Europe, now sinks deeply into the minds of kings, and abides there amidst all the agitation of more unsettled thoughts.

Our contemporaries are therefore much less divided than is commonly supposed; they are constantly disputing as to the hands in which supremacy is to be vested, but they readily agree upon the duties and the rights of that supremacy. The notion they all form of government is that of a sole, simple, providential, and creative power. All secondary opinions in politics are unsettled; this one remains fixed, invariable and consistent. It is adopted by statesmen and political philosophers; it is eagerly laid hold of by the multitude; those who govern and those who are governed agree to pursue it with equal ardour: it is the foremost notion of their minds, it seems connatural with their feelings. It originates therefore in no caprice of the human intellect, but it is a necessary condition of the present state of mankind.

Chapter III

That the Sentiments of Democratic Nations Accord with Their Opinions in Leading Them to Concentrate Political Power

If it be true that, in ages of equality, men readily adopt the notion of a great central power, it cannot be doubted on the other hand that their habits and sentiments predispose them to recognise such a power and to give it their support. This may be demonstrated in a few words, as the greater part of the reasons, to which the fact may be attributed, have been previously stated.[1] As the men who inhabit democratic countries have no superiors, no inferiors, and no habitual or necessary partners in their undertakings, they readily fall back upon themselves and consider themselves as beings apart. I had occasion to point this out at considerable length in treating of individualism. Hence such men can never, without an effort, tear themselves from their private affairs to engage in public business; their natural bias leads them to abandon the latter to the sole visible and permanent representative of the interests of the community, that is to say, to the State. Not only are they naturally wanting in a taste for public business, but they have frequently no time to attend to it. Private life is so busy in democratic periods, so excited, so full of wishes and of work, that hardly any energy or leisure remains to each individual for public life. I am the last man to contend that these propensities are unconquerable, since my chief object in writing this book has been to combat them. I only maintain that at the present day a secret power is fostering them in the human heart, and that if they are not checked they will wholly overgrow it.

I have also had occasion to show how the increasing love of well-being, and the fluctuating character of property, cause democratic nations to dread all violent disturbance. The love of public tranquillity is frequently the only passion which these nations retain, and it becomes more active and powerful amongst them in proportion as all other passions droop and die. This naturally disposes the members of the community constantly to give or to surrender additional rights to the central power, which alone seems to be interested in defending them by the same means that it uses to defend itself. As in ages of equality no man is compelled to lend his assistance to his fellow-men, and none has any right to expect much support from them, every one is at once independent and powerless. These two conditions, which must never be either separately considered or confounded together, inspire the citizen of a democratic country with very contrary propensities. His independence fills him with self-reliance and pride amongst his equals; his debility makes him feel from time to time the want of some outward assistance, which he cannot expect from any of them, because they are all impotent and unsympathizing. In this predicament he naturally turns his eyes to that imposing power which alone rises above the level of universal depression. Of that power his wants and especially his desires continually remind him, until he ultimately views it as the sole and necessary support of his own weakness.[2] This may more completely explain what frequently takes place in democratic countries, where the very men who are so impatient of superiors patiently submit to a master, exhibiting at once their pride and their servility.

The hatred which men bear to privilege increases in proportion as privileges become more scarce and less considerable, so that democratic passions would seem to burn most fiercely at the very time when they have least fuel. I have already given the reason of this phenomenon. When all conditions are unequal, no inequality is so great as to offend the eye; whereas the slightest dissimilarity is odious in the midst of general uniformity: the more complete is this uniformity, the more insupportable does the sight of such a difference become. Hence it is natural that the love of equality should constantly increase together with equality itself, and that it should grow by what it feeds upon. This never-dying, ever-kindling hatred, which sets a democratic people against the smallest privileges, is peculiarly favourable to the gradual concentration of all political rights in the hands of the representative of the State alone. The sovereign, being necessarily and incontestably above all the citizens, excites not their envy, and each of them thinks that he strips his equals of the prerogative which he concedes to the crown. The man of a democratic age is extremely reluctant to obey his neighbour who is his equal; he refuses to acknowledge in such a person ability superior to his own; he mistrusts his justice, and is jealous of his power; he fears and he contemns him; and he loves continually to remind him of the common dependence in which both of them stand to the same master. Every central power which follows its natural tendencies courts and encourages the principle of equality; for equality singularly facilitates, extends, and secures the influence of a central power.

In like manner it may be said that every central government worships uniformity: uniformity relieves it from inquiry into an infinite number of small details which must be attended to if rules were to be adapted to men, instead of indiscriminately subjecting men to rules: thus the government likes what the citizens like, and naturally hates what they hate. These common sentiments, which, in democratic nations, constantly unite the sovereign and every member of the community in one and the same conviction, establish a secret and lasting sympathy between them. The faults of the government are pardoned for the sake of its tastes; public confidence is only reluctantly withdrawn in the midst even of its excesses and its errors, and it is restored at the first call. Democratic nations often hate those in whose hands the central power is vested; but they always love that power itself.

Thus, by two separate paths, I have reached the same conclusion. I have shown that the principle of equality suggests to men the notion of a sole, uniform, and strong government: I have now shown that the principle of equality imparts to them a taste for it. To governments of this kind the nations of our age are therefore tending. They are drawn thither by the natural inclination of mind and heart; and in order to reach that result, it is enough that they do not check themselves in their course. I am of opinion, that, in the democratic ages which are opening upon us, individual independence and local liberties will ever be the produce of artificial contrivance; that centralization will be the natural form of government.[3]

Chapter IV

Of Certain Peculiar and Accidental Causes Which Either Lead a People to Complete Centralization of Government, or Which Divert Them from It

If all democratic nations are instinctively led to the centralization of government, they tend to this result in an unequal manner. This depends on the particular circumstances which may promote or prevent the natural consequences of that state of society—circumstances which are exceedingly numerous; but I shall only advert to a few of them. Amongst men who have lived free long before they became equal, the tendencies derived from free institutions combat, to a certain extent, the propensities superinduced by the principle of equality; and although the central power may increase its privileges amongst such a people, the private members of such a community will never entirely forfeit their independence. But when the equality of conditions grows up amongst a people which has never known, or has long ceased to know, what freedom is (and such is the case upon the continent of Europe), as the former habits of the nation are suddenly combined, by some sort of natural attraction, with the novel habits and principles engendered by the state of society, all powers seem spontaneously to rush to the centre. These powers accumulate there with astonishing rapidity, and the State instantly attains the utmost limits of its strength, whilst private persons allow themselves to sink as suddenly to the lowest degree of weakness.

The English who emigrated three hundred years ago to found a democratic commonwealth on the shores of the New World, had all learned to take a part in public affairs in their mother country; they were conversant with trial by jury; they were accustomed to liberty of speech and of the press—to personal freedom, to the notion of rights and the practice of asserting them. They carried with them to America these free institutions and manly customs, and these institutions preserved them against the encroachments of the State. Thus amongst the Americans it is freedom which is old—equality is of comparatively modern date. The reverse is occurring in Europe, where equality, introduced by absolute power and under the rule of kings, was already infused into the habits of nations long before freedom had entered into their conceptions.

I have said that amongst democratic nations the notion of government naturally presents itself to the mind under the form of a sole and central power, and that the notion of intermediate powers is not familiar to them. This is peculiarly applicable to the democratic nations which have witnessed the triumph of the principle of equality by means of a violent revolution. As the classes which managed local affairs have been suddenly swept away by the storm, and as the confused mass which remains has as yet neither the organisation nor the habits which fit it to assume the administration of these same affairs, the State alone seems capable of taking upon itself all the details of government, and centralization becomes, as it were, the unavoidable state of the country. Napoleon deserves neither praise nor censure for having centred in his own hands almost all the administrative power of France; for, after the abrupt disappearance of the nobility and the higher rank of the middle classes, these powers devolved on him of course: it would have been almost as difficult for him to reject as to assume them. But no necessity of this kind has ever been felt by the Americans, who, having passed through no revolution, and having governed themselves from the first, never had to call upon the State to act for a time as their guardian. Thus the progress of centralization amongst a democratic people depends not only on the progress of equality, but on the manner in which this equality has been established.

At the commencement of a great democratic revolution, when hostilities have but just broken out between the different classes of society, the people endeavours to centralize the public administration in the hands of the government, in order to wrest the management of local affairs from the aristocracy. Towards the close of such a revolution, on the contrary, it is usually the conquered aristocracy that endeavours to make over the management of all affairs to the State, because such an aristocracy dreads the tyranny of a people which has become its equal, and not unfrequently its master. Thus it is not always the same class of the community which strives to increase the prerogative of the government; but as long as the democratic revolution lasts there is always one class in the nation, powerful in numbers or in wealth, which is induced, by peculiar passions or interests, to centralize the public administration, independently of that hatred of being governed by one's neighbour, which is a general and permanent feeling amongst democratic nations. It may be remarked, that at the present day the lower orders in England are striving with all their might to destroy local independence, and to transfer the administration from all points of the circumference to the centre; whereas the higher classes are endeavouring to retain this administration within its ancient boundaries. I venture to predict that a time will come when the very reverse will happen.

These observations explain why the supreme power is always stronger, and private individuals weaker, amongst a democratic people which has passed through a long and arduous struggle to reach a state of equality than amongst a democratic community in which the citizens have been equal from the first. The example of the Americans completely demonstrates the fact. The inhabitants of the United States were never divided by any privileges; they have never known the mutual relation of master and inferior, and as they neither dread nor hate each other, they have never known the necessity of calling in the supreme power to manage their affairs. The lot of the Americans is singular: they have derived from the aristocracy of England the notion of private rights and the taste for local freedom; and they have been able to retain both the one and the other, because they have had no aristocracy to combat.

If at all times education enables men to defend their independence, this is most especially true in democratic ages. When all men are alike, it is easy to found a sole and all-powerful government, by the aid of mere instinct. But men require much intelligence, knowledge, and art to organise and to maintain secondary powers under similar circumstances, and to create amidst the independence and individual weakness of the citizens such free associations as may be in a condition to struggle against tyranny without destroying public order.

Hence the concentration of power and the subjection of individuals will increase amongst democratic nations, not only in the same proportion as their equality, but in the same proportion as their ignorance. It is true, that in ages of imperfect civilisation the government is frequently as wanting in the knowledge required to impose a despotism upon the people as the people are wanting in the knowledge required to shake it off; but the effect is not the same on both sides. However rude a democratic people may be, the central power which rules it is never completely devoid of cultivation, because it readily draws to its own uses what little cultivation is to be found in the country, and, if necessary, may seek assistance elsewhere. Hence, amongst a nation which is ignorant as well as democratic, an amazing difference cannot fail speedily to arise between the intellectual capacity of the ruler and that of each of his subjects. This completes the easy concentration of all power in his hands: the administrative function of the State is perpetually extended, because the State alone is competent to administer the affairs of the country. Aristocratic nations, however unenlightened they may be, never afford the same spectacle, because in them instruction is nearly equally diffused between the monarch and the leading members of the community.

The Pasha who now rules in Egypt, found the population of that country composed of men exceedingly ignorant and equal, and he has borrowed the science and ability of Europe to govern that people. As the personal attainments of the sovereign are thus combined with the ignorance and democratic weakness of his subjects, the utmost centralisation has been established without impediment, and the Pasha has made the country his manufactory, and the inhabitants his workmen.

I think that extreme centralisation of government ultimately enervates society, and thus after a length of time weakens the government itself; but I do not deny that a centralised social power may be able to execute great undertakings with facility in a given time and on a particular point. This is more especially true of war, in which success depends much more on the means of transferring all the resources of a nation to one single point, than on the extent of those resources. Hence it is chiefly in war that nations desire and frequently require to increase the powers of the central government. All men of military genius are fond of centralisation, which increases their strength; and all men of centralising genius are fond of war, which compels nations to combine all their powers in the hands of the government. Thus the democratic tendency which leads men unceasingly to multiply the privileges of the State, and to circumscribe the rights of private persons, is much more rapid and constant amongst those democratic nations which are exposed by their position to great and frequent wars, than amongst all others.

I have shown how the dread of disturbance and the love of well-being insensibly lead democratic nations to increase the functions of central government, as the only power which appears to be intrinsically sufficiently strong, enlightened, and secure, to protect them from anarchy. I would now add, that all the particular circumstances which tend to make the state of a democratic community agitated and precarious, enhance this general propensity, and lead private persons more and more to sacrifice their rights to their tranquility. A people is therefore never so disposed to increase the functions of central government as at the close of a long and bloody revolution, which, after having wrested property from the hands of its former possessors, has shaken all belief, and filled the nation with fierce hatreds, conflicting interests, and contending factions. The love of public tranquillity becomes at such times an indiscriminating passion, and the members of the community are apt to conceive a most inordinate devotion to order.

I have already examined several of the incidents which may concur to promote the centralisation of power, but the principal cause still remains to be noticed. The foremost of the incidental causes which may draw the management of all affairs into the hands of the ruler in democratic countries, is the origin of that ruler himself, and his own propensities. Men who live in the ages of equality are naturally fond of central power, and are willing to extend its privileges; but if it happens that this same power faithfully represents their own interests, and exactly copies their own inclinations, the confidence they place in it knows no bounds, and they think that whatever they bestow upon it is bestowed upon themselves.

The attraction of administrative powers to the centre will always be less easy and less rapid under the reign of kings who are still in some way connected with the old aristocratic order, than under new princes, the children of their own achievements, whose birth, prejudices, propensities, and habits appear to bind them indissolubly to the cause of equality. I do not mean that princes of aristocratic origin who live in democratic ages do not attempt to centralise; I believe they apply themselves to that object as diligently as any others. For them, the sole advantages of equality lie in that direction; but their opportunities are less great, because the community, instead of volunteering compliance with their desires, frequently obeys them with reluctance. In democratic communities the rule is that centralisation must increase in proportion as the sovereign is less aristocratic. When an ancient race of kings stands at the head of an aristocracy, as the natural prejudices of the sovereign perfectly accord with the natural prejudices of the nobility, the vices inherent in aristocratic communities have a free course, and meet with no corrective. The reverse is the case when the scion of a feudal stock is placed at the head of a democratic people. The sovereign is constantly led, by his education, his habits, and his associations, to adopt sentiments suggested by the inequality of conditions, and the people tend as constantly, by their social condition, to those manners which are engendered by equality. At such times it often happens that the citizens seek to control the central power far less as a tyrannical than as an aristocratical power, and that they persist in the firm defence of their independence, not only because they would remain free, but especially because they are determined to remain equal. A revolution which overthrows an ancient regal family, in order to place men of more recent growth at the head of a democratic people, may temporarily weaken the central power; but however anarchical such a revolution may appear at first, we need not hesitate to predict that its final and certain consequence will be to extend and to secure the prerogatives of that power. The foremost or indeed the sole condition which is required in order to succeed in centralising the supreme power in a democratic community, is to love equality, or to get men to believe you love it. Thus the science of despotism, which was once so complex, is simplified, and reduced as it were to a single principle.

CHAPTER V

That amongst the European Nations of Our Time the Power of Governments Is Increasing, although the Persons Who Govern Are Less Stable

On reflecting upon what has already been said, the reader will be startled and alarmed to find that in Europe everything seems to conduce to the indefinite extension of the prerogatives of government, and to render all that enjoyed the rights of private independence more weak, more subordinate, and more precarious. The democratic nations of Europe have all the general and permanent tendencies which urge the Americans to the centralisation of government, and they are moreover exposed to a number of secondary and incidental causes with which the Americans are unacquainted. It would seem as if every step they make towards equality brings them nearer to despotism. And indeed if we do but cast our looks around, we shall be convinced that such is the fact. During the aristocratic ages which preceded the present time, the sovereigns of Europe had been deprived of, or had relinquished, many of the rights inherent in their power. Not a hundred years ago, amongst the greater part of European nations, numerous private persons and corporations were sufficiently independent to administer justice, to raise and maintain troops, to levy taxes, and frequently even to make or interpret the law. The State has everywhere resumed to itself alone these natural attributes of sovereign power; in all matters of government the State tolerates no intermediate agent between itself and the people, and in general business it directs the people by its own immediate influence. I am far from blaming this concentration of power, I simply point it out.

At the same period a great number of secondary powers existed in Europe, which represented local interests and administered local affairs. Most of these local authorities have already disappeared; all are speedily tending to disappear, or to fall into the most complete dependence. From one end of Europe to the other the privileges of the nobility, the liberties of cities, and the powers of provincial bodies, are either destroyed or upon the verge of destruction. Europe has endured, in the course of the last half-century, many revolutions and counter-revolutions which have agitated it in opposite directions: but all these perturbations resemble each other in one respect—they have all shaken or destroyed the secondary powers of government. The local privileges which the French did not abolish in the countries they conquered, have finally succumbed to the policy of the princes who conquered the French. Those princes rejected all the innovations of the French Revolution except centralisation: that is the only principle they consented to receive from such a source. My object is to remark, that all these various rights, which have been successively wrested, in our time, from classes, corporations, and individuals, have not served to raise new secondary powers on a more democratic basis, but have uniformly been concentrated in the hands of the sovereign. Everywhere the State acquires more and more direct control over the humblest members of the community, and a more exclusive power of governing each of them in his smallest concerns.[1] Almost all the charitable establishments of Europe were formerly in the hands of private persons or of corporations; they are now almost all dependent on the supreme government, and in many countries are actually administered by that power. The State almost exclusively undertakes to supply bread to the hungry, assistance and shelter to the sick, work to the idle, and to act as the sole reliever of all kinds of misery. Education, as well as charity, is become in most countries at the present day a national concern. The State receives, and often takes, the child from the arms of the mother, to hand it over to official agents: the State undertakes to train the heart and to instruct the mind of each generation. Uniformity prevails in the courses of public instruction as in everything else; diversity, as well as freedom, is disappearing day by day. Nor do I hesitate to affirm, that amongst almost all the Christian nations of our days, Catholic as well as Protestant, religion is in danger of falling into the hands of the government. Not that rulers are over-jealous of the right of settling points of doctrine, but they get more and more hold upon the will of those by whom doctrines are expounded; they deprive the clergy of their property, and pay them by salaries; they divert to their own use the influence of the priesthood, they make them their own ministers—often their own servants—and by this alliance with religion they reach the inner depths of the soul of man.[2]

But this is as yet only one side of the picture. The authority of government has not only spread, as we have just seen, throughout the sphere of all existing powers, till that sphere can no longer contain it, but it goes further, and invades the domain heretofore reserved to private independence. A multitude of actions, which were formerly entirely beyond the control of the public administration, have been subjected to that control in our time, and the number of them is constantly increasing. Amongst aristocratic nations the supreme government usually contented itself with managing and superintending the community in whatever directly and ostensibly concerned the national honour; but in all other respects the people were left to work out their own free will. Amongst these nations the government often seemed to forget that there is a point at which the faults and the sufferings of private persons involve the general prosperity, and that to prevent the ruin of a private individual must sometimes be a matter of public importance. The democratic nations of our time lean to the opposite extreme. It is evident that most of our rulers will not content themselves with governing the people collectively: it would seem as if they thought themselves responsible for the actions and private condition of their subjects—as if they had undertaken to guide and to instruct each of them in the various incidents of life, and to secure their happiness quite independently of their own consent. On the other hand private individuals grow more and more apt to look upon the supreme power in the same light; they invoke its assistance in all their necessities, and they fix their eyes upon the administration as their mentor or their guide.

I assert that there is no country in Europe in which the public administration has not become, not only more centralised, but more inquisitive and more minute: it everywhere interferes in private concerns more than it did; it regulates more undertakings, and undertakings of a lesser kind; and it gains a firmer footing every day about, above, and around all private persons, to assist, to advise, and to coerce them. Formerly a sovereign lived upon the income of his lands, or the revenue of his taxes; this is no longer the case now that his wants have increased as well as his power. Under the same circumstances which formerly compelled a prince to put on a new tax, he now has recourse to a loan. Thus the State gradually becomes the debtor of most of the wealthier members of the community, and centralises the largest amounts of capital in its own hands. Small capital is drawn into its keeping by another method. As men are intermingled and conditions become more equal, the poor have more resources, more education, and more desires; they conceive the notion of bettering their condition, and this teaches them to save. These savings are daily producing an infinite number of small capitals, the slow and gradual produce of labour, which are always increasing. But the greater part of this money would be unproductive if it remained scattered in the hands of its owners. This circumstance has given rise to a philanthropic institution, which will soon become, if I am not mistaken, one of our most important political institutions. Some charitable persons conceived the notion of collecting the savings of the poor and placing them out at interest. In some countries these benevolent associations are still completely distinct from the State; but in almost all they manifestly tend to identify themselves with the government; and in some of them the government has superseded them, taking upon itself the enormous task of centralising in one place, and putting out at interest on its own responsibility, the daily savings of many millions of the working classes. Thus the State draws to itself the wealth of the rich by loans, and has the poor man's mite at its disposal in the savings-banks. The wealth of the country is perpetually flowing around the government and passing through its hands; the accumulation increases in the same proportion as the equality of conditions; for in a democratic country the State alone inspires private individuals with confidence, because the State alone appears to be endowed with strength and durability.[3] Thus the sovereign does not confine himself to the management of the public treasury; he interferes in private money-matters; he is the superior, and often the master, of all the members of the community; and, in addition to this, he assumes the part of their steward and paymaster.

The central power not only fulfils of itself the whole of the duties formerly discharged by various authorities—extending those duties, and surpassing those authorities—but it performs them with more alertness, strength, and independence than it displayed before. All the governments of Europe have in our time singularly improved the science of administration: they do more things, and they do everything with more order, more celerity, and at less expense; they seem to be constantly enriched by all the experience of which they have stripped private persons. From day to day the princes of Europe hold their subordinate officers under stricter control, and they invent new methods for guiding them more closely, and inspecting them with less trouble. Not content with managing everything by their agents, they undertake to manage the conduct of their agents in everything; so that the public administration not only depends upon one and the same power, but it is more and more confined to one spot and concentrated in the same hands. The government centralises its agency whilst it increases its prerogative—hence a twofold increase of strength.

In examining the ancient constitution of the judicial power, amongst most European nations, two things strike the mind—the independence of that power, and the extent of its functions. Not only did the courts of justice decide almost all differences between private persons, but in very many cases they acted as arbiters between private persons and the State. I do not here allude to the political and administrative offices which courts of judicature had in some countries usurped, but the judicial office common to them all. In most of the countries of Europe, there were, and there still are, many private rights, connected for the most part with the general right of property, which stood under the protection of the courts of justice, and which the State could not violate without their sanction. It was this semi-political power which mainly distinguished the European courts of judicature from all others; for all nations have had judges, but all have not invested their judges with the same privileges. Upon examining what is now occurring amongst the democratic nations of Europe which are called free, as well as amongst the others, it will be observed that new and more dependent courts are everywhere springing up by the side of the old ones, for the express purpose of deciding, by an extraordinary jurisdiction, such litigated matters as may arise between the government and private persons. The elder judicial power retains its independence, but its jurisdiction is narrowed; and there is a growing tendency to reduce it to be exclusively the arbiter between private interests. The number of these special courts of justice is continually increasing, and their functions increase likewise. Thus the government is more and more absolved from the necessity of subjecting its policy and its rights to the sanction of another power. As judges cannot be dispensed with, at least the State is to select them, and always to hold them under its control; so that, between the government and private individuals, they place the effigy of justice rather than justice itself. The State is not satisfied with drawing all concerns to itself, but it acquires an ever-increasing power of deciding on them all without restriction and without appeal.[4]

There exists amongst the modern nations of Europe one great cause, independent of all those which have already been pointed out, which perpetually contributes to extend the agency or to strengthen the prerogative of the supreme power, though it has not been sufficiently attended to: I mean the growth of manufactures, which is fostered by the progress of social equality. Manufactures generally collect a multitude of men of the same spot, amongst whom new and complex relations spring up. These men are exposed by their calling to great and sudden alternations of plenty and want, during which public tranquillity is endangered. It may also happen that these employments sacrifice the health, and even the life, of those who gain by them, or of those who live by them. Thus the manufacturing classes require more regulation, superintendence, and restraint than the other classes of society, and it is natural that the powers of government should increase in the same proportion as those classes.

This is a truth of general application; what follows more especially concerns the nations of Europe. In the centuries which preceded that in which we live, the aristocracy was in possession of the soil, and was competent to defend it: landed property was therefore surrounded by ample securities, and its possessors enjoyed great independence. This gave rise to laws and customs which have been perpetuated, notwithstanding the subdivision of lands and the ruin of the nobility; and, at the present time, landowners and agriculturists are still those amongst the community who most easily escape from the control of the supreme power. In these same aristocratic ages, in which all the sources of our history are to be traced, personal property was of small importance, and those who possessed it were despised and weak: the manufacturing class formed an exception in the midst of those aristocratic communities; as it had no certain patronage, it was not outwardly protected, and was often unable to protect itself. Hence a habit sprung up of considering manufacturing property as something of a peculiar nature, not entitled to the same deference, and not worthy of the same securities as property in general; and manufacturers were looked upon as a small class in the bulk of the people, whose independence was of small importance, and who might with propriety be abandoned to the disciplinary passions of princes. On glancing over the codes of the middle ages, one is surprised to see, in those periods of personal independence, with what incessant royal regulations manufactures were hampered, even in their smallest details: on this point centralisation was as active and as minute as it can ever be. Since that time a great revolution has taken place in the world; manufacturing property, which was then only in the germ, has spread till it covers Europe: the manufacturing class has been multiplied and enriched by the remnants of all other ranks; it has grown and is still perpetually growing in number, in importance, in wealth. Almost all those who do not belong to it are connected with it at least on some one point; after having been an exception in society, it threatens to become the chief, if not the only, class; nevertheless the notions and political precedents engendered by it of old still cling about it. These notions and these precedents remain unchanged, because they are old, and also because they happen to be in perfect accordance with the new notions and general habits of our contemporaries. Manufacturing property then does not extend its rights in the same ratio as its importance. The manufacturing classes do not become less dependent, whilst they become more numerous; but, on the contrary, it would seem as if despotism lurked within them and naturally grew with their growth.[5] As a nation becomes more engaged in manufactures, the want of roads, canals, harbours, and other works of a semi-public nature, which facilitate the acquisition of wealth, is more strongly felt; and as a nation becomes more democratic, private individuals are less able, and the State more able, to execute works of such magnitude. I do not hesitate to assert that the manifest tendency of all governments at the present time is to take upon themselves alone the execution of these undertakings; by which means they daily hold in closer dependence the population which they govern.

On the other hand, in proportion as the power of a State increases, and its necessities are augmented, the State consumption of manufactured produce is always growing larger, and these commodities are generally made in the arsenals or establishments of the government. Thus, in every kingdom, the ruler becomes the principal manufacturer; he collects and retains in his service a vast number of engineers, architects, mechanics, and handicraftsmen. Not only is he the principal manufacturer, but he tends more and more to become the chief, or rather the master of all other manufacturers. As private persons become more powerless by becoming more equal, they can effect nothing in manufactures without combination; but the government naturally seeks to place these combinations under its own control.

It must be admitted that these collective beings, which are called combinations, are stronger and more formidable than a private individual can ever be, and that they have less of the responsibility of their own actions; whence it seems reasonable that they should not be allowed to retain so great an independence of the supreme government as might be conceded to a private individual.

Rulers are the more apt to follow this line of policy, as their own inclinations invite them to it. Amongst democratic nations it is only by association that the resistance of the people to the government can ever display itself: hence the latter always looks with ill-favour on those associations which are not in its own power; and it is well worthy of remark, that amongst democratic nations, the people themselves often entertain a secret feeling of fear and jealousy against these very associations, which prevents the citizens from defending the institutions of which they stand so much in need. The power and the duration of these small private bodies, in the midst of the weakness and instability of the whole community, astonish and alarm the people; and the free use which each association makes of its natural powers is almost regarded as a dangerous privilege. All the associations which spring up in our age are, moreover, new corporate powers, whose rights have not been sanctioned by time; they come into existence at a time when the notion of private rights is weak, and when the power of government is unbounded; hence it is not surprising that they lose their freedom at their birth. Amongst all European nations there are some kinds of associations which cannot be formed until the State has examined their by-laws, and authorised their existence. In several others, attempts are made to extend this rule to all associations; the consequences of such a policy, if it were successful, may easily be foreseen. If once the sovereign had a general right of authorising associations of all kinds upon certain conditions, he would not be long without claiming the right of superintending and managing them, in order to prevent them from departing from the rules laid down by himself. In this manner, the State, after having reduced all who are desirous of forming associations into dependence, would proceed to reduce into the same condition all who belong to associations already formed—that is to say, almost all the men who are now in existence. Governments thus appropriate to themselves, and convert to their own purposes, the greater part of this new power which manufacturing interests have in our time brought into the world. Manufacturers govern us—they govern manufactures.

I attach so much importance to all that I have just been saying, that I am tormented by the fear of having impaired my meaning in seeking to render it more clear. If the reader thinks that the examples I have adduced to support my observations are insufficient or ill-chosen—if he imagines that I have anywhere exaggerated the encroachments of the supreme power, and, on the other hand, that I have underrated the extent of the sphere which still remains open to the exertions of individual independence, I entreat him to lay down the book for a moment, and to turn his mind to reflect for himself upon the subjects I have attempted to explain. Let him attentively examine what is taking place in France and in other countries—let him inquire of those about him–let him search himself, and I am much mistaken if he does not arrive, without my guidance, and by other paths, at the point to which I have sought to lead him. He will perceive that for the last half-century, centralisation has everywhere been growing up in a thousand different ways. Wars, revolutions, conquests, have served to promote it: all men have laboured to increase it. In the course of the same period, during which men have succeeded each other with singular rapidity at the head of affairs, their notions, interests, and passions have been infinitely diversified; but all have by some means or other sought to centralise. This instinctive centralisation has been the only settled point amidst the extreme mutability of their lives and of their thoughts.

If the reader, after having investigated these details of human affairs, will seek to survey the wide prospect as a whole, he will be struck by the result. On the one hand the most settled dynasties shaken or overthrown—the people everywhere escaping by violence from the sway of their laws—abolishing or limiting the authority of their rulers or their princes—the nations, which are not in open revolution, restless at least, and excited—all of them animated by the same spirit of revolt: and on the other hand, at this very period of anarchy, and amongst these untractable nations, the incessant increase of the prerogative of the supreme government, becoming more centralised, more adventurous, more absolute, more extensive—the people perpetually falling under the control of the public administration—led insensibly to surrender to it some further portion of their individual independence, till the very men, who from time to time upset a throne and trample on a race of kings, bend more and more obsequiously to the slightest dictate of a clerk. Thus two contrary revolutions appear in our days to be going on; the one continually weakening the supreme power, the other as continually strengthening it: at no other period in our history has it appeared so weak or so strong.

But upon a more attentive examination of the state of the world, it appears that these two revolutions are intimately connected together, that they originate in the same source, and that after having followed a separate course, they lead men at last to the same result. I may venture once more to repeat what I have already said or implied in several parts of this book: great care must be taken not to confound the principle of equality itself, with the revolution which finally establishes that principle in the social condition and the laws of a nation: here lies the reason of almost all the phenomena which occasion our astonishment. All the old political powers of Europe, the greatest as well as the least, were founded in ages of aristocracy, and they more or less represented or defended the principles of inequality and of privilege. To make the novel wants and interests, which the growing principle of equality introduced, preponderate in government, our contemporaries had to overturn or to coerce the established powers. This led them to make revolutions, and breathed into many of them, that fierce love of disturbance and independence, which all revolutions, whatever be their object, always engender. I do not believe that there is a single country in Europe in which the progress of equality has not been preceded or followed by some violent changes in the state of property and persons; and almost all these changes have been attended with much anarchy and licence, because they have been made by the least civilised portion of the nation against that which is most civilised. Hence proceeded the twofold contrary tendencies which I have just pointed out. As long as the democratic revolution was glowing with heat, the men who were bent upon the destruction of old aristocratic powers hostile to that revolution, displayed a strong spirit of independence; but as the victory or the principle of equality became more complete, they gradually surrendered themselves to the propensities natural to that condition of equality, and they strengthened and centralised their governments. They had sought to be free in order to make themselves equal; but in proportion as equality was more established by the aid of freedom, freedom itself was thereby rendered of more difficult attainment.

These two states of a nation have sometimes been contemporaneous: the last generation in France showed how a people might organise a stupendous tyranny in the community, at the very time when they were baffling the authority of the nobility and braving the power of all kings—at once teaching the world the way to win freedom, and the way to lose it. In our days men see that constituted powers are dilapidated on every side—they see all ancient authority gasping away, all ancient barriers tottering to their fall, and the judgment of the wisest is troubled at the sight: they attend only to the amazing revolution which is taking place before their eyes, and they imagine that mankind is about to fall into perpetual anarchy: if they looked to the final consequences of this revolution, their fears would perhaps assume a different shape. For myself, I confess that I put no trust in the spirit of freedom which appears to animate my contemporaries. I see well enough that the nations of this age are turbulent, but I do not clearly perceive that they are liberal; and I fear lest, at the close of those perturbations which rock the base of thrones, the domination of sovereigns may prove more powerful than it ever was before.

Chapter VI

What Sort of Despotism Democratic Nations Have to Fear

I had remarked during my stay in the United States, that a democratic state of society, similar to that of the Americans, might offer singular facilities for the establishment of despotism; and I perceived, upon my return to Europe, how much use had already been made by most of our rulers, of the notions, the sentiments, and the wants engendered by this same social condition, for the purpose of extending the circle of their power. This led me to think that the nations of Christendom would perhaps eventually undergo some sort of oppression like that which hung over several of the nations of the ancient world. A more accurate examination of the subject, and five years of further meditations, have not diminished my apprehensions, but they have changed the object of them. No sovereign ever lived in former ages so absolute or so powerful as to undertake to administer by his own agency, and without the assistance of intermediate powers, all the parts of a great empire: none ever attempted to subject all his subjects indiscriminately to strict uniformity of regulation, and personally to tutor and direct every member of the community. The notion of such an undertaking never occurred to the human mind; and if any man had conceived it, the want of information, the imperfection of the administrative system, and above all, the natural obstacles caused by the inequality of conditions, would speedily have checked the execution of so vast a design. When the Roman emperors were at the height of their power, the different nations of the empire still preserved manners and customs of great diversity; although they were subject to the same monarch, most of the provinces were separately administered; they abounded in powerful and active municipalities; and although the whole government of the empire was centred in the hands of the emperor alone, and he always remained, upon occasions, the supreme arbiter in all matters, yet the details of social life and private occupations lay for the most part beyond his control. The emperors possessed, it is true, an immense and unchecked power, which allowed them to gratify all their whimsical tastes, and to employ for that purpose the whole strength of the State. They frequently abused that power arbitrarily to deprive their subjects of property or of life: their tyranny was extremely onerous to the few, but it did not reach the greater number; it was fixed to some few main objects, and neglected the rest; it was violent, but its range was limited.

But it would seem that if despotism were to be established amongst the democratic nations of our days, it might assume a different character; it would be more extensive and more mild; it would degrade men without tormenting them. I do not question, that in an age of instruction and equality like our own, sovereigns might more easily succeed in collecting all political power into their own hands, and might interfere more habitually and decidedly within the circle of private interests, than any sovereign of antiquity could ever do. But this same principle of equality which facilitates despotism, tempers its rigour. We have seen how the manners of society become more humane and gentle in proportion as men become more equal and alike. When no member of the community has much power or much wealth, tyranny is, as it were, without opportunities and a field of action. As all fortunes are scanty, the passions of men are naturally circumscribed—their imagination limited, their pleasures simple. This universal moderation moderates the sovereign himself, and checks within certain limits the inordinate extent of his desires.

Independently of these reasons drawn from the nature of the state of society itself, I might add many others arising from causes beyond my subject; but I shall keep within the limits I have laid down to myself. Democratic governments may become violent and even cruel at certain periods of extreme effervescence or of great danger: but these crises will be rare and brief. When I consider the petty passions of our contemporaries, the mildness of their manners, the extent of their education, the purity of their religion, the gentleness of their morality, their regular and industrious habits, and the restraint which they almost all observe in their vices no less than in their virtues, I have no fear that they will meet with tyrants in their rulers, but rather guardians.[1] I think then that the species of oppression by which democratic nations are menaced is unlike anything which ever before existed in the world: our contemporaries will find no prototype of it in their memories. I am trying myself to choose an expression which will accurately convey the whole of the idea I have formed of it, but in vain; the old words despotism and tyranny are inappropriate: the thing itself is new; and since I cannot name it, I must attempt to define it.

I seek to trace the novel features under which despotism may appear in the world. The first thing that strikes the observation is an innumerable multitude of men all equal and alike, incessantly endeavouring to procure the petty and paltry pleasures with which they glut their lives. Each of them, living apart, is as a stranger to the fate of all the rest—his children and his private friends constitute to him the whole of mankind; as for the rest of his fellow-citizens, he is close to them, but he sees them not—he touches them, but he feels them not; he exists but in himself and for himself alone; and if his kindred still remain to him, he may be said at any rate to have lost his country. Above this race of men stands an immense and tutelary power, which takes upon itself alone to secure their gratifications, and to watch over their fate. That power is absolute, minute, regular, provident, and mild. It would be like the authority of a parent, if, like that authority, its object was to prepare men for manhood; but it seeks on the contrary to keep them in perpetual childhood: it is well content that the people should rejoice, provided they think of nothing but rejoicing. For their happiness such a government willingly labours, but it chooses to be the sole agent and the only arbiter of that happiness: it provides for their security, foresees and supplies their necessities, facilitates their pleasures, manages their principal concerns, directs their industry, regulates the descent of property, and subdivides their inheritances—what remains, but to spare them all the care of thinking and all the trouble of living? Thus it every day renders the exercise of the free agency of man less useful and less frequent; it circumscribes the will within a narrower range, and gradually robs a man of all the uses of himself. The principle of equality has prepared men for these things: it has predisposed men to endure them, and oftentimes to look on them as benefits.

After having thus successively taken each member of the community in its powerful grasp, and fashioned them at will, the supreme power then extends its arm over the whole community. It covers the surface of society with a network of small complicated rules, minute and uniform, through which the most original minds and the most energetic characters cannot penetrate, to rise above the crowd. The will of man is not shattered, but softened, bent, and guided: men are seldom forced by it to act, but they are constantly restrained from acting: such a power does not destroy, but it prevents existence; it does not tyrannize, but it compresses, enervates, extinguishes, and stupefies a people, till each nation is reduced to be nothing better than a flock of timid and industrious animals, of which the government is the shepherd.

I have always thought that servitude of the regular, quiet, and gentle kind which I have just described, might be combined more easily than is commonly believed with some of the outward forms of freedom; and that it might even establish itself under the wing of the sovereignty of the people. Our contemporaries are constantly excited by two conflicting passions; they want to be led, and they wish to remain free: as they cannot destroy either one or the other of these contrary propensities, they strive to satisfy them both at once. They devise a sole, tutelary, and all-powerful form of government, but elected by the people. They combine the principle of centralisation and that of popular sovereignty; this gives them a respite; they console themselves for being in tutelage by the reflection that they have chosen their own guardians. Every man allows himself to be put in leading-strings, because he sees that it is not a person or a class of persons, but the people at large that holds the end of his chain. By this system the people shake off their state of dependence just long enough to select their master, and then relapse into it again. A great many persons at the present day are quite contented with this sort of compromise between administrative despotism and the sovereignty of the people; and they think they have done enough for the protection of individual freedom when they have surrendered it to the power of the nation at large. This does not satisfy me: the nature of him I am to obey signifies less to me than the fact of extorted obedience.

I do not however deny that a constitution of this kind appears to me to be infinitely preferable to one, which, after having concentrated all the powers of government, should vest them in the hands of an irresponsible person or body of persons. Of all the forms which democratic despotism could assume, the latter would assuredly be the worst. When the sovereign is elective, or narrowly watched by a legislature which is really elective and independent, the oppression which he exercises over individuals is sometimes greater, but it is always less degrading; because every man, when he is oppressed and disarmed, may still imagine, that whilst he yields obedience it is to himself he yields it, and that it is to one of his own inclinations that all the rest give way. In like manner I can understand that when the sovereign represents the nation, and is dependent upon the people, the rights and the power of which every citizen is deprived, not only serve the head of the state, but the state itself; and that private persons derive some return from the sacrifice of their independence which they have made to the public. To create a representation of the people in every centralised country, is therefore, to diminish the evil which extreme centralisation may produce, but not to get rid of it. I admit that by this means room is left for the intervention of individuals in the more important affairs; but it is not the less suppressed in the smaller and more private ones. It must not be forgotten that it is especially dangerous to enslave men in the minor details of life. For my own part, I should be inclined to think freedom less necessary in great things than in little ones, if it were possible to be secure of the one without possessing the other. Subjection in minor affairs breaks out every day, and is felt by the whole community indiscriminately. It does not drive men to resistance, but it crosses them at every turn, till they are led to surrender the exercise of their will. Thus their spirit is gradually broken and their character enervated; whereas that obedience, which is exacted on a few important but rare occasions, only exhibits servitude at certain intervals, and throws the burden of it upon a small number of men. It is in vain to summon a people, which has been rendered so dependent on the central power, to choose from time to time the representatives of that power; this rare and brief exercise of their free choice, however important it may be, will not prevent them from gradually losing the faculties of thinking, feeling, and acting for themselves, and thus gradually falling below the level of humanity.[2] I add that they will soon become incapable of exercising the great and only privilege which remains to them. The democratic nations which have introduced freedom into their political constitution, at the very time when they were augmenting the despotism of their administrative constitution, have been led into strange paradoxes. To manage those minor affairs in which good sense is all that is wanted—the people are held to be unequal to the task but when the government of the country is at stake, the people are invested with immense powers; they are alternately made the playthings of their ruler, and his masters—more than kings, and less than men. After having exhausted all the different modes of election, without finding one to suit their purpose, they are still amazed, and still bent on seeking further; as if the evil they remark did not originate in the constitution of the country far more than in that of the electoral body. It is, indeed, difficult to conceive how men who have entirely given up the habit of self-government should succeed in making a proper choice of those by whom they are to be governed; and no one will ever believe that a liberal, wise, and energetic government can spring from the suffrages of a subservient people. A constitution, which should be republican in its head and ultra-monarchical in all its other parts, has ever appeared to me to be a short-lived monster. The vices of rulers and the ineptitude of the people would speedily bring about its ruin; and the nation, weary of its representatives and of itself, would create freer institutions, or soon return to stretch itself at the feet of a single master.

Chapter VII

Continuation of the Preceding Chapters

I believe that it is easier to establish an absolute and despotic government amongst a people in which the conditions of society are equal, than amongst any other; and I think that if such a government were once established amongst such a people, it would not only oppress men, but would eventually strip each of them of several of the highest qualities of humanity. Despotism therefore appears to me peculiarly to be dreaded in democratic ages. I should have loved freedom, I believe, at all times, but in the time in which we live I am ready to worship it. On the other hand, I am persuaded that all who shall attempt, in the ages upon which we are entering, to base freedom upon aristocratic privilege, will fail—that all who shall attempt to draw and to retain authority within a single class, will fail. At the present day no ruler is skilful or strong enough to found a despotism, by reestablishing permanent distinctions of rank amongst his subjects: no legislator is wise or. powerful enough to preserve free institutions, if he does not take equality for his first principle and his watchword. All those of our contemporaries who would establish or secure the independence and the dignity of their fellow-men, must show themselves the friends of equality; and the only worthy means of showing themselves as such, is to be so: upon this depends the success of their holy enterprise. Thus the question is not how to reconstruct aristocratic society, but how to make liberty proceed out of that democratic state of society in which God has placed us.

These two truths appear to me simple, clear, and fertile in consequences; and they naturally lead me to consider what kind of free government can be established amongst a people in which social conditions are equal.

It results from the very constitution of democratic nations and from their necessities, that the power of government amongst them must be more uniform, more centralised, more extensive, more searching, and more efficient than in other countries. Society at large is naturally stronger and more active, individuals more subordinate and weak; the former does more, the latter less; and this is inevitably the case. It is not therefore to be expected that the range of private independence will ever be as extensive in democratic as in aristocratic countries—nor is this to be desired; for, amongst aristocratic nations, the mass is often sacrificed to the individual, and the prosperity of the greater number to the greatness of the few. It is both necessary and desirable that the government of a democratic people should be active and powerful: and our object should not be to render it weak or indolent, but solely to prevent it from abusing its aptitude and its strength.

The circumstance which most contributed to secure the independence of private persons in aristocratic ages, was, that the supreme power did not affect to take upon itself alone the government and administration of the community; those functions were necessarily partially left to the members of the aristocracy: so that as the supreme power was always divided, it never weighed with its whole weight and in the same manner on each individual. Not only did the government not perform everything by its immediate agency; but as most of the agents who discharged its duties derived their power not from, the State, but from the circumstance of their birth, they were not perpetually under its control. The government could not make or unmake them in an instant, at pleasure, nor bend them in strict uniformity to its slightest caprice—this was an additional guarantee of private independence. I readily admit that recourse cannot be had to the same means at the present time: but I discover certain democratic expedients which may be substituted for them. Instead of vesting in the government alone all the administrative powers of which corporations and nobles have been deprived, a portion of them may be entrusted to secondary public bodies, temporarily composed of private citizens: thus the liberty of private persons will be more secure, and their equality will not be diminished.

The Americans, who care less for words than the French, still designate by the name of County the largest of their administrative districts: but the duties of the count or lord-lieutenant are in part performed by a provincial assembly. At a period of equality like our own it would be unjust and unreasonable to institute hereditary officers; but there is nothing to prevent us from substituting elective public officers to a certain extent. Election is a democratic expedient which insures the independence of the public officer in relation to the government, as much and even more than hereditary rank can insure it amongst aristocratic nations. Aristocratic countries abound in wealthy and influential persons who are competent to provide for themselves, and who cannot be easily or secretly oppressed: such persons restrain a government within general habits of moderation and reserve. I am very well aware that democratic countries contain no such persons naturally; but something analogous to them may be created by artificial means. I firmly believe that an aristocracy cannot again be founded in the world; but I think that private citizens, by combining together, may constitute bodies of great wealth, influence, and strength, corresponding to the persons of an aristocracy. By this means many of the greatest political advantages of aristocracy would be obtained without its injustice or its dangers. An association for political, commercial, or manufacturing purposes, or even for those of science and literature, is a powerful and enlightened member of the community, which cannot be disposed of at pleasure, or oppressed without remonstrance; and which, by defending its own rights against the encroachments of the government, saves the common liberties of the country.

In periods of aristocracy every man is always bound so closely to many of his fellow-citizens, that he cannot be assailed without their coming to his assistance. In ages of equality every man naturally stands alone; he has no hereditary friends whose cooperation he may demand—no class upon whose sympathy he may rely: he is easily got rid of, and he is trampled on with impunity. At the present time, an oppressed member of the community has therefore only one method of self-defence—he may appeal to the whole nation; and if the whole nation is deaf to his complaint, he may appeal to mankind: the only means he has of making this appeal is by the press. Thus the liberty of the press is infinitely more valuable amongst democratic nations than amongst all others; it is the only cure for the evils which equality may produce. Equality sets men apart and weakens them; but the press places a powerful weapon within every man's reach, which the weakest and loneliest of them all may use. Equality deprives a man of the support of his connexions; but the press enables him to summon all his fellow-countrymen and all his fellow-men to his assistance. Printing has accelerated the progress of equality, and it is also one of its best correctives.

I think that men living in aristocracies may, strictly speaking, do without the liberty of the press: but such is not the case with those who live in democratic countries. To protect their personal independence I trust not to great political assemblies, to parliamentary privilege, or to the assertion of popular sovereignty. All these things may, to a certain extent, be reconciled with personal servitude—but that servitude cannot be complete if the press is free: the press is the chiefest democratic instrument of freedom.

Something analogous may be said of the judicial power. It is a part of the essence of judicial power to attend to private interests, and to fix itself with predilection on minute objects submitted to its observation; another essential quality of judicial power is never to volunteer its assistance to the oppressed, but always to be at the disposal of the humblest of those who solicit it; their complaint, however feeble they may themselves be, will force itself upon the ear of justice and claim redress, for this is inherent in the very constitution of the courts of justice. A power of this kind is therefore peculiarly adapted to the wants of freedom, at a time when the eye and finger of the government are constantly intruding into the minutest details of human actions, and when private persons are at once too weak to protect themselves, and too much isolated for them to reckon upon the assistance of their fellows. The strength of the courts of law has ever been the greatest security which can be offered to personal independence; but this is more especially the case in democratic ages: private rights and interests are in constant danger, if the judicial power does not grow more extensive and more strong to keep pace with the growing equality of conditions.

Equality awakens in men several propensities extremely dangerous to freedom, to which the attention of the legislator ought constantly to be directed. I shall only remind the reader of the most important amongst them. Men living in democratic ages do not readily comprehend the utility of forms: they feel an instinctive contempt for them—I have elsewhere shown for what reasons. Forms excite their contempt and often their hatred; as they commonly aspire to none but easy and present gratifications, they rush onwards to the object of their desires, and the slightest delay exasperates them. This same temper, carried with them into political life, renders them hostile to forms, which perpetually retard or arrest them in some of their projects. Yet this objection which the men of democracies make to forms is the very thing which renders forms so useful to freedom; for their chief merit is to serve as a barrier between the strong and the weak, the ruler and the people, to retard the one, and give the other time to look about him. Forms become more necessary in proportion as the government becomes more active and more powerful, whilst private persons are becoming more indolent and more feeble. Thus democratic nations naturally stand more in need of forms than other nations, and they naturally respect them less. This deserves most serious attention. Nothing is more pitiful than the arrogant disdain of most of our contemporaries for questions of form; for the smallest questions of form have acquired in our time an importance which they never had before: many of the greatest interests of mankind depend upon them. I think that if the statesmen of aristocratic ages could sometimes contemn forms with impunity, and frequently rise above them, the statesmen to whom the government of nations is now confided ought to treat the very least among them with respect, and not neglect them without imperious necessity. In aristocracies the observance of forms was superstitious; amongst us they ought to be kept with a deliberate and enlightened deference.

Another tendency, which is extremely natural to democratic nations and extremely dangerous, is that which leads them to despise and undervalue the rights of private persons. The attachment which men feel to a right, and the respect which they display for it, is generally proportioned to its importance, or to the length of time during which they have enjoyed it. The rights of private persons amongst democratic nations are commonly of small importance, of recent growth, and extremely precarious—the consequence is that they are often sacrificed without regret, and almost always violated without remorse. But it happens that at the same period and amongst the same nations in which men conceive a natural contempt for the rights of private persons, the rights of society at large are naturally extended and consolidated: in other words, men become less attached to private rights at the very time at which it would be most necessary to retain and to defend what little remains of them. It is therefore most especially in the present democratic ages, that the true friends of the liberty and the greatness of man ought constantly to be on the alert to prevent the power of government from lightly sacrificing the private rights of individuals to the general execution of its designs. At such times no citizen is so obscure that it is not very dangerous to allow him to be oppressed—no private rights are so unimportant that they can be surrendered with impunity to the caprices of a government. The reason is plain:—if the private right of an individual is violated at a time when the human mind is fully impressed with the importance and the sanctity of such rights, the injury done is confined to the individual whose right is infringed; but to violate such a right, at the present day, is deeply to corrupt the manners of the nation and to put the whole community in jeopardy, because the very notion of this kind of right constantly tends amongst us to be impaired and lost.

There are certain habits, certain notions, and certain vices which are peculiar to a state of revolution, and which a protracted revolution cannot fail to engender and to propagate, whatever be, in other respects, its character, its purpose, and the scene on which it takes place. When any nation has, within a short space of time, repeatedly varied its rulers, its opinions, and its laws, the men of whom it is composed eventually contract a taste for change, and grow accustomed to see all changes effected by sudden violence. Thus they naturally conceive a contempt for forms which daily prove ineffectual; and they do not support without impatience the dominion of rules which they have so often seen infringed. As the ordinary notions of equity and morality no longer suffice to explain and justify all the innovations daily begotten by a revolution, the principle of public utility is called in, the doctrine of political necessity is conjured up, and men accustom themselves to sacrifice private interests without scruple, and to trample on the rights of individuals in order more speedily to accomplish any public purpose.

These habits and notions, which I shall call revolutionary, because all revolutions produce them, occur in aristocracies just as much as amongst democratic nations; but amongst the former they are often less powerful and always less lasting, because there they meet with habits, notions, defects, and impediments, which counteract them: they consequently disappear as soon as the revolution is terminated, and the nation reverts to its former political courses. This is not always the case in democratic countries, in which it is ever to be feared that revolutionary tendencies, becoming more gentle and more regular, without entirely disappearing from society, will be gradually transformed into habits of subjection to the administrative authority of the government. I know of no countries in which revolutions are more dangerous than in democratic countries; because, independently of the accidental and transient evils which must always attend them, they may always create some evils which are permanent and unending. I believe that there are such things as justifiable resistance and legitimate rebellion: I do not therefore assert, as an absolute proposition, that the men of democratic ages ought never to make revolutions; but I think that they have especial reason to hesitate before they embark in them, and that it is far better to endure many grievances in their present condition than to have recourse to so perilous a remedy.

I shall conclude by one general idea, which comprises not only all the particular ideas which have been expressed in the present chapter, but also most of those which it is the object of this book to treat of. In the ages of aristocracy which preceded our own, there were private persons of great power, and a social authority of extreme weakness. The outline of society itself was not easily discernible, and constantly confounded with the different powers by which the community was ruled. The principal efforts of the men of those times were required to strengthen, aggrandise, and secure the supreme power; and on the other hand, to circumscribe individual independence within narrower limits, and to subject private interests to the interests of the public. Other perils and other cares await the men of our age. Amongst the greater part of modem nations, the government, whatever may be its origin, its constitution, or its name, has become almost omnipotent, and private persons are falling, more and more, into the lowest stage of weakness and dependence. In olden society everything was different; unity and uniformity were nowhere to be met with. In modern society everything threatens to become so much alike, that the peculiar characteristics of each individual will soon be entirely lost in the general aspect of the world. Our forefathers were ever prone to make an improper use of the notion, that private rights ought to be respected; and we are naturally prone on the other hand to exaggerate the idea that the interest of a private individual ought always to bend to the interest of the many. The political world is metamorphosed: new remedies must henceforth be sought for new disorders. To lay down extensive, but distinct and settled limits, to the action of the government; to confer certain rights on private persons, and to secure to them the undisputed enjoyment of those rights; to enable individual man to maintain whatever independence, strength, and original power he still possesses; to raise him by the side of society at large, and uphold him in that position—these appear to me the main objects of legislators in the ages upon which we are now entering. It would seem as if the rulers of our time sought only to use men in order to make things great; I wish that they would try a little more to make great men; that they would set less value on the work, and more upon the workman; that they would never forget that a nation cannot long remain strong when every man belonging to it is individually weak, and that no form or combination of social polity has yet been devised, to make an energetic people out of a community of pusillanimous and enfeebled citizens.

I trace amongst our contemporaries two contrary notions which are equally injurious. One set of men can perceive nothing in the principle of equality but the anarchical tendencies which it engenders: they dread their own free agency—they fear themselves. Other thinkers, less numerous but more enlightened, take a different view: beside that track which starts from the principle of equality to terminate in anarchy, they have at last discovered the road which seems to lead men to inevitable servitude. They shape their souls beforehand to this necessary condition; and, despairing of remaining free, they already do obeisance in their hearts to the master who is soon to appear. The former abandon freedom, because they think it dangerous; the latter, because they hold it to be impossible. If I had entertained the latter conviction, I should not have written this book, but I should have confined myself to deploring in secret the destiny of mankind. I have sought to point out the dangers to which the principle of equality exposes the independence of man, because I firmly believe that these dangers are the most formidable, as well as the least foreseen, of all those which futurity holds in store: but I do not think that they are insurmountable. The men who live in the democratic ages upon which we are entering have naturally a taste for independence: they are naturally impatient of regulation, and they are wearied by the permanence even of the condition they themselves prefer. They are fond of power; but they are prone to despise and hate those who wield it, and they easily elude its grasp by their own mobility and insignificance. These propensities will always manifest themselves, because they originate in the groundwork of society, which will undergo no change: for a long time they will prevent the establishment of any despotism, and they will furnish fresh weapons to each succeeding generation which shall struggle in favour of the liberty of mankind. Let us then look forward to the future with that salutary fear which makes men keep watch and ward for freedom, not with that faint and idle terror which depresses and enervates the heart.

Chapter VIII

General Survey of the Subject

Before I close forever the theme that has detained me so long, I would fain take a parting survey of all the various characteristics of modern society, and appreciate at last the general influence to be exercised by the principle of equality upon the fate of mankind; but I am stopped by the difficulty of the task, and in presence of so great an object my sight is troubled, and my reason fails. The society of the modern world which I have sought to delineate, and which I seek to judge, has but just come into existence. Time has not yet shaped it into perfect form: the great revolution by which it has been created is not yet over: and amidst the occurrences of our time, it is almost impossible to discern what will pass away with the revolution itself, and what will survive its close. The world which is rising into existence is still half encumbered by the remains of the world which is waning into decay; and amidst the vast perplexity of human affairs, none can say how much of ancient institutions and former manners will remain, or how much will completely disappear. Although the revolution which is taking place in the social condition, the laws, the opinions, and the feelings of men, is still very far from being terminated, yet its results already admit of no comparison with anything that the world has ever before witnessed. I go back from age to age up to the remotest antiquity; but I find no parallel to what is occurring before my eyes: as the past has ceased to throw its light upon the future, the mind of man wanders in obscurity.

Nevertheless, in the midst of a prospect so wide, so novel and so confused, some of the more prominent characteristics may already be discerned and pointed out. The good things and the evils of life are more equally distributed in the world: great wealth tends to disappear, the number of small fortunes to increase; desires and gratifications are multiplied, but extraordinary prosperity and irremediable penury are alike unknown. The sentiment of ambition is universal, but the scope of ambition is seldom vast. Each individual stands apart in solitary weakness; but society at large is active, provident, and powerful: the performances of private persons are insignificant, those of the state immense. There is little energy of character; but manners are mild, and laws humane. If there be few instances of exalted heroism or of virtues of the highest, brightest, and purest temper, men's habits are regular, violence is rare, and cruelty almost unknown. Human existence becomes longer, and property more secure: life is not adorned with brilliant trophies, but it is extremely easy and tranquil. Few pleasures are either very refined or very coarse; and highly-polished manners are as uncommon as great brutality of tastes. Neither men of great learning, nor extremely ignorant communities, are to be met with; genius becomes more rare, information more diffused. The human mind is impelled by the small efforts of all mankind combined together, not by the strenuous activity of certain men. There is less perfection, but more abundance, in all the productions of the arts. The ties of race, of rank, and of country are relaxed; the great bond of humanity is strengthened. If I endeavour to find out the most general and the most prominent of all these different characteristics, I shall have occasion to perceive, that what is taking place in men's fortunes manifests itself under a thousand other forms. Almost all extremes are softened or blunted: all that was most prominent is superseded by some mean term, at once less lofty and less low, less brilliant and less obscure, than what before existed in the world.

When I survey this countless multitude of beings, shaped in each other's likeness, amidst whom nothing rises and nothing falls, the sight of such universal uniformity saddens and chills me, and I am tempted to regret that state of society which has ceased to be. When the world was full of men of great importance and extreme insignificance, of great wealth and extreme poverty, of great learning and extreme ignorance, I turned aside from the latter to fix my observation on the former alone, who gratified my sympathies. But I admit that this gratification arose from my own weakness: it is because I am unable to see at once all that is around me, that I am allowed thus to select and separate the objects of my predilection from among so many others. Such is not the case with that Almighty and Eternal Being, whose gaze necessarily includes the whole of created things, and who surveys distinctly, though at once, mankind and man. We may naturally believe that it is not the singular prosperity of the few, but the greater well-being of all, which is most pleasing in the sight of the Creator and Preserver of men. What appears to me to be man's decline, is to His eye advancement; what afflicts me is acceptable to

Him. A state of equality is perhaps less elevated, but it is more just; and its justice constitutes its greatness and its beauty. I would strive then to raise myself to this point of the Divine contemplation, and thence to view and to judge the concerns of men.

No man, upon the earth, can as yet affirm, absolutely and generally, that the new state of the world is better than its former one; but it is already easy to perceive that this state is different. Some vices and some virtues were so inherent in the constitution of an aristocratic nation, and are so opposite to the character of a modern people, that they can never be infused into it; some good tendencies and some bad propensities which were unknown to the former, are natural to the latter; some ideas suggest themselves spontaneously to the imagination of the one, which are utterly repugnant to the mind of the other. They are like two distinct orders of human beings, each of which has its own merits and defects, its own advantages and its own evils. Care must therefore be taken not to judge the state of society, which is now coming into existence, by notions derived from a state of society which no longer exists; for as these states of society are exceedingly different in their structure, they cannot be submitted to a just or fair comparison. It would be scarcely more reasonable to require of our own contemporaries the peculiar virtues which originated in the social condition of their forefathers, since that social condition is itself fallen, and has drawn into one promiscuous ruin the good and evil which belonged to it.

But as yet these things are imperfectly understood. I find that a great number of my contemporaries undertake to make a certain selection from amongst the institutions, the opinions, and the ideas which originated in the aristocratic constitution of society as it was: a portion of these elements they would willingly relinquish, but they would keep the remainder and transplant them into their new world. I apprehend that such men are wasting their time and their strength in virtuous but unprofitable efforts. The object is not to retain the peculiar advantages which the inequality of conditions bestows upon mankind, but to secure the new benefits which equality may supply. We have not to seek to make ourselves like our progenitors, but to strive to work out that species of greatness and happiness which is our own. For myself, who now look back from this extreme limit of my task, and discover from afar, but at once, the various objects which have attracted my more attentive investigation upon my way, I am full of apprehensions and of hopes. I perceive mighty dangers which it is possible to ward off—mighty evils which may be avoided or alleviated; and I cling with a firmer hold to the belief, that for democratic nations to be virtuous and prosperous they require but to will it. I am aware that many of my contemporaries maintain that nations are never their own masters here below, and that they necessarily obey some insurmountable and unintelligent power, arising from anterior events, from their race, or from the soil and climate of their country. Such principles are false and cowardly; such principles can never produce aught but feeble men and pusillanimous nations. Providence has not created mankind entirely independent or entirely free. It is true that around every man a fatal circle is traced, beyond which he cannot pass; but within the wide verge of that circle he is powerful and free: as it is with man, so with communities. The nations of our time cannot prevent the conditions of men from becoming equal; but it depends upon themselves whether the principle of equality is to lead them to servitude or freedom, to knowledge or barbarism, to prosperity or to wretchedness.

Appendices

Appendix A.—Vol. I. p. 16.

For information concerning all the countries of the West which have not been visited by Europeans, consult the account of two expeditions undertaken at the expense of Congress by Major Long. This traveller particularly mentions, on the subject of the great American desert, that a line may be drawn nearly parallel to the 20th degree of longitude[1] (meridian of Washington), beginning from the Red River and ending at the River Platte. From this imaginary line to the Rocky Mountains, which bound the valley of the Mississippi on the West, lie immense plains, which are almost entirely covered with sand, incapable of cultivation, or scattered over with masses of granite. In summer, these plains are quite destitute of water, and nothing is to be seen on them but herds of buffaloes and wild horses. Some hordes of Indians are also found there, but in no great numbers. Major Long was told that in travelling northwards from the River Platte you find the same desert lying constantly on the left; but he was unable to ascertain the truth of this report. ('Long's Expedition,' vol. ii. p. 361.) However worthy of confidence may be the narrative of Major Long, it must be remembered that he only passed through the country of which he speaks, without deviating widely from the line which he had traced out for his journey.

Appendix B.—Vol. I. p. 17.

South America, in the region between the tropics, produces an incredible profusion of climbing plants, of which the Flora of the Antilles alone presents us with forty different species. Among the most graceful of these shrubs is the Passionflower, which, according to Descourtiz, grows with such luxuriance in the Antilles, as to climb trees by means of the tendrils with which it is provided, and form moving bowers of rich and elegant festoons, decorated with blue and purple flowers, and fragrant with perfume. (Vol. I. p. 265.) The *Mimosa scandens* (Acacia à grandes gousses) is a creeper of enormous and rapid growth, which climbs from tree to tree, and sometimes covers more than half a league. (Vol. iii. p. 227.)

Appendix C.—Vol. I. p. 18.

The languages which are spoken by the Indians of America, from the Pole to Cape Horn, are said to be all formed upon the same model, and subject to the same grammatical rules; whence it may fairly be concluded that all the Indian nations sprang from the same stock. Each tribe of the American continent speaks a different dialect; but the number of languages, properly so called, is very small, a fact which tends to prove that the nations of the New World had not a very remote origin. Moreover, the languages of America have a great degree of regularity, from which it seems probable that the tribes which employ them had not undergone any great revolutions, or been incorporated, voluntarily or by constraint, with foreign nations. For it is generally the union of several languages into one which produces grammatical irregularities. It is not long since the American languages, especially those of the North, first attracted the serious attention of philologists, when the discovery was made that this idiom of a barbarous people was the product of a complicated system of ideas and very learned combinations. These languages were found to be very rich, and great pains had been taken at their formation to render them agreeable to the ear. The grammatical system of the Americans differs from all others in several points, but especially in the following:—

Some nations of Europe, amongst others the Germans, have the power of combining at pleasure different expressions, and thus giving a complex sense to certain words. The Indians have given a most surprising extension to this power, so as to arrive at the means of connecting a great number of ideas with a single term. This will be easily understood with the help of an example quoted by Mr. Duponceau, in the 'Memoirs of the Philosophical Society of America.' 'A Delaware woman playing with a cat or a young dog,' says this writer, 'is heard to pronounce the word *kuligatschis,* which is thus composed: *k* is the sign of the second person, and signifies "thou" or "thy;" *uli* is a part of the word *wulit,* which signifies "beautiful," "pretty;" *gat* is another fragment of the word *wichgat,* which means "paw;" and, lastly, *schis* is a diminutive giving the idea of smallness. Thus in one word the Indian woman has expressed "Thy pretty little paw." 'Take another example of the felicity with which the savages of America have composed their words. A young man of Delaware is called *pilapé.* This word is formed from *pilsit,* chaste, innocent; and *lenapé,* man; viz. man in his purity and innocence. This facility of combining words is most remarkable in the strange formation of their verbs. The most complex action is often expressed by a single verb, which serves to convey all the shades of an idea by the modification of its construction. Those who may wish to examine more in detail this subject, which I have only glanced at superficially, should read:—

1. The correspondence of Mr. Duponceau and the Rev. Mr. Hecwelder relative to the Indian languages, which is to be found in the first volume of the 'Memoirs of the Philosophical Society of America,' published at Philadelphia, 1819, by Abraham Small; vol. i. p. 356–464.

2. The 'Grammar of the Delaware or the Lenape Language,' by Geiberger, and the preface of Mr. Duponceau. All these are in the same collection; vol. iii.

3. An excellent account of these works, which is at the end of the sixth volume of the 'American Encyclopaedia.'

Appendix D.—Vol. I. p. 19.

See in Charlevoix, vol. i. p. 235, the history of the first war which the French inhabitants of Canada carried on, in 1610, against the Iroquois. The latter, armed with bows and arrows, offered a desperate resistance to the French and their allies. Charlevoix is not a great painter, yet he exhibits clearly enough, in this narrative, the contrast between the European manners and those of savages, as well as the different way in which the two races of men understood the sense of honour. When the French, says he, seized upon the beaver skins which covered the Indians who had fallen, the Hurons, their allies, were greatly offended at this proceeding; but without hesitation they set to work in their usual manner, inflicting horrid cruelties upon the prisoners, and devouring one of those who had been killed, which made the Frenchmen shudder. The barbarians prided themselves upon a scrupulousness which they were surprised at not finding in our nation, and could not understand that there was less to reprehend in the stripping of dead bodies than in the devouring of their flesh like wild beasts. Charlevoix, in another place (vol. i. p. 230), thus describes the first torture of which Champlain was an eye-witness, and the return of the Hurons into their own village. Having proceeded about eight leagues, says he, our allies halted; and having singled out one of their captives, they reproached him with all the cruelties which he had practised upon the warriors of their nation who had fallen into his hands, and told him that he might expect to be treated in like manner; adding, that if he had any spirit he would prove it by singing. He immediately chanted forth his death-song, and then his war-song, and all the songs he knew 'but in a very mournful strain,' says Champlain, who was not then aware that all savage music has a melancholy character. The tortures which succeeded, accompanied by all the horrors which we shall mention hereafter, terrified the French, who made every effort to put a stop to them, but in vain. The following night, one of the Hurons having dreamt that they were pursued, the retreat was changed to a real flight, and the savages never stopped until they were out of the reach of danger.

The moment they perceived the cabins of their own village, they cut themselves long sticks, to which they fastened the scalps which had fallen to their share, and carried them in triumph. At this sight, the women swam to the canoes, where they received the bloody scalps from the hands of their husbands, and tied them round their necks. The warriors offered one of these horrible trophies to Champlain; they also presented him with some bows and arrows—the only spoils of the Iroquois which they had ventured to seize—entreating him to show them to the King of France. Champlain lived a whole winter quite alone among these barbarians, without being under any alarm for his person or property.

Appendix E.—Vol. I. p. 29.

Although the Puritanical strictness which presided over the establishment of the English colonies in America is now much relaxed, remarkable traces of it are still found in their habits and their laws. In 1792, at the very time when the anti-Christian republic of France began its ephemeral existence, the legislative body of Massachusetts promulgated the following law, to compel the citizens to observe the Sabbath. We give the preamble and the principal articles of this law, which is worthy of the reader's attention: 'Whereas,' says the legislator, 'the observation of the Sunday is an affair of public interest; inasmuch as it produces a necessary suspension of labour, leads men to reflect upon the duties of life, and the errors to which human nature is liable, and provides for the public and private worship of God, the creator and governor of the universe, and for the performance of such acts of charity as are the ornament and comfort of Christian societies: Whereas—irreligious or light-minded persons, forgetting the duties which the Sabbath imposes, and the benefits which these duties confer on society, are known to profane its sanctity, by following their pleasures or their affairs; this way of acting being contrary to their own interest as Christians, and calculated to annoy those who do not follow their example; being also of great injury to society at large, by spreading a taste for dissipation and dissolute manners; Be it enacted and ordained by the Governor, Council, and Representatives convened in General Court of Assembly, that all and every person and persons shall on that day carefully apply themselves to the duties of religion and piety, that no tradesman or labourer shall exercise his ordinary calling, and that no game or recreation shall be used on the Lord's Day, upon pain of forfeiting ten shillings.

'That no one shall travel on that day, or any part thereof, under pain of forfeiting twenty shillings; that no vessel shall leave a harbour of the colony; that no persons shall keep outside the meeting-house during the time of public worship, or profane the time by playing or talking, on penalty of five shillings.

'Public-houses shall riot entertain any other than strangers or lodgers, under penalty of five shillings for every person found drinking and abiding therein.

'Any person in health, who, without sufficient reason, shall omit to worship God in public during three months, shall be condemned to a fine often shillings.

'Any person guilty of misbehaviour in a place of public worship, shall be fined from five to forty shillings.

'These laws are to be enforced by the tything-men of each township, who have authority to visit public-houses on the Sunday. The innkeeper who shall refuse them admittance, shall be fined forty shillings for such offence.

'The tything-men are to stop travellers, and require of them their reason for being on the road on Sunday; anyone refusing to answer, shall be sentenced to pay a fine not exceeding five pounds sterling. If the reason given by the traveller be not deemed by the tything-man sufficient, he may bring the traveller before the justice of the peace of the district.' (Law of the 8th March, 1792; 'General Laws of Massachusetts,' vol. i. p. 410.)

On the 11th March, 1797, a new law increased the amount of fines, half of which was to be given to the informer. (Same collection, vol. ii. p. 525.) On the 16th February, 1816, a new law confirmed these same measures. (Same collection, vol. ii. p. 405.) Similar enactments exist in the laws of the State of New York, revised in 1827 and 1828. (See 'Revised Statutes,' Part I. chapter 20, p. 675.) In these it is declared that no one is allowed on the Sabbath to sport, to fish, to play at games, or to frequent houses where liquor is sold. No one can travel, except in case of necessity.

And this is not the only trace which the religious strictness and austere manners of the first emigrants have left behind them in the American laws. In the "Revised Statutes of the State of New York," vol. i. p. 662, is the following clause:—

'Whoever shall win or lose in the space of twenty-four hours, by gaming or betting, the sum of twenty-five dollars, shall be found guilty of a misdemeanour, and upon conviction shall be condemned to pay a fine equal to at least five times the value of the sum lost or won; which shall be paid to the inspector of the poor of the township. He that loses twenty-five dollars or more may bring an action to recover them; and if he neglects to do so the inspector of the poor may prosecute the winner, and oblige him to pay into the poor's box both the sum he has gained and three times as much besides.' The laws we quote from are of recent date; but they are unintelligible without going back to the very origin of the colonies. I have no doubt that in our days the penal part of these laws is very rarely applied. Laws preserve their inflexibility, long after the manners of a nation have yielded to the influence of time. It is still true, however, that nothing strikes a foreigner on his arrival in America more forcibly than the regard paid to the Sabbath. There is one, in particular, of the large American cities, in which all social movements begin to be suspended even on Saturday evening. You traverse its streets at the hour at which you expect men in the middle of life to be engaged in business, and young people in pleasure; and you meet with solitude and silence. Not only have all ceased to work, but they appear to have ceased to exist. Neither the movements of industry are heard, nor the accents of joy, nor even the confused murmur which arises from the midst of a great city. Chains are hung across the streets in the neighborhood of the churches; the half-closed shutters of the houses scarcely admit a ray of sun into the dwellings of the citizens. Now and then you perceive a solitary individual who glides silently along the deserted streets and lanes. Next day, at early dawn, the rolling of carriages, the noise of hammers, the cries of the population, begin to make themselves heard again. The city is awake. An eager crowd hastens towards the resort of commerce and industry; everything around you bespeaks motion, bustle, hurry. A feverish activity succeeds to the lethargic stupor of yesterday; you might almost suppose that they had but one day to acquire wealth and to enjoy it.

Appendix F.—Vol. I. p. 33.

It is unnecessary for me to say, that in the chapter which has just been read, I have not had the intention of giving a history of America. My only object was to enable the reader to appreciate the influence which the opinions and manners of the first emigrants had exercised upon the fate of the different colonies, and of the Union in general. I have therefore confined myself to the quotation of a few detached fragments. I do not know whether I am deceived, but it appears to me that, by pursuing the path which I have merely pointed out, it would be easy to present such pictures of the American republics as would not be unworthy the attention of the public, and could not fail to suggest to the statesman matter for reflection. Not being able to devote myself to this labour, I am anxious to render it easy to others; and, for this purpose, I subjoin a short catalogue and analysis of the works which seem to me the most important to consult.

At the head of the general documents which it would be advantageous to examine I place the work entitled 'An Historical Collection of State Papers, and other authentic Documents, intended as Materials for a History of the United States of America,' by Ebenezer Hasard. The first volume of this compilation, which was printed at Philadelphia in 1792, contains a literal copy of all the charters granted by the Crown of England to the emigrants, as well as the principal acts of the colonial governments, during the commencement of their existence. Amongst other authentic documents, we here find a great many relating to the affairs of New England and

Virginia during this period. The second volume is almost entirely devoted to the acts of the Confederation of 1643. This Federal compact, which was entered into by the colonies of New England with the view of resisting the Indians, was the first instance of union afforded by the Anglo-Americans. There were besides many other confederations of the same nature, before the famous one of 1776, which brought about the independence of the colonies.

Each colony has, besides, its own historic monuments, some of which are extremely curious; beginning with Virginia, the State which was first peopled. The earliest historian of Virginia was its founder, Captain John Smith. Captain Smith has left us an octavo volume, entitled 'The generall Historie of Virginia and New England, by Captain John Smith, sometymes Governor in those Countryes, and Admirall of New England;' printed at London in 1627. The work is adorned with curious maps and engravings of the time when it appeared; the narrative extends from the year 1584 to 1626. Smith's work is highly and deservedly esteemed. The author was one of the most celebrated adventurers of a period of remarkable adventure; his book breathes that ardour for discovery, that spirit of enterprise which characterised the men of his time, when the manners of chivalry were united to zeal for commerce, and made subservient to the acquisition of wealth. But Captain Smith is most remarkable for uniting to the virtues which characterised his contemporaries several qualities to which they were generally strangers; his style is simple and concise, his narratives bear the stamp of truth, and his descriptions are free from false ornament. This author throws most valuable light upon the state and condition of the Indians at the time when North America was first discovered.

The second historian to consult is Beverley, who commences his narrative with the year 1585, and ends it with 1700. The first part of his book contains historical documents, properly so called, relative to the infancy of the colony. The second affords a most curious picture of the state of the Indians at this remote period. The third conveys very clear ideas concerning the manners, social conditions, laws, and political customs of the Virginians in the author's lifetime. Beverley was a native of Virginia, which occasions him to say at the beginning of his book, that he entreats his readers not to exercise their critical severity upon it, since, having been born in the Indies, he does not aspire to purity of language. Notwithstanding this colonial modesty, the author shows throughout his book the impatience with which he endures the supremacy of the mother-country. In this work of Beverley are also found numerous traces of that spirit of civil liberty which animated the English colonies of America at the time when he wrote. He also shows the dissensions which existed among them, and retarded their independence. Beverley detests his Catholic neighbours of Maryland even more than he hates the English government: his style is simple, his narrative interesting, and apparently trustworthy.

I saw in America another work which ought to be consulted, entitled 'The History of Virginia,' by William Stith. This book affords some curious details, but I thought it long and diffuse.

The most ancient as well as the best document to be consulted on the history of Carolina, is a work in small quarto, entitled "The History of Carolina," by John Lawson, printed at London in 1718. This work contains, in the first part, a journey of discovery in the west of Carolina; the account of which, given in the form of a journal, is in general confused and superficial; but it contains a very striking description of the mortality caused among the savages of that time both by the small-pox and the immoderate use of brandy; with a curious picture of the corruption of manners prevalent amongst them, which was increased by the presence of Europeans. The second part of Lawson's book is taken up with a description of the physical condition of Carolina, and its productions. In the third part, the author gives an interesting account of the manners, customs, and government of the Indians at that period. There is a good deal of talent and originality in this part of the work. Lawson concludes his history with a copy of the charter granted to the Carolinas in the reign of Charles II. The general tone of this work is light, and often licentious, forming a perfect contrast to the solemn style of the works published at the same period in New England. Lawson's History is extremely scarce in America, and cannot be procured in Europe. There is, however, a copy of it in the Royal Library at Paris.

From the southern extremity of the United States, I pass at once to the northern limit; as the intermediate space was not peopled till a later period. I must first point out a very curious compilation, entitled 'Collection of the Massachusetts Historical Society,' printed for the first time at Boston in 1792, and reprinted in 1806. The Collection of which I speak, and which is continued to the present day, contains a great number of very valuable documents relating to the history of the different States in New England. Among them are letters which have never been published, and authentic pieces which had been buried in provincial archives. The whole work of Gookin, concerning the Indians, is inserted there.

I have mentioned several times in the chapter to which this note relates, the work of Nathaniel Norton entitled 'New England's Memorial;' sufficiently, perhaps, to prove that it deserves the attention of those who would be conversant with the history of New England. This book is in 8vo., and was reprinted at Boston in 1826.

The most valuable and important authority which exists upon the history of New England, is the work of the Rev. Cotton Mather, entitled 'Magnalia Christi

Americana, or the Ecclesiastical History of New England, 1620–1698, 2 vols. 8vo, reprinted at Hartford, United States, in 1820.'[2] The author divided his work into seven books. The first presents the history of the events which prepared and brought about the establishment of New England. The second contains the lives of the first governors and chief magistrates who presided over the country. The third is devoted to the lives and labours of the evangelical ministers who, during the same period, had the care of souls. In the fourth the author relates the institution and progress of the University of Cambridge (Massachusetts). In the fifth he describes the principles and the discipline of the Church of New England. The sixth is taken up in retracing certain facts, which, in the opinion of Mather, prove the merciful interposition of Providence in behalf of the inhabitants of New England. Lastly, in the seventh, the author gives an account of the heresies and the troubles to which the Church of New England was exposed. Cotton Mather was an evangelical minister who was born at Boston, and passed his life there. His narratives are distinguished by the same ardour and religious zeal which led to the foundation of the colonies of New England. Traces of bad taste sometimes occur in his manner of writing; but he interests, because he is full of enthusiasm. He is often intolerant, still oftener credulous, but he never betrays an intention to deceive. Sometimes his book contains fine passages, and true and profound reflections, such as the following:—'Before the arrival of the Puritans,' says he (vol. i. chap, iv.), 'there were more than a few attempts of the English to people and improve the parts of New England which were to the northward of New Plymouth; but the designs of those attempts being aimed no higher than the advancement of some worldly interests, a constant series of disasters has confounded them, until there was a plantation erected upon the nobler designs of Christianity: and that plantation though it has had more adversaries than perhaps any one upon earth, yet, having obtained help from God, it continues to this day.' Mather occasionally relieves the austerity of his descriptions with images full of tender feeling: after having spoken of an English lady whose religious ardour had brought her to America with her husband, and who soon after sank under the fatigues and privations of exile, he adds, 'As for her virtuous husband, Isaac Johnson,

He tryed
To live without her, liked it not, and dyed' (vol. i.)

Mather's work gives an admirable picture of the time and country which he describes. In his account of the motives which led the Puritans to seek an asylum beyond seas, he says:—

'The God of Heaven served, as it were, a summons upon the spirits of his people in the English nation, stirring up the spirits of thousands which never saw the faces of each other, with a most unanimous inclination to leave all the pleasant accommodations of their native country, and go over a terrible ocean, into a more terrible desert, for the pure enjoyment of all his ordinances. It is now reasonable that, before we pass any further, the reasons of his undertaking should be more exactly made known unto posterity, especially unto the posterity of those that were the undertakers, lest they come at length to forget and neglect the true interest of New England. Wherefore I shall now transcribe some of them from a manuscript, wherein they were then tendered unto consideration:

'General Considerations for the Plantation of New England.

'First, It will be a service unto the Church of great consequence, to carry the Gospel unto those parts of the world, and raise a bulwark against the kingdom of Antichrist, which the Jesuits labour to rear up in all parts of the world.

'Secondly, All other Churches of Europe have been brought under desolations; and it may be feared that the like judgments are coming upon us; and who knows but God hath provided this place to be a refuge for many whom he means to save out of the general destruction?

'Thirdly, The land grows weary of her inhabitants, insomuch that man, which is the most precious of all creatures, is here more vile and base than the earth he treads upon; children, neighbours, and friends, especially the poor, are counted the greatest burdens, which, if things were right, would be the chiefest of earthly blessings.

'Fourthly, We are grown to that intemperance in all excess of riot, as no mean estate almost will suffice a man to keep sail with his equals, and he that fails in it must live in scorn and contempt: hence it comes to pass, that all arts and trades are carried in that deceitful manner and unrighteous course, as it is almost impossible for a good upright man to maintain his constant charge and live comfortably in them.

'Fifthly, The schools of learning and religion are so corrupted, as (besides the unsupportable charge of education) most children, even the best, wittiest, and of the fairest hopes, are perverted, corrupted, and utterly overthrown by the multitude of evil examples and licentious behaviours in these seminaries.

'Sixthly, The whole earth is the Lord's garden, and he hath given it to the sons of Adam, to be tilled and improved by them: why, then, should we stand starving here for places of habitation, and in the meantime suffer whole countries, as profitable for the use of man, to lie waste without any improvement?

'Seventhly, What can be a better or nobler work, and more worthy of a Christian, than to erect and support a reformed particular Church in its infancy, and unite our forces with such a company of faithful people, as by timely assistance may grow stronger and prosper; but for want of it, may be put to great hazards, if not be wholly ruined?

'Eighthly, If any such as are known to be godly, and live in wealth and prosperity here, shall forsake all this to join with this reformed Church, and with it run the hazard of an hard and mean condition, it will be an example of great use, both for the removing of scandal and to give more life unto the faith of God's people in their prayers for the plantation, and also to encourage others to join the more willingly in it.'

Further on, when he declares the principles of the Church of New England with respect to morals, Mather inveighs with violence against the custom of drinking healths at table, which he denounces as a pagan and abominable practice. He proscribes with the same rigour all ornaments for the hair used by the female sex, as well as their custom of having the arms and neck uncovered. In another part of his work he relates several instances of witchcraft which had alarmed New England. It is plain that the visible action of the devil in the affairs of this world appeared to him an incontestable and evident fact.

This work of Cotton Mather displays, in many places, the spirit of civil liberty and political independence which characterised the times in which he lived. Their principles respecting government are discoverable at every page. Thus, for instance, the inhabitants of Massachusetts, in the year 1630, ten years after the foundation of Plymouth, are found to have devoted Pound 400 sterling to the establishment of the University of Cambridge. In passing from the general documents relative to the history of New England to those which describe the several States comprised within its limits, I ought first to notice 'The History of the Colony of Massachusetts,' by Hutchinson, Lieutenant-Governor of the Massachusetts Province, 2 vols. 8vo. The History of Hutchinson, which I have several times quoted in the chapter to which this note relates, commences in the year 1628, and ends in 1750. Throughout the work there is a striking air of truth and the greatest simplicity of style: it is full of minute details. The best history to consult concerning Connecticut is that of Benjamin Trumbull, entitled 'A Complete History of Connecticut, Civil and Ecclesiastical,' 1630–1764; 2 vols. 8vo., printed in 1818 at New Haven. This history contains a clear and calm account of all the events which happened in Connecticut during the period given in the title. The author drew from the best sources, and his narrative bears the stamp of truth. All that he says of the early days of Connecticut is extremely curious. See especially the Constitution of 1639, vol. i. ch. vi. p. 100; and also the Penal Laws of Connecticut, vol. i. ch. vii. p. 123.

'The History of New Hampshire,' by Jeremy Belknap, is a work held in merited estimation. It was printed at Boston in 1792, in 2 vols. 8vo. The third chapter of the first volume is particularly worthy of attention for the valuable details it affords on the political and religious principles of the Puritans, on the causes of their emigration, and on their laws. The following curious quotation is given from a sermon delivered in 1663:—'It concerneth New England always to remember that they are a plantation religious, not a plantation of trade. The profession of the purity of doctrine, worship, and discipline, is written upon her forehead. Let merchants, and such as are increasing cent. per cent., remember this, that worldly gain was not the end and design of the people of New England, but religion. And if any man among us make religion as twelve, and the world as thirteen, such an one hath not the spirit of a true New Englishman.' The reader of Belknap will find in his work more general ideas, and more strength of thought, than are to be met with in the American historians even to the present day.

Among the Central States which deserve our attention for their remote origin, New York and Pennsylvania are the foremost. The best history we have of the former is entitled 'A History of New York,' by William Smith, printed at London in 1757. Smith gives us important details of the wars between the French and English in America. His is the best account of the famous confederation of the Iroquois.

With respect to Pennsylvania, I cannot do better than point out the work of Proud, entitled 'The History of Pennsylvania, from the original Institution and Settlement of that Province, under the first Proprietor and Governor, William Penn, in 1681, till after the year 1742,' by Robert Proud, 2 vols. 8vo., printed at Philadelphia in 1797. This work is deserving of the especial attention of the reader; it contains a mass of curious documents concerning Penn, the doctrine of the Quakers, and the character, manners, and customs of the first inhabitants of Pennsylvania. I need not add that among the most important documents relating to this State are the works of Penn himself, and those of Franklin.

Appendix G.—Vol. I. p. 38.

We read in Jefferson's Memoirs as follows:—

'At the time of the first settlement of the English in Virginia, when land was to be had for little or nothing, some provident persons having obtained large grants of it, and being desirous of maintaining the splendour of their families, entailed their property upon their descendants. The transmission of these estates from generation to generation, to men who bore the same name, had the effect of raising up a distinct class of families, who, possessing by law the privilege of perpetuating their wealth, formed by these means a sort of patrician order, distinguished by the grandeur and luxury of their establishments. From this order it was that the King usually chose his councillors of state.'[3]

In the United States, the principal clauses of the English law respecting descent have been universally rejected. The first rule that we follow, says Mr. Kent, touching inheritance, is the following:—If a man dies intestate, his property goes to his heirs in a direct line. If he has but one heir or heiress, he or she succeeds to the whole. If there are several heirs of the same degree, they divide the inheritance equally amongst them, without distinction of sex. This rule was prescribed for the first time in the State of New York by a statute of the 23rd of February, 1786. (See 'Revised Statutes,' vol. iii. Appendix, p. 48.) It has since then been adopted in the "Revised Statutes" of the same State. At the present day this law holds good throughout the whole of the United States, with the exception of the State of Vermont, where the male heir inherits a double portion. (Kent's Commentaries, vol. iv. p. 370.) Mr. Kent, in the same work, vol. iv. p. 1–22, gives a historical account of American legislation on the subject of entail: by this we learn that, previous to the revolution, the colonies followed the English law of entail. Estates tail were abolished in Virginia in 1776, on a motion of Mr. Jefferson. They were suppressed in New York in 1786, and have since been abolished in North Carolina, Kentucky, Tennessee, Georgia, and Missouri. In Vermont, Indiana, Illinois, South Carolina, and Louisiana, entail was never introduced. Those States which thought proper to preserve the English law of entail, modified it in such a way as to deprive it of its most aristocratic tendencies. 'Our general principles on the subject of government,' says Mr. Kent, 'tend to favour the free circulation of property.'

It cannot fail to strike the French reader who studies the law of inheritance, that on these questions the French legislation is infinitely more democratic even than the American. The American law makes an equal division of the father's property, but only in the case of his will not being known; 'for every man,' says the law, 'in the State of New York ('Revised Statutes,' vol. iii. Appendix, p. 51), has entire liberty, power, and authority, to dispose of his property by will, to leave it entire, or divided in favour of any persons he chooses as his heirs, provided he do not leave it to a political body or any corporation.' The French law obliges the testator to divide his property equally, or nearly so, among his heirs. Most of the American republics still admit of entails, under certain restrictions; but the French law prohibits entail in all cases. If the social condition of the Americans is more democratic than that of the French, the laws of the latter are the most democratic of the two. This may be explained more easily than at first appears to be the case.

In France, democracy is still occupied in the work of destruction; in America, it reigns quietly over the ruins it has made.

Appendix H.—Vol. I. p. 44.

Summary Of The Qualifications Of Voters In The United States As They Existed In 1832.

All the States agree in granting the right of voting at the age of twenty-one. In all of them it is necessary to have resided for a certain time in the district where the vote is given. This period varies from three months to two years.

As to the qualification: in the State of Massachusetts it is necessary to have an income of three pounds sterling or a capital of sixty pounds. In Rhode Island, a man must possess landed property to the amount of 133 dollars.

In Connecticut, he must have a property which gives an income of seventeen dollars. A year of service in the militia also gives the elective privilege.

In New Jersey, an elector must have a property of fifty pounds a year.

In South Carolina and Maryland, the elector must possess fifty acres of land.

In Tennessee, he must possess some property.

In the States of Mississippi, Ohio, Georgia, Virginia, Pennsylvania, Delaware, New York, the only necessary qualification for voting is that of paying the taxes; and in most of the States, to serve in the militia is equivalent to the payment of taxes.

In Maine and New Hampshire any man can vote who is not on the pauper list.

Lastly, in the States of Missouri, Alabama, Illinois, Louisiana, Indiana, Kentucky, and Vermont, the conditions of voting have no reference to the property of the elector.

I believe there is no other State besides that of North Carolina in which different conditions are applied to the voting for the Senate and the electing the House of Representatives. The electors of the former, in this case, should possess in property fifty acres of land; to vote for the latter, nothing more is required than to pay taxes.

Appendix I.—Vol. I. p. 71.

The small number of Custom-house officers employed in the United States, compared with the extent of the coast, renders smuggling very easy; notwithstanding which, it is less practised than elsewhere, because everybody endeavours to repress it. In America there is no police for the prevention of fires, and such accidents are more frequent than in Europe; but in general they are more speedily extinguished, because the surrounding population is prompt in lending assistance.

Appendix K.—Vol. I. p. 72.

It is incorrect to assert that centralisation was produced by the French Revolution; the revolution brought it to perfection, but did not create it. The mania for centralisation and government regulations dates from the time when jurists began to take a share in the Government, in the time of Philippe-le-Bel; ever since which period they have been on the increase. In the year 1775, M. de Malesherbes, speaking in the name of the Cour des Aides, said to Louis XIV:—[4]

'. . . Every corporation and every community of citizens retained the right of administering its own affairs; a right which not only forms part of the primitive constitution of the kingdom, but has a still higher origin; for it is the right of nature, and of reason. Nevertheless, your subjects, Sire, have been deprived of it; and we cannot refrain from saying that in this respect your government has fallen into puerile extremes. From the time when powerful ministers made it a political principle to prevent the convocation of a national assembly, one consequence has succeeded another, until the deliberations of the inhabitants of a village are declared null when they have not been authorised by the Intendant. Of course, if the community has an expensive undertaking to carry through, it must remain under the control of the sub-delegate of the Intendant, and, consequently, follow the plan he proposes, employ his favourite workmen, pay them according to his pleasure; and if an action at law is deemed necessary, the Intendant's permission must be obtained. The cause must be pleaded before this first tribunal, previous to its being carried into a public court; and if the opinion of the Intendant is opposed to that of the inhabitants, or if their adversary enjoys his favour, the community is deprived of the power of defending its rights. Such are the means, Sire, which have been exerted to extinguish the municipal spirit in France; and to stifle, if possible, the opinions of the citizens. The nation may be said to lie under an interdict, and to be in wardship under guardians.' What could be said more to the purpose at the present day, when the revolution has achieved what are called its victories in centralisation?

In 1789, Jefferson wrote from Paris to one of his friends:—'There is no country where the mania for over-governing has taken deeper root than in France, or been the source of greater mischief.' (Letter to Madison, 28th August, 1789.) The feet is, that for several centuries past the central power of France has done everything it could to extend central administration; it has acknowledged no other limits than its own strength. The central power to which the revolution gave birth made more rapid advances than any of its predecessors, because it was stronger and wiser than they had been; Louis XIV. committed the welfare of such communities to the caprice of an intendant; Napoleon left them to that of the Minister. The same principle governed both, though its consequences were more or less remote.

Appendix L.—Vol. I. p. 75.

The immutability of the Constitution of France is a necessary consequence of the laws of that country. To begin with the most important of all the laws, that which decides the order of succession to the Throne; what can be more immutable in its principle than a political order founded upon the natural succession of father to son? In 1814, Louis XVIII. had established the perpetual law of hereditary succession in favour of his own family. The individuals who regulated the consequences of the revolution of 1830 followed his example; they merely established the perpetuity of the law in favour of another family. In this respect they imitated the Chancellor Meaupou, who, when he erected the new Parliament upon the ruins of the old, took care to declare in the same ordinance that the rights of the new magistrates should be as inalienable as those of their predecessors had been. The laws of 1830, like those of 1814, point out no way of changing the Constitution: and it is evident that the ordinary means of legislation are insufficient for this purpose. As the King, the Peers, and the Deputies, all derive their authority from the Constitution, these three powers united cannot alter a law by virtue of which alone they govern. Out of the pale of the Constitution they are nothing: where, when, could they take their stand to effect a change in its provisions? The alternative is clear: either their efforts are powerless against the Charter, which continues to exist in spite of them, in which case they only reign in the name of the Charter; or they succeed in changing the Charter, and then, the law by which they existed being annulled, they themselves cease to exist. By destroying the Charter, they destroy themselves. This is much more evident in the laws of 1830 than in those of 1814. In 1814, the royal prerogative took its stand above and beyond the Constitution; but in 1830, it was avowedly created by, and dependent on, the Constitution. A part, therefore, of the French Constitution is immutable, because it is united to the destiny of a family; and the body of the Constitution is equally immutable, because there appear to be no legal means of changing it. These remarks are not applicable to England. That country having no written Constitution, who can assert when its Constitution is changed?

Appendix M.—Vol. I. p. 75.

The most esteemed authors who have written upon the English Constitution agree with each other in establishing the omnipotence of the Parliament. Delolme says: 'It is a fundamental principle with the English lawyers, that Parliament can do everything except making a woman a man, or a man a woman.' Blackstone expresses himself more in detail, if not more energetically, than Delolme, in the following terms:—'The power and jurisdiction of Parliament, says Sir Edward Coke (4 Inst. 36), is so transcendent and absolute that it cannot be confined, either for causes or persons, within any bounds.' And of this High Court, he adds, may be truly said, "Si antiquitatem spectes, est vetustissima; si dignitatem, est honoratissima; si jurisdictionem, est capacissima." It hath sovereign and uncontrollable authority in the making, confirming, enlarging, restraining, abrogating, repealing, reviving, and expounding of laws, concerning matters of all possible denominations; ecclesiastical or temporal; civil, military, maritime, or criminal; this being the place where that absolute despotic power which must, in all Governments, reside somewhere, is intrusted by the Constitution of these kingdoms. All mischiefs and grievances, operations and remedies, that transcend the ordinary course of the laws, are within the reach of this extraordinary tribunal. It can regulate or new-model the succession to the Crown; as was done in the reign of Henry VIII. and William III. It can alter the established religion of the land; as was done in a variety of instances in the reigns of King Henry VIII. and his three children. It can *change and create afresh even the Constitution of the kingdom,* and of parliaments themselves; as was done by the Act of Union and the several statutes for triennial and septennial elections. It can, in short, do everything that is not naturally impossible to be done; and, therefore some have not scrupled to call its power, by a figure rather too bold, the omnipotence of Parliament.'

Appendix N.—Vol. I. p. 84.

There is no question upon which the American Constitutions agree more fully than upon that of political jurisdiction. All the Constitutions which take cognisance of this matter, give to the House of Delegates the exclusive right of impeachment; excepting only the Constitution of North Carolina, which grants the same privilege to grand juries. (Article 23.) Almost all the Constitutions give the exclusive right of pronouncing sentence to the Senate, or to the Assembly which occupies its place.

The only punishments which the political tribunals can inflict are removal, or the interdiction of public functions for the future. There is no other Constitution but that of Virginia (p. 152), which enables them to inflict every kind of punishment. The crimes which are subject to political jurisdiction are, in the Federal Constitution (Section 4, Art. 1); in that of Indiana (Art. 3, paragraphs 23 and 24); of New York (Art. 5); of Delaware (Art. 5), high treason, bribery, and other high crimes or offences. In the Constitution of Massachusetts (Chap. I, Section 2); that of North

Carolina (Art. 23); of Virginia (p. 252), misconduct and maladministration. In the Constitution of New Hampshire (p. 105), corruption, intrigue, and maladministration. In Vermont (Chap. 2, Art. 24), maladministration. In South Carolina (Art. 5); Kentucky (Art. 5); Tennessee (Art. 4); Ohio (Art. 1, §23, 24); Louisiana (Art. 5); Mississippi (Art. 5); Alabama (Art. 6); Pennsylvania (Art. 4), crimes committed in the non-performance of official duties. In the States of Illinois, Georgia, Maine, and Connecticut, no particular offences are specified.

Appendix O.—Vol. I. p. 133.

It is true that the powers of Europe may carry on maritime wars with the Union; but there is always greater facility and less danger in supporting a maritime than a continental war. Maritime warfare only requires one species of effort. A commercial people which consents to furnish its Government with the necessary funds, is sure to possess a fleet. And it is far easier to induce a nation to part with its money, almost unconsciously, than to reconcile it to sacrifices of men and personal efforts. Moreover, defeat by sea rarely compromises the existence or independence of the people which endures it. As for continental wars, it is evident that the nations of Europe cannot be formidable in this way to the American Union. It would be very difficult to transport and maintain in America more than 25,000 soldiers; an army which may be considered to represent a nation of about 2,000,000 of men. The most populous nation of Europe contending in this way against the Union, is in the position of a nation of 2,000,000 of inhabitants at war with one of 12,000,000. Add to this, that America has all its resources within reach, whilst the European is at 4,000 miles distance from his; and that the immensity of the American continent would of itself present an insurmountable obstacle to its conquest.

Appendix P.—Vol. I. p. 146.

The first American journal appeared in April, 1704, and was published at Boston. See 'Collection of the Historical Society of Massachusetts,' vol. vi. p. 66. It would be a mistake to suppose that the periodical press has always been entirely free in the American colonies: an attempt was made to establish something analogous to a censorship and preliminary security. Consult the Legislative Documents of Massachusetts of the 14th of January, 1722. The Committee appointed by the General Assembly (the Legislative body of the province) for the purpose of examining into circumstances connected with a paper entitled 'The New England Courier,' expresses its opinion that 'the tendency of the said journal is to turn religion into derision and bring it into contempt; that it mentions the sacred writers in a profane and irreligious manner; that it puts malicious interpretations upon the conduct of the ministers of the Gospel; and that the Government of his Majesty is insulted, and the peace and tranquillity of the province disturbed by the said journal. The Committee is consequently of opinion that the printer and publisher, James Franklin, should be forbidden to print and publish the said journal or any other work in future, without having previously submitted it to the Secretary of the province; and that the justices of the peace for the county of Suffolk should be commissioned to require bail of the said James Franklin for his good conduct during the ensuing year.' The suggestion of the Committee was adopted and passed into a law, but the effect of it was null, for the journal eluded the prohibition by putting the name of Benjamin Franklin instead of James Franklin at the bottom of its columns, and this manœuvre was supported by public opinion.

Appendix Q.—Vol. I. p. 225.

The Federal Constitution has introduced the jury into the tribunals of the Union in the same way as the States had introduced it into their own several courts; but as it has not established any fixed rules for the choice of jurors, the Federal Courts select them from the ordinary jury-list which each State makes for itself. The laws of the States must therefore be examined for the theory of the formation of juries. See 'Story's Commentaries on the Constitution,' B. hi. chap. 38, p. 654–659; 'Sergeant's Constitutional Law,' p. 165. See also the Federal Laws of the years 1789, 1800, and 1802, upon the subject. For the purpose of thoroughly understanding the American principles with respect to the formation of juries, I examined the laws of States at a distance from one another, and the following observations were the result of my inquiries. In America, all the citizens who exercise the elective franchise have the right of serving upon a jury. The great State of New York, however, has made a slight difference between the two privileges, but in a spirit quite contrary to that of the laws of France; for in the State of New York there are fewer persons eligible as jurymen than there are electors. It may be said in general that the right of forming part of a jury, like the right of electing, is open to all the citizens: the exercise of this right, however, is not put indiscriminately into any hands. Every year a body of municipal or county magistrates—called *selectmen* in New England, *supervisors* in New York, *trustees* in Ohio, and *sheriffs of the parish* in Louisiana—choose for each county a certain number of citizens who have the right of serving as jurymen, and who are supposed to be capable of exercising their functions. These magistrates, being themselves elective, excite no distrust; their powers, like those of most republican magistrates, are very extensive and very arbitrary, and they frequently make use of them to remove unworthy or incompetent jurymen. The names of the jurymen thus chosen are transmitted to the

County Court; and the jury who have to decide any affair are drawn by lot from the whole list of names. The Americans have contrived in every way to make the common people eligible to the jury, and to render the service as little onerous as possible. The sessions are held in the chief town of every county, and the jury are indemnified for their attendance either by the State or the parties concerned. They receive in general a dollar per day, besides their travelling expenses. In America, the being placed upon the jury is looked upon as a burden, but it is a burden which is very supportable. See 'Brevard's Digest of the Public Statute Law of South Carolina,' vol. i. pp. 446 and 454, vol. ii. pp. 218 and 338; 'The General Laws of Massachusetts, revised and published by authority of the Legislature,' vol. ii. pp. 187 and 331; 'The Revised Statutes of the State of New York,' vol. ii. pp. 411, 643, 717, 720; 'The Statute Law of the State of Tennessee,' vol. i. p. 209; 'Acts of the State of Ohio,' pp. 95 and 210; and 'Digeste général des Actes de la Legislature de la Louisiane.'

Appendix R.—Vol. I. p. 227.

If we attentively examine the constitution of the jury as introduced into civil proceedings in England, we shall readily perceive that the jurors are under the immediate control of the judge. It is true that the verdict of the jury, in civil as well as in criminal cases, comprises the question of fact and the question of right in the same reply; thus—a house is claimed by Peter as having been purchased by him: this is the fact to be decided. The defendant puts in a plea of incompetency on the part of the vendor: this is the legal question to be resolved. But the jury do not enjoy the same character of infallibility in civil cases, according to the practice of the English courts, as they do in criminal cases. The judge may refuse to receive the verdict; and even after the first trial has taken place, a second or new trial may be awarded by the Court. See 'Blackstone's Commentaries,' book iii. ch. 24.

Appendix S.—Vol. II. p. 541.

I find in my travelling-journal a passage which may serve to convey a more complete notion of the trials to which the women of America, who consent to follow their husbands into the wilds, are often subjected. This description has nothing to recommend it to the reader but its strict accuracy.

'... From time to time we come to fresh clearings; all these places are alike; I shall describe the one at which we have halted tonight, for it will serve to remind me of all the others.

'The bell which the pioneers hang round the necks of their cattle, in order to find them again in the woods, announced our approach to a clearing, when we were yet a long way off; and we soon afterwards heard the stroke of the hatchet, hewing down the trees of the forest. As we came nearer, traces of destruction marked the presence of civilised man; the road was strewn with shattered boughs; trunks of trees, half consumed by fire, or cleft by the wedge, were still standing in the track we were following. We continued to proceed till we reached a wood in which all the trees seemed to have been suddenly struck dead; in the height of summer their boughs were as leafless as in winter; and upon closer examination we found that a deep circle had been cut round the bark, which by stopping the circulation of the sap, soon kills the tree. We were informed that this is commonly the first thing a pioneer does; as he cannot in the first year cut down all the trees which cover his new parcel of land, he sows Indian corn under their branches, and puts the trees to death in order to prevent them from injuring his crop. Beyond this field, at present imperfectly traced out, we suddenly came upon the cabin of its owner, situated in the centre of a plot of ground more carefully cultivated than the rest, but where man was still waging unequal warfare with the forest; there the trees were cut down, but their roots were not removed, and the trunks still encumbered the ground which they so recently shaded. Around these dry blocks, wheat, suckers of trees, and plants of every kind, grow and intertwine in all the luxuriance of wild, untutored Nature. Amidst this vigorous and various vegetation stands the house of the pioneer, or, as they call it, the log-house. Like the ground about it, this rustic dwelling bore marks of recent and hasty labour; its length seemed not to exceed thirty feet, its height fifteen; the walls as well as the roof were formed of rough trunks of trees, between which a little moss and clay had been inserted to keep out the cold and rain.

'As night was coming on, we determined to ask the master of the log-house for a lodging. At the sound of our footsteps, the children who were playing amongst the scattered branches sprang up and ran towards the house, as if they were frightened at the sight of man; whilst two large dogs, almost wild, with ears erect and outstretched nose, came growling out of their hut, to cover the retreat of their young masters. The pioneer himself made his appearance at the door of his dwelling; he looked at us with a rapid and inquisitive glance, made a sign to the dogs to go into the house, and set them the example, without betraying either curiosity or apprehension at our arrival.

'We entered the log house: the inside is quite unlike that of the cottages of the peasantry of Europe: it contains more than is superfluous, less than is necessary. A single window with a muslin blind; on a hearth of trodden clay an immense fire, which lights the whole structure; above the hearth a good rifle, a deer's skin, and plumes of eagles' feathers; on the right hand of the chimney a map of the United States, raised and shaken by the wind through the crannies in the wall; near the map, upon a shelf formed of a roughly hewn plank, a few volumes of books—a Bible, the six first books of Milton, and two of Shakespeare's plays; along the wall, trunks instead of closets; in the centre of the room a rude table, with legs of green wood, and with the bark still upon them, looking as if they grew out of the ground on which they stood; but on this table a tea-pot of British ware, silver spoons, cracked tea-cups, and some newspapers.

'The master of this dwelling has the strong angular features and lank limbs peculiar to the native of New England. It is evident that this man was not born in the solitude in which we have met with him: his physical constitution suffices to show that his earlier years were spent in the midst of civilised society, and that he belongs to that restless, calculating, and adventurous race of men, who do with the utmost coolness things only to be accounted for by the ardour of the passions, and who endure the life of savages for a time, in order to conquer and civilise the back-woods.

'When the pioneer perceived that we were crossing his threshold, he came to meet us and shake hands, as is their custom; but his face was quite unmoved; he opened the conversation by inquiring what was going on in the world; and when his curiosity was satisfied, he held his peace, as if he were tired by the noise and importunity of mankind. When we questioned him in our turn, he gave us all the information we required; he then attended sedulously, but without eagerness, to our personal wants. Whilst he was engaged in providing thus kindly for us, how came it that in spite of ourselves we felt our gratitude die upon our lips? It is that our host whilst he performs the duties of hospitality, seems to be obeying an irksome necessity of his condition: he treats it as a duty imposed upon him by his situation, not as a pleasure. By the side of the hearth sits a woman with a baby on her lap: she nods to us without disturbing herself. Like the pioneer, this woman is in the prime of life; her appearance would seem superior to her condition, and her apparel even betrays a lingering taste for dress; but her delicate limbs appear shrunken, her features are drawn in, her eye is mild and melancholy; her whole physiognomy bears marks of a degree of religious resignation, a deep quiet of all passions, and some sort of natural and tranquil firmness, ready to meet all the ills of life, without fearing and without braving them. Her children cluster about her, full of health, turbulence, and energy: they are true children of the wilderness; their mother watches them from time to time with mingled melancholy and joy: to look at their strength and her languor, one might imagine that the life she has given them has exhausted her own, and still she regrets not what they have cost her. The house inhabited by these emigrants has no internal partition or loft. In the one chamber of which it consists, the whole family is gathered for the night. The dwelling is itself a little world—an ark of civilisation amidst an ocean of foliage: a hundred steps beyond it the primeval forest spreads its shades, and solitude resumes its sway.'

Appendix T.—Vol. II. p. 543.

It is not the equality of conditions which makes men immoral and irreligious; but when men, being equal, are at the same time immoral and irreligious, the effects of immorality and irreligion easily manifest themselves outwardly, because men have but little influence upon each other, and no class exists which can undertake to keep society in order. Equality of conditions never engenders profligacy of morals, but it sometimes allows that profligacy to show itself.

Appendix U.—Vol. II. p. 558.

Setting aside all those who do not think at all, and those who dare not say what they think, the immense majority of the Americans will still be found to appear satisfied with the political institutions by which they are governed; and, I believe, really to be so. I look upon this state of public opinion as an indication, but not as a demonstration, of the absolute excellence of American laws. The pride of a nation, the gratification of certain ruling passions by the law, a concourse of circumstances, defects which escape notice, and more than all the rest, the influence of a majority which shuts the mouth of all cavillers, may long perpetuate the delusions of a people as well as those of a man. Look at England throughout the eighteenth century. No nation was ever more prodigal of self-applause, no people was ever more self-satisfied; then every part of its constitution was right—everything, even to its most obvious defects, was irreproachable: at the present day a vast number of Englishmen seem to have nothing better to do than to prove that this constitution was faulty in many respects. Which was right?—the English people of the last century, or the English people of the present day? The same thing has occurred in France. It is certain that during the reign of Louis XIV. the great bulk of the nation was devotedly attached to the form of government which, at that time, governed the community. But it is a vast error to suppose that there was anything degraded in the character of the French of that age. There might be some sort of servitude in France at that time, but assuredly there was no servile spirit among the people. The writers of that age felt a species of genuine enthusiasm in extolling the power of their king; and there was no peasant so obscure in his hovel as not to take a pride in the glory of his sovereign, and to die cheerfully with the cry 'Vive le Roi!' upon his lips. These very same forms of loyalty are now odious to the French people. Which are wrong?—the French of the age of Louis XIV., or their descendants of the present day?

Our judgment of the laws of a people must not then be founded exclusively upon its inclinations, since those inclinations change from age to age; but upon more elevated principles and a more general experience. The love which a people may show for its law proves only this:—that we should not be in too great a hurry to change them.

Appendix V.—Vol. II. p. 597.

In the chapter to which this note relates I have pointed out one source of danger: I am now about to point out another kind of peril, more rare indeed, but far more formidable if it were ever to make its appearance. If the love of physical gratification and the taste for well-being, which are naturally suggested to men by a state of equality, were to get entire possession of the mind of a democratic people, and to fill it completely, the manners of the nation would become so totally opposed to military tastes, that perhaps even the army would eventually acquire a love of peace, in spite of the peculiar interest which leads it to desire war. Living in the midst of a state of general relaxation, the troops would ultimately think it better to rise without efforts, by the slow but commodious advancement of a peace establishment, than to purchase more rapid promotion at the cost of all the toils and privations of the field. With these feelings, they would take up arms without enthusiasm, and use them without energy; they would allow themselves to be led to meet the foe, instead of marching to attack him. It must not be supposed that this pacific state of the army would render it adverse to revolutions; for revolutions, and especially military revolutions, which are generally very rapid, are attended indeed with great dangers, but not with protracted toil; they gratify ambition at less cost than war; life only is at stake, and the men of democracies care less for their lives than for their comforts. Nothing is more dangerous for the freedom and the tranquillity of a people than an army afraid of war, because, as such an army no longer seeks to maintain its importance and its influence on the field of battle, it seeks to assert them elsewhere. Thus it might happen that the men of whom a democratic army consists should lose the interests of citizens without acquiring the virtues of soldiers; and that the army should cease to be fit for war without ceasing to be turbulent. I shall here repeat what I have said in the text: the remedy for these dangers is not to be found in the army, but in the country: a democratic people which has preserved the manliness of its character will never be at a loss for military prowess in its soldiers.

Appendix W.—Vol. II. p. 609.

Men connect the greatness of their idea of unity with means, God with ends: hence this idea of greatness, as men conceive it, leads us into infinite littleness. To compel all men to follow the same course towards the same object is a human notion;—to introduce infinite variety of action, but so combined that all these acts lead by a multitude of different courses to the accomplishment of one great design, is a conception of the Deity. The human idea of unity is almost always barren; the divine idea pregnant with abundant results. Men think they manifest their greatness by simplifying the means they use; but it is the purpose of God which is simple—his means are infinitely varied.

Appendix X.—Vol. II. p. 611.

A democratic people is not only led by its own tastes to centralise its government, but the passions of all the men by whom it is governed constantly urge it in the same direction. It may easily be foreseen that almost all the able and ambitious members of a democratic community will labour without ceasing to extend the powers of government, because they all hope at some time or other to wield those powers. It is a waste of time to attempt to prove to them that extreme centralisation may be injurious to the State, since they are centralising for their own benefit. Amongst the public men of democracies there are hardly any but men of great disinterestedness or extreme mediocrity who seek to oppose the centralisation of government: the former are scarce, the latter powerless.

Appendix Y.—Vol. II. p. 626.

I have often asked myself what would happen if, amidst the relaxation of democratic manners, and as a consequence of the restless spirit of the army, a military government were ever to be founded amongst any of the nations of the present age. I think that even such a government would not differ very much from the outline I have drawn in the chapter to which this note belongs, and that it would retain none of the fierce characteristics of a military oligarchy. I am persuaded that, in such a case, a sort of fusion would take place between the habits of official men and those of the military service. The administration would assume something of a military character, and the army some of the usages of the civil administration. The result would be a regular, clear, exact, and absolute system of government; the people would become the reflection of the army, and the community be drilled like a garrison.

Appendix Z.—Vol. II. p. 629.

It cannot be absolutely or generally affirmed that the greatest danger of the present age is licence or tyranny, anarchy or despotism. Both are equally to be feared; and the one may as easily proceed as the other from the selfsame cause, namely, *general* that *apathy,* which is the consequence of what I have termed Individualism: it is because this apathy exists, that the executive government, having mustered a few troops, is able to commit acts of oppression one day, and the next day a party, which has mustered some thirty men in its ranks, can also commit acts of oppression. Neither one nor the other can found anything to last; and the causes which enable them to succeed easily, prevent them from succeeding long: they rise because nothing opposes them, and they sink because nothing supports them. The proper object therefore of our most strenuous resistance, is far less either anarchy or despotism than the apathy which may almost indifferently beget either the one or the other.

Notes

First Part

Note on This Reeve Edition

1. In addition to the current standard, the Phillips Bradley edition, George Lawrence and Harvey Mansfield have editions of *Democracy in America* in print, and James T. Schleifer is currently working on an edition for Liberty Fund (Indianapolis, Indiana).
2. Published in 1862 and 1945, respectively.
3. Alexis de Tocqueville, *Democracy in America,* ed. Henry Steele Commager (Oxford University Press, 1947), p. vi.
4. Bowen, Appendix III, in Alexis de Tocqueville, *Democracy in America,* ed. Phillips Bradley (Vintage, 1990), p. 466.
5. Ibid., pp. 476–77.
6. Ibid., p. 478.
7. Letter from Tocqueville to Reeve, May 23, 1840, reprinted in John Knox Laughton, *Memoirs of the Life and Correspondence of Henry Reeve* (Longmans, Green, and Co., 1898), p. 115.
8. Russell Kirk, *The Conservative Mind: From Burke to Eliot,* 7th rev. ed. (Regnery Publishing, Inc.), p. 186.
9. Ibid., p. 187.

Chapter I: Exterior Form of North America

1. Darby's 'View of the United States.'
2. The Red River.
3. Warden's 'Description of the United States.'
4. See Appendix, A.
5. Make Brun tells us (vol. v. p. 726) that the water of the Caribbean Sea is so transparent that corals and fish are discernible at a depth of sixty fathoms. The ship seemed to float in air, the navigator became giddy as his eye penetrated through the crystal flood, and beheld submarine gardens, or beds of shells, or gilded fishes gliding among tufts and thickets of seaweed.
6. See Appendix, B.
7. With the progress of discovery some resemblance has been found to exist between the physical conformation, the language, and the habits of the Indians of North America, and those of the Tongous, Mantchous, Mongols, Tartars, and other wandering tribes of Asia. The land occupied by these tribes is not very distant from Behring's Strait, which allows of the supposition, that at a remote period they gave inhabitants to the desert continent of America. But this is a point which has not yet been clearly elucidated by science. See Malte Brun, vol. v.; the works of Humboldt; Fischer, 'Conjecture sur l'Origine des Americains;' Adair, 'History of the American Indians.'
8. See Appendix, C.

9. We learn from President Jefferson's 'Notes upon Virginia,' p. 148, that among the Iroquois, when attacked by a superior force, aged men refused to fly or to survive the destruction of their country; and they braved death like the ancient Romans when their capital was sacked by the Gauls. Further on, p. 150, he tells us that there is no example of an Indian who, having fallen into the hands of his enemies, begged for his life; on the contrary, the captive sought to obtain death at the hands of his conquerors by the use of insult and provocation.
10. See 'Histoire de la Louisiane,' by Lepage Dupratz; Charlevoix, 'Histoire de la Nouvelle France'; 'Lettres du Rev. G. Hecwelder;' 'Transactions of the American Philosophical Society,' v. 1; Jefferson's 'Notes on Virginia,' pp. 135–190. What is said by Jefferson is of especial weight, on account of the personal merit of the writer, of his peculiar position, and of the matter-of-fact age in which he lived.
11. See Appendix, D.

Chapter II: Origin of the Anglo-Americans, and Its Importance in Relation to Their Future Condition.

1. The Charter granted by the Crown of England in 1609 stipulated, amongst other conditions, that the adventurers should pay to the Crown a fifth of the produce of all gold and silver mines. See Marshall's 'Life of Washington,' vol. i. pp. 18–66.
2. A large portion of the adventurers, says Stith (History of Virginia), were unprincipled young men of family, whom their parents were glad to ship off, discharged servants, fraudulent bankrupts, or debauchees; and others of the same class, people more apt to pillage and destroy than to assist the settlement, were the seditious chiefs, who easily led this band into every kind of extravagance and excess. See for the history of Virginia the following works: 'History of Virginia, from the First Settlements in the year 1624,' by Smith; 'History of Virginia,' by William Stith; 'History of Virginia, from the Earliest Period,' by Beverley.
3. It was not till some time later that a certain number of rich English capitalists came to fix themselves in the colony.
4. Slavery was introduced about the year 1620 by a Dutch vessel which landed twenty negroes on the banks of the river James. See Chalmer.
5. The States of New England are those situated to the east of the Hudson; they are now six in number: 1, Connecticut; 2, Rhode Island; 3, Massachusetts; 4, Vermont; 5, New Hampshire; 6, Maine.
6. 'New England's Memorial,' p. 13. Boston, 1826. See also 'Hutchinson's History,' vol. ii. p. 440.
7. [The emigrants were, for the most part, godly Christians from the North of England, who had quitted their native country because they were 'studious of reformation, and entered into covenant to walk with one another according to the primitive pattern of the Word of God.' They emigrated to Holland, and settled in the city of Leyden in 1610, where they abode, being lovingly respected by the Dutch, for many years: they left it in 1620 for several reasons, the last of which was, that their posterity would in a few generations become Dutch, and so lose their interest in the English nation; they being desirous rather to enlarge His Majesty's dominions, and to live under their natural prince.—*Translator's Note.*]
8. This rock is become an object of veneration in the United States. I have seen bits of it carefully preserved in several towns of the Union. Does not this sufficiently show how entirely all human power and greatness is in the soul of man? Here is a stone which the feet of a few outcasts pressed for an instant, and this stone becomes famous; it is treasured by a great nation, its very dust is shared as a relic: and what is become of the gateways of a thousand palaces?
9. The emigrants who founded the State of Rhode Island in 1638, those who landed at New Haven in 1637, the first settlers in Connecticut in 1639, and the founders of Providence in

1640, began in like manner by drawing up a social contract, which was acceded to by all the interested parties. See 'Pitkin's History,' pp. 42 and 47.

10. This was the case in the State of New York.
11. Maryland, the Carolinas, Pennsylvania, and New Jersey were in this situation. See 'Pitkin's History,' vol. i. pp. 11–31.
12. See the work entitled *'Historical Collection of State Papers and other authentic Documents intended as materials for a History of the United States of America, by Ebenezer Hasard. Philadelphia,* 1792,' for a great number of documents relating to the commencement of the colonies, which are valuable from their contents and their authenticity: amongst them are the various charters granted by the King of England, and the first acts of the local governments. See also the analysis of all these charters given by Mr. Story, Judge of the Supreme Court of the United States, in the Introduction to his Commentary on the Constitution of the United States. It results from these documents that the principles of representative government and the external forms of political liberty were introduced into all the colonies at their origin. These principles were more fully acted upon in the North than in the South, but they existed everywhere.
13. See 'Pitkin's History,' p. 35. See the 'History of the Colony of Massachusetts Bay,' by Hutchinson, vol. i. p. 9.
14. See 'Pitkin's History,' pp. 42, 47.
15. The inhabitants of Massachusetts had deviated from the forms which are preserved in the criminal and civil procedure of England; in 1650 the decrees of justice were not yet headed by the royal style. See Hutchinson, vol. i. p. 452.
16. Code of 1650, p. 28. Hartford, 1830.
17. See also in 'Hutchinson's History,' vol. i. pp. 435, 456, the analysis of the penal code adopted in 1648 by the Colony of Massachusetts: this code is drawn up on the same principles as that of Connecticut.
18. Adultery was also punished with death by the law of Massachusetts: and Hutchinson, vol. i. p. 441, says that several persons actually suffered for this crime. He quotes a curious anecdote on this subject, which occurred in the year 1663. A married woman had had criminal intercourse with a young man; her husband died, and she married the lover. Several years had elapsed, when the public began to suspect the previous intercourse of this couple: they were thrown into prison, put upon trial, and very narrowly escaped capital punishment.
19. Code of 1650, p. 48. It seems sometimes to have happened that the judges superadded these punishments to each other, as is seen in a sentence pronounced in 1643 (p. 114, New Haven Antiquities), by which Margaret Bedford, convicted of loose conduct, was condemned to be whipped, and afterwards to marry Nicholas Jemmings, her accomplice.
20. 'New Haven Antiquities,' p. 104. See also 'Hutchinson's History,' for several causes equally extraordinary.
21. Code of 1650, pp. 50, 57.
22. Code of 1650, p. 64.
23. Ibid., p. 44.
24. This was not peculiar to Connecticut. See, for instance, the law which, on September 13, 1644, banished the Anabaptists from the State of Massachusetts. (Historical Collection of State Papers, vol. i. p. 538.) See also the law against the Quakers, passed on the 14th of October, 1656: 'Whereas,' says the preamble, 'an accursed race of heretics called Quakers has sprung up,' &c. The clauses of the statute inflict a heavy fine on all captains of ships who should import Quakers into the country. The Quakers who may be found there shall be whipped and imprisoned with hard labour. Those members of the sect who should defend their opinions shall be first fined, then imprisoned, and finally driven out of the province.—Historical Collection of State Papers, vol. i. p. 630.

25. By the penal law of Massachusetts, any Catholic priest who should set foot in the colony after having been once driven out of it was liable to capital punishment.
26. Code of 1650, p. 96.
27. New England's Memorial, p. 316. See Appendix, E.
28. Constitution of 1638, p. 17.
29. In 1641 the General Assembly of Rhode Island unanimously declared that the government of the State was a democracy, and that the power was vested in the body of free citizens, who alone had the right to make the laws and to watch their execution.—Code of 1650, p. 70.
30. 'Pitkin's History,' p. 47.
31. Constitution of 1638, p. 12.
32. Code of 1650, p. 80.
33. Ibid., p. 78.
34. Ibid., p. 49.
35. See 'Hutchinson's History,' vol. i. p. 455.
36. Code of 1650, p. 86.
37. Ibid., p. 40.
38. Ibid., p. 90.
39. Mather's 'Magnalia Christi Americana,' vol. ii. p. 13. This speech was made by Winthrop; he was accused of having committed arbitrary actions during his magistracy, but after having made the speech of which the above is a fragment, he was acquitted by acclamation, and from that time forwards he was always reelected governor of the State. See Marshal, vol. i. p. 166.
40. See Appendix, F.
41. Crimes no doubt exist for which bail is inadmissible, but they are few in number.
42. See Blackstone; and Delolme, book I chap. x.

Chapter III: Social Condition of the Anglo-Americans

1. I understand by the law of descent all those laws whose principal object is to regulate the distribution of property after the death of its owner. The law of entail is of this number; it certainly prevents the owner from disposing of his possessions before his death; but this is solely with the view of preserving them entire for the heir. The principal object, therefore, of the law of entail is to regulate the descent of property after the death of its owner: its other provisions are merely means to this end.
2. I do not mean to say that the small proprietor cultivates his land better, but he cultivates it with more ardor and care; so that he makes up by his labor for his want of skill.
3. Land being the most stable kind of property, we find, from time to time, rich individuals who are disposed to make great sacrifices in order to obtain it, and who willingly forfeit a considerable part of their income to make sure of the rest. But these are accidental cases. The preference for landed property is no longer found habitually in any class but among the poor. The small landowner, who has less information, less imagination, and fewer passions than the great one, is generally occupied with the desire of increasing his estate: and it often happens that by inheritance, by marriage, or by the chances of trade, he is gradually furnished with the means. Thus, to balance the tendency which leads men to divide their estates, there exists another, which incites them to add to them. This tendency, which is sufficient to prevent estates from being divided *ad infinitum,* is not strong enough to create great territorial possessions, certainly not to keep them up in the same family.
4. See Appendix, G.
5. [This may have been true in 1832, but is not so in 1874, when great cities like Chicago and San Francisco have sprung up in the Western States. But as yet the Western States exert no powerful influence on American society.—*Translator's Note.*]

Chapter IV: The Principle of the Sovereignty of the People in America

1. See Appendix, H.

Chapter V: Necessity of Examining the Condition of the States before That of the Union at Large

1. In 1830 there were 305 townships in the State of Massachusetts, and 610,014 inhabitants, which gives an average of about 2,000 inhabitants to each township.
2. The same rules are not applicable to the great towns, which generally have a mayor, and a corporation divided into two bodies; this, however, is an exception which requires the sanction of a law.—See the Act of February 22, 1822, for appointing the authorities of the City of Boston. It frequently happens that small towns as well as cities are subject to a peculiar administration. In 1832, 104 townships in the State of New York were governed in this manner.—*Williams's Register.*
3. Three selectmen are appointed in the small townships, and nine in the large ones. See 'The Town Officer,' p. 186. See also the principal laws of the State of Massachusetts relative to the selectmen: Act of February 20, 1786, vol. i. p. 219; February 24, 1796, vol. i. p. 488; March 7, 1801, vol. ii. p. 45; June 16, 1795, vol. i. p. 475; March 12, 1808, vol. ii. p. 186; February 28, 1787, vol. i. p. 302; June 22, 1797, vol. i. p. 539.
4. See Laws of Massachusetts, vol. i. p. 150, Act of the 25th March, 1786.
5. All these magistrates actually exist; their different functions are all detailed in a book called 'The Town-Officer,' by Isaac Goodwin, Worcester, 1827; and in the 'Collection of the General Laws of Massachusetts,' 3 vols., Boston, 1823.
6. See the Act of February 14, 1821, Laws of Massachusetts, vol. i. p. 551.
7. See the Act of February 20, 1819, Laws of Massachusetts, vol. ii. p. 494.
8. The council of the Governor is an elective body.
9. See 'The Town-Officer,' especially at the words SELECTMEN, ASSESSORS, COLLECTORS, SCHOOLS, SURVEYORS OF HIGHWAYS. I take one example in a thousand: the State prohibits travelling on the Sunday; the *tything-men,* who are town-officers, are specially charged to keep watch and to execute the law. See the Laws of Massachusetts, vol. i. p. 410. The selectmen draw up the lists of electors for the election of the Governor, and transmit the result of the ballot to the Secretary of the State. See Act of February 24, 1796: Id., vol. i. p. 488.
10. Thus, for instance, the selectmen authorize the construction of drains, point out the proper sites for slaughter-houses and other trades which are a nuisance to the neighborhood. See the Act of June 7, 1785: Id., vol. i. p. 193.
11. The selectmen take measures for the security of the public in case of contagious diseases, conjointly with the justices of the peace. See Act of June 22, 1797, vol. i. p. 539.
12. I say *almost,* for there are various circumstances in the annals of a township which are regulated by the justice of the peace in his individual capacity, or by the justices of the peace assembled in the chief town of the county; thus licenses are granted by the justices. See the Act of February 28, 1787, vol. i. p. 297.
13. Thus licenses are only granted to such persons as can produce a certificate of good conduct from the selectmen. If the selectmen refuse to give the certificate, the party may appeal to the justices assembled in the Court of Sessions, and they may grant the license. See Act of March 12, 1808, vol. ii. p. 186. The townships have the right to make by-laws, and to enforce them by fines which are fixed by law; but these by-laws must be approved by the Court of Sessions. See Act of March 23, 1786, vol. i. p. 254.
14. In Massachusetts the county magistrates are frequently called upon to investigate the acts of the town magistrates; but it will be shown further on that this investigation is a consequence, not of their administrative, but of their judicial power.
15. The town committees of schools are obliged to make an annual report to the Secretary of the State on the condition of the school. See Act of March 10, 1827, vol. iii. p. 183.
16. We shall hereafter learn what a Governor is: I shall content myself with remarking in this place that he represents the executive power of the whole State.

17. See the Constitution of Massachusetts, Chap. II. sect. 1. § 9; Chap. III. § 3.
18. Thus, for example, a stranger arrives in a township from a country where a contagious disease prevails, and he falls ill. Two justices of the peace can, with the assent of the selectmen, order the sheriff of the county to remove and take care of him.—Act of June 22, 1797, vol. i. p. 540. In general the justices interfere in all the important acts of the administration, and give them a semi-judicial character.
19. I say *the greater number,* because certain administrative misdemeanours are brought before ordinary tribunals. If, for instance, a township refuses to make the necessary expenditure for its schools or to name a school-committee, it is liable to a heavy fine. But this penalty is pronounced by the Supreme Judicial Court or the Court of Common Pleas. See Act of March 10, 1827, Laws of Massachusetts, vol. iii. p. 190. Or when a township neglects to provide the necessary war-stores.—Act of February 21, 1822: Id., vol. ii. p. 570.
20. In their individual capacity the justices of the peace take a part in the business of the counties and townships.
21. These affairs may be brought under the following heads:—1. The erection of prisons and courts of justice. 2. The county budget, which is afterwards voted by the State. 3. The distribution of the taxes so voted. 4. Grants of certain patents. 5. The laying down and repairs of the country roads.
22. Thus, when a road is under consideration, almost all difficulties are disposed of by the aid of the jury.
23. See Act of 20th February, 1786, Laws of Massachusetts, vol. i. p. 217.
24. There is an indirect method of enforcing the obedience of a township. Suppose that the funds which the law demands for the maintenance of the roads have not been voted, the town surveyor is then authorized, *ex officio,* to levy the supplies. As he is personally responsible to private individuals for the state of the roads, and indictable before the Court of Sessions, he is sure to employ the extraordinary right which the law gives him against the township. Thus by threatening the officer the Court of Sessions exacts compliance from the town. See Act of 5th March, 1787: Id., vol. i. p. 305.
25. Laws of Massachusetts, vol. ii. p. 45.
26. If, for instance, a township persists in refusing to name its assessors, the Court of Sessions nominates them; and the magistrates thus appointed are invested with the same authority as elected officers. See the Act quoted above, 20th Feb. 1787.
27. I say the Court of Sessions, because in common courts there is a magistrate who exercises some of the functions of a public prosecutor.
28. The grand-jurors are, for instance, bound to inform the court of the bad state of the roads.—Laws of Massachusetts, vol. i. p. 308.
29. If, for instance, the treasurer of the county holds back his accounts.—Laws of Massachusetts, vol. i. p. 406.
30. Thus, if a private individual breaks down or is wounded in consequence of the badness of a road, he can sue the township or the county for damages at the sessions.—Laws of Massachusetts, vol. i. p. 309.
31. In cases of invasion or insurrection, if the town-officers neglect to furnish the necessary stores and ammunition for the militia, the township may be condemned to a fine of from 200 to 500 dollars. It may readily be imagined that in such a case it might happen that no one cared to prosecute; hence the law adds that all the citizens may indict offences of this kind, and that half of the fine shall belong to the plaintiff. See Act of 6th March, 1810, vol. ii. p. 236. The same clause is frequently to be met with in the law of Massachusetts. Not only are private individuals thus incited to prosecute the public officers, but the public officers are encouraged in the same manner to bring the disobedience of private individuals to justice. If a citizen refuses to perform the work which has been assigned to him upon a road,

the road-surveyor may prosecute him, and he receives half the penalty for himself. See the Laws above quoted, vol. i. p. 308.

32. For details see the Revised Statutes of the State of New York, part i. chap. xi. vol. i. pp. 336–364, entitled, 'Of the Powers, Duties, and Privileges of Towns.' See in the Digest of the Laws of Pennsylvania, the words ASSESSORS COLLECTOR, CONSTABLES, OVERSEER OF THE POOR, SUPERVISORS OF HIGHWAYS, and in the Acts of a general nature of the State of Ohio, the Act of 25th February, 1834, relating to townships, p. 412; besides the peculiar dispositions relating to divers town-officers, such as Township's Clerk, Trustees, Overseers of the Poor, Fence Viewers, Appraisers of Property, Township's Treasurer, Constables, Supervisors of Highways.
33. See the Revised Statutes of the State of New York, part i. chap. xi. vol. i. p. 340. *Id.* chap. xii. p. 366; also in the Acts of the State of Ohio, an act relating to county commissioners, 25th February, 1824, p. 263. See the Digest of the Laws of Pennsylvania, at the words COUNTY-RATES and LEVIES, p. 170. In the State of New York each township elects a representative, who has a share in the administration of the county as well as in that of the township.
34. In some of the Southern States the county courts are charged with all the details of the administration. See the Statutes of the State of Tennessee, *arts.* JUDICIARY, TAXES, &c.
35. For instance, the direction of public instruction centres in the hands of the Government. The legislature names the members of the University, who are denominated Regents; the Governor and Lieutentant-Governor of the State are necessarily of the number. Revised Statutes, vol. i. p. 455. The Regents of the University annually visit the colleges and academies, and make their report to the legislature. Their superintendence is not inefficient, for several reasons: the colleges in order to become corporations stand in need of a charter, which is only granted on the recommendation of the Regents; every year funds are distributed by the State for the encouragement of learning, and the Regents are the distributors of this money. See Chap. xv. 'Public Instruction,' Revised Statutes, vol. i. p. 455. The school-commissioners are obliged to send an annual report to the Superintendent of the Republic—Id. p. 488. A similar report is annually made to the same person on the number and condition of the poor.—Id. p. 631.
36. If any one conceives himself to be wronged by the school-commissioners (who are town-officers), he can appeal to the superintendent of the primary schools, whose decision is final.—Revised Statutes, vol. i. p. 487. Provisions similar to those above cited are to be met with from time to time in the laws of the State of New York; but in general these attempts at centralization are weak and unproductive. The great authorities of the State have the right of watching and controlling the subordinate agents, without that of rewarding or punishing them. The same individual is never empowered to give an order and to punish disobedience; he has therefore the right of commanding, without the means of exacting compliance. In 1830 the Superintendent of Schools complained in his Annual Report addressed to the legislature that several school-commissioners had neglected, notwithstanding his application, to furnish him with the accounts which were due. He added that if this omission continued he should be obliged to prosecute them, as the law directs, before the proper tribunals.
37. Thus the district-attorney is directed to recover all fines below the sum of fifty dollars, unless such a right has been specially awarded to another magistrate.—Revised Statutes, vol. i. p. 383.
38. Several traces of centralization may be discovered in Massachusetts; for instance, the committees of the town-schools are directed to make an annual report to the Secretary of State. See Laws of Massachusetts, vol. i. p. 367.
39. In Massachusetts the Senate is not invested with any administrative functions.
40. As in the State of New York.

41. Practically speaking, it is not always the Governor who executes the plans of the Legislature; it often happens that the latter, in voting a measure, names special agents to superintend the execution of it.
42. In some of the States the justices of the peace are not elected by the Governor.
43. [The war of 1862 cruelly belied this statement, and in the course of the struggle the North alone called two millions and a half of men to arms; but to the honor of the United States it must be added that, with the cessation of the contest, this army disappeared as rapidly as it had been raised.—*Translator's Note.*]
44. The authority which represents the State ought not, I think, to waive the right of inspecting the local administration, even when it does not interfere more actively. Suppose, for instance, that an agent of the Government was stationed at some appointed spot in the country, to prosecute the misdemeanours of the town and county officers, would not a more uniform order be the result, without in any way compromising the independence of the township? Nothing of the kind, however, exists in America: there is nothing above the county-courts, which have, as it were, only an incidental cognizance of the offences they are meant to repress.
45. China appears to me to present the most perfect instance of that species of well-being which a completely central administration may furnish to the nations among which it exists. Travellers assure us that the Chinese have peace without happiness, industry without improvement, stability without strength, and public order without public morality. The condition of society is always tolerable, never excellent. I am convinced that, when China is opened to European observation, it will be found to contain the most perfect model of a central administration which exists in the universe.
46. A writer of talent, who, in the comparison which he has drawn between the finances of France and those of the United States, has proved that ingenuity cannot always supply the place of a knowledge of facts, very justly reproaches the Americans for the sort of confusion which exists in the accounts of the expenditure in the townships; and after giving the model of a Departmental Budget in France, he adds: 'We are indebted to centralization, that admirable invention of a great man, for the uniform order and method which prevail alike in all the municipal budgets, from the largest town to the humblest commune.' Whatever may be my admiration of this result, when I see the communes of France, with their excellent system of accounts, plunged into the grossest ignorance of their true interests, and abandoned to so incorrigible an apathy that they seem to vegetate rather than to live; when, on the other hand, I observe the activity, the information, and the spirit of enterprise which keep society in perpetual labour, in those American townships whose budgets are drawn up with small method and with still less uniformity, I am struck by the spectacle; for to my mind the end of a good government is to ensure the welfare of a people, and not to establish order and regularity in the midst of its misery and its distress. I am therefore led to suppose that the prosperity of the American townships and the apparent confusion of their accounts, the distress of the French communes and the perfection of their budget, may be attributable to the same cause. At any rate I am suspicious of a benefit which is united to so many evils, and I am not averse to an evil which is compensated by so many benefits.
47. See Appendix, I.
48. See Appendix, K.

Chapter VI: Judicial Power in the United States, and Its Influence on Political Society

1. See Appendix L.
2. See Appendix M.
3. [The fifth article of the original Constitution of the United States provides the mode in which amendments of the Constitution may be made. Amendments must be proposed by

two-thirds of both Houses of Congress, and ratified by the Legislatures of three-fourths of the several States. Fifteen amendments of the Constitution have been made at different times since 1789, the most important of which are the thirteenth, fourteenth, and fifteenth, framed and ratified after the Civil War. The original Constitution of the United States, followed by these fifteen amendments, is printed at the end of this edition.—*Translator's Note,* 1874.]

Chapter VII: Political Jurisdiction in the United States

1. [As it existed under the constitutional monarchy down to 1848.]
2. Chapter I. sect. ii. § 8.
3. See the Constitutions of Illinois, Maine, Connecticut, and Georgia.
4. See Appendix, N. [The impeachment of President Andrew Johnson in 1868—which was resorted to by his political opponents solely as a means of turning him out of office, for it could not be contended that he had been guilty of high crimes and misdemeanours, and he was in fact honourably acquitted and reinstated in office—is a striking confirmation of the truth of this remark.—*Translator's Note,* 1874.]

Chapter VIII: The Federal Constitution

1. See the articles of the first confederation formed in 1778. This constitution was not adopted by all the States until 1781. See also the analysis given of this constitution in 'The Federalist' from No. 15 to No. 22, inclusive, and Story's 'Commentaries on the Constitution of the United States,' pp. 85–115.
2. Congress made this declaration on February 21, 1787.
3. It consisted of fifty-five members; Washington, Madison, Hamilton, and the two Morrises were amongst the number.
4. It was not adopted by the legislative bodies, but representatives were elected by the people for this sole purpose; and the new constitution was discussed at length in each of these assemblies.
5. See the Amendment to the Federal Constitution; 'Federalist,' No. 32; Story, p. 711; Kent's 'Commentaries,' vol. i. p. 364. It is to be observed that whenever the *exclusive* right of regulating certain matters is not reserved to Congress by the Constitution, the States may take up the affair until it is brought before the National Assembly. For instance, Congress has the right of making a general law on bankruptcy, which, however, it neglects to do. Each State is then at liberty to make a law for itself. This point has been established by discussion in the law-courts, and may be said to belong more properly to jurisprudence.
6. The action of this court is indirect, as we shall hereafter show.
7. It is thus that 'The Federalist,' No. 45, explains the division of supremacy between the Union and the States: 'The powers delegated by the Constitution to the Federal Government are few and defined. Those which are to remain in the State Governments are numerous and indefinite. The former will be exercised principally on external objects, as war, peace, negotiation, and foreign commerce. The powers reserved to the several States will extend to all the objects which, in the ordinary course of affairs, concern the internal order and prosperity of the State.' I shall often have occasion to quote 'The Federalist' in this work. When the bill which has since become the Constitution of the United States was submitted to the approval of the people, and the discussions were still pending, three men, who had already acquired a portion of that celebrity which they have since enjoyed.—John Jay, Hamilton, and Madison—formed an association with the intention of explaining to the nation the advantages of the measure which was proposed. With this view they published a series of articles in the shape of a journal, which now form a complete treatise. They entitled their journal 'The Federalist,' a name which has been retained in the work. 'The Federalist' is an excellent book, which ought to be familiar to the statesmen of all countries, although it especially concerns America.

8. See Constitution, sect. 8; 'Federalist,' Nos. 41 and 42; Kent's 'Commentaries,' vol. i. p. 207; Story, pp. 358–382; Ibid. pp. 409–426.
9. Several other privileges of the same kind exist, such as that which empowers the Union to legislate on bankruptcy, to grant patents, and other matters in which its intervention is clearly necessary.
10. Even in these cases its interference is indirect. The Union interferes by means of the tribunals, as will be hereafter shown.
11. Federal Constitution, sect. 10, art. I.
12. Constitution, sects. 8, 9, and 10; 'Federalist,' Nos. 30–36, inclusive, and 41–44; Kent's 'Commentaries,' vol. i. pp. 207 and 381; Story, pp. 329 and 514.
13. [In this chapter the author points out the essence of the conflict between the seceding States and the Union which caused the civil war of 1861.]
14. Every ten years Congress fixes anew the number of representatives which each State is to furnish. The total number was 69 in 1789, and 240 in 1833. (See 'American Almanac,' 1834, p. 194.) The Constitution decided that there should not be more than one representative for every 30,000 persons; but no minimum was fixed on. The Congress has not thought fit to augment the number of representatives in proportion to the increase of population. The first Act which was passed on the subject (April 14, 1792: see 'Laws of the United States,' by Story, vol. i. p. 235) decided that there should be one representative for every 33,000 inhabitants. The last Act, which was passed in 1832, fixes the proportion at one for 48,000. The population represented is composed of all the free men and of three-fifths of the slaves.
15. [The last Act of apportionment, passed February 2, 1872, fixes the representation at one to 134,684 inhabitants. There are now (1875) 283 members of the lower House of Congress, and 9 for the States at large, making in all 292 members. The old States have of course lost the representatives which the new States have gained.—*Translator's Note.*]
16. See 'The Federalist,' Nos. 52–56, inclusive; Story, pp. 199—314; Constitution of the United States, sects. 2 and 3.
17. See 'The Federalist,' Nos. 67–77; Constitution of the United States, art. 2; Story, p. 315, pp. 615–780; Kent's 'Commentaries,' p. 255.
18. The Constitution had left it doubtful whether the President was obliged to consult the Senate in the removal as well as in the appointment of Federal officers. 'The Federalist' (No. 77) seemed to establish the affirmative; but in 1789 Congress formally decided that, as the President was responsible for his actions, he ought not to be forced to employ agents who had forfeited his esteem. See Kent's 'Commentaries,' vol. i. p. 289.
19. [This comparison applied to the Constitutional King of France and to the powers he held under the Charter of 1830, till the overthrow of the monarchy in 1848.—*Translator's Note.*]
20. The sums annually paid by the State to these officers amount to 200,000,000 francs (eight millions sterling).
21. This number is extracted from the 'National Calendar' for 1833. The 'National Calendar' is an American almanac which contains the names of all the Federal officers. It results from this comparison that the King of France has eleven times as many places at his disposal as the President, although the population of France is not much more than double that of the Union.
22. [I have not the means of ascertaining the number of appointments now at the disposal of the President of the United States, but his patronage and the abuse of it have largely increased since 1833.—*Translator's Note,* 1875.]
23. [This, however, may be a great danger. The period during which Mr. Buchanan retained office, after the election of Mr. Lincoln, from November, 1860, to March, 1861, was that which enabled the seceding States of the South to complete their preparations for the civil war, and the Executive Government was paralysed. No greater evil could befall a nation.—*Translator's Note.*]

24. As many as it sends members to Congress. The number of electors at the election of 1833 was 288. (See 'The National Calendar,' 1833.)
25. The electors of the same State assemble, but they transmit to the central government the list of their individual votes, and not the mere result of the vote of the majority.
26. In this case it is the majority of the States, and not the majority of the members, which decides the question; so that New York has not more influence in the debate than Rhode Island. Thus the citizens of the Union are first consulted as members of one and the same community; and, if they cannot agree, recourse is had to the division of the States, each of which has a separate and independent vote. This is one of the singularities of the Federal Constitution which can only be explained by the jar of conflicting interests.
27. Jefferson, in 1801, was not elected until the thirty-sixth time of balloting.
28. [General Grant is now (1874) the eighteenth President of the United States.]
29. [Not always. The election of President Lincoln was the signal of civil war.—*Translator's Note.*]
30. See Chapter VI., entitled 'Judicial Power in the United States.' This chapter explains the general principles of the American theory of judicial institutions. See also 'The Federal Constitution,' Art. 3. See 'The Federalists,' Nos. 78–83, inclusive; and a work entitled 'Constitutional Law,' being a view of the practice and jurisdiction of the courts of the United States, by Thomas Sergeant. See Story, pp. 134, 162, 489, 511, 581, 668; and the organic law of the 24th September, 1789, in the 'Collection of the Laws of the United States,' by Story, vol. i. p. 53.
31. Federal laws are those which most require courts of justice, and those at the same time which have most rarely established them. The reason is that confederations have usually been formed by independent States, which entertained no real intention of obeying the central Government, and which very readily ceded the right of command to the federal executive, and very prudently reserved the right of non-compliance to themselves.
32. The Union was divided into districts, in each of which a resident Federal judge was appointed, and the court in which he presided was termed a 'District Court.' Each of the judges of the Supreme Court annually visits a certain portion of the Republic, in order to try the most important causes upon the spot; the court presided over by this magistrate is styled a 'Circuit Court.' Lastly, all the most serious cases of litigation are brought before the Supreme Court, which holds a solemn session once a year, at which all the judges of the Circuit Courts must attend. The jury was introduced into the Federal Courts in the same manner, and in the same cases, as into the courts of the States.
33. It will be observed that no analogy exists between the Supreme Court of the United States and the French Cour de Cassation, since the latter only hears appeals on questions of law. The Supreme Court decides upon the evidence of the fact as well as upon the law of the case, whereas the Cour de Cassation does not pronounce a decision of its own, but refers the cause to the arbitration of another tribunal. See the law of the 24th September, 1789, 'Laws of the United States,' by Story, vol. i. p. 53.
34. In order to diminish the number of these suits, it was decided that in a great many Federal causes the courts of the States should be empowered to decide conjointly with those of the Union, the losing party having then a right of appeal to the Supreme Court of the United States. The Supreme Court of Virginia contested the right of the Supreme Court of the United States to judge an appeal from its decisions, but unsuccessfully. See 'Kent's Commentaries,' vol. i. p. 300, pp. 370 *et seq.;* Story's 'Commentaries,.' p. 646; and 'The Organic Law of the United States,' vol. i. p. 35.
35. The Constitution also says that the Federal courts shall decide 'controversies between a State and the citizens of another State.' And here a most important question of a constitutional nature arose, which was, whether the jurisdiction given by the Constitution in cases in which a State is a party extended to suits brought *against* a State as well as *by* it, or was

exclusively confined to the latter. The question was most elaborately considered in the case of *Chisholm v. Georgia,* and was decided by the majority of the Supreme Court in the affirmative. The decision created general alarm among the States, and an amendment was proposed and ratified by which the power was entirely taken away, so far as it regards suits brought *against* a State. See Story's 'Commentaries,' p. 624, or in the large edition § 1677.

36. As for instance, all cases of piracy.
37. This principle was in some measure restricted by the introduction of the several States as independent powers into the Senate, and by allowing them to vote separately in the House of Representatives when the President is elected by that body. But these are exceptions, and the contrary principle is the rule.
38. It is perfectly clear, says Mr. Story ('Commentaries,' p. 503, or in the large edition § 1379), that any law which enlarges, abridges, or in any manner changes the intention of the parties, resulting from the stipulations in the contract, necessarily impairs it. He gives in the same place a very long and careful definition of what is understood by a contract in Federal jurisprudence. A grant made by the State to a private individual, and accepted by him, is a contract, and cannot be revoked by any future law. A charter granted by the State to a company is a contract, and equally binding to the State as to the grantee. The clause of the Constitution here referred to insures, therefore, the existence of a great part of acquired rights, but not of all. Property may legally be held, though it may not have passed into the possessor's hands by means of a contract; and its possession is an acquired right, not guaranteed by the Federal Constitution.
39. A remarkable instance of this is given by Mr. Story (p. 508, or in the large edition § 1388): 'Dartmouth College in New Hampshire had been founded by a charter granted to certain individuals before the American Revolution, and its trustees formed a corporation under this charter. The legislature of New Hampshire had, without the consent of this corporation, passed an act changing the organization of the original provincial charter of the college, and transferring all the rights, privileges, and franchises from the old charter trustees to new trustees appointed under the act. The constitutionality of the act was contested, and, after solemn arguments, it was deliberately held by the Supreme Court that the provincial charter was a contract within the meaning of the Constitution (Art. I. Section 10), and that the emendatory Act was utterly void, as impairing the obligation of that charter. The college was deemed, like other colleges of private foundation, to be a private eleemosynary institution, endowed by its charter with a capacity to take property unconnected with the Government. Its funds were bestowed upon the faith of the charter, and those funds consisted entirely of private donations. It is true that the uses were in some sense public, that is, for the general benefit, and not for the mere benefit of the corporators; but this did not make the corporation a public corporation. It was a private institution for general charity. It was not distinguishable in principle from a private donation, vested in private trustees, for a public charity, or for a particular purpose of beneficence. And the State itself, if it had bestowed funds upon a charity of the same nature, could not resume those funds.'
40. See Chapter VI. on 'Judicial Power in America.'
41. See Kent's 'Commentaries,' vol. i. p. 387.
42. [The number of States has now risen to 46 (1874), besides the district of Columbia.]
43. At this time Alexander Hamilton, who was one of the principal founders of the Constitution, ventured to express the following sentiments in 'The Federalist,' No. 71:—
44. 'There are some who would be inclined to regard the servile pliancy of the Executive to a prevailing current, either in the community or in the Legislature, as its best recommendation. But such men entertain very crude notions, as well of the purposes for which government was instituted as of the true means by which the public happiness may be promoted. The Republican principle demands that the deliberative sense of the community

should govern the conduct of those to whom they entrust the management of their affairs; but it does not require an unqualified complaisance to every sudden breeze of passion, or to every transient impulse which the people may receive from the arts of men who natter their prejudices to betray their interests. It is a just observation, that the people commonly *intend the public good.* This often applies to their very errors. But their good sense would despise the adulator who should pretend that they always *reason right* about the *means* of promoting it. They know from experience that they sometimes err; and the wonder is that they so seldom err as they do, beset, as they continually are, by the wiles of parasites and sycophants; by the snares of the ambitious, the avaricious, the desperate; by the artifices of men who possess their confidence more than they deserve it, and of those who seek to possess rather than to deserve it. When occasions present themselves in which the interests of the people are at variance with their inclinations, it is the duty of persons whom they have appointed to be the guardians of those interests to withstand the temporary delusion, in order to give them time and opportunity for more cool and sedate reflection. Instances might be cited in which a conduct of this kind has saved the people from very fatal consequences of their own mistakes, and has procured lasting monuments of their gratitude to the men who had courage and magnanimity enough to serve them at the peril of their displeasure.'

45. This was the case in Greece, when Philip undertook to execute the decree of the Amphictyons; in the Low Countries, where the province of Holland always gave the law; and, in our own time, in the Germanic Confederation, in which Austria and Prussia assume a great degree of influence over the whole country, in the name of the Diet.
46. Such has always been the situation of the Swiss Confederation, which would have perished ages ago but for the mutual jealousies of its neighbours.
47. I do not speak of a confederation of small republics, but of a great consolidated Republic.
48. See the Mexican Constitution of 1824.
49. [This is precisely what occurred in 1862, and the following paragraph describes correctly the feelings and notions of the South. General Lee held that his primary allegiance was due, not to the Union, but to Virginia.]
50. For instance, the Union possesses by the Constitution the right of selling unoccupied lands for its own profit. Supposing that the State of Ohio should claim the same right in behalf of certain territories lying within its boundaries, upon the plea that the Constitution refers to those lands alone which do not belong to the jurisdiction of any particular State, and consequently should choose to dispose of them itself, the litigation would be carried on in the names of the purchasers from the State of Ohio and the purchasers from the Union, and not in the names of Ohio and the Union. But what would become of this legal fiction if the Federal purchaser was confirmed in his right by the courts of the Union, whilst the other competitor was ordered to retain possession by the tribunals of the State of Ohio?
51. Kent's 'Commentaries,' vol. i. p. 244. I have selected an example which relates to a time posterior to the promulgation of the present Constitution. If I had gone back to the days of the Confederation, I might have given still more striking instances. The whole nation was at that time in a state of enthusiastic excitement; the Revolution was represented by a man who was the idol of the people; but at that very period Congress had, to say the truth, no resources at all at its disposal. Troops and supplies were perpetually wanting. The best-devised projects failed in the execution, and the Union, which was constantly on the verge of destruction, was saved by the weakness of its enemies far more than by its own strength. [All doubt as to the powers of the Federal Executive was, however, removed by its efforts in the civil war, and those powers were largely extended.]
52. [War broke out between the United States and Mexico in 1846, and ended in the conquest of an immense territory, including California.]
53. See Appendix O.

Chapter X: Parties in the United States

1. [It is scarcely necessary to remark that in more recent times the signification of these terms has changed. The Republicans are the representatives of the old Federalists, and the Democrats of the old Republicans.—*Trans, note* (1861).]
2. [The divisions of North and South have since acquired a far greater degree of intensity, and the South, though conquered, still presents a formidable spirit of opposition to Northern government.]

Chapter XI: Liberty of the Press in the United States

1. They only write in the papers when they choose to address the people in their own name; as, for instance, when they are called upon to repel calumnious imputations, and to correct a misstatement of facts.
1. See Appendix, P.
1. It may, however, be doubted whether this rational and self-guiding conviction arouses as much fervour or enthusiastic devotedness in men as their first dogmatical belief.

Chapter XIII: Government of the Democracy in America

1. I here use the word *Magistrates* in the widest sense in which it can be taken; I apply it to all the officers to whom the execution of the laws is intrusted.
2. See the Act of 27th February, 1813. 'General Collection of the Laws of Massachusetts,' vol. ii. p. 331. It should be added that the jurors are afterwards drawn from these lists by lot.
3. See Act of 28th February, 1787. 'General Collection of the Laws of Massachusetts,' vol. i. p. 302.
4. It is needless to observe that I speak here of the democratic form of government as applied to a people, not merely to a tribe.
5. The word *poor* is used here, and throughout the remainder of this chapter, in a relative, not in an absolute sense. Poor men in America would often appear rich in comparison with the poor of Europe; but they may with propriety by styled poor in comparison with their more affluent countrymen.
6. The gross receipts of the Treasury of the United States in 1832 were about 28 millions of dollars; in 1870 they had risen to 411 millions of dollars. The gross expenditure in 1832 was 30 millions of dollars; in 1870, 309 millions of dollars.
7. The easy circumstances in which secondary functionaries are placed in the United States result also from another cause, which is independent of the general tendencies of democracy; every kind of private business is very lucrative, and the State would not be served at all if it did not pay its servants. The country is in the position of a commercial undertaking, which is obliged to sustain an expensive competition, notwithstanding its tastes for economy.
8. The State of Ohio, which contains a million of inhabitants, gives its Governor a salary of only 1,200 dollars (£ 260) a year.
9. To render this assertion perfectly evident, it will suffice to examine the scale of salaries of the agents of the Federal Government. I have added the salaries attached to the corresponding officers in France under the constitutional monarchy to complete the comparison.

United States.

Treasury Department.

Messenger	$700 (150*l.*)
Clerk with lowest salary	$1,000 (217*l.*)
Clerk with highest salary	$1,600 (347*l.*)
Chief Clerk	$2,000 (434*l.*)
Secretary of State	$6,000 (1,300*l.*)
The President	$25,000 (5,400*l.*)

France.

Minisstère des Finances.

Huissier	1,500 fr. (60*l.*)
Clerk with lowest salary,	1,000 to 1,800 fr. (40*l.* to 72*l.*)
Clerk with highest salary	3,200 to 8,600 fr. (128*l.* to 144*l.*)
Secrétaire-général	20,000 fr. (800*l.*)
The Minister	80,000 fr. (3,200*l.*)
The King	12,000,000 fr. (480,000*l.*)

I have perhaps done wrong in selecting France as my standard of comparison. In France the democratic tendencies of the nation exercise an ever-increasing influence upon the Government, and the Chambers show a disposition to raise the low salaries and to lower the principal ones. Thus, the Minister of Finance, who received 160,000 fr. under the Empire, receives 80,000 fr. in 1835: the Directeurs-Généraux of Finance, who then received 50,000 fr., now receive only 20,000 fr. [This comparison is based on the state of things existing in France and the United States in 1831. It has since materially altered in both countries, but not so much as to impugn the truth of the author's observation.]

10. See the American budgets for the cost of indigent citizens and gratuitous instruction. In 1831 50,000*l.* were spent in the State of New York for the maintenance of the poor, and at least 200,000*l.* were devoted to gratuitous instruction. (Williams's 'New York Annual Register,' 1832, pp. 205 and 243.) The State of New York contained only 1,900,000 inhabitants in the year 1830, which is not more than double the amount of population in the Department du Nord in France.

11. The Americans, as we have seen, have four separate budgets, the Union, the States, the Counties, and the Townships having each severally their own. During my stay in America I made every endeavour to discover the amount of the public expenditure in the townships and counties of the principal States of the Union, and I readily obtained the budget of the larger townships, but I found it quite impossible to procure that of the smaller ones. I possess, however, some documents relating to county expenses, which, although incomplete, are still curious. I have to thank Mr. Richards, Mayor of Philadelphia, for the budgets of thirteen of the counties of Pennsylvania, viz., Lebanon, Centre, Franklin, Fayette, Montgomery, Luzerne, Dauphin, Buder, Alleghany, Columbia, Northampton, Northumberland, and Philadelphia, for the year 1830. Their population at that time consisted of 495,207 inhabitants. On looking at the map of Pennsylvania, it will be seen that these thirteen counties are scattered in every direction, and so generally affected by the causes which usually influence the condition of a country, that they may easily be supposed to furnish a correct average of the financial state of the counties of Pennsylvania in general; and thus, upon reckoning that the expenses of these counties amounted in the year 1830 to about 72,330*l.*, or nearly 3*s.* for each inhabitant, and calculating that each of them contributed in the same year about 10*s.* 2*d.* towards the Union, and about 3*s.* to the State of Pennsylvania, it appears that they each contributed as their share of all the public expenses (except those of the townships) the sum of 16*s.* 2*d.* This calculation is doubly incomplete, as it applies only to a single year and to one part of the public charges; but it has at least the merit of not being conjectural.

12. Those who have attempted to draw a comparison between the expenses of France and America have at once perceived that no such comparison could be drawn between the total expenditure of the two countries; but they have endeavoured to contrast detached portions of this expenditure. It may readily be shown that this second system is not at all less defective than the first. If I attempt to compare the French budget with the budget of the Union, it must be remembered that the latter embraces much fewer objects than the central Government of the former country, and that the expenditure must consequently be much smaller. If I contrast the budgets of the Departments with those of the States which constitute the Union, it must be observed that, as the power and control exercised by the States is much greater than that which is exercised by the Departments, their expenditure

is also more considerable. As for the budgets of the counties, nothing of the kind occurs in the French system of finances; and it is, again, doubtful whether the corresponding expenses should be referred to the budget of the State or to those of the municipal divisions. Municipal expenses exist in both countries, but they are not always analogous. In America the townships discharge a variety of offices which are reserved in France to the Departments or to the State. It may, moreover, be asked what is to be understood by the municipal expenses of America. The organization of the municipal bodies or townships differs in the several States. Are we to be guided by what occurs in New England or in Georgia, in Pennsylvania or in the State of Illinois? A kind of analogy may very readily be perceived between certain budgets in the two countries; but as the elements of which they are composed always differ more or less, no fair comparison can be instituted between them. [The same difficulty exists, perhaps to a greater degree at the present time, when the taxation of America has largely increased.—1874.]

13. Even if we knew the exact pecuniary contributions of every French and American citizen to the coffers of the State, we should only come at a portion of the truth. Governments do not only demand supplies of money, but they call for personal services, which may be looked upon as equivalent to a given sum. When a State raises an army, besides the pay of the troops, which is furnished by the entire nation, each soldier must give up his time, the value of which depends on the use he might make of it if he were not in the service. The same remark applies to the militia; the citizen who is in the militia devotes a certain portion of valuable time to the maintenance of the public peace, and he does in reality surrender to the State those earnings which he is prevented from gaining. Many other instances might be cited in addition to these. The governments of France and of America both levy taxes of this kind, which weigh upon the citizens; but who can estimate with accuracy their relative amount in the two countries?
14. See the details in the Budget of the French Minister of Marine; and for America, the National Calendar of 1833, p. 228. [But the public debt of the United States in 1870, caused by the Civil War, amounted to 2,480,672,427 dollars; that of France was more than doubled by the extravagance of the Second Empire and by the war of 1870.]
15. [That is precisely what has since occurred.]
16. One of the most singular of these occurrences was the resolution which the Americans took of temporarily abandoning the use of tea. Those who know that men usually cling more to their habits than to their life will doubtless admire this great though obscure sacrifice which was made by a whole people.
17. [The Civil War showed that when the necessity arose the American people, both in the North and in the South, are capable of making the most enormous sacrifices, both in money and in men.]
18. 'The President,' says the Constitution, Art. II, sect. 2, Section 2, 'shall have power, by and with the advice and consent of the Senate, to make treaties, provided two-thirds of the senators present concur.' The reader is reminded that the senators are returned for a term of six years, and that they are chosen by the legislature of each State.
19. See the fifth volume of Marshall's 'Life of Washington.' 'In a government constituted like that of the United States,' he says, 'it is impossible for the chief magistrate, however firm he may be, to oppose for any length of time the torrent of popular opinion; and the prevalent opinion of that day seemed to incline to war. In fact, in the session of Congress held at the time, it was frequently seen that Washington had lost the majority in the House of Representatives.' The violence of the language used against him in public was extreme, and in a political meeting they did not scruple to compare him indirectly to the treacherous Arnold. 'By the opposition,' says Marshall, 'the friends of the administration were declared to be an aristocratic and corrupt faction, who, from a desire to introduce monarchy, were hostile to France and under the influence of Britain; that they were a paper nobility,

whose extreme sensibility at every measure which threatened the funds, induced a tame submission to injuries and insults, which the interests and honor of the nation required them to resist.'

Chapter XIV: What the Real Advantages Are Which American Society Derives from the Government of the Democracy

1. [The legislation of England for the forty years is certainly not fairly open to this criticism, which was written before the Reform Bill of 1832, and accordingly Great Britain has thus far escaped and surmounted the perils and calamities to which she seemed to be exposed.]
1. [This, too, has been amended by much larger provisions for the amusements of the people in public parks, gardens, museums, &c; and the conduct of the people in these places of amusement has improved in the same proportion.]
1. At the time of my stay in the United States the Temperance Societies already consisted of more than 270,000 members, and their effect had been to diminish the consumption of fermented liquors by 500,000 gallons per annum in the State of Pennsylvania alone.
1. The same remark was made at Rome under the first Caesars. Montesquieu somewhere alludes to the excessive despondency of certain Roman citizens who, after the excitement of political life, were all at once flung back into the stagnation of private life.

Chapter XV: Unlimited Power of the Majority in the United States, and Its Consequences

1. We observed, in examining the Federal Constitution, that the efforts of the legislators of the Union had been diametrically opposed to the present tendency. The consequence has been that the Federal Government is more independent in its sphere than that of the States. But the Federal Government scarcely ever interferes in any but external affairs; and the governments of the State are in reality the authorities which direct society in America.
2. The legislative acts promulgated by the State of Massachusetts alone, from the year 1780 to the present time, already fill three stout volumes; and it must not be forgotten that the collection to which I allude was published in 1823, when many old laws which had fallen into disuse were omitted. The State of Massachusetts, which is not more populous than a department of France, may be considered as the most stable, the most consistent, and the most sagacious in its undertakings of the whole Union.
3. No one will assert that a people cannot forcibly wrong another people; but parties may be looked upon as lesser nations within a greater one, and they are aliens to each other: if, therefore, it be admitted that a nation can act tyrannically towards another nation, it cannot be denied that a party may do the same towards another party.
4. A striking instance of the excesses which may be occasioned by the despotism of the majority occurred at Baltimore in the year 1812. At that time the war was very popular in Baltimore. A journal which had taken the other side of the question excited the indignation of the inhabitants by its opposition. The populace assembled, broke the printing-presses, and attacked the houses of the newspaper editors. The militia was called out, but no one obeyed the call; and the only means of saving the poor wretches who were threatened by the frenzy of the mob was to throw them into prison as common malefactors. But even this precaution was ineffectual; the mob collected again during the night, the magistrates again made a vain attempt to call out the militia, the prison was forced, one of the newspaper editors was killed upon the spot, and the others were left for dead; the guilty parties were acquitted by the jury when they were brought to trial.
5. This power may be centred in an assembly, in which case it will be strong without being stable; or it may be centred in an individual, in which case it will be less strong, but more stable.

6. I presume that it is scarcely necessary to remind the reader here, as well as throughout the remainder of this chapter, that I am speaking, not of the Federal Government, but of the several governments of each State, which the majority controls at its pleasure.
7. 15th March, 1789.

Chapter XVI: Causes Which Mitigate the Tyranny of the Majority in the United States

1. See chapter VI. on the 'Judicial Power in the United States.'
2. The investigation of trial by jury as a judicial institution, and the appreciation of its effects in the United States, together with the advantages the Americans have derived from it, would suffice to form a book, and a book upon a very useful and curious subject. The State of Louisiana would in particular afford the curious phenomenon of a French and English legislation, as well as a French and English population, which are gradually combining with each other. See the 'Digeste des Lois de la Louisiane,' in two volumes; and the 'Traite surles Regies des Actions civiles,' printed in French and English at New Orleans in 1830.
3. All the English and American jurists are unanimous upon this head. Mr. Story, judge of the Supreme Court of the United States, speaks, in his 'Treatise on the Federal Constitution,' of the advantages of trial by jury in civil cases:—'The inestimable privilege of a trial by jury in civil cases—a privilege scarcely inferior to that in criminal cases, which is counted by all persons to be essential to political and civil liberty. . . .' (Story, book iii. ch. xxxviii.)
4. If it were our province to point out the utility of the jury as a judicial institution in this place, much might be said, and the following arguments might be brought forward amongst others:—
5. By introducing the jury into the business of the courts you are enabled to diminish the number of judges, which is a very great advantage. When judges are very numerous, death is perpetually thinning the ranks of the judicial functionaries, and laying places vacant for newcomers. The ambition of the magistrates is therefore continually excited, and they are naturally made dependent upon the will of the majority, or the individual who fills up the vacant appointments; the officers of the court then rise like the officers of an army. This state of things is entirely contrary to the sound administration of justice, and to the intentions of the legislator. The office of a judge is made inalienable in order that he may remain independent: but of what advantage is it that his independence should be protected if he be tempted to sacrifice it of his own accord? When judges are very numerous many of them must necessarily be incapable of performing their important duties, for a great magistrate is a man of no common powers; and I am inclined to believe that a half-enlightened tribunal is the worst of all instruments for attaining those objects which it is the purpose of courts of justice to accomplish. For my own part, I had rather submit the decision of a case to ignorant jurors directed by a skilful judge than to judges a majority of whom are imperfectly acquainted with jurisprudence and with the laws.
6. An important remark must, however, be made. Trial by jury does unquestionably invest the people with a general control over the actions of citizens, but it does not furnish means of exercising this control in all cases, or with an absolute authority. When an absolute monarch has the right of trying offences by his representatives, the fate of the prisoner is, as it were, decided beforehand. But even if the people were predisposed to convict, the composition and the non-responsibility of the jury would still afford some chances favourable to the protection of innocence.
7. [This may be true to some extent of special juries, but not of common juries. The author seems not to have been aware that the qualifications of jurors in England vary exceedingly.]
8. See Appendix, Q.
9. See Appendix, R.

10. The Federal judges decide upon their own authority almost all the questions most important to the country.

Chapter XVII: Principal Causes Which Tend to Maintain the Democratic Republic in the United States

1. The United States have no metropolis, but they already contain several very large cities. Philadelphia reckoned 161,000 inhabitants and New York 202,000 in the year 1830. The lower orders which inhabit these cities constitute a rabble even more formidable than the populace of European towns. They consist of freed blacks in the first place, who are condemned by the laws and by public opinion to a hereditary state of misery and degradation. They also contain a multitude of Europeans who have been driven to the shores of the New World by their misfortunes or their misconduct; and these men inoculate the United States with all our vices, without bringing with them any of those interests which counteract their baneful influence. As inhabitants of a country where they have no civil rights, they are ready to turn all the passions which agitate the community to their own advantage; thus, within the last few months serious riots have broken out in Philadelphia and in New York. Disturbances of this kind are unknown in the rest of the country, which is nowise alarmed by them, because the population of the cities has hitherto exercised neither power nor influence over the rural districts. Nevertheless, I look upon the size of certain American cities, and especially on the nature of their population, as a real danger which threatens the future security of the democratic republics of the New World; and I venture to predict that they will perish from this circumstance unless the government succeeds in creating an armed force, which, whilst it remains under the control of the majority of the nation, will be independent of the town population, and able to repress its excesses.

   [The population of the city of New York had risen, in 1870, to 942,292, and that of Philadelphia to 674,022. Brooklyn, which may be said to form part of New York city, has a population of 396,099, in addition to that of New York. The frequent disturbances in the great cities of America, and the excessive corruption of their local governments—over which there is no effectual control–are amongst the greatest evils and dangers of the country.]
2. [The number of foreign immigrants into the United States in the last fifty years (from 1820 to 1871) is stated to be 7,556,007. Of these, 4,104,553 spoke English that is, they came from Great Britain, Ireland, or the British colonies; 2,643,069 came from Germany or northern Europe; and about half a million from the south of Europe.]
3. In New England the estates are exceedingly small, but they are rarely subjected to further division.
4. [It is difficult to ascertain with accuracy the amount of the Roman Catholic population of the United States, but in 1868 an able writer in the *Edinburgh Review* (vol. cxxvii. p. 521) affirmed that the whole Catholic population of the United States was then about 4,000,000, divided into 43 dioceses, with 3,795 churches, under the care of 45 bishops and 2,317 clergymen. But this rapid increase is mainly supported by immigration from the Catholic countries of Europe.]
5. The New York *Spectator* of August 23, 1831, relates the fact in the following terms:—'The Court of Common Pleas of Chester county (New York) a few days since rejected a witness who declared his disbelief in the existence of God. The presiding judge remarked that he had not before been aware that there was a man living who did not believe in the existence of God; that this belief constituted the sanction of all testimony in a court of justice, and that he knew of no cause in a Christian country where a witness had been permitted to testify without such belief.'

6. Unless this term be applied to the functions which many of them fill in the schools. Almost all education is entrusted to the clergy.
7. See the 'Constitution of New York,' art. 7, §7 4:—

   'And whereas the ministers of the Gospel are, by their profession, dedicated to the service of God and the care of souls, and ought not to be diverted from the great duties of their functions: therefore no minister of the Gospel, or priest of any denomination whatsoever, shall at any time hereafter, under any pretence or description whatever, be eligible to, or capable of holding, any civil or military office or place within this State.'

   See also the 'Constitutions of North Carolina,' art. 31. Virginia. South Carolina, art. 1. §7 23. Kentucky, art. 2. §7 26. Tennessee, art. 8. §7 1. Louisiana, art. 2. §7 22.
8. [This cannot be said with truth of the country of Kent, Story, and Wheaton.]
9. [In the Northern States the number of persons destitute of instruction is inconsiderable, the largest number being 241,152 in the State of New York (according to Spaulding's 'Handbook of American Statistics' for 1874); but in the South no less than 1,516,339 whites and 2,671,396 coloured persons are returned as 'illiterate.']
10. I travelled along a portion of the frontier of the United States in a sort of cart which was termed the mail. We passed, day and night, with great rapidity along the roads which were scarcely marked out, through immense forests; when the gloom of the woods became impenetrable the coachman lighted branches of fir, and we journeyed along by the light they cast. From time to time we came to a hut in the midst of the forest, which was a post-office. The mail dropped an enormous bundle of letters at the door of this isolated dwelling, and we pursued our way at full gallop, leaving the inhabitants of the neighbouring log-houses to send for their share of the treasure.

    [When the author visited America the locomotive and the railroad were scarcely invented, and not yet introduced in the United States. It is superfluous to point out the immense effect of those inventions in extending civilisation and developing the resources of that vast continent. In 1831 there were 51 miles of railway in the United States; in 1872 there were 60,000 miles of railway.]
11. In 1832 each inhabitant of Michigan paid a sum equivalent to 1 fr. 22 cent. (French money) to the post-office revenue, and each inhabitant of the Floridas paid 1 fr. 5 cent. (See *National Calendar,* 1833, p. 244.) In the same year each inhabitant of the Departement du Nord paid 1 fr. 4 cent, to the revenue of the French post-office. (See the *Compte rendu de l'administration des Finances,* 1833, p. 623.) Now the State of Michigan only contained at that time 7 inhabitants per square league and Florida only 5: the public instruction and the commercial activity of these districts is inferior to that of most of the States in the Union, whilst the Departement du Nord, which contains 3,400 inhabitants per square league, is one of the most enlightened and manufacturing parts of France.
12. I remind the reader of the general signification which I give to the word *manners,* namely, the moral and intellectual characteristics of social man taken collectively.
13. [A remark which, since the great Civil War of 1862, ceases to be applicable.]
14. [This prediction of the return of France to imperial despotism, and of the true character of that despotic power, was written in 1832, and realized to the letter in 1852.]

Chapter XVIII: The Present and Probable Future Condition of the Three Races Which Inhabit the Territory of the United States

1. The native of North America retains his opinions and the most insignificant of his habits with a degree of tenacity which has no parallel in history. For more than two hundred years the wandering tribes of North America have had daily intercourse with the whites, and they have never derived from them either a custom or an idea. Yet the Europeans have exercised a powerful influence over the savages: they have made them more licentious, but not more European. In the summer of 1831 I happened to be beyond Lake Michigan, at a place called

Green Bay, which serves as the extreme frontier between the United States and the Indians on the north-western side. Here I became acquainted with an American officer, Major H., who, after talking to me at length on the inflexibility of the Indian character, related the following fact:—'I formerly knew a young Indian,' said he, 'who had been educated at a college in New England, where he had greatly distinguished himself, and had acquired the external appearance of a member of civilised society. When the war broke out between ourselves and the English in 1810, I saw this young man again; he was serving in our army, at the head of the warriors of his tribe, for the Indians were admitted amongst the ranks of the Americans, upon condition that they would abstain from their horrible custom of scalping their victims. On the evening of the battle of* * *, C. came and sat himself down by the fire of our bivouac. I asked him what had been his fortune that day: he related his exploits; and growing warm and animated by the recollection of them, he concluded by suddenly opening the breast of his coat, saying, "You must not betray me—see here!" And I actually beheld,' said the Major, 'between his body and his shirt, the skin and hair of an English head, still dripping with gore.'

2. In the thirteen original States there are only 6,273 Indians remaining. (See Legislative Documents, 20th Congress, No. 117, p. 90.) [The decrease is now far greater, and is verging on extinction. See note at p. 361.]

3. Messrs. Clarke and Cass, in their Report to Congress the 4th of February, 1829, p. 23, expressed themselves thus:—'The time when the Indians generally could supply themselves with food and clothing, without any of the articles of civilised life, has long since passed away. The more remote tribes, beyond the Mississippi, who live where immense herds of buffalo are yet to be found and who follow those animals in their periodical migrations, could more easily than any others recur to the habits of their ancestors, and live without the white man or any of his manufactures. But the buffalo is constantly receding. The smaller animals, the bear, the deer, the beaver, the otter, the muskrat, &c, principally minister to the comfort and support of the Indians; and these cannot be taken without guns, ammunition, and traps. Among the North-Western Indians particularly, the labour of supplying a family with food is excessive. Day after day is spent by the hunter without success, and during this interval his family must subsist upon bark or roots, or perish. Want and misery are around them and among them. Many die every winter from actual starvation.'

The Indians will not live as Europeans live, and yet they can neither subsist without them, nor exactly after the fashion of their fathers. This is demonstrated by a fact which I likewise give upon official authority. Some Indians of a tribe on the banks of Lake Superior had killed a European; the American Government interdicted all traffic with the tribe to which the guilty parties belonged, until they were delivered up to justice. This measure had the desired effect.

4. 'Five years ago,' (says Volney in his 'Tableau des États-Unis,' p. 370) 'in going from Vincennes to Kaskaskia, a territory which now forms part of the State of Illinois, but which at the time I mention was completely wild (1797), you could not cross a prairie without seeing herds of from four to five hundred buffaloes. There are now none remaining; they swam across the Mississippi to escape from the hunters, and more particularly from the bells of the American cows.'

5. The truth of what I here advance may be easily proved by consulting the tabular statement of Indian tribes inhabiting the United States and their territories. (Legislative Documents, 20th Congress, No. 117, p. 90–105.) It is there shown that the tribes in the centre of America are rapidly decreasing, although the Europeans are still at a considerable distance from them.

6. 'The Indians,' say Messrs. Clarke and Cass in their Report to Congress, p. 15, 'are attached to their country by the same feelings which bind us to ours; and, besides, there are certain superstitious notions connected with the alienation of what the Great Spirit gave to their ancestors, which operate strongly upon the tribes who have made few or no cessions, but

which are gradually weakened as our intercourse with them is extended. "We will not sell the spot which contains the bones of our fathers," is almost always the first answer to a proposition for a sale.'

7. See, in the Legislative Documents of Congress (Doc. 117), the narrative of what takes place on these occasions. This curious passage is from the above-mentioned Report, made to Congress by Messrs. Clarke and Cass in February, 1829. Mr. Cass is now the Secretary at War.

   'The Indians,' says the Report, 'reach the treaty-ground poor and almost naked. Large quantities of goods are taken there by the traders, and are seen and examined by the Indians. The women and children become importunate to have their wants supplied, and their influence is soon exerted to induce a sale. Their improvidence is habitual and unconquerable. The gratification of his immediate wants and desires is the ruling passion of an Indian. The expectation of future advantages seldom produces much effect. The experience of the past is lost, and the prospects of the future disregarded. It would be utterly hopeless to demand a cession of land, unless the means were at hand of gratifying their immediate wants; and when their condition and circumstances are fairly considered, it ought not to surprise us that they are so anxious to relieve themselves.'

8. On the 19th of May, 1830, Mr. Edward Everett affirmed before the House of Representatives, that the Americans had already acquired by treaty, to the east and west of the Mississippi, 230,000,000 of acres. In 1808 the Ovages gave up 48,000,000 acres for an annual payment of 1,000 dollars. In 1818 the Quapaws yielded up 29,000,000 acres for 4,000 dollars. They reserved for themselves a territory of 1,000,000 acres for a hunting-ground. A solemn oath was taken that it should be respected: but before long it was invaded like the rest.

   Mr. Bell, in his 'Report of the Committee on Indian Affairs,' February 24, 1830, has these words:—'To pay an Indian tribe what their ancient hunting-grounds are worth to them, after the game is fled or destroyed, as a mode of appropriating wild lands claimed by Indians, has been found more convenient, and certainly it is more agreeable to the forms of justice, as well as more merciful, than to assert the possession of them by the sword. Thus the practice of buying Indian titles is but the substitute which humanity and expediency have imposed, in place of the sword, in arriving at the actual enjoyment of property claimed by the right of discovery, and sanctioned by the natural superiority allowed to the claims of civilised communities over those of savage tribes. Up to the present time so invariable has been the operation of certain causes, first in diminishing the value of forest lands to the Indians, and secondly in disposing them to sell readily, that the plan of buying their right of occupancy has never threatened to retard, in any perceptible degree, the prosperity of any of the States.' (Legislative Documents, 21st Congress, No. 227, p. 6.)

9. This seems, indeed, to be the opinion of almost all American statesmen. 'Judging of the future by the past,' says Mr. Cass, 'we cannot err in anticipating a progressive diminution of their numbers, and their eventual extinction, unless our border should become stationary, and they be removed beyond it, or unless some radical change should take place in the principles of our intercourse with them, which it is easier to hope for than to expect.'

10. Amongst other warlike enterprises, there was one of the Wampanaogs, and other confederate tribes, under Metacom in 1675, against the colonists of New England; the English were also engaged in war in Virginia in 1622.

11. See the 'Histoire de la Nouvelle France," by Charlevoix, and the work entitled 'Lettres edifiantes.'

12. 'In all the tribes,' says Volney, in his 'Tableau des États-Unis,' p. 423, 'there still exists a generation of old warriors, who cannot forbear, when they see their countrymen using the hoe, from exclaiming against the degradation of ancient manners, and asserting that the savages owe their decline to these innovations; adding, that they have only to return to their primitive habits in order to recover their power and their glory.'

13. The following description occurs in an official document:—'Until a young man has been engaged with an enemy, and has performed some acts of valour, he gains no consideration, but is regarded nearly as a woman. In their great war-dances all the warriors in succession strike the post, as it is called, and recount their exploits. On these occasions their auditory consists of the kinsmen, friends, and comrades of the narrator. The profound impression which his discourse produces on them is manifested by the silent attention it receives, and by the loud shouts which hail its termination. The young man who finds himself at such a meeting without anything to recount is very unhappy; and instances have sometimes occurred of young warriors, whose passions had been thus inflamed, quitting the war-dance suddenly, and going off alone to seek for trophies which they might exhibit, and adventures which they might be allowed to relate.'
14. These nations are now swallowed up in the States of Georgia, Tennessee, Alabama, and Mississippi. There were formerly in the South four great nations (remnants of which still exist), the Choctaws, the Chickasaws, the Creeks, and the Cherokees. The remnants of these four nations amounted, in 1830, to about 75,000 individuals. It is computed that there are now remaining in the territory occupied or claimed by the Anglo-American Union about 300,000 Indians. (See Proceedings of the Indian Board in the City of New York.) The official documents supplied to Congress make the number amount to 313,130. The reader who is curious to know the names and numerical strength of all the tribes which inhabit the Anglo-American territory should consult the documents I refer to. (Legislative Documents, 20th Congress, No. 117, p. 90–105.) [In the Census of 1870 it is stated that the Indian population of the United States is only 25,731, of whom 7,241 are in California.]
15. I brought back with me to France one or two copies of this singular publication.
16. See in the Report of the Committee on Indian Affairs, 21st Congress, No. 227, p. 23, the reasons for the multiplication of Indians of mixed blood among the Cherokees. The principal cause dates from the War of Independence. Many Anglo-Americans of Georgia, having taken the side of England, were obliged to retreat among the Indians, where they married.
17. Unhappily the mixed race has been less numerous and less influential in North America than in any other country. The American continent was peopled by two great nations of Europe, the French and the English. The former were not slow in connecting themselves with the daughters of the natives, but there was an unfortunate affinity between the Indian character and their own: instead of giving the tastes and habits of civilised life to the savages, the French too often grew passionately fond of the state of wild freedom they found them in. They became the most dangerous of the inhabitants of the desert, and won the friendship of the Indian by exaggerating his vices and his virtues. M. de Senonville, the governor of Canada, wrote thus to Louis XIV. in 1685:—'It has long been believed that in order to civilise the savages we ought to draw them nearer to us. But there is every reason to suppose we have been mistaken. Those which have been brought into contact with us have not become French, and the French who have lived among them are changed into savages, affecting to dress and live like them.' ('History of New France,' by Charlevoix, vol. ii., p. 345.) The Englishman, on the contrary, continuing obstinately attached to the customs and the most insignificant habits of his forefathers, has remained in the midst of the American solitudes just what he was in the bosom of European cities; he would not allow of any communication with savages whom he despised, and avoided with care the union of his race with theirs. Thus while the French exercised no salutary influence over the Indians, the English have always remained alien from them.
18. There is in the adventurous life of the hunter a certain irresistible charm, which seizes the heart of man and carries him away in spite of reason and experience. This is plainly shown by the memoirs of Tanner. Tanner is a European who was carried away at the age of six by the Indians, and has remained thirty years with them in the woods. Nothing can be

conceived more appalling that the miseries which he describes. He tells us of tribes without a chief, families without a nation to call their own, men in a state of isolation, wrecks of powerful tribes wandering at random amid the ice and snow and desolate solitudes of Canada. Hunger and cold pursue them; every day their life is in jeopardy. Amongst these men, manners have lost their empire, traditions are without power. They become more and more savage. Tanner shared in all these miseries; he was aware of his European origin; he was not kept away from the whites by force; on the contrary, he came every year to trade with them, entered their dwellings, and witnessed their enjoyments; he knew that whenever he chose to return to civilised life he was perfectly able to do so—and he remained thirty years in the deserts. When he came into civilised society he declared that the rude existence which he described, had a secret charm for him which he was unable to define: he returned to it again and again: at length he abandoned it with poignant regret; and when he was at length fixed among the Whites, several of his children refused to share his tranquil and easy situation. I saw Tanner myself at the lower end of Lake Superior; he seemed to me to be more like a savage than a civilised being. His book is written without either taste or order; but he gives, even unconsciously, a lively picture of the prejudices, the passions, the vices, and, above all, of the destitution in which he lived.

19. The destructive influence of highly civilised nations upon others which are less so, has been exemplified by the Europeans themselves. About a century ago the French founded the town of Vincennes upon the Wabash, in the middle of the desert; and they lived there in great plenty until the arrival of the American settlers, who first ruined the previous inhabitants by their competition, and afterwards purchased their lands at a very low rate. At the time when M. de Volney, from whom I borrow these details, passed through Vincennes, the number of the French was reduced to a hundred individuals, most of whom were about to pass over to Louisiana or to Canada. These French settlers were worthy people, but idle and uninstructed: they had contracted many of the habits of savages. The Americans, who were perhaps their inferiors, in a moral point of view, were immeasurably superior to them in intelligence: they were industrious, well informed, rich, and accustomed to govern their own community.

I myself saw in Canada, where the intellectual difference between the two races is less striking, that the English are the masters of commerce and manufacture in the Canadian country, that they spread on all sides, and confine the French within limits which scarcely suffice to contain them. In like manner, in Louisiana, almost all activity in commerce and manufacture centres in the hands of the Anglo-Americans.

But the case of Texas is still more striking: the State of Texas is a part of Mexico, and lies upon the frontier between that country and the United States. In the course of the last few years the Anglo-Americans have penetrated into this province, which is still thinly peopled; they purchase land, they produce the commodities of the country, and supplant the original population. It may easily be foreseen that if Mexico takes no steps to check this change, the province of Texas will very shortly cease to belong to that Government.

If the different degrees—comparatively so slight—which exist in European civilisation produce results of such magnitude, the consequences which must ensue from the collision of the most perfect European civilisation with Indian savages may readily be conceived.

20. See in the Legislative Documents (21st Congress, No. 89) instances of excesses of every kind committed by the Whites upon the territory of the Indians, either in taking possession of a part of their lands, until compelled to retire by the troops of Congress, or carrying off their cattle, burning their houses, cutting down their corn, and doing violence to their persons. It appears, nevertheless, from all these documents that the claims of the natives are constantly protected by the Government from the abuse of force. The Union has a representative agent continually employed to reside among the Indians; and the report of the Cherokee agent, which is among the documents I have referred to, is almost always favourable to the Indians.

'The intrusion of Whites,' he says, 'upon the lands of the Cherokees would cause ruin to the poor, helpless, and inoffensive inhabitants.' And he further remarks upon the attempt of the State of Georgia to establish a division line for the purpose of limiting the boundaries of the Cherokees, that the line drawn having been made by the Whites, and entirely upon *ex parte* evidence of their several rights, was of no validity whatever.

21. In 1829 the State of Alabama divided the Creek territory into counties, and subjected the Indian population to the power of European magistrates.

 In 1830 the State of Mississippi assimilated the Choctaws and Chickasaws to the white population, and declared that any of them that should take the title of chief would be punished by a fine of 1000 dollars and a year's imprisonment. When these laws were enforced upon the Choctaws, who inhabited that district, the tribe assembled, their chief communicated to them the intentions of the Whites, and read to them some of the laws to which it was intended that they should submit; and they unanimously declared that it was better at once to retreat again into the wilds.

22. The Georgians, who are so much annoyed by the proximity of the Indians, inhabit a territory which does not at present contain more than seven inhabitants to the square mile. In France there are one hundred and sixty-two inhabitants to the same extent of country.

23. In 1818 Congress appointed commissioners to visit the Arkansas Territory, accompanied by a deputation of Creeks, Choctaws, and Chickasaws. This expedition was commanded by Messrs. Kennedy, M'Coy, Wash Hood, and John Bell. See the different reports of the commissioners, and their journal, in the Documents of Congress, No. 87, House of Representatives.

24. The fifth article of the treaty made with the Creeks in August, 1790, is in the following words:—'The United States solemnly guarantee to the Creek nation all their land within the limits of the United States.'

 The seventh article of the treaty concluded in 1791 with the Cherokees says:—'The United States solemnly guarantee to the Cherokee nation all their lands not hereby ceded.' The following article declared that if any citizen of the United States or other settler not of the Indian race should establish himself upon the territory of the Cherokees, the United States would withdraw their protection from that individual, and give him up to be punished as the Cherokee nation should think fit.

25. This does not prevent them from promising in the most solemn manner to do so. See the letter of the President addressed to the Creek Indians, March 23, 1829 (Proceedings of the Indian Board, in the city of New York, p. 5.) 'Beyond the great river Mississippi, where a part of your nation has gone, your father has provided a country large enough for all of you, and he advises you to remove to it. There your white brothers will not trouble you; they will have no claim to the land, and you can live upon it, you and all your children, as long as the grass grows, or the water runs, in peace and plenty. *It will be yours forever.'*

 The Secretary of War, in a letter written to the Cherokees, April 18, 1829, (see the same work, p. 6), declares to them that they cannot expect to retain possession of the lands at that time occupied by them, but gives them the most positive assurance of uninterrupted peace if they would remove beyond the Mississippi: as if the power which could not grant them protection then, would be able to afford it them hereafter!

26. To obtain a correct idea of the policy pursued by the several States and the Union with respect to the Indians, it is necessary to consult, 1st, 'The Laws of the Colonial and State Governments relating to the Indian Inhabitants.' (See the Legislative Documents, 21st Congress, No. 319.) 2nd, The Laws of the Union on the same subject, and especially that of March 30, 1802. (See Story's 'Laws of the United States.') 3rd, The Report of Mr. Cass, Secretary of War, relative to Indian Affairs, November 29, 1823.

27. December 18, 1829.

28. The honour of this result is, however, by no means due to the Spaniards. If the Indian tribes had not been tillers of the ground at the time of the arrival of the Europeans, they would unquestionably have been destroyed in South as well as in North America.
29. See, amongst other documents, the report made by Mr. Bell in the name of the Committee on Indian Affairs, February 24, 1830, in which is most logically established and most learnedly proved, that 'the fundamental principle, that the Indians had no right by virtue of their ancient possession either of will or sovereignty, has never been abandoned either expressly or by implication.' In perusing this report, which is evidently drawn up by an experienced hand, one is astonished at the facility with which the author gets rid of all arguments founded upon reason and natural right, which he designates as abstract and theoretical principles. The more I contemplate the difference between civilised and uncivilised man with regard to the principles of justice, the more I observe that the former contests the justice of those rights which the latter simply violates.
30. It is well known that several of the most distinguished authors of antiquity, and amongst them Æsop and Terence, were, or had been slaves. Slaves were not always taken from barbarous nations, and the chances of war reduced highly civilised men to servitude.
31. To induce the whites to abandon the opinion they have conceived of the moral and intellectual inferiority of their former slaves, the negroes must change; but as long as this opinion subsists, to change is impossible.
32. See 'Beverley's History of Virginia.' See also in 'Jefferson's Memoirs' some curious details concerning the introduction of negroes into Virginia, and the first Act which prohibited the importation of them in 1778.
33. The number of slaves was less considerable in the North, but the advantages resulting from slavery were not more contested there than in the South. In 1740, the Legislature of the State of New York declared that the direct importation of slaves ought to be encouraged as much as possible, and smuggling severely punished in order not to discourage the fair trader. ('Kent's Commentaries,' vol. ii. p. 206.) Curious researches, by Belknap, upon slavery in New England, are to be found in the 'Historical Collection of Massachusetts,' vol. iv. p. 193. It appears that negroes were introduced there in 1630, but that the legislation and manners of the people were opposed to slavery from the first; see also, in the same work, the manner in which public opinion, and afterwards the laws, finally put an end to slavery.
34. Not only is slavery prohibited in Ohio, but no free negroes are allowed to enter the territory of that State, or to hold property in it. See the Statutes of Ohio.
35. The activity of Ohio is not confined to individuals, but the undertakings of the State are surprisingly great; a canal has been established between Lake Erie and the Ohio, by means of which the valley of the Mississippi communicates with the river of the North, and the European commodities which arrive at New York may be forwarded by water to New Orleans across five hundred leagues of continent.
36. The exact numbers given by the census of 1830 were: Kentucky, 688,844; Ohio, 937,679. [In 1870 the population of Ohio was 2,665,260, that of Kentucky, 1,321,011.]
37. Independently of these causes, which, wherever free workmen abound, render their labour more productive and more economical than that of slaves, another cause may be pointed out which is peculiar to the United States: the sugar-cane has hitherto been cultivated with success only upon the banks of the Mississippi, near the mouth of that river in the Gulf of Mexico. In Louisiana the cultivation of the sugar-cane is exceedingly lucrative, and nowhere does a labourer earn so much by his work, and, as there is always a certain relation between the cost of production and the value of the produce, the price of slaves is very high in Louisiana. But Louisiana is one of the Confederated States, and slaves may be carried thither from all parts of the Union; the price given for slaves in New Orleans consequently raises the value of slaves in all the other markets. The consequence of this is, that in the

countries where the land is less productive, the cost of slave-labour is still very considerable, which gives an additional advantage to the competition of free labour.

38. A peculiar reason contributes to detach the two last-mentioned States from the cause of slavery. The former wealth of this part of the Union was principally derived from the cultivation of tobacco. This cultivation is specially carried on by slaves; but within the last few years the market-price of tobacco has diminished, whilst the value of the slaves remains the same. Thus the ratio between the cost of production and the value of the produce is changed. The natives of Maryland and Virginia are therefore more disposed than they were thirty years ago, to give up slave-labour in the cultivation of tobacco, or to give up slavery and tobacco at the same time.

39. The States in which slavery is abolished usually do what they can to render their territory disagreeable to the negroes as a place of residence; and as a kind of emulation exists between the different States in this respect, the unhappy blacks can only choose the least of the evils which beset them.

40. There is a very great difference between the mortality of the blacks and of the whites in the States in which slavery is abolished; from 1820 to 1831 only one out of forty-two individuals of the white population died in Philadelphia; but one negro out of twenty-one individuals of the black population died in the same space of time. The mortality is by no means so great amongst the negroes who are still slaves. (See 'Emmerson's Medical Statistics,' p. 28.)

41. This is true of the spots in which rice is cultivated; rice-grounds, which are unwholesome in all countries, are particularly dangerous in those regions which are exposed to the beams of a tropical sun. Europeans would not find it easy to cultivate the soil in that part of the New World if it must necessarily be made to produce rice; but may they not subsist without rice-grounds?

42. These States are nearer to the equator than Italy and Spain, but the temperature of the continent of America is very much lower than that of Europe.

43. The Spanish Government formerly caused a certain number of peasants from the Açores to be transported into a district of Louisiana called Attakapas, by way of experiment. These settlers still cultivate the soil without the assistance of slaves, but their industry is so languid as scarcely to supply their most necessary wants.

44. We find it asserted in an American work, entitled 'Letters on the Colonisation Society,' by Mr. Carey, 1833, 'That for the last forty years the black race has increased more rapidly than the white race in the State of South Carolina; and that if we take the average population of the five States of the South into which slaves were first introduced, viz., Maryland, Virginia, South Carolina, North Carolina, and Georgia, we shall find that from 1790 to 1830 the whites have augmented in the proportion of 80 to 100, and the blacks in that of 112 to 100.'

In the United States, in 1830, the population of the two races stood as follows:—

States where slavery is abolished, 6,565,434 whites; 120,520 blacks. Slave States, 3,960,814 whites; 2,208,102 blacks. [In 1870 the United States contained a population of 33,589,377 whites, and 4,880,009 negroes.]

45. This opinion is sanctioned by authorities infinitely weightier than anything that I can say: thus, for instance, it is stated in the 'Memoirs of Jefferson' (as collected by M. Conseil), 'Nothing is more clearly written in the book of destiny than the emancipation of the blacks; and it is equally certain that the two races will never live in a state of equal freedom under the same government, so insurmountable are the barriers which nature, habit, and opinions have established between them.'

46. If the British West India planters had governed themselves, they would assuredly not have passed the Slave Emancipation Bill which the mother-country has recently imposed upon them.

47. This society assumed the name of 'The Society for the Colonisation of the Blacks.' See its annual reports; and more particularly the fifteenth. See also the pamphlet, to which

allusion has already been made, entitled 'Letters on the Colonisation Society, and on its probable Results,' by Mr. Carey, Philadelphia, 1833.

48. This last regulation was laid down by the founders of the settlement; they apprehended that a state of things might arise in Africa similar to that which exists on the frontiers of the United States, and that if the negroes, like the Indians, were brought into collision with a people more enlightened than themselves, they would be destroyed before they could be civilized.

49. Nor would these be the only difficulties attendant upon the undertaking; if the Union undertook to buy up the negroes now in America, in order to transport them to Africa, the price of slaves, increasing with their scarcity, would soon become enormous; and the States of the North would never consent to expend such great sums for a purpose which would procure such small advantages to themselves. If the Union took possession of the slaves in the Southern States by force, or at a rate determined by law, an insurmountable resistance would arise in that part of the country. Both alternatives are equally impossible.

50. In 1830 there were in the United States 2,010,327 slaves and 319,439 free blacks, in all 2,329,766 negroes: which formed about one-fifth of the total population of the United States at that time.

51. [This chapter is no longer applicable to the condition of the Negro Race in the United States, since the abolition of slavery was the result, though not the object, of the great Civil War, and the negroes have been raised to the condition not only of freedmen, but of citizens; and in some States they exercise a preponderating political power by reason of their numerical majority. Thus, in South Carolina there were in 1870, 289,667 Whites and 415,814 Blacks. But the emancipation of the slaves has not solved the problem, how two races so different and so hostile are to live together in peace in one country on equal terms. That problem is as difficult, perhaps more difficult than ever; and to this difficulty the author's remarks are still perfectly applicable.]

52. [This chapter is one of the most curious and interesting portions of the work, because it embraces almost all the constitutional and social questions which were raised by the great secession of the South and decided by the results of the Civil War. But it must be confessed that the sagacity of the author is sometimes at fault in these speculations, and did not save him from considerable errors, which the course of events has since made apparent. He held that 'the legislators of the Constitution of 1789 were not appointed to constitute the government of a single people, but to regulate the association of several States; that the Union was formed by the voluntary agreement of the States, and in uniting together they have not forfeited their nationality, nor have they been reduced to the condition of one and the same people.' Whence he inferred that 'if one of the States chose to withdraw its name from the contract, it would be difficult to disprove its right of doing so; and that the Federal Government would have no means of maintaining its claims directly, either by force or by right.' This is the Southern theory of the Constitution, and the whole case of the South in favour of secession. To many Europeans, and to some American (Northern) jurists, this view appeared to be sound; but it was vigorously resisted by the North, and crushed by force of arms.

The author of this book was mistaken in supposing that the 'Union was a vast body which presents no definite object to patriotic feeling.' When the day of trial came, millions of men were ready to lay down their lives for it. He was also mistaken in supposing that the Federal Executive is so weak that it requires the free consent of the governed to enable it to subsist, and that it would be defeated in a struggle to maintain the Union against one or more separate States. In 1861 nine States, with a population of 8,753,000, seceded, and maintained for four years a resolute but unequal contest for independence, but they were defeated.

Lastly, the author was mistaken in supposing that a community of interests would always prevail between North and South sufficiently powerful to bind them together. He overlooked the influence which the question of slavery must have on the Union the moment that the majority of the people of the North declared against it. In 1831, when the author visited America, the anti-slavery agitation had scarcely begun; and the fact of Southern slavery was accepted by men of all parties, even in the States where there were no slaves: and that was unquestionably the view taken by all the States and by all American statesmen at the time of the adoption of the Constitution, in 1789. But in the course of thirty years a great change took place, and the North refused to perpetuate what had become the 'peculiar institution' of the South, especially as it gave the South a species of aristocratic preponderance. The result was the ratification, in December, 1865, of the celebrated 13th article or amendment of the Constitution, which declared that 'neither slavery nor involuntary servitude—except as a punishment for crime—shall exist within the United States.' To which was soon afterwards added the 15th article, 'The right of citizens to vote shall not be denied or abridged by the United States, or by any State, on account of race, colour, or previous servitude.' The emancipation of several millions of negro slaves without compensation, and the transfer to them of political preponderance in the States in which they outnumber the white population, were acts of the North totally opposed to the interests of the South, and which could only have been carried into effect by conquest.—*Translator's Note.*]

53. See the conduct of the Northern States in the war of 1812. 'During that war,' says Jefferson in a letter to General Lafayette, 'four of the Eastern States were only attached to the Union, like so many inanimate bodies to living men.'

54. The profound peace of the Union affords no pretext for a standing army; and without a standing army a government is not prepared to profit by a favourable opportunity to conquer resistance, and take the sovereign power by surprise. [This note, and the paragraph in the text which precedes, have been shown by the results of the Civil War to be a misconception of the writer.]

55. Thus the province of Holland in the republic of the Low Countries, and the Emperor in the Germanic Confederation, have sometimes put themselves in the place of the union, and have employed the Federal authority to their own advantage.

56. See 'Darby's View of the United States,' p. 435. [In 1870 the number of States and Territories had increased to 38, the population to 38,558,710, and the area of the States, 3,500,000 square miles.—*Translator's Note.*]

57. It is scarcely necessary for me to observe that by the expression *Anglo-Americans,* I only mean to designate the great majority of the nation; for a certain number of isolated individuals are of course to be met with holding very different opinions.

58. Census of 1790 3,929,328.
    ________1830 12,856,165.
    ________1860 31,443,321.
    ________1870 38,555,983.

59. This indeed is only a temporary danger. I have no doubt that in time society will assume as much stability and regularity in the West as it has already done upon the coast of the Atlantic Ocean.

60. Pennsylvania contained 431,373 inhabitants in 1790 [and 3,521,951 in 1870.]

61. The area of the State of New York is about 46,000 square miles. See 'Carey and Lea's American Geography,' p. 142.

62. If the population continues to double every twenty-two years, as it has done for the last two hundred years, the number of inhabitants in the United States in 1852 will be twenty millions; in 1874, forty-eight millions; and in 1896, ninety-six millions. This may still be the case even if the lands on the western slope of the Rocky Mountains should be found to be unfit for cultivation. The territory which is already occupied can easily contain this number of inhabitants. One hundred millions of men disseminated over the surface of the twenty-four States, and the three dependencies, which constitute the Union, would only

give 762 inhabitants to the square league; this would be far below the mean population of France, which is 1,063 to the square league; or of England, which is 1,457; and it would even be below the population of Switzerland, for that country, notwithstanding its lakes and mountains, contains 783 inhabitants to the square league. See 'Malte Brun,' vol. vi. p. 92.

[The actual result has fallen somewhat short of these calculations, in spite of the vast territorial acquisitions of the United States: but in 1874 the population is probably about forty-two millions.]

63. See Legislative Documents, 20th Congress, No. 117, p. 105.
64. 3,672,317; census of 1830.
65. The distance from Jefferson, the capital of the State of Missouri, to Washington is 1,019 miles. ('American Almanac,' 1831, p. 48.)
66. The following statements will suffice to show the difference which exists between the commerce of the South and that of the North:—

    In 1829 the tonnage of all the merchant vessels belonging to Virginia, the two Carolinas, and Georgia (the four great Southern States), amounted to only 5,243 tons. In the same year the tonnage of the vessels of the State of Massachusetts alone amounted to 17,322 tons. (See Legislative Documents, 21st Congress, 2nd session, No. 140, p. 244.) Thus the State of Massachusetts had three times as much shipping as the four above-mentioned States. Nevertheless the area of the State of Massachusetts is only 7,335 square miles, and its population amounts to 610,014 inhabitants [1,457,351 in 1870]; whilst the area of the four other States I have quoted is 210,000 square miles, and their of population 3,047,767. Thus the area of the State of Massachusetts forms only one-thirtieth part the area of the four States; and its population is five times smaller than theirs. (See 'Darby's View of the United States.') Slavery is prejudicial to the commercial prosperity of the South in several different ways; by diminishing the spirit of enterprise amongst the whites, and by preventing them from meeting with as numerous a class of sailors as they require. Sailors are usually taken from the lowest ranks of the population. But in the Southern States these lowest ranks are composed of slaves, and it is very difficult to employ them at sea. They are unable to serve as well as a white crew, and apprehensions would always be entertained of their mutinying in the middle of the ocean, or of their escaping in the foreign countries at which they might touch.
67. 'Darby's View of the United States,' p. 444.
68. It may be seen that in the course of the last ten years (1820–1830) the population of one district, as, for instance, the State of Delaware, has increased in the proportion of 5 per cent.; whilst that of another, as the territory of Michigan, has increased 250 per cent. Thus the population of Virginia had augmented 13 per cent., and that of the border State of Ohio 61 per cent., in the same space of time. The general table of these changes, which is given in the 'National Calendar,' displays a striking picture of the unequal fortunes of the different States.
69. It has just been said that in the course of the last term the population of Virginia has increased 13 per cent.; and it is necessary to explain how the number of representatives for a State may decrease, when the population of that State, far from diminishing, is actually upon the increase. I take the State of Virginia, to which I have already alluded, as my term of comparison. The number of representatives of Virginia in 1823 was proportionate to the total number of the representatives of the Union, and to the relation which the population bore to that of the whole Union: in 1833 the number of representatives of Virginia was likewise proportionate to the total number of the representatives of the Union, and to the relation which its population, augmented in the course of ten years, bore to the augmented population of the Union in the same space of time. The new number of Virginian representatives will then be to the old number, on the one hand, as the new number of all the representatives is to the old number; and, on the other hand, as the augmentation of the population of Virginia

is to that of the whole population of the country. Thus, if the increase of the population of the lesser country be to that of the greater in an exact inverse ratio of the proportion between the new and the old numbers of all the representatives, the number of the representatives of Virginia will remain stationary; and if the increase of the Virginian population be to that of the whole Union in a feebler ratio than the new number of the representatives of the Union to the old number, the number of the representatives of Virginia must decrease. [Thus, to the 43rd Congress in 1874, Virginia and West Virginia send only twelve representatives.]

70. See the report of its committee to the Convention which proclaimed the nullification of the tariff in South Carolina.
71. The population of a country assuredly constitutes the first element of its wealth. In the ten years (1820–1830) during which Virginia lost two of its representatives in Congress, its population increased in the proportion of 13.7 per cent.; that of Carolina in the proportion of 15 per cent.; and that of Georgia, 15.5 per cent. (See the 'American Almanac,' 1832, p. 162) But the population of Russia, which increases more rapidly than that of any other European country, only augments in ten years at the rate of 9.5 per cent.; in France, at the rate of 7 per cent.; and in Europe in general, at the rate of 4.7 per cent. (See 'Malte Brun,' vol. vi. p. 95)
72. It must be admitted, however, that the depreciation which has taken place in the value of tobacco, during the last fifty years, has notably diminished the opulence of the Southern planters: but this circumstance is as independent of the will of their Northern brethren as it is of their own.
73. In 1832, the district of Michigan, which only contains 31,639 inhabitants, and is still an almost unexplored wilderness, possessed 940 miles of mail-roads. The territory of Arkansas, which is still more uncultivated, was already intersected by 1,938 miles of mail-roads. (See the report of the General Post Office, 30th November, 1833.) The postage of newspapers alone in the whole Union amounted to 254,796 dollars.
74. In the course often years, from 1821 to 1831, 271 steamboats have been launched upon the rivers which water the valley of the Mississippi alone. In 1829, 259 steamboats existed in the United States. (See Legislative Documents, No. 140, p. 274.)
75. [Since 1861 the movement is certainly in the opposite direction, and the Federal power has largely increased, and tends to further increase.]
76. See in the Legislative Documents, already quoted in speaking of the Indians, the letter of the President of the United States to the Cherokees, his correspondence on this subject with his agents, and his messages to Congress.
77. The first act of session was made by the State of New York in 1780; Virginia, Massachusetts, Connecticut, South and North Carolina, followed this example at different times, and lastly, the act of cession of Georgia was made as recently as 1802.
78. It is true that the President refused his assent to this law; but he completely adopted it in principle. (See Message of 8th December, 1833.)
79. The present Bank of the United States was established in 1816, with a capital of 35,000,000 dollars; its charter expires in 1836. Last year Congress passed a law to renew it, but the President put his veto upon the bill. The struggle is still going on with great violence on either side, and the speedy fall of the Bank may easily be foreseen. [It was soon afterwards extinguished by General Jackson.]
80. See principally for the details of this affair, the Legislative Documents, 22nd Congress, 2nd Session. No. 30.
81. That is to say, the majority of the people; for the opposite party, called the Union party, always formed a very strong and active minority. Carolina may contain about 47,000 electors; 30,000 were in favour of nullification, and 17,000 opposed to it.
82. This decree was preceded by a report of the committee by which it was framed, containing the explanation of the motives and object of the law. The following passage occurs in it, p. 34:—'When the rights reserved by the Constitution to the different States are deliberately

violated, it is the duty and the right of those States to interfere, in order to check the progress of the evil; to resist usurpation, and to maintain, within their respective limits, those powers and privileges which belong to them as *independent sovereign States*. If they were destitute of this right, they would not be sovereign. South Carolina declares that she acknowledges no tribunal upon earth above her authority. She has indeed entered into a solemn compact of union with the other States; but she demands, and will exercise, the right of putting her own construction upon it; and when this compact is violated by her sister States, and by the Government which they have created, she is determined to avail herself of the unquestionable right of judging what is the extent of the infection, and what are the measures best fitted to obtain justice.'

83. Congress was finally decided to take this step by the conduct of the powerful State of Virginia, whose legislature offered to serve as mediator between the Union and South Carolina. Hitherto the latter State had appeared to be entirely abandoned, even by the States which had joined in her remonstrances.
84. This law was passed on the 2nd March, 1833.
85. This bill was brought in by Mr. Clay, and it passed in four days through both Houses of Congress by an immense majority.
86. The total value of goods imported during the year which ended on the 30th September, 1832, was 101,129,266 dollars. The value of the cargoes of foreign vessels did not amount to 10,731,039 dollars, or about one-tenth of the entire sum.
87. The value of goods exported during the same year amounted to 87,176,943 dollars; the value of goods exported by foreign vessels amounted to 21,036,183 dollars, or about one quarter of the whole sum. ('Williams's Register,' 1833, p. 398.)
88. The tonnage of the vessels which entered all the ports of the Union in the years 1829, 1830, and 1831, amounted to 3,307,719 tons, of which 544,571 tons were foreign vessels; they stood, therefore, to the American vessels in a ratio of about 16 to 100. ('National Calendar,' 1833, p. 304.) The tonnage of the English vessels which entered the ports of London, Liverpool, and Hull, in the years 1820, 1826, and 1831, amounted to 443,800 tons. The foreign vessels which entered the same ports during the same years amounted to 159,431 tons. The ratio between them was, therefore, about 36 to 100. ('Companion to the Almanac,' 1834, p. 169.) In the year 1832 the ratio between the foreign and British ships which entered the ports of Great Britain was 29 to 100. [These statements relate to a condition of affairs which has ceased to exist; the civil war and the heavy taxation of the United States entirely altered the trade and navigation of the country.]
89. Materials are, generally speaking, less expensive in America than in Europe, but the price of labour is much higher.
90. It must not be supposed that English vessels are exclusively employed in transporting foreign produce into England, or British produce to foreign countries; at the present day the merchant shipping of England may be regarded in the light of a vast system of public conveyances, ready to serve all the producers of the world, and to open communications between all peoples. The maritime genius of the Americans prompts them to enter into competition with the English.
91. Part of the commerce of the Mediterranean is already carried on by American vessels.
92. The foremost of these circumstances is, that nations which are accustomed to free institutions and municipal government are better able than any others to found prosperous colonies. The habit of thinking and governing for oneself is indispensable in a new country, where success necessarily depends, in a great measure, upon the individual exertions of the settlers.
93. [This was speedily accomplished, and ere long both Texas and California formed part of the United States. The Russian settlements were acquired by purchase.]
94. The United States already extend over a territory equal to one-half of Europe. The area of Europe is 500,000 square leagues, and its population 205,000,000 of inhabitants. ('Malte Brun,' liv. 114. vol. vi. p. 4.)

[This computation is given in French leagues, which were in use when the author wrote. Twenty years later, in 1850, the superficial area of the United States had been extended to 3,306,865 square miles of territory, which is about the area of Europe.]

95. See 'Malte Brun,' liv. 116, vol. vi. p. 92.
96. This would be a population proportionate to that of Europe, taken at a mean rate of 410 inhabitants to the square league.
97. Russia is the country in the Old World in which population increases most rapidly in proportion.

Second Part, First Book

Chapter V: Of the Manner in Which Religion in the United States Avails Itself of Democratic Tendencies

1. In all religions there are some ceremonies which are inherent in the substance of the faith itself, and in these nothing should, on any account, be changed. This is especially the case with Roman Catholicism, in which the doctrine and the form are frequently so closely united as to form one point of belief.

Chapter XIII: Literary Characteristics of Democratic Ages

1. All this is especially true of the aristocratic countries which have been long and peacefully subject to a Monarchical Government. When liberty prevails in an aristocracy, the higher ranks are constantly obliged to make use of the lower classes; and when they use, they approach them. This frequently introduces something of a democratic spirit into an aristocratic community. There springs up, moreover, in a privileged body, governing with energy and an habitually bold policy, a taste for stir and excitement which must infallibly affect all literary performances.

Second Part, Second Book

Chapter II: Of Individualism in Democratic Countries

1. [I adopt the expression of the original, however strange it may seem to the English ear, partly because it illustrates the remark on the introduction of general terms into democratic language which was made in a preceding chapter, and partly because I know of no English word exactly equivalent to the expression. The chapter itself defines the meaning attached to it by the author.—*Translator's Note.*]

Chapter VI: Of the Relation between Public Associations and Newspapers

1. I say a *democratic people:* the administration of an aristocratic people may be the reverse of centralised, and yet the want of newspapers be little felt, because local powers are then vested in the hands of a very small number of men, who either act apart, or who know each other and can easily meet and come to an understanding.

Chapter VII: Connection of Civil and Political Associations

1. This is more especially true when the executive government has a discretionary power of allowing or prohibiting associations. When certain associations are simply prohibited by law, and the courts of justice have to punish infringements of that law, the evil is far less considerable. Then every citizen knows beforehand pretty nearly what he has to expect. He judges himself before he is judged by the law, and, abstaining from prohibited associations, he embarks in those which are legally sanctioned. It is by these restrictions that all free nations have always admitted that the right of association might be limited. But if the legislature should invest a man with a power of ascertaining beforehand which associations are dangerous and which are useful, and should authorise him to destroy all associations in the bud or allow them to be formed, as nobody would be able to foresee in what cases associations might be established and in what cases they would be put down, the spirit of

association would be entirely paralysed. The former of these laws would only assail certain associations; the latter would apply to society itself, and inflict an injury upon it. I can conceive that a regular government may have recourse to the former, but I do not concede that any government has the right of enacting the latter.

Chapter XIX: That Almost All the Americans Follow Industrial Callings

1. It has often been remarked that manufacturers and mercantile men are inordinately addicted to physical gratifications, and this has been attributed to commerce and manufactures; but that is, I apprehend, to take the effect for the cause. The taste for physical gratifications is not imparted to men by commerce or manufactures, but it is rather this taste which leads men to embark in commerce and manufactures, as a means by which they hope to satisfy themselves more promptly and more completely. If commerce and manufactures increase the desire of well-being, it is because every passion gathers strength in proportion as it is cultivated, and is increased by all the efforts made to satiate it. All the causes which make the love of worldly welfare predominate in the heart of man are favourable to the growth of commerce and manufactures. Equality of conditions is one of those causes; it encourages trade, not directly by giving men a taste for business, but indirectly by strengthening and expanding in their minds a taste for prosperity.
2. Some aristocracies, however, have devoted themselves eagerly to commerce, and have cultivated manufactures with success. The history of the world might furnish several conspicuous examples. But, generally speaking, it may be affirmed that the aristocratic principle is not favourable to the growth of trade and manufactures. Monied aristocracies are the only exception to the rule. Amongst such aristocracies there are hardly any desires which do not require wealth to satisfy them; the love of riches becomes, so to speak, the high road of human passions, which is crossed by or connected with all lesser tracks. The love of money and the thirst for that distinction which attaches to power, are then so closely intermixed in the same souls, that it becomes difficult to discover whether men grow covetous from ambition, or whether they are ambitious from covetousness. This is the case in England, where men seek to get rich in order to arrive at distinction, and seek distinctions as a manifestation of their wealth. The mind is then seized by both ends, and hurried into trade and manufactures, which are the shortest roads that lead to opulence.
3. This, however, strikes me as an exceptional and transitory circumstance. When wealth is become the only symbol of aristocracy, it is very difficult for the wealthy to maintain sole possession of political power, to the exclusion of all other men. The aristocracy of birth and pure democracy are at the two extremes of the social and political state of nations: between them monied aristocracy finds its place. The latter approximates to the aristocracy of birth by conferring great privileges on a small number of persons; it so far belongs to the democratic element, that these privileges may be successively acquired by all. It frequently forms a natural transition between these two conditions of society, and it is difficult to say whether it closes the reign of aristocratic institutions, or whether it already opens the new era of democracy.

Second Part: Third Book

Chapter I: That Manners Are Softened as Social Conditions Become More Equal

1. To feel the point of this joke the reader should recollect that Madame de Grignan was Gouvernante de Provence.

Chapter V: How Democracy Affects the Relation of Masters and Servants

1. If the principal opinions by which men are guided are examined closely and in detail, the analogy appears still more striking, and one is surprised to find amongst them, just as much as amongst the haughtiest scions of a feudal race, pride of birth, respect for their ancestry

and their descendants, disdain of their inferiors, a dread of contact, a taste for etiquette, precedents, and antiquity.

Chapter VIII: Influence of Democracy on Kindred

1. The Americans, however, have not yet thought fit to strip the parent, as has been done in France, of one of the chief elements of parental authority, by depriving him of the power of disposing of his property at his death. In the United States there are no restrictions on the powers of a testator. In this respect, as in almost all others, it is easy to perceive, that if the political legislation of the Americans is much more democratic than that of the French, the civil legislation of the latter is infinitely more democratic than that of the former. This may easily be accounted for. The civil legislation of France was the work of a man who saw that it was his interest to satisfy the democratic passions of his contemporaries in all that was not directly and immediately hostile to his own power. He was willing to allow some popular principles to regulate the distribution of property and the government of families, provided they were not to be introduced into the administration of public affairs. Whilst the torrent of democracy overwhelmed the civil laws of the country, he hoped to find an easy shelter behind its political institutions. This policy was at once both adroit and selfish; but a compromise of this kind could not last; for in the end political institutions never fail to become the image and expression of civil society; and in this sense it may be said that nothing is more political in a nation than its civil legislation.

Chapter X: The Young Woman in the Character of a Wife

1. See Appendix, S.

Chapter XI: That the Equality of Conditions Contributes to the Maintenance of Good Morals in America

1. See Appendix, T.
2. The literature of Europe sufficiently corroborates this remark. When a European author wishes to depict in a work of imagination any of these great catastrophes in matrimony which so frequently occur amongst us, he takes care to bespeak the compassion of the reader by bringing before him ill-assorted or compulsory marriages. Although habitual tolerance has long since relaxed our morals, an author could hardly succeed in interesting us in the misfortunes of his characters, if he did not first palliate their faults. This artifice seldom fails: the daily scenes we witness prepare us long beforehand to be indulgent. But American writers could never render these palliations probable to their readers; their customs and laws are opposed to it; and as they despair of rendering levity of conduct pleasing, they cease to depict it. This is one of the causes to which must be attributed the small number of novels published in the United States.

Chapter XVI: Why the National Vanity of the Americans Is More Restless and Captious Than That of the English

1. See Appendix, U.

Chapter XVIII: Of Honour in the United States and in Democratic Communities

1. The word "honour" is not always used in the same sense either in French or English. 1. It first signifies the dignity, glory, or reverence which a man receives from his kind; and in this sense a man is said *to acquire honor*; 2. Honor signifies the aggregate of those rules by the assistance of which this dignity, glory, or reverence is obtained. Thus we say that *a man has always strictly obeyed the laws of honor;* or *a man has violated his honor.* In this chapter the word is always used in the latter sense.
2. Even the word *patrie* was not used by the French writers until the sixteenth century.

3. I speak here of the Americans inhabiting those States where slavery does not exist; they alone can be said to present a complete picture of democratic society.

Chapter XX: The Trade of Place-Hunting in Certain Democratic Countries

1. [As a matter of fact, more recent experience has shown that place-hunting is quite as intense in the United States as in any country in Europe. It is regarded by the Americans themselves as one of the great evils of their social condition, and it powerfully affects their political institutions. But the American who seeks a place seeks not so much a means of subsistence as the distinction which office and public employment confer. In the absence of any true aristocracy, the public service creates a spurious one, which is as much an object of ambition as the distinctions of rank in aristocratic countries.—*Translator's Note.*]

Chapter XXI: Why Great Revolutions Will Become More Rare

1. If I inquire what state of society is most favourable to the great revolutions of the mind, I find that it occurs somewhere between the complete equality of the whole community and the absolute separation of ranks. Under a system of castes generations succeed each other without altering men's positions; some have nothing more, others nothing better, to hope for. The imagination slumbers amidst this universal silence and stillness, and the very idea of change fades from the human mind. When ranks have been abolished and social conditions are almost equalized, all men are in ceaseless excitement, but each of them stands alone, independent and weak. This latter state of things is excessively different from the former one; yet it has one point of analogy—great revolutions of the human mind seldom occur in it. But between these two extremes of the history of nations is an intermediate period—a period as glorious as it is agitated—when the conditions of men are not sufficiently settled for the mind to be lulled in torpor, when they are sufficiently unequal for men to exercise a vast power on the minds of one another, and when some few may modify the convictions of all. It is at such times that great reformers start up, and new opinions suddenly change the face of the world.

Chapter XXIII: Which Is the Most Warlike and Most Revolutionary Class in Democratic Armies

1. The position of officers is indeed much more secure amongst democratic nations than elsewhere; the lower the personal standing of the man, the greater is the comparative importance of his military grade, and the more just and necessary is it that the enjoyment of that rank should be secured by the laws.

Chapter XXIV: Causes Which Render Democratic Armies Weaker Than Other Armies at the Outset of a Campaign, and More Formidable in Protracted Warfare

1. See Appendix, V.

Chapter XXVI: Some Considerations on War in Democratic Communities

1. It is scarcely necessary for me to observe that the dread of war displayed by the nations of Europe is not solely attributable to the progress made by the principle of equality amongst them; independently of this permanent cause several other accidental causes of great weight might be pointed out, and I may mention before all the rest the extreme lassitude which the wars of the Revolution and the Empire have left behind them.
2. This is not only because these nations have the same social condition, but it arises from the very nature of that social condition which leads men to imitate and identify themselves with each other. When the members of a community are divided into castes and classes, they not only differ from one another, but they have no taste and no desire to be alike; on the contrary, everyone endeavours, more and more, to keep his own opinions undisturbed, to

retain his own peculiar habits, and to remain himself. The characteristics of individuals are very strongly marked. When the state of society amongst a people is democratic—that is to say, when there are no longer any castes or classes in the community, and all its members are nearly equal in education and in property—the human mind follows the opposite direction. Men are much alike, and they are annoyed, as it were, by any deviation from that likeness: far from seeking to preserve their own distinguishing singularities, they endeavour to shake them off, in order to identify themselves with the general mass of the people, which is the sole representative of right and of might to their eyes. The characteristics of individuals are nearly obliterated. In the ages of aristocracy even those who are naturally alike strive to create imaginary differences between themselves: in the ages of democracy even those who are not alike seek only to become so, and to copy each other—so strongly is the mind of every man always carried away by the general impulse of mankind. Something of the same kind may be observed between nations: two nations having the same aristocratic social condition, might remain thoroughly distinct and extremely different, because the spirit of aristocracy is to retain strong individual characteristics; but if two neighbouring nations have the same democratic social condition, they cannot fail to adopt similar opinions and manners, because the spirit of democracy tends to assimilate men to each other.

3. It should be borne in mind that I speak here of sovereign and independent democratic nations, not of confederate democracies; in confederacies, as the preponderating power always resides, in spite of all political fictions, in the state governments, and not in the federal government, civil wars are in fact nothing but foreign wars in disguise.

Second Part, Fourth Book

Chapter III: That the Sentiments of Democratic Nations Accord with Their Opinions in Leading Them to Concentrate Political Power

1. See Appendix, W.
2. In democratic communities nothing but the central power has any stability in its position or any permanence in its undertakings. All the members of society are in ceaseless stir and transformation. Now it is in the nature of all governments to seek constantly to enlarge their sphere of action; hence it is almost impossible that such a government should not ultimately succeed, because it acts with a fixed principle and a constant will, upon men, whose position, whose notions, and whose desires are in continual vacillation. It frequently happens that the members of the community promote the influence of the central power without intending it. Democratic ages are periods of experiment, innovation, and adventure. At such times there are always a multitude of men engaged in difficult or novel undertakings, which they follow alone, without caring for their fellow-men. Such persons may be ready to admit, as a general principle, that the public authority ought not to interfere in private concerns; but, by an exception to that rule, each of them craves for its assistance in the particular concern on which he is engaged, and seeks to draw upon the influence of the government for his own benefit, though he would restrict it on all other occasions. If a large number of men apply this particular exception to a great variety of different purposes, the sphere of the central power extends insensibly in all directions, although each of them wishes it to be circumscribed. Thus a democratic government increases its power simply by the fact of its permanence. Time is on its side; every incident befriends it; the passions of individuals unconsciously promote it; and it may be asserted, that the older a democratic community is, the more centralised will its government become.
3. See Appendix, X.

Chapter V: That amongst the European Nations of Our Time the Power of Governments Is Increasing, although the Persons Who Govern Are Less Stable

1. This gradual weakening of individuals in relation to society at large may be traced in a thousand ways. I shall select from amongst these examples one derived from the law of wills. In aristocracies it is common to profess the greatest reverence for the last testamentary dispositions of a man; this feeling sometimes even became superstitious amongst the older nations of Europe: the power of the State, far from interfering with the caprices of a dying man, gave full force to the very least of them, and insured to him a perpetual power. When all living men are enfeebled, the will of the dead is less respected: it is circumscribed within a narrow range, beyond which it is annulled or checked by the supreme power of the laws. In the middle ages, testamentary power had, so to speak, no limits: amongst the French at the present day, a man cannot distribute his fortune amongst his children without the interference of the State; after having domineered over a whole life, the law insists upon regulating the very last act of it.
2. In proportion as the duties of the central power are augmented, the number of public officers by whom that power is represented must increase also. They form a nation in each nation; and as they share the stability of the government, they more and more fill up the place of an aristocracy.

   In almost every part of Europe the government rules in two ways; it rules one portion of the community by the fear which they entertain of its agents, and the other by the hope they have of becoming its agents.
3. On the one hand the taste for worldly welfare is perpetually increasing, and on the other the government gets more and more complete possession of the sources of that welfare. Thus men are following two separate roads to servitude: the taste for their own welfare withholds them from taking a part in the government, and their love of that welfare places them in closer dependence upon those who govern.
4. A strange sophism has been made on this head in France. When a suit arises between the government and a private person, it is not to be tried before an ordinary judge—in order, they say, not to mix the administrative and the judicial powers; as if it were not to mix those powers, and to mix them in the most dangerous and oppressive manner, to invest the government with the office of judging and administering at the same time.
5. I shall quote a few facts in corroboration of this remark. Mines are the natural sources of manufacturing wealth: as manufactures have grown up in Europe, as the produce of mines has become of more general importance, and good mining more difficult from the subdivision of property which is a consequence of the equality of conditions, most governments have asserted a right of owning the soil in which the mines lie, and of inspecting the works; which has never been the case with any other kind of property. Thus mines, which were private property, liable to the same obligations and sheltered by the same guarantees as all other landed property, have fallen under the control of the State. The State either works them or farms them; the owners of them are mere tenants, deriving their rights from the State; and, moreover, the State almost everywhere claims the power of directing their operations: it lays down rules, enforces the adoption of particular methods, subjects the mining adventurers to constant superintendence, and, if refractory, they are ousted by a government court of justice, and the government transfers their contract to other hands; so that the government not only possesses the mines, but has all the adventurers in its power. Nevertheless, as manufactures increase, the working of old mines increases also; new ones are opened, the mining population extends and grows up; day by day governments augment their subterranean dominions, and people them with their agents.

Chapter VI: What Sort of Despotism Democratic Nations Have to Fear

1. See Appendix, Y.
2. See Appendix, Z.

Appendices

1. The 20th degree of longitude, according to the meridian of Washington, agrees very nearly with the 97th degree on the meridian of Greenwich.
2. A folio edition of this work was published in London in 1702.
3. This passage is extracted and translated from M. Conseil's work upon the life of Jefferson, entitled 'Mélanges Politiques et Philosophiques de Jefferson.'
4. See 'Memoires pour servir a l'Histoire du Droit Public de la France en matiere d'impôts,' p. 654, printed at Brussels in 1779.